SOLID STATE
NUCLEAR TRACK DETECTORS

Volume 2

HONORARY COMMITTEE

H. MATTHÖFER, Bundesministerium für Forschung und Technologie
Prof. Dr. H. MAIER, Bayer. Staatsminister für Unterricht und Kultus
G. KAHN-ACKERMANN, Secretary General of the Council of Europe
G. KRONAWITTER, Oberbürgermeister der Landeshauptstadt München
Prof.Dr.H. GLUBRECHT, Deputy Director General of the IAEA
Prof.Dr.U. GRIGULL, Präsident der Technischen Universität München
Prof.Dr.R. WITTENZELLNER, Wissenschaftlich -Technischer Geschäftsführer der GSF, Neuherberg/München

ORGANIZING COMMITTEE

F. GRANZER, Institut f. Angewandte Physik, Universität Frankfurt
G. HAASE, Institut f. Wissenschaftliche Photographie, Technische Universität München
W. JACOBI, Institut f. Strahlenschutz, GSF - Neuherberg
J.P. MASSUE, Committee on Science and Technology, Council of Europe, Strasbourg
H.G. PARETZKE, Institut f. Strahlenschutz, GSF - Neuherberg
E. SCHOPPER, Institut f. Kernphysik, Universität Frankfurt
F. ZÖRGIEBEL, Institut f. Wissenschaftliche Photographie, Technische Universität München

ADVISORY SCIENTIFIC COMMITTEE

J. HEBERT (Canada); A.J. HERTZ (CERN); O.C. ALLKOFER, W. ENGE (Federal Republic of Germany); H. FRANCOIS, R. SCHMITT (France); P.H. FOWLER (Great Britain); L. MEDVECZKY, G. SOMOGYI (Hungary); M. NICOLAE (Romania); V. GANDIA (Spain); K. KRISTIANSSON (Sweden); E.V. BENTON, R.L. FLEISCHER, R. KATZ, M. SHAPIRO (USA); C.S. BOGOMOLOV, N.A. PERFILOV, G. ZDANOV (USSR).

THE CONGRESS WAS SPONSORED BY

Gesellschaft für Strahlen- und Umweltforschung mbH (GSF), München
Council of Europe, Parlamentary Assembly, Strasbourg.

IT WAS HELD IN CO-OPERATION WITH

International Atomic Energy Agency (IAEA), Vienna.

SOLID STATE NUCLEAR TRACK DETECTORS

*Proceedings of the 9th International Conference,
Neuherberg/Munchen, 30th September - 6th October 1976*

Edited by

F. GRANZER
Institut für Angewandte Physik, Frankfurt/Main

H. PARETZKE
Institut für Strahlenschutz, GSF-Neuherberg

E. SCHOPPER
Institut für Kernphysik, Frankfurt/Main

IN TWO VOLUMES

Volume 2

PERGAMON PRESS
OXFORD · NEW YORK · TORONTO · SYDNEY · PARIS · FRANKFURT

U.K.	Pergamon Press Ltd., Headington Hill Hall, Oxford OX3 0BW, England
U.S.A.	Pergamon Press Inc., Maxwell House, Fairview Park, Elmsford, New York 10523, U.S.A.
CANADA	Pergamon of Canada Ltd., 75 The East Mall, Toronto, Ontario, Canada
AUSTRALIA	Pergamon Press (Aust.) Pty. Ltd., 19a Boundary Street, Rushcutters Bay, N.S.W. 2011, Australia
FRANCE	Pergamon Press SARL, 24 rue des Ecoles, 75240 Paris, Cedex 05, France
FEDERAL REPUBLIC OF GERMANY	Pergamon Press GmbH, 6242 Kronberg-Taunus, Pferdstrasse 1, Federal Republic of Germany

Copyright © 1978 Pergamon Press Ltd.

All Rights Reserved. No part of this publication may be reproduced, stored in a retrieval system or transmitted in any form or by any means: electronic, electrostatic, magnetic tape, mechanical, photocopying, recording or otherwise, without permission in writing from the publisher.

First edition 1978

British Library Cataloguing in Publication Data

International Conference on Solid State Nuclear
Track Detectors, 9th, Munich, 1976
Solid state nuclear track detectors.
1. Solids, Effect of radiation on - Congresses
2. Particle tracks (Nuclear physics) - Congresses
I. Title II. Granzer, F III. Paretzke, H
IV. Schopper, E
539.7'7   QC176.8.R3   77-30630
ISBN 0-08-021659-5

In order to make this volume available as economically and as rapidly as possible the authors' typescripts have been reproduced in their original forms. This method unfortunately has its typographical limitations but it is hoped that they in no way distract the reader.

*Printed in Great Britain by William Clowes & Sons Limited
London, Beccles and Colchester*

CONTENTS

VOLUME 1

Preface	xix
List of participants	xxiii

OPENING LECTURE

Chairman: E. Schopper

R.L. Fleischer (Invited Paper)	The past and future roles of solid state nuclear track detectors	3

Session 1: TRACK THEORY

Chairman: C.S. Bogomolov

R. Katz (Invited Paper)	Track structure theory in radiobiology and in radiation detection	27
H.G. Paretzke	On primary damage and secondary electron damage in heavy ion tracks in plastics	87
K. Rössler	Computer simulation of collision cascades in some non-metallic solids	99
A. Ali, S.A. Durrani	A statistical model for track formation	101

Chairman: E.V. Benton

G. Somogyi, R. Scherzer, K. Grabisch, W. Enge	A spatial track formation model and its use for calculating etch-pit parameters of light nuclei	103
G. Somogyi, K. Grabisch, R. Scherzer, W. Enge	Revision of the concept of registration threshold in plastic track detectors	119
G. Siegmon, H.J. Köhnen, K.P. Bartholomä, W. Enge	The dependence of the mass-identification scale on different track formation models	137

L. Larsson, F.E. Pinkerton, R. Katz	Supralinearity of nuclear research emulsions	145
L. Larsson, F.E. Pinkerton, R. Katz, E.V. Benton	Particle tracks in supralinear nuclear research emulsions	175

Session 2: NEW DETECTOR SYSTEMS

Chairman: C.B. Childs

G. Haase, E. Schopper, F. Granzer	Solid state nuclear track detectors: Track forming, stabilizing and development processes	199
P.J. Ouseph	Photochromic radiation detector	215
V.D. Dmitriev, N.P. Kocherov, N.R. Novikova, N.A. Perfilov	Properties of silver chloride track detectors	217
E.D. Avdonina, V.M. Belous, A.L. Kartuzhanski, T.E. Kehva, B.T. Plachenov	Luminescent investigation of mechanism of silver halides radiolysis	221
J.L. Gisclon, J.F. Jal, J. Dupuy	Fundamental analysis of defects induced by energetic ions implantation in AgCl	229
H. Schmidt, G. Haase, F. Zörgiebel	Investigations on the photographic elementary process in AgCl-single crystal foils	237
J. Gourcy, M. Monnin, J. Faïn	Charged particles detection: The Graft-and-Dye Method	243
B. Žižić, S. Božin, R. Ristić	Effects of alpha-particles on ammonium dihydrogen phosphate (ADP) single crystals	251

Session 3: TRACKS IN PLASTICS

Chairman: H.G. Paretzke

G. Somogyi (Invited Paper)	Processing of plastic track detectors	255
G. Somogyi	A study of the basic properties of electrochemical track etching	285
A. Chambaudet, A. Bernas, J. Roncin	On the formation of heavy ion latent tracks in polymeric track detectors	301
A. Chambaudet, Ph. Romary	On the variation of some heavy ion track characteristics with the polymeric detector crystallinity	307
J.L. Decossas, J.P. Moliton, J.C. Vareille, J.L. Teyssier, B. Delaunay	Contribution to the research on cellulose diacetate as a solid state track detector	317
D. Hildebrand, G. Reitz, H. Buecker	Some aspects of the etching behaviour of cellulose nitrate as track detector	325
F.H. Ruddy, H.B. Knowles, S.C. Luckstead, G.E. Tripard	Etch induction time in cellulose nitrate: A new particle identification parameter	333
D.D. Peterson, M. Tran, E.V. Benton	Measurement of activation energies for annealing and etching C-12 and Ne-20 ion tracks in cellulose nitrate and lexan polycarbonate plastics	349
M. Balcázar-García, S.A. Durrani	^{3}He and ^{4}He spectroscopy using plastic solid state nuclear track detectors	351
C.B. Besant, A.Y. Qaqish, B.B. Varga	Detection efficiency and range measurements of alphas and protons in cellulose nitrate	363

Session 4: TRACKS IN CRYSTALS AND GLASSES

Chairman: M. Monnin

P.F. Green, S.A. Durrani (Invited Paper)	Annealing studies of tracks in crystals	375

A. Sigrist, R. Balzer	Investigations on the formation of tracks in crystals	387
L.L. Kashkarov, V.S. Pestov	Calibration of parameters of tracks in silicate minerals with the aid of irradiation with accelerated Zn and Kr ions	393
E. Dartige, J.P. Duraud, Y. Langevin	Thermal annealing of iron tracks in muscovite, labradorite, and olivine	395

Chairman: G. Haase

L.L. Kashkarov, V.L. Koshkin, A.N. Troshin	Influence of the ultrasound upon the etching of tracks in silicate matter	401
H.A. Khan, R. A. Akber	Fission fragment etch pit formation in mica, quartz and feldspar crystals under varying etching conditions	403
G. Fiedler, J. Aschenbach, W. Otto, T. Rautenberg, U. Steinhauser, G. Siegert	Further developments with glass detectors for heavy ions and fission fragments	417
T. Werba, F. Granzer	Electron microscopic investigations of tracks of Cf^{252}-fission fragments in quartz glass	425
H. A. Khan, R. A. Akber, A. Waheed, P. Chaudhry, S. Mubarakmand	New etchants for soda lime glass track detectors	439
A. Aframian	Track retaining properties of quartz for high temperature in-core neutron fluence measurements	447

Session 5: TRACKS IN EMULSIONS

Chairman: R. Katz

R.V. Rechenmann, E. Wittendorp, B. Senger (Invited Paper)	A new approach for the detection of charged particles by photographic recording systems - first applications in corpuscular physics, biology and electron microscopy	463

Contents

P.J. McNulty, R.C. Filz	Width measurements on neon nitrogen tracks in Ilford G-5 emulsion	501
C.S. Bogomolov, V.A. Ditlov	The determination of nuclear charges by the method of grain counting in heavy particle tracks	511
C.S. Bogomolov, I.F. Razorenova, I.A. Ruditskaya, L.S. Khruleva, G.I. Kozinets, V.V. Fetisov	A study of recording possibilities of emulsions in autoradiography conditions	523
V.I. Zakharov, N.R. Novikova, N.A. Perfilov	On the use of alpha-naphtol in nuclear photography	533
K.M. Romanovskaya, J.P. Savateeva, E.N. Tolkacheva	Increasing of registering capacity of nuclear emulsion for autoradiography	541
A.B. Akopova, N.V. Magradze, L.V. Melkumyan, Y.P. Prokhorenko	The separation of heavy ion tracks in nuclear emulsions by means of the pulsed electric field	547
A.V. Apanasenko, G.B. Zhdanov, K.A. Kotelnikov, J.A. Smorodin, K.S. Bogomolov, V.A. Mylzeva, I.N. Sokolovskaya, L.S. Khruleva, E.I. Chikunova, R.S. Gadiullin, V.O. Ivanov, G.G. Kuznetzova	Thin nuclear film R-2T-50 for cosmic rays	553
L.G. Baranova	The development of 1200 micron glass mounted emulsion	565
V.I. Baranov	Color identification of distortions in nuclear emulsions	567

Session 6: EVALUATION TECHNIQUES

Chairman: C. O'Ceallaigh

L. Tommasino, N. Klein, P. Solomon	Fission fragment detection by thin-film capacitors. I. Breakdown counter	571
N. Klein, P. Solomon, L. Tommasino	Fission fragment detection by thin-film capacitors. II. Current pulse counter and mechanisms	581
E. Piesch, J. Jasiak	Automatic spark counting of fast neutron-induced recoil particles in polymers	587
G. Somogyi, L. Medveczky, I. Hunyadi, B. Nyako	Automatic spark counting of alpha-tracks in plastic foils	599
J.U. Schott, E. Schopper, R. Staudte	A high precision video-electronic measuring system for use with solid state track detectors	615
D. Azimi-Garakani, J.G. Williams	Automatic fission track counting using Quantimet 720	625
S. Di Liberto, P. Ginobbi	Automatic device for measurements of heavy ions tracks in plastics	635
W. Abmayr, P. Gais, H.G. Paretzke, K. Rodenacker, G. Schwarzkopf	Real-time automatic evaluation of solid state nuclear track detectors with an on-line TV-device	643
A. Aframian, S.A. Durrani	Semiautomatic evaluation of fast neutron fluences in plastic SSNTDs using a double-beam microdensitometer	651

Session 7: APPLICATIONS IN EARTH SCIENCES AND RADIOCHEMISTRY

Chairman: M. Juric

R.L. Fleischer, O.G. Raabe	Fragmentation of respirable PuO_2 particles in water by alpha decay. A mode of "Dissolution"	663

J.H. Roberts, V.P. Kafalenos, T.J. Yule	Characterization of aerosols containing fissionable elements using solid-state track recorders	669
R. Coppens, P. Richard, S. Bashir	Utilization of alpha-autoradiography of rocks in the investigation of the radioactive equilibrium	677
A.E. Liehu	Geologic analysis by track etch method	689
T.U. Jensen, W. Enge, H. Erlenkeuser, H. Willkomm	Age determination of sediments by Pb-210 using a plastic detector technique	697
T. Nakanishi, M. Sakanoue, B. Sansoni	Application of fission track technique to determine the $^{240}Pu/^{239}Pu$ isotope ratio	705
J.C. Dran, J.P. Duraud, Y. Langevin, J.C. Petit	Fission track dating of quartz grains from the Oklo Uranium Ore Deposit	707
A. Daniş, M. Oncescu, I. Purica, E.G. Badea	Several considerations on radiocolloidal and pseudoradiocolloidal states of the fissionable element solutions	715
A. Daniş, E.G. Badea, M. Oncescu, I. Purica	The fissionable materials used in neutron detection by fission track method	719
A. Daniş	On the nature and distribution of the fissionable element impurities in minerals and soils	725
K. Thiel, G. Damm	A sensitive mapping technique for bismuth using α-particle tracks	729

VOLUME 2

Session 8: APPLICATIONS IN DOSIMETRY AND RADIOGRAPHY

Chairman: R.L. Fleischer

E.V. Benton, C.A. Tobias, R.P. Henke, M.R. Cruty (Invited Paper)	Heavy-particle radiography with plastic nuclear track detectors	739

E. Bagge, E. Dühmke, W. Ehge, W. Hunger, U. Roose, R. Scherzer	Fast neutron radiography for extended objects by a plastic detector technique	753
I.Y. Khadduri	A neutron radiography facility on the IRT-2000 reactor	761
M. Nicolae	Possibilities and limits of using nuclear track detectors in dosimetry and microdosimetry	771
H.B. Knowles, F.H. Ruddy, G.E. Tripard, G.M. West, M.M. Kligerman	Status report: Direct track detector dosimetry in negative pion beams	787

Chairman: F. Granzer

H.A. Khan, R.A. Akber (Invited Paper)	The measurement of radon by alpha-sensitive plastic track detectors for use in uranium exploration	803
H. A. Khan, R. A. Akber, A. Waheed, M. Afzal, P. Chaudhary, S. Mubarakmand, F. I. Nagi	The use of CA80-15 and LR-115 cellulose nitrate track detectors for discrimination between radon and thoron	815
H.G. Paretzke	Results of an international alpha particle registration intercomparison with solid state nuclear track detectors	821
R. Antanasijević, Z. Todorović, A. Stamatović D. Miocinović	Factors affecting the sensitivity of plastic detectors to ionizing alpha particle registration	831
J. Simonivić, J. Vuković, R. Antanasijević	Alpha autoradiography by cellulose nitrate layer	835

Chairman: H. Francois

F. Spurný, K. Turek (Invited Paper)	Neutron dosimetry by means of solid state nuclear track detectors	839
F. Spurný, K. Turek	On the energetical dependence of polymer solid state nuclear track detectors as fast neutron dosimeters	863
U. Lotz, E. Pitt, A. Scharmann, B. Vitt	Fast neutron dosimetry by track detection with cellulose nitrate films	875
I.Y. Khadduri, I.K. Al-Haddad	A neutron dose meter using cellulose nitrate film	883
H.B. Knowles, F.H. Ruddy, G.E. Tripard, H. Bichsel, J. Eenmaa, J.E. Smathers	Status report: Direct track detector dosimetry in fast neutron beams	891
G.M. Hassib, J.W.N. Tuyn, J. Dutrannois	On the electrochemical etching of neutron-induced tracks in plastics and its application to personnel neutron dosimetry	905
H. Schraube, H.G. Paretzke	Neutron fluence measurements with solid state nuclear track detectors. Results of an international intercomparison	917

Chairman: L. Medveczky

H.A. Khan, R.A. Akber, G. Hussain	The development and applications of plastic track detectors for neutron and gamma dose measurements	931
M.A. Kenawy, M. El-Fiki, S. El-Konsol, M.A. Fadel, A.M. Basha	Detection of fast and slow neutrons by etch pit method of nuclear track registration in plastics	943

xiv   Contents

J. Dutrannois, J.W.N. Tuyn	Application of solid state nuclear track detectors for personnel monitoring around high energy accelerators	953
D. Haşegan, A. Drăgu, M. Nicolae, A. Apostol	Personal dosimeter for high energy corpuscular radiation	967
C. Heilmann, H. Francois, C. Jacquot	Dosimetry with activated emulsion	977

Session 9: APPLICATIONS IN COSMIC RAY PHYSICS

Chairman: A.J. Herz

P.H. Fowler (Invited Paper)	Ultra heavy cosmic ray nuclei - analysis and results	983
P.H. Fowler, C. Alexander, V.M. Clapham, D.L. Henshaw, C. O'Ceallaigh, D. O'Sullivan, A. Thompson	High resolution study of nucleonic cosmic rays with $Z \geq 34$	1007
P.H. Fowler, D.L. Henshaw, C. O'Ceallaigh, D. O'Sullivan, A. Thompson	Measurement of the cosmic ray element abundances between $\simeq 300$ and $\simeq 750$ MeV/N in the region from nickel to krypton using Lexan track detectors	1017
R.C. Filz	Search for trapped Van Allen belt particles heavier than hydrogen in satellite-exposed nuclear emulsions	1023
W. Krätschmer	Lunar and meteoritic mineral track detectors and the composition of the galactic cosmic radiation	1025
R.K. Bull, S.A. Durrani	Studies of fresh and fossil tracks in meteoritic hypersthene	1031

Chairman: P.H. Fowler

W. Enge (Invited Paper)	Isotopic composition of cosmic ray nuclei	1039

G. Siegmon, K.-P. Bartholomä, W. Enge	Composition of Fe-isotopes in cosmic rays	1059
R. Beaujean, H. Sagebiel, W. Enge	Isotopic composition of low energy cosmic ray particles with charges Z=5-8	1069
R. Scherzer, W. Enge, R. Beaujean, S. Hertzman, K. Kristiansson, K. Söderström	Study on cosmic ray iron isotopes in an emulsion-plastic detector	1075
N.S. Ivanova, D.G. Baranov, V.V. Varyukhin, Yu.F. Gagarin, V.N. Kulinkov, V.E. Myshkin, K.M. Romanovskaya, I.G. Khilyuto, E.A. Yakubovsky	Using nuclear emulsions and plastics in a long exposure satellite experiment	1081
J. Sequeiros, J. Medina, A. Durã, M. Ortega, A. Vidal-Quadras, F. Fernandez, R.T. Thorne	Low energy heavy cosmic ions charge discrimination with plastic detectors	1087
B. Sojka, H. Röhrs	Some results of charge spectrum measurements with a rocket equipment using plastic detectors	1089
J. Tripier, M. Debeauvais	Calibration of two plastic detectors and application on study of heavy cosmic rays	1091

Session 10: APPLICATIONS IN NUCLEAR PHYSICS

Chairman: E. Schopper

I. Otterlund (Invited Paper)	Applications of SSNTDs in high energy physics	1107
A. Waheed, M. Jurić	Fast stable particles from light nuclei of emulsion following the interaction of 1.5 GeV/c K^- meson	1129
A. Waheed, M. Jurić, V. Zlatarov	Probability distribution of K^-p and K^-n channels in the interaction of 1.5 GeV/c K^- meson in light emulsion nuclei	1137
M.K. Jurić, S.B. Drndarević	Hypernuclei in some hammer-like events recorded in a photonuclear emulsion irradiated with stopping K^- mesons	1145

M. Brun, H. Annoni	Method for a determination of low momentum antineutron flux	1153

Chairman: I. Otterlund

P.S. Young, K. Fukui, Y.V. Rao	Some aspects of 400 GeV proton interactions in nuclear emulsions	1155
Y.V. Rao, K. Fukui, P.S. Young	The scattering constant for multiply-charged particles in emulsions	1163
K. Grabisch, R. Beaujean, R. Scherzer, W. Enge	Spallation products induced by energetic neutrons in plastic detector material	1171
M. Debeauvais, J. Tripier, S. Jokic	Fission cross sections of heavy nuclei induced by 300 GeV protons with the help of plastic detector	1179
B. Grabež, Ž. Todorović, A. Antanasijević	Fission of Bi, Pb, and Au induced by 0.65, 1.74 and 4.12 GeV alpha particles	1187
R. Beaujean, W. Enge	Fragmentation and isotope measurements on accelerator neon and argon particles of 280 MeV/Nuc	1197

Chairman: R. Schmitt

P. Vater, H.J. Becker, R. Brandt, H. Freiesleben (Invited Paper)	Multi-fragment decay reactions induced by heavy ions and studied with Mica track detectors	1207
S. Mubarakmand, P. Chaudhry, K. Rashid, R. A. Akber, H. A. Khan	The measurement of helium-ion-induced fission crossection of uranium by glass track detectors	1231
V.A. Nikolaev	Application of the track diameter measurement method in nuclear physics	1235
G. Somogyi, I. Hunyadi, E. Koltay, L. Zolnai	On the detection of low-energy ^{4}He, ^{12}C, ^{14}N, ^{16}O ions in PC foils and its use in nuclear reaction measurements	1245

Contents

Combined Session 11: SPACE BIOPHYSICS

Chairman: J.P. Massue

O.C. Allkofer (Invited Paper)	Dosimetric significance of cosmic radiation in the altitude of supersonic transports and in free space	1265
R. Facius, G. Hölz, B. Toth, H. Bücker	Radiobiological investigations of cosmic HZE-particles with visual track detectors in the Biostack experiment	1283
M. Schäfer, H. Bücker, R. Facius, D. Hildebrand	High precision localization methods for HZE-particles	1291
U. Scheidemann, H. Bücker, R. Facius, C. Thomas	Determination of the trajectories of HZE particles in seeds of <u>Arabidopsis thaliana</u> by use of plastic detectors	1299
E.H. Graul, W. Rüther	Radiobiological studies on biological systems of animals exposed to the heavy nuclei of cosmic galactic radiation	1301

Chairman: O.C. Allkofer

W. Heinrich	Calculation of LET-spectra of heavy cosmic ray nuclei at various absorber depths	1313
R. Pfohl, R. Kaiser, J.P. Massue, H. Francois	Dosimetry of cosmic particles in nuclear emulsions for the Apollo 16, 17, and Apollo-Soyouz-Test-Project experiments (1972 - 1975)	1325
P.J. McNulty, V.P. Pease, V.P. Bond, R.C. Filz, P.L. Rothwell	Particle induced visual phenomena in space	1335

Author Index	1345
List of proceedings of former conferences	1349

Session 8

Applications in Dosimetry and Radiography

Chairmen: R. L. Fleischer
F. Granzer
H. Francois
L. Medveczky

HEAVY-PARTICLE RADIOGRAPHY WITH PLASTIC NUCLEAR TRACK DETECTORS

E. V. Benton*, C. A. Tobias**, R. P. Henke*
and M. R. Cruty**

*University of San Francisco, San Francisco, California, U.S.A.
**University of California, Berkeley, California, U.S.A.

ABSTRACT

A nearly monoenergetic beam of heavy particles produced by the Berkeley Bevatron or Bevalac can be used as a very sensitive tool for the measurement of the mass thicknesses of a stopping material. If such a beam is stopped in a stack of plastic nuclear track detectors (PNTD) after having passed through a specimen, the processed detectors reveal a high-contrast radiograph which shows very slight gradations in the mass thickness of the specimen. The PNTD are the logical choice of detector because of their threshold registration characteristics. They register each particle only near its stopping point but with nearly 100% efficiency.

1. INTRODUCTION

The potential use of monoenergetic charged particle beams as a sensitive "thickness gauge" in the detection of small density variations has been recognized for many years. It was first reduced to practice about 1959 be Belanger (School of Medicine at Ottawa), who made use of alpha particles from polonium 210 to image thin tissue slices. The principles of heavy-ion radiography with plastic nuclear track detectors (PNTD) were developed by Tobias and Benton [1,2] in 1972. In that same year, the first heavy-particle radiographs of small animals were produced by Benton, Henke, and Tobias utilizing O^{16} particles from the Bevatron recorded on plastics through the use of the track-etch technique [3].

The scheme of producing radiographs in PNTD with heavy-particle beams is quite straight forward. A beam of nearly monoenergetic heavy nuclei is caused to spread to lateral size of the order of the size of the specimen under study. The beam then impinges on the specimen and is stopped in a stack of PNTD. Where the specimen is thicker or has a higher stopping power (approximately proportional to the electron density) the particles stop earlier in the stack; where it is thinner or has a lower stopping power, the particles stop in a layer of the stack which is further downstream. The thickness of the required stack is minimized by adjusting the beam energy either within the accelerator itself or with an energy degrading absorber to place the particles passing through the maximum thickness of the specimen near the upstream (front) side of the stack. In most situations, since it is the internal structure of the subject which is being studied, the effects of the external geometry are minimized by placing the subject in a water bath. The external walls of this bath, through which the beam passes, are especially designed to be flat and parallel.

This exposure situation is illustrated in Fig. 1. In this case the specimen is a phantom made of Lucite with a spherical cavity filled with a 5% lower stopping power sucrose solution. At each lateral point the beam stops in the shaded band. The band is displaced downstream for those particles passing through the cavity.

Heavy-particle radiography is of interest to radiologists because it is complementary to conventional X-ray radiography and in many cases produces radiographs of much higher contrast. Conventional X-ray radiography is well suited to subjects with regions of highly varying effective atomic number and therefore X-ray absorption coefficient. If the subject is homogeneous with respect to atomic number, however, heavy-particle radiography can provide a much more sensitive measure of the variation in the electron density because a very small displacement of the beam stopping point leads to relatively large variation in the registered track fluences in the various layers of the stack. This fact makes heavy-particle radiography eminently suited to the detection of abnormalities in soft tissue regions, such as internal organs and breasts.

2. IMAGE DETECTION WITH PLASTIC NUCLEAR TRACK DETECTORS

The role of the stack of PNTD is to measure the residual energy or range of the beam at each point after it has passed through the subject. There are alternative ways of achieving the same end but they each appear to have limitations as compared to PNTD. For example, active electronic counting systems are quite expensive and must laterally scan the emerging beam to achieve the same resolution. Because of scanning requirement and the need to limit beam intensity because of the limited counting speed, the exposure time can be quite long. Other passive detectors, such as film, do not provide high accuracy in the measurement of the stopping point because they do not have the required high detection threshold of the PNTD. Also, conventional film does not reveal individual tracks as PNTD do. It should be mentioned, however, that PNTD are only applicable to imaging with highly ionizing particles. This means $Z \geq 6$ for thick subjects (≥ 1 cm in thickness) and $Z \geq 2$ for very thin subjects.

The working requirement on the detector stack is that each particle be recorded in one and only one layer. This simultaneously maximizes the signal to dose ratio and eliminates the redundancy which leads to a statistical correlation between the signal in one layer and the next. It is also desirable to have the band of stopping particles span at least two layers of the stack. Thus the relative track densities on the involved layers provides a determination of the average stopping point which is much finer than the thickness of a single layer. These constraints imply that the detector sensitivity and thickness (or spacing if inert spacers are used) must be matched to the beam type and energy. Our most usual exposure configuration meets these requirements almost exactly. We expose stacks of 254 μm layers of cellulose nitrate to beams of 250 MeV/amu C^{12} ions. The registration range (maximum residual range at which the particle will register) of the ions is approximately one layer, and the range straggling of the beam is approximately ±1.6 layers (standard deviation).

The detectors are processed in 6.25N NaOH at 60°C for a time sufficient to produce 10-20% overlap of the tracks. A shorter etch would reduce the optical visibility of the image. A longer etch would lead to increased clumping of the tracks with a corresponding reduction in the statistics and an increased graininess.

The detectors are usually "read out" by photographing each of the layers individually with dark-field illumination. An alternate procedure is to photographically synthesize the images on several or all of the layers to both reduce the graininess and to combine the unique information on the various layers into a single image. The synthesis technique that we used which achieves the ultimate in flexibility and quantitativeness is to digitize each of the layers of the stack and then to computer process and display the final radiograph.

3. RESULTS

Initially, radiographs of various phantoms and resolution testers where used to make quantitative physical measurement of capabilities and limitations of heavy-particle radiography. Next a number of radiographs of small animals were made. An example of such a radiograph is shown in Fig. 2a, which is the radiograph of a rat.

A comparison X ray is shown in Fig. 2b. It can be seen that the X ray is sensitive to high-Z features such as bones and particularly three metal suture clips. In contrast the heavy-particle radiograph quite conspicuously registers such features as air bubbles. It should also be noted that the heavy-particle radiograph is only the result from a single layer. It is necessary to either inspect or synthesize all of the layers to obtain all of the information contained in the radiographic stack.

The program of heavy-particle diagnostic radiology was begun with studies of excised tissues. To date we have radiographed specimens of many of the internal organs as well as breasts and diseased portions of the vascular system. An example of the radiograph of a freshly excised breast specimen is shown in Fig. 3. In this case, three layers have been photographically synthesized to improve statistics. The tumor is quite evident. In contrast, the conventional X ray of the same specimen, shown in Fig. 4, shows little indication of the tumor.

The last step to date in our radiography program has been the radiography of patients with known or suspected malignancies. These have included a patient with a giant cell tumor of the distal femur as well as several mammography patients. An example of a heavy-particle mammogram is shown in Fig. 5. In this case the digitized data from approximately 20 layers of a radiography stack has been synthesized by our PDP-11 computer to yield the average stopping layer at each point on a 204 × 156 point array. Each array element represents an area of approximately 0.6 mm square. The computer display represents the $z(x,y)$ surface as it would be seen with light shining from the top of the figure at an angle of 70 deg with the normal. The depressed areas such as the skin line are produced by high-stopping-power regions and the raised areas such as the fatty tissue immediately under the skin are produced by low-stopping-power regions in the subject.

4. SUMMARY

Heavy-particle radiography presents some unique capabilities which complement and significantly add to the capabilities of conventional X-ray radiography. In particular very small differences in electron density can be detected even in the case where the atomic composition is essentially constant. In addition, these differences are revealed with remarkable contrast as compared to X rays. An added bonus of heavy-particle radiography is that the radiographs can be made to yield quite quantitative data about the subject.

Plastic nuclear track detectors appear to be the most suitable recording medium for heavy-particle radiography using particles with $Z \geq 6$. This stems from the near 100% efficiency of the detectors as well as their threshold characteristics, which maximize signal to noise and contrast and provide the minimum possible graininess for the particle type and fluence. In addition, PNTD have no particle rate dependences such as electronic counters. They are also inexpensive and easy to handle.

Given sufficient support for the building of medical accelerators, heavy-particle radiography could become a very useful tool in diagnostic radiology. Because of its sensitivity to soft tissue abnormalities, it would make possible the detection of malignancies which are much smaller than those now visible with X-ray techniques and at a comparable or lower dose.

5. ACKNOWLEDGMENTS

The authors wish to thank Drs. K. Woodruff, H. Genant, and E. Sickles for their valuable contribution in performing heavy-particle mammography.

REFERENCES

[1] E. V. Benton, R. P. Henke, and C. A. Tobias, Science 182, 474 (1973); E. V. Benton, R. P. Henke, and C. A. Tobias, in: LBL-2016 (Lawrence Berkeley Laboratory, Berkeley, California, 1973), p. 1.

[2] C. A. Tobias, E. V. Benton, R. P. Henke, and M. R. Cruty, "Developments in heavy particle radiography" (in preparation).

[3] C. A. Tobias and E. V. Benton, "Additional techniques for heavy-ion radiography and laminography", UCID-3593, Lawrence Berkeley Laboratory, Berkeley, California (1972), 15 p; E. V. Benton, R. P. Henke, C. A. Tobias, and M. R. Cruty, "Radiography with heavy particles", LBL-2887 (Lawrence Berkeley Laboratory, Berkeley, California, 1975), 67 p.

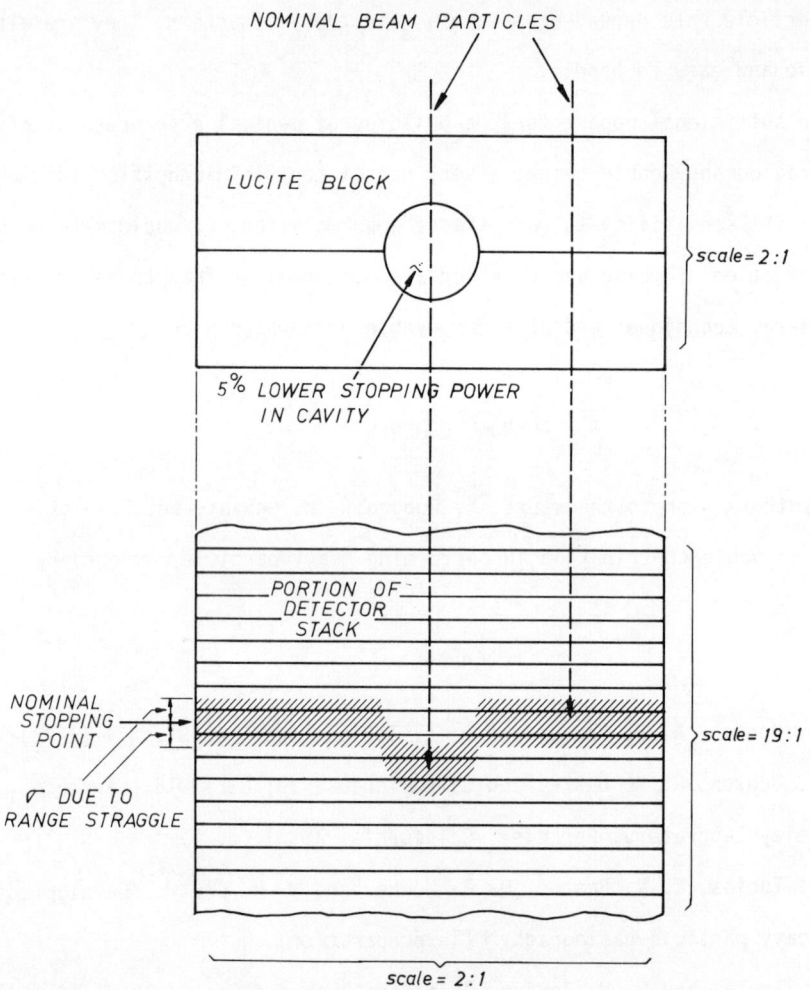

Figure 1

An illustration of imaging with a heavy-particle beam, using a multi-layered stack of thin plastic track detectors for image registration. The lower stopping power of the cavity in the Lucite block results in a downstream displacement of beam particles traversing the cavity, relative to particles not traversing the cavity. The spread in the beam stopping points, inversely related to beam particle atomic mass number (as $A^{-0.5}$), is a result of range straggling. Note: The shaded area represents the stopping point region in the detector; the diagonal lines do not represent tracks. Actual particle tracks are perpendicular to the layer surface.

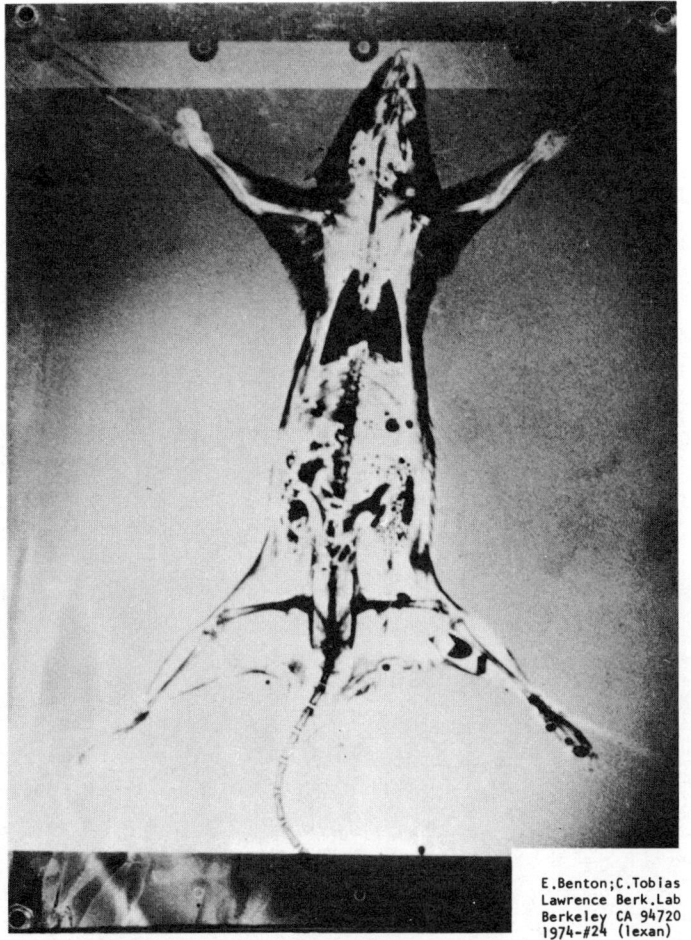

Figure 2a

A photograph of one layer of a 400 MeV/amu neon radiograph of a rat with Lexan used as the detector. The beam fluence is 10^6 particles/cm^2. The rat is immersed in a 2" wide Lucite water bath. Organs, such as the stomach and liver, can be discerned as well as small bone features such as the patella. The average dose to the rat was approximately 9 rads.

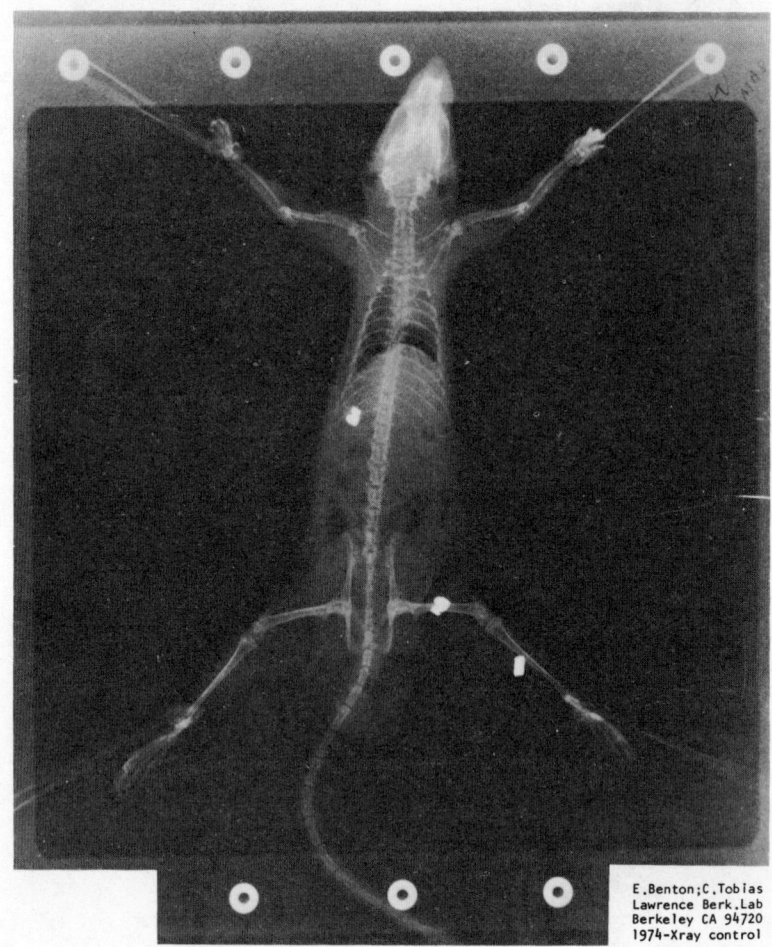

Figure 2b

A photograph of an X-ray radiograph of the rat in Fig. 2. The X-ray was made in air at 45 kv, 200 mas. The three bright spots on the radiograph are metal suture clamps.

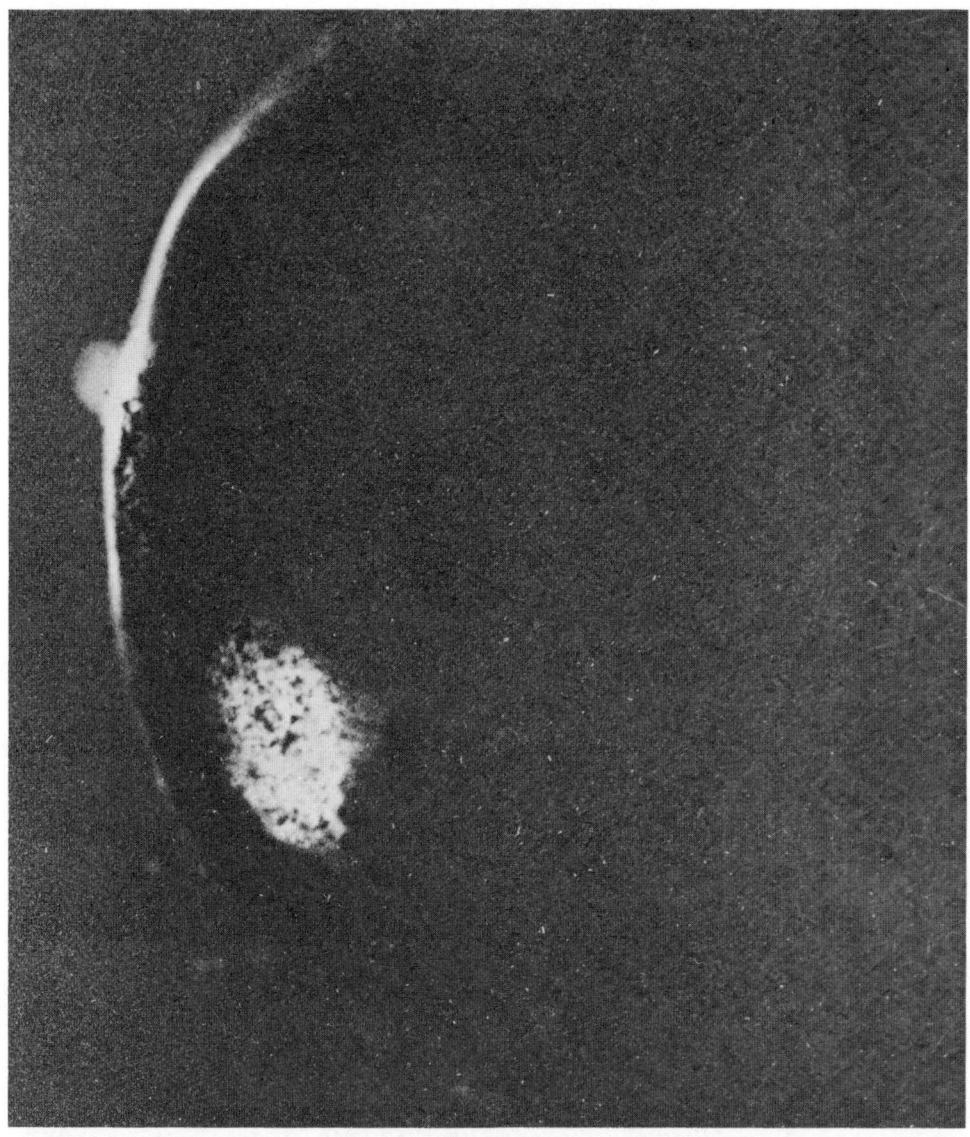

Figure 3

A composite photograph synthesizing three successive cellulose nitrate layers from a multilayered stack used to obtain a 400 MeV/amu oxygen radiograph of a freshly excised breast specimen. This composite represents our initial attempt at photographic synthesis of images from individual layers of a multilayered detector stack.

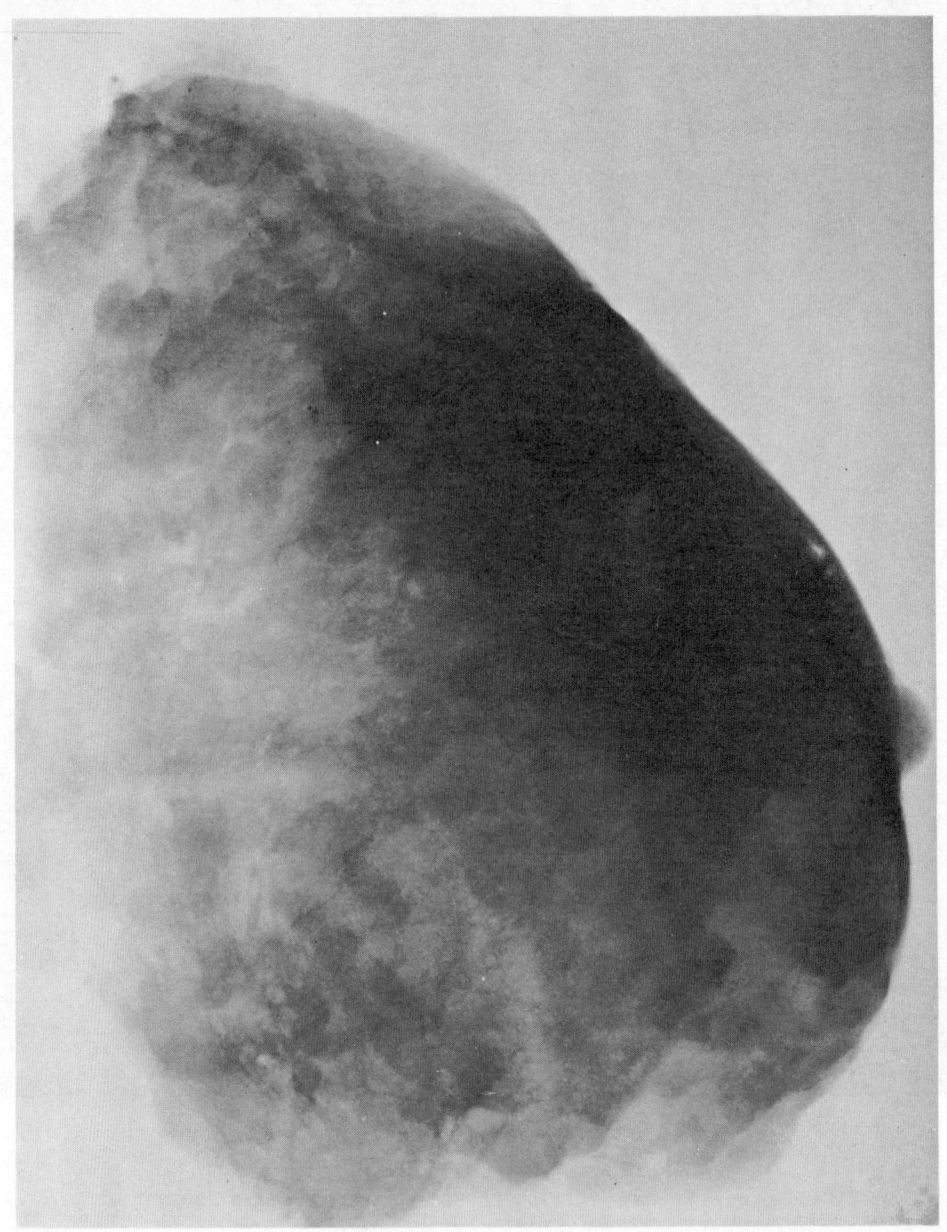

Figure 4

A photograph of an X-ray radiograph of specimen shown in Fig. 3 in air at 37 kv, 250 mas.

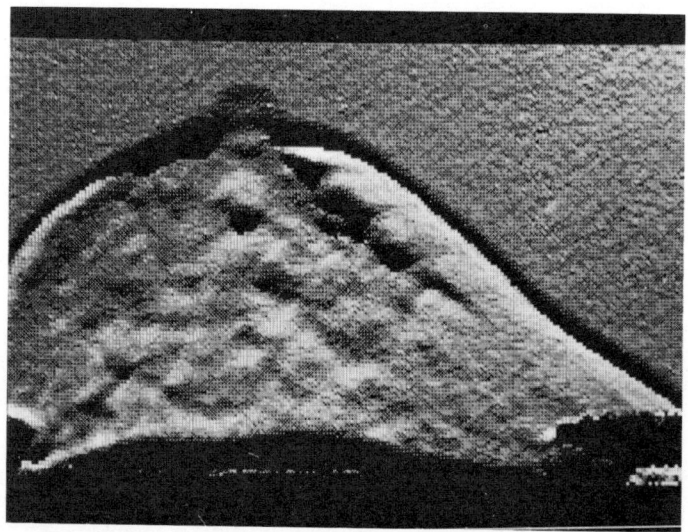

Figure 5

Simulated three-dimensional display of the synthesized radiograph of the right breast of patient B18. The computer represents the z(x,y) surface as it would be seen with light shining from the top of the figure at an angle of 70 deg with the normal.

FAST NEUTRON RADIOGRAPHY FOR EXTENDED OBJECTS BY A PLASTIC DETECTOR TECHNIQUE

E. Bagge*, E. Dühmke**, W. Enge*, W. Hunger*,
U. Roose*** and R. Scherzer*

*Institut für Reine und Angewandte Kernphysik, University of Kiel,
23 Kiel, West Germany
**Radiologische Universitätsklinik, University of Kiel, 23 Kiel,
West Germany
***INT der Frauenhofergesellschaft, 23 Kiel, West Germany

Abstract

Slow neutron radiography is restricted by the inability of slow neutrons to penetrate extended (up to 15 cm) objects. Therefore using 14.9 MeV-neutrons we tested the Kodak CA 80-15 plastic detector with regard to its applicability in fast neuton radiography. The efficiency of this detector as well as its blackenand its contrasting properties were examined in order to achieve an optical image. A conversion factor of 10^4 neutrons per track was measured and minimum contrast necessary for visibility was found to be 9 per cent. Furthermore, some experimental radiograms are shown and possible applications of fast neutron radiography are discussed.

Introduction

Radiography using slow neutrons was widely explored and quite successfully applied to many scientific and technical objects (1) because of different attenuation as compared to X-rays.

Extended objects, however, need fast neutrons because of the inability of slow neutrons to penetrate such an amount of matter. Although attenuation differences of inspected elements decrease with increasing energy it seems to be possible to analyse lighter elements masked by heavier elements (e.g. airpockets masked by bones or tissue (2) or the inside of steel-shielded objects). This expectation stimulated us to start studying fast neutron radiography by means of plastic detectors which were introduced to neutron detection by Fleischer, Price and Walker (3) and since then has become a powerful tool in radiographic field (4).

In a previous study (5) we performed radiographic experiments using fission neutrons (about 1 MeV mean energy). In continuation of this work we attend to the application of 14.9 MeV-neutrons.

In analogy to optical photography we consider the equivalent terms like sensitivity - efficiency, density - blackening and contrast.

Efficiency

Studies concerning the efficiency of the plastic CA 80-15 for the detection of fast neutrons have been made using 14.9 MeV-neutrons generated in the $t(d,n)\alpha$ reaction. By means of an α-counter the source strength was known within 5 per cent. In comparison with the microscopical track counting a conversion factor of $k = 10^{-4} \pm 20\%$ tracks per neutron was determined, considering track formation on both surfaces of the detector sheet. Making use of the plastic components and the neutron cross sections the estimated value of k yields the same order of magnitude. The measured conversion factor shows a weak dependency on the etching conditions.

We tried to improve this conversion factor in order to reduce the radiation dose. The testing of several radiator materials yielded following results:

(a) Heavy radiator elements like ^{232}Th, ^{235}U and ^{238}U produce a background track density which may conceal the effect to be measured. These background tracks originate from spontaneous fissions and are thus increased with increasing exposure time.
(b) Light elements like Li,Be,B,C do not cause measurable variations of the track density.

Further studies are in progress.

Blackening

At a first step a measure for a quantitative treatment of macroscopic blackening induced by etched tracks had to be established. We believe that as an adequate approximation for the impression of the human eye the signal of a photomultiplier coupled to light microscope can be taken. Thus blackening of an irradiated and etched plastic foil in contrast to a nonirradiated foil is measured by the relative difference of multiplier signals which are proportional to intensity of transmission light.

Fast neutron radiography

The next step was to seek for microscopic parameters which influence the blackening. These parameters can be subdivided into three categories (Fig.1) according to the track density, the size of these tracks and the microdistribution of blackness of the single track. The relation between these parameters and macroscopic blackening is demonstrated in Fig.2 and Fig.3. The well known influence of etching time on the track size is shown in Fig.4. This results in a blackness growing curve of Fig.3 according to the t^2-law, since for perpendicular incident particles only the part of the track at the surface contributes to blackening. Deviation from this curve for higher blackening is due to overlapping tracks. The influence of the track number density can be seen from calibration curves (Fig.2), where

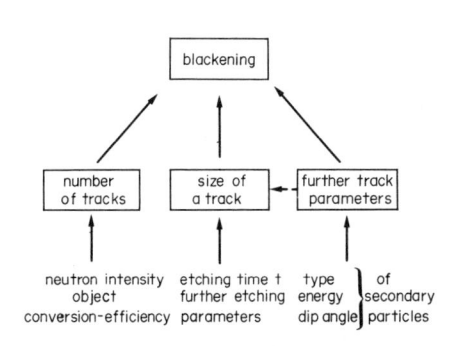

Fig.1 : Influence of parameters on blackening, which is caused by secondary particles.

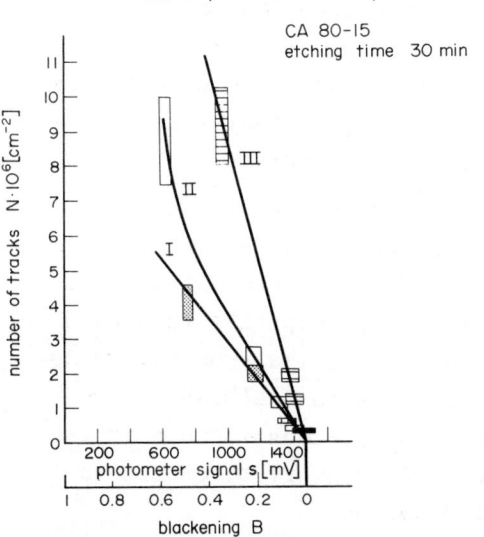

Fig.2 : Calibration curves for different types of blackening tracks

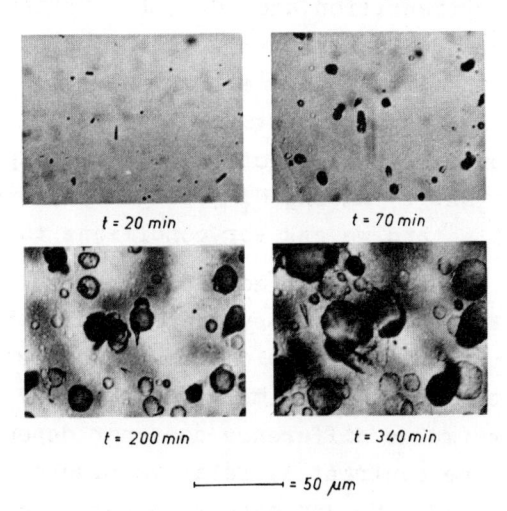

Fig.4 : Increase of track size with etching time

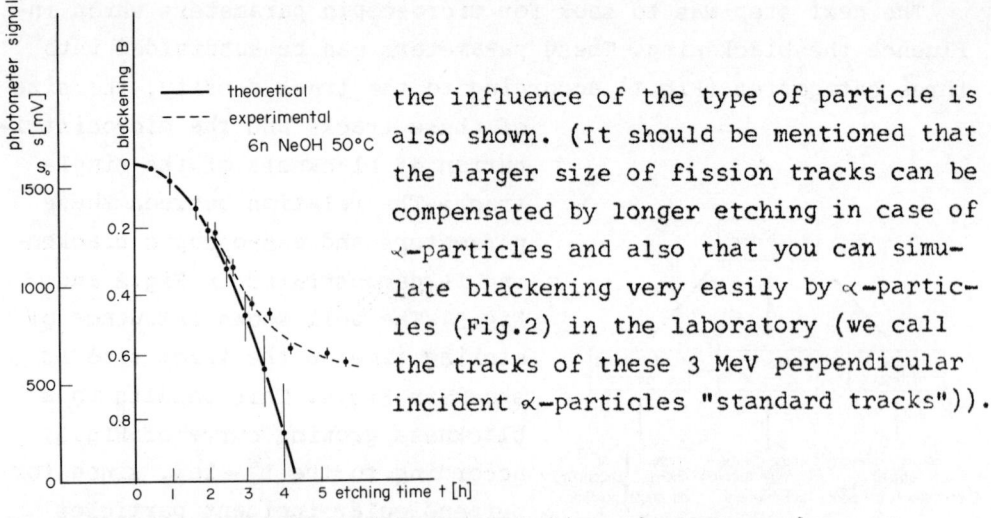

the influence of the type of particle is also shown. (It should be mentioned that the larger size of fission tracks can be compensated by longer etching in case of α-particles and also that you can simulate blackening very easily by α-particles (Fig.2) in the laboratory (we call the tracks of these 3 MeV perpendicular incident α-particles "standard tracks")).

Fig.3 : Growing of blackening with etching time. Error bars on calculated curve indicate for a few points the uncertainty which results from experimentally determined parameters.

Contrast

The visual contrast of radiograms recorded in plastics is based on compairing areas which are different in light reflection or transition according to different blackening induced by etched tracks. Since this visual contrast is influenced by irradiation condition (e.g. neutron energy), object (desired) and detector treatment we restrict to the last aspect neglecting parameters outside the detector. Thus we compare two areas of plastic (index 1 and 2) having N_1 (background) and N_2 ($N_2 > N_1$) latent track density, and we ask for conditions to get visible contrast.

Let us first consider the light reception of the human eye. From the psychophysical law of Weber (6) we learn that the perceptible threshold stimulus difference ΔS_{crit} is proportional to the absolute stimulus S, i.e. the relative perceptible threshold stimulus difference does not depend on S ($\Delta S_{crit}/S = $ const). We define contrast as relative blackening difference $C := (B_2-B_1)/B_2 \sim (N_2-N_1)/N_2$ and expect to get a relation $C_{crit} = $ const. Since it is not possible to "see" blackness we have to convert blackness to light intensity using dark field illumination.

To study the dependence of critical contrast C_{crit} on blackening we used several test foils showing different background blackening B_1, which was measured by photomultiplier. We provided

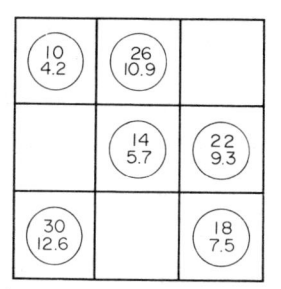

Fig.5 : Example of a test foil provided with contrast spots by standard tracks.
The upper number means irradiation time in min, the lower number means contrast C in per cent.

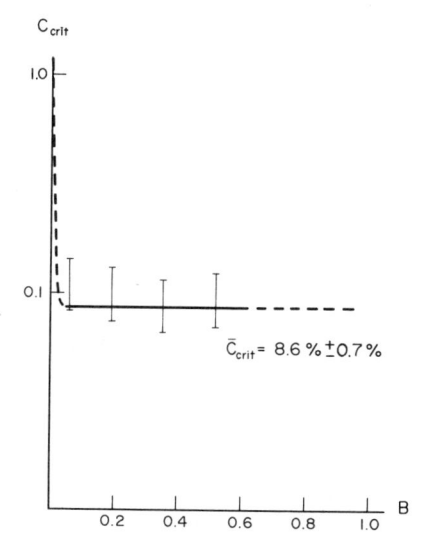

Fig.6 : Critical contrast in dependence of blackening. The dotted part of the curve has not been measured, it has been drawn according to plausibility considerations.

these foils with contrast spots of variing blackening B_2 in random order (Fig.5) by our α-particle source ($\emptyset$ = 3 mm). The examination of the foils by several persons with regard to visibility of these spots was accomplished by help of a small darkfield illumination box which achieved objective and reproducable experimental conditions. The results (Fig.6) show a critical contrast of about 9 per cent, which remains constant within the considered range of background blackening B_1 (see law of Weber).

Further experiments will be performed to prove if smaller differences of contrast than C_{crit} can be observed photometrically.

Applications

Knowing C_{crit} we constructed a paraffin dummy (Fig.7) whose alternating steps should be visible in a radiogram except in one case. This expectation was verified by our experimental observations (Fig.8).

If a radiographic object guarantees a contrast $(1 - N_1/N_2) > C_{crit}$ the image will be visible in the radiogram. One only needs a sufficient etching time, while further etching does not improve the contrast. Overlapping of tracks due to "overetching" diminishes the contrast because $(B_2 - B_1)/B_2 \to 0$. In this contents it is interesting to mention the analogy to over- resp. underdevelopment in photography.

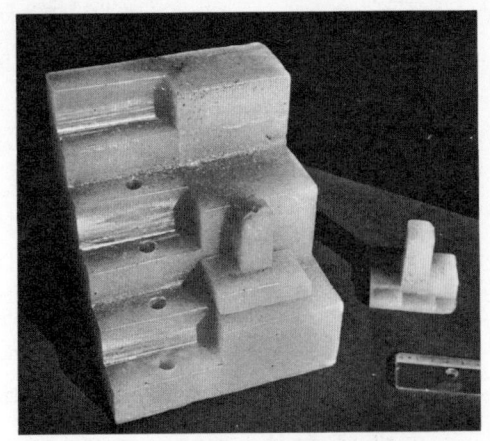

Fig.7 : Paraffin dummy for 14.9 MeV neutrons (left) and 1 MeV (right, used in former studies (5)).

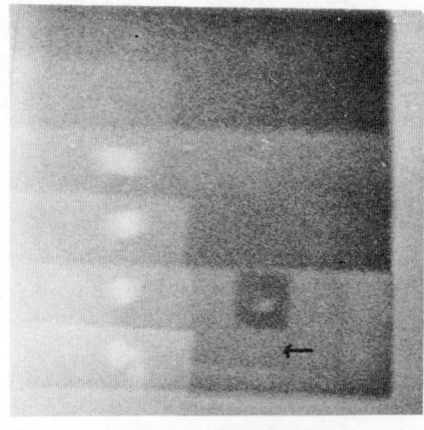

Fig.8 : Neutron radiogram of the 14.9 MeV dummy. Indicating arrow marks invisible step ($C = 4\%$).

Fig.9 : Neutron radiogram of a 1 cm steel shielded paraffin block containing a table tennis ball (total thickness 12 cm).

Because of the good penetration capability of the 14.9 MeV neutrons (interacting mean free path in $H_2O = 13$ cm) it is practicable to image hidden air-pockets inside of objects of this size. Fig.9 shows the neutron radiogram of a 1 cm – steel shielded paraffin block where the implanted table tennis ball is visible. Using a 120 kV X-ray equipment the ball could not be discovered.

In Fig.10 and Fig.11 one can see the neutron radiogram and the corresponding X-ray radiograph of a roe-head. The calcium containing bone tissue appears as transparent as soft tissue but gas hollows contrast to soft tissue as well as in X-ray radiography. Hence for medical application fundamentally follows the possibility to image pathological processes in soft tissue perhaps better, but in any case free from disturbances by superpositionend skeleton.

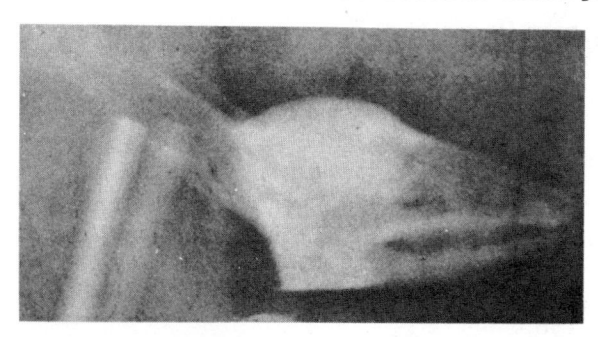

Fig.10 : Neutron radiogram of a roe head.

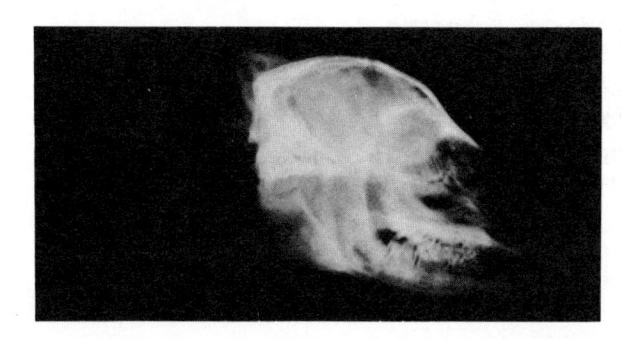

Fig.11 : X-ray radiograph corresponding to Fig.10 (120 kV , autom. exposure).

The neutron fluence at the roe head amounted $2.4 \cdot 10^9$ n/cm^2. Using an energy dose of $5.19 \cdot 10^{-9}$ rd per n / cm^2 (7) and a RBE-factor of 6 the skin dose yields to 75 rem and lies therefore according to size between diagnostical and therapeutical single dose.

Outlook

Up to now we only tested the detector and its imaging properties. For the future the examination of contrasting materials as well as of other converter matter is planned in order to achieve a reduction of radiation dose.

Furthermore, calculations and experimental inspections of materials and medical objects will be carried out to study the contrast properties of these objects. The findings of these studies should exhibit the range of use for fast neutron radiography in medical problems and nondestructive testing.

This work was supported partly by the
Deutsche Forschungsgemeinschaft.

References

(1) Berger, H.,1971, Ann. Rev. Nucl. Sci.
(2) Parks,P.B.,Brown,M.,Biomedical Sciences Instrumentation 6,118-126,Pittsburgh,Instrum.Soc.Am.
(3) Fleischer,R.L.,Price,P.B.,Walker,R.M.,1965, Nucl.Sci.Eng. 22,153-156
(4) Berger,H.,1973,Nuclear Technology 19,188-198
(5) Bagge,E.,Dühmke,E.,Enge,W.,Scherzer,R.,1975, Atomkernenergie,26,198-200
(6) Rohracher,H.,1971,"Einführung in die Psychologie", 113 ff.,Urban+Schwarzenberg
(7) Angerstein,W.,1975,"Lexikon der radiologischen Technik in der Medizin",Thieme,275

A NEUTRON RADIOGRAPHY FACILITY ON THE IRT-2000 REACTOR

I. Y. Khadduri

Nuclear Research Institute, Tuwaitha, Baghdad, Iraq

A neutron radiography facility has been constructed on the thermal neutron channel of the IRT-2000 reactor. A collimated thermal neutron beam exposure area of 10 cm diameter is obtained with an L/D ratio of 48.8. The film used is cellulose nitrate coated with lithium tetraborate which is insensitive to gamma and beta radiation. Some pictures with good contrast and resolution have been obtained. Pictures of parts of an IRT-2000 reactor fuel pin have also been recorded.

1. Introduction

The use of neutron beams for radiographic purposes is a relatively new method of non-destructive testing. A major advantage of using neutrons for radiography[1,2] is that certain material combinations are easier, and even possible, to radiograph with neutrons whereas problems would arise with X-radiography. This is primarily due to the different attenuation mechanisms inherent in the two methods. The attenuation of neutrons in a material is a function of the nuclear absorption and scattering cross sections and is independant of the charge of the nuclide, in contradistinction to the situation with X and γ-rays where the charge, which increase monotically with the atomic number, is the important factor in their attenuation.

For example, the comparatively high attenuation of thermal neutrons in light and medium elements such as hydrogen, lithium, boron, cadmium and several rare earth elements provides for their detection with neutrongraphy even when they may be combined with some high atomic weight materials such as steel, lead, or uranium. Such a material combination would be difficult to X-radiograph due to the strong attenuation of the X-rays in the heavy material. Furthermore, the sharp differences in the attenuation of thermal neutrons often found between neihboring materials in the periodic table, and even among the isotopes of one material, offer an advantage for neutrongraphy discrimination between materials that have similar X-ray attenuation characteristics. In addition, neutrongraphy of highly radioactive samples, such as reactor fuel pins, is possible with the use of certain neutron sensitive films that can discriminate against gamma and beta radiation or alternatively the "transfer method" is used by which the converter foil (e.g. dysprosium) is neutron activated on its own and is then taken away from the radioactive sample and the neutron irradiation area and is placed on the recording film that would print the image due to the decaying activity of the converter.

As one of the objects of this project was to neutron radiography a fuel pin of the IRT-2000 reactor fuel cassette, a recently available cellulose nitrate film, which is insensitive to gamma rays, was used and the details of which will be outlined below.

2. The collimated thermal neutron beam

The IRT-2000 reactor, a 2 MW swimming pool type reactor, served as the neutron source. One of its experimental channels, a graphite moderated thermal channel, was used for neutron radiography purposes, see figure (1). The diameter of the entrance opening (D) of the thermal channel is 15 cm. A lead plug is inserted at the beginning of the channel in order to reduce the gamma contribution. Four converging collimators, made of lithium floride and lead, are installed along the channel within the biological shield wall which narrows the beam to 40 mm. diameter. A further collimator constructed of several layers of lead and borated parafin and situated within the concrete shielding block further reduces the neutron beam diameter to 20 mm[3]. Beyond that, a lead cave 50 cm x 50 cm x 320 cm long 10 cm. thick was constructed. It was lined on the inside with 1 mm. thick cadmium sheet in order to absorb the scattered neutrons. At the end of the lead cave, a 20 cm diameter lead door, 10 cm thick, that is operated manually, allowed access to the film and object holders, see figure (2). A neutron beam catcher consisting of lithium floride, lead, and 10% boric acid solution is placed at the end of the lead cave. A prompt neutron activation analysis assembly is situated near the begining of the cave sharing the use of the thermal neutron beam.

The thermal neutron flux inside the lead cave at 50 cm. beyond the 20 mm. diameter collimator was measured to be 1.5×10^8 n/cm^2/sec. at 2 MW reactor power[3]. The total distance (L) from the entrance opening of the thermal channel to the film was 732 cm., and hence, the neutron beam transport system L/D ratio is 48.8. A neutron beam exposure area diameter of about 10 cm. can be obtained at the position of the film.

3. The photographic procedure

The film used in this work is the specially treated cellulose nitrate film (KODAK CA 80-15 type B)*. These films consist of 100 μm thick lightly rose-tinted cellulose nitrate which is light insensitive and is primarily intended for recording α particles or fission fragments. The film is coated on both sides with lithium tetraborate which is dispersed in a water soluble binder that is easily washed off with ordinary water after irradiation and prior to the etching procedure. The lithium tetraborate serves as the (n,α) converter with the following reactions, ^{6}Li (n,α) T and ^{10}B(n,α)^{7}Li. The tracks of the α particles that penetrate the cellulose nitrate are greatly enhanced into pitted holes upon etching in a strong alkaline solution[4]. Several million tracks/cm^2 are needed for a visual image.

The procedure for taking a neutron radiograph is to place the object in front of and in contact as much as possible with the cellulose nitrate film. An exposure time of 20 minutes at 2 MW reactor power is sufficient to produce an image with good contrast. The film is then removed and washed gently with

----

* The film can be obtained from : KODAK - PATHE, 30 Rue des Vignerons, 94300 Vincennes, FRANCE.

water in order to remove the lithium tetraborate coating. Care must be taken not to wipe the coating with the fingers lest the film is scratched but to let the running water to slowly dissolve the coating. The film is then submerged in a 2.7N KOH solution (150 grams of KOH Analar pellets in one litre of distilled water) at 30°C in a constant temperture water bath for 100 minutes. It is then dipped into a 2% acetic acid solution and rinsed with water for one hour and dried with a hot air hand blower.

The film can then be viewed for the picture of the object either directly or it can be considered as a "positive" from which a negative is obtained (industrial R film) and then printed on photographic paper (G1 low contrast). Alternatively, the cellulose nitrate film itself can be considered as a "negative" and printed directly on photographic paper (G4 Kodabromide extra contrast).

4. Results and conclusion

Neutron radiographs of a clock and of a collection of a hand lighter, a bullet, and an electronic valve were obtained as shown in figure 3. One can notice the good resolution in the details of the clock especially in the teeth of the wheels of the clock. Of interest also is the strong neutron attenuation effect of the light organic material of the coton and filament of the lighter and the powder of the bullet.

A major aim of this work was to begin acquiring an experience in neutron radiography of nuclear fuel pins and irradiated assemblies and loops. An initial attempt was undertaken to radiograph the 3K-10 10% enriched and unirradiated fuel pin of the IRT-2000 reactor fuel cassette. After preliminary calculations on the effects of a 20 minutes neutron exposure of the fuel, it was found that the temperature increase of the pin would be negligible (an increase of 0.1°C) and that the gamma activity of the pin would yield about 50 mRem/hr per 10 cm of the fuel at a distance of 30 cm. which is tolerable; (it was found, in fact, to be 40 mRem/hr).

Figure 4 shows the pictures of two neutron radiographs of the fuel pin, one is an end section and the other a midsection of it. Several points on the physical properties of the pin were concluded:

1. The aluminum cladding of the fuel pin is 1.5 mm thick and not 1 mm thick as originally supplied by the designer.
2. The nuclear fuel is apparently composed of one single rod and not of several pellets.
3. It has been reported[*] that the fuel pin of a similar kind was found to contain a graphite plug inside one of the aluminum end sections of the pin, and it was thought that the 3K-10 fuel pin might also contain such a graphite plug. But neutron radiography of the two end sections of the fuel pin did not show any evidence of such a plug.

The present facility allows for taking a picture of up to 10 cm diameter exposure. A larger exposure area can be achieved by removing the last collimator in the concrete shielding block near the channel exit. It is predicted that with the end collimator removed, a 20 cm diameter area can be achieved at the expense of worsened resolution. Alternatively, one can place the object and film at shorter distances to the last collimator thus obtaining better resolution but with a smaller area of exposure.

Further work on this project would be to employ the transfer method (using a dysprosium foil) with the aim of neutron radiographing a spent radioactive fuel pin.

The author acknowledges with gratitude the assistance of R. Barbalat in obtaining some of the negatives from the cellulose nitrate films and S. Zamel for providing the positive prints.

---

[*] Oral Communication, Atomic Energy Commission, Egypt.

References

1. Hawksworth, M. R. and Walker, J., "Review: Radiography with Neutrons". J. of Materials Science, Vol. 4, (1969), pp. 817-835.

2. Berger, H., "Neutron Radiography", Annual Review of Nuclear Science, Vol. 21, (1971), pp. 335-364.

3. Jafar, J. D. et al., "Measurement of the $^{35}Cl(n,\gamma)^{36}Cl$ using a Three Crystal Pair and Anti-Compton Spectrometer". Iraqi Atomic Energy Commission, Report No. Ph-2 (1970).

4. Barbier, J., "Transactions of the American Nuclear Society," Vol. 13, (1970) p. 530.

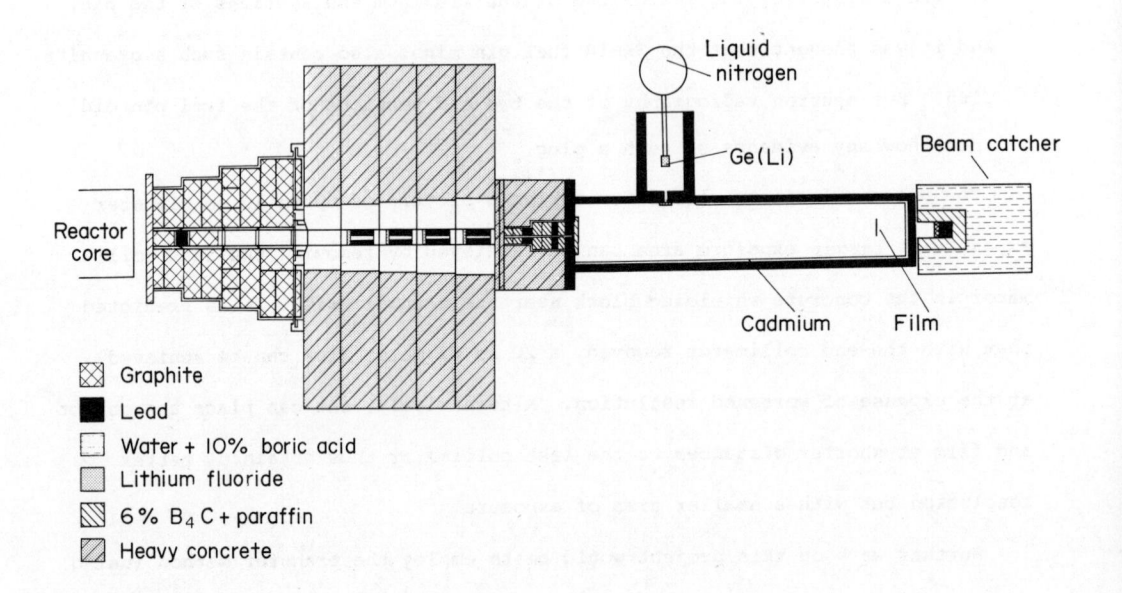

Fig. 1. Neutron Beam Transport System.

Figure 2 : The lead cave for neutron radiography.

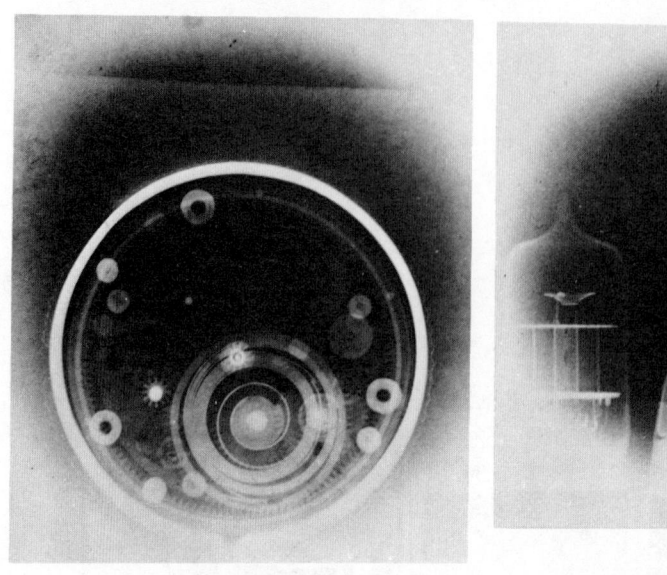

Figure 3 : Pictures of neutron radiographs with cellulose nitrate film of a) a clock and b) a collection of a hand lighter, a bullet and an electronic valve.

A neutron radiography facility 769

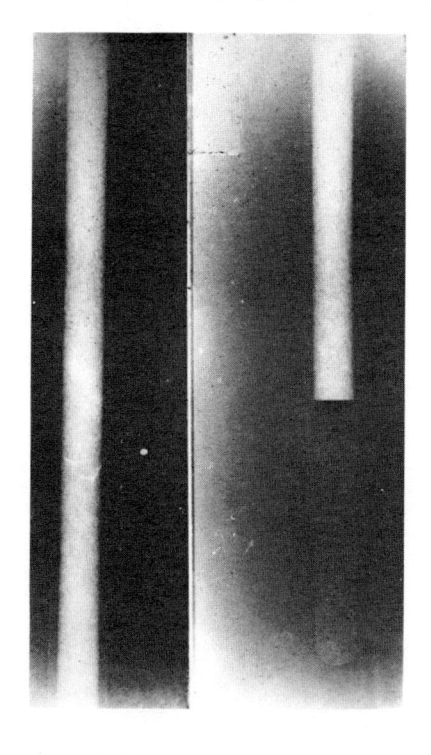

(a)   (b)

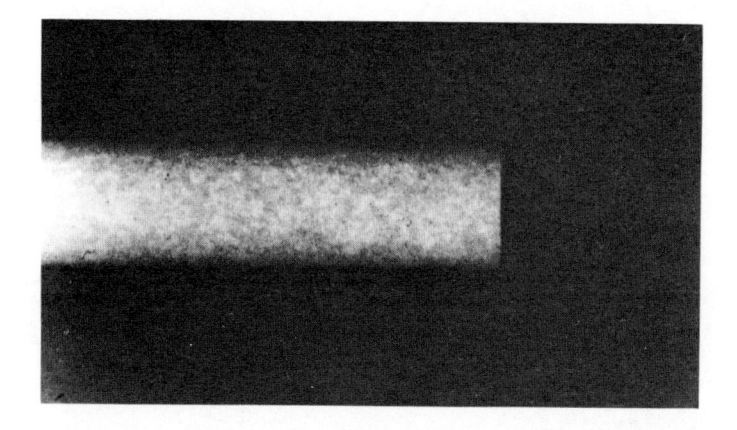

(c)

Figure 4 : Pictures of neutron radiographs of sections
of an 3K-10 fuel pin, a) a midsection of it,
b) an end section, and c) an end section (x3).

POSSIBILITIES AND LIMITS OF USING NUCLEAR TRACK DETECTORS FOR DOSIMETRY AND MICRODOSIMETRY

M. Nicolae

Institute for Atomic Physics, Bucharest, Romania

Abstract

The mean characteristics of the nuclear track detectors suitable for application in dosimetry and microdosimetry are summarly analysed; some sources of systematic errors as the procedure and the tedious character of experimental data aquisition are mentioned as factors which can limitate the utilization of these detecting systems in dosimetry. Some experimental results obtained at the Institute of Atomic Physics from Bucharest are then presentend which underline the advantages of the method for some special applications in dosimetry. It concers these situations which necessitate: high spatial resolution as energy distribution measurements in small volumes in microdosimetry and biological applications; time integrating capacity as for low radioactivity measurements; particle-particle or particle-gamma ray discrimination capacity as neutron dosimetry.

1. Introduction

Dosimetry of to-day makes use of a lot of methods which are in competition as concerns the fulfiling of the basic requirements of dosimetry as: suitable sensitivity, time stability to the environmental factors, good accuracy and reproductibility, low prices easly handling etc.

The photographic method generally used till 10 - 15 years ago for routine personal dosimetry was obliged to restrict its application area in competition with the solid state luminescences detectors method and the electronic pocket alarm dosimeters.

The last years, a lot of scientists start to use the method of Solid state track detectors for neutron dosimetry.

Some attitudes of reserve are observed as concerns the routine use of the method due, at first, to the tedious character of measurement.

Therefore is very important to perform an objective analysis in view of detach these features of the nuclear track detector method and these situations when that becomes preferable owing to its advantages.

In this paper we will tray to answer partially to this problem, unassuming to be exhausted.

2. Mean feature of the nuclear track detectors

The nuclear track detectors, it means nuclear emulsion and solid state nuclear track detectors, become very convenient for a lot of scientific or routine applications due to the following positive features:

1) possibility to registrate single nuclear particle tracks;

2) possibility to analyse the particle parameters as: mass energy, charge by performing measurements along the tracks;

3) particle-particle or particle-gamma ray discrimination capacity;

4) spatial integrating capacity;

5) time integrating capacity, which enables high sensitivity;

6) spatial resolution capacity which enables particle energy distribution measurements on small volumes (microvolumes);

7) tissue equivalence (for plastic track detectors).

Beside the above mentionned positive features we must to count some unfavorable characteristics as:

1) the tedious character of obtaining the experimental data;

2) a chemical processing is necessary to render visible the particle tracks what alter the primary information and can generate some systematical errors.

Therefore, it seems resonable to recommend the application of the nuclear track detectors in dosimetry for such purposes which necessitate:

1) nuclear particle discrimination in mixt radiation field as for instance neutron dosimetry;

2) high sensitivity, by using the time integrating capacity of the detector, what becomes very important for low α-radioactivity measurements;

3) high spatial resolution what is very important in microdosimetry for studyng the spatial distribution of energy absorbed in small volumes.

The method can be successfuly used in radiobiological investigation dealing with heavy corpuscular radiations or in medical dosimetry.

In this paper we tray to support these conclusions with some of our applications when the method of track detection gave excellent results.

3. Some application of nuclear track detectors in dosimetry

3.1. Determination of the internal α-radioactive contamination by spute autoradiography

The problem to detect the internal contamination through the respiratory tractus tp the personal working with α-radioactive materials - from nuclear industry and uranium mines is of great importance.

Taking into account the high LET of α-particles, the radioactive contamination with α-radioactive elements is more dangerous than that due to β ray emiters.

The α-particle energy is delivered into a very small tissue volume, practically one cilinder with the length equal to the particle path and the diameter from some tens of microns till 1 - 2 um.

In such conditions, the conventional (macroscopic) methods of dosimetry cann't supply correct informations.

On the other hand, no routine procedures are available

for α-radioactive internal contamination control, excepting of associated γ-ray spectrometry.

We studied a method for α-radioactive contamination control by using the α-track autoradiography with nuclear emulsions.[1]

Briefly, our method consists of the following. Small amount of spute from subjects supposed to be α-radioactive contaminated is droped-out and uniformly spreaded on the center of a small glass plate, on an areea of about 2 cm^2. After drying, the plate is covered by an 100 um coat of nuclear emulsion IFA EN3. The samples are kept for 30 days at 3 - 5°C in dry athosphere for exposure and then the emulsion is processed. On the processed samples, the α-track density is measured. Some times, microscopical particles of radioactive mineral as can be seen in fig. 1 (a, b, c) were observed. Together with the contamined samples 5 - 6 control samples, containing spute from a healthful men were processed.

The level of internal contamination is estimated by the formula:
$$C = \frac{N - \bar{N}_M}{\sqrt{\sigma_1^2 + \sigma_2^2}}$$ were

N is the α-track density on the measured sample;
$\bar{N}_M$ is the mean value of the α-track density on the control samples;
σ_1 and σ_2 - the standard deviation of the α-track density on the control samples, and on the measured pattern, respectively.

The internal contamination is considered significant when $\frac{N - \bar{N}_M}{\sigma} \geq 2$.

The radioactive concentration on the sample will be:
$c = k(N - \bar{N}_M)$ where k is an experimental parameter - which relates the unity of radioactive concentration on the track density.

In tab. 1 we give some results obtained by using the method for routine works. After two years of application it was proved extremly sensitive and very specifical.

The calibration of the method is in course in order to determine the local dose equivalent which corresponds to the small radioactive mineral particles observed on some samples.

We estimate this kind of application to be an evidence. on the advantage of the nuclear emulsion as analytical and integrating detector of high sensitivity.

3.2. Fast neutron dosimetry

A great number of papers dealing with the application of the solid state detectors in dosimetry are related to the neutron fluence measurements [2, 3, 4, 5].

In this paper will be breefly described the spectrodosimeter for neutrons more in detail presented in [6]. It consists from a plastic holder as can be seen in fig. 2 which contains inside four (n,α) targets, as presented in fig. 3 each of them being in good contact with a cellulose nitrate foil used as α-track detector. The (n,α) targets are small pellets or foil of: LiF, Al, C, B_2O_3 respectively fixed in the walls of the holder. In the center of the holder is fixed a single foil of cellulose nitrate as control detector for α and recoil tracks generated from the cellulose nitrate foil. After exposure to fast neutrons, the cellulose nitrate detectors are processed and then the α track density is measured on each of them corresponding to a given target and at the same time on the control one.

A computer programme was established in order to calculate the neutron equivalent dose and the energy spectrum as function of the α-track density on each sheet, by using the method of model spectra [7].

In fig. 4 we give the energy spectum determined for a 14 MeV neutron beam by using this method. As can be seen the agreement between the experimental and the theoretical spectra is rather good. We expect some improvement of the method in order to use it for routine work in personal dosimetry. Its great advantage is to alow the estimation of the energy spectra and therefpre a correct determination of the neutron equivalent dose by using the I.C.R.U. quality factors.

3.3. Study on the absorbed dose spatial distribution by using radiochromatic detectors

As was already mentioned, the method of track detectors is considered rather tedious for routine personal dosimetry. In order to combine some advantages of the method as its high sensitivity and spatial resolution with the requirements of routine works the possibility of using some plastic track detectors as radiochroms for medical dosimetry, was investigated in [8]

The radiochromatic plastics, already widely used in megarad dosimetry, are not enough sensitive for medical dosimetry despite to be very convenient for such purpose from other point of view.

We studied the possibility to improve to a great extent the sensitivity of the method by eching the plastics after exposure.

The method, described in detail in other paper [9] consists of the following. Cellulose nitrate sheets of 40 - 60 um coloured in violet - blue or yelow with cyanine dyes were exposed to α particles or to neutrons for doses between few rads till some hundred rads. No changes in the colour was observed by spectrophotometric measurements after exposure.

Such changes become visible only above one thousand rads. But if the coloured cellulose nitrate is etched after exposure, visible changes in transparency can be observed even by eye at small doses, as few rads.

Some direct relation between the optical transparency and the dose were observed as presented in fig. 5.

The images observed by eye or at slow magnification have high resolution and very good location of the particle beam can be performed what is of great importance for medical dosimetry.

We estime the method to be useful for neutron therapy when stacks of a great number of cellulose nitrate foils can be used.

After a standard etching the cellulose nitrate sheets are observed by a microphotometer in order to measure the transparency change.

The spatial distribution of the absorbed dose can be easy determined as function of the transparency distribution on the measured sheet.

3.4. Microdosimetry of α and β radioactive airborne particles

As is known, the radioactive airborne particles have an esential contribution to the internal contamination of the personnel. Because the energy of α or B particles is deposed in a very small tissue volume, a correct estimation of the local tissue dose can be performed only it the average size and the activity of such airborne particles are known.

We studied an autoradigraphic method intended to this purpose.

The method which is presented in detail in other paper [10] consists in principle of the following.

From the contamined atmosphere radioactive airborne particles are prevented on a filter tape by some conventional way. Then the filter tape is bronght in good contact with a plate of nuclear emulsion of 100 μm. After some delay of time, of 1 - 2 hours till about 24 hours, necessary for exposure, the emulsion is developed and observed to the microscope.

Images like those given in fig. 6 and 7 can be seen, which correspond to B radioactive airborne particles of Au^{198} and P^{32} respectively.

By an photometric microscope Leitz MPV-1 were measured the optical transparency and size distribution of such images given by single radioactive airborne particles.

Similar measurements were performed on the standard pattern having known activity from the same radionuclide.

By using these data the activity of each observed radioactive airborne particle of Au^{198} and P^{32} was calculated as presented in tab. 2 and 3.

As can be seen, activities situated between $10^{-13}-10^{-14} Ci$ for each particle were measured. From these activities the tissue absorbed dose in small volumes comparable with the α-particle or effective electron path can be calculated.

In conclusion we estime this method offer new possibilities by respect to the conventional radiometric methods by making use of the high sensitivity and spatial resolution of the nuclear emulsion.

4. Conclusions

1. In competition with other methods used today in dosimetry, the nuclear track detectors are imposing for some kind of applications due to their remarcable features: time integrating capacity, high spatial resolution, possibility to render visible single particle path, possibility to measure energy or dose distribution in small tissue volumes.

2. A large field of application can be found for nuclear track detectors in microdosimetry, medicine and radiobiology due to the above mentioned features.

3. New combined methods which make use of some rapid kind of measurement like microphotometric or spectrophotometric procedures instead of microscope observation are intended to improve the tedious character of data aquisition what is of great importance for routine work in dosimetry.

TABLE 1
AUTORADIOGRAPHIC DETERMINATION OF RADIOACTIVE CONTAMINATION - SOME EXPERIMENTAL DATA

	Sample No	Mean density tracks/cm^2	$N - N_M$	$\dfrac{N - N_M}{\sqrt{\sigma_1^2 + \sigma_2^2}}$	Remarks on contamination
	Control 1	160			
	Control 2	167			
	Control 3	192			
	Control 4	162			
	Control 5	137			
	Control 6	152			
	Control (mean value)	162±25			
1.	603	211 \ 223	61	2,44	Significant
2.	603	235 /			
3.	604	221 \ 213	51	2,55	Significant
4.	604	205 /			
5.	607	212 \ 199	37	1,85	Nonsignificant
6.	607	187 /			
7.	611	160	355	17,7	Microscopic mineral particles observed
8.	611	517 (185)*	(10)		
9.	612	212 \ 234	72	2,88	Significant
10.	612	257 /			
11.	613	1098	936	37,4	Microscopic mineral particles observed (high contamination)
12.	613	1489 (414)*	1327	53,1	
13.	617	216 \ 206	44	1,76	Nonsignificant
14.	617	196 /			
15.	626	270 \ 292	130	5,20	Significant
16.	626	294 /			
17.	632	230	95	3,8	Significant
18.	632	284			
19.	634	277 \ 268	106	4,2	Significant
20.	634	259 /			
21.	635	292	109	4,3	Significant
22.	635	251			

* The values given in brakets are determined without taking in consideration the α particles generated by the microscopic mineral particles observed on the same pattern.

Tab. 2

Measured activities of Au^{198} airborne particles

Nr. crt.	d_{um}	T	D/mm^2	D/p	A $D/\delta \cdot p$	A Ci/p
1.	16,1	4	$1,25 \cdot 10^6$	$2,54 \cdot 10^2$	$3,37 \cdot 10^{-3}$	$0,91 \cdot 10^{-13}$
2.	18,4	5	$1,20 \cdot 10^6$	$3,18 \cdot 10^2$	$4,22 \cdot 10^{-3}$	$1,14 \cdot 10^{-13}$
3.	23	5	$1,20 \cdot 10^6$	$4,98 \cdot 10^2$	$6,59 \cdot 10^{-3}$	$1,78 \cdot 10^{-13}$
4.	36,8	2	$1,30 \cdot 10^6$	$13,82 \cdot 10^2$	$18,28 \cdot 10^{-3}$	$4,94 \cdot 10^{-13}$
5.	6,44	3	$1,28 \cdot 10^6$	$41,67$	$5,51 \cdot 10^{-4}$	$1,49 \cdot 10^{-14}$
6.	7,59	4	$1,25 \cdot 10^6$	$56,53$	$7,48 \cdot 10^{-4}$	$2,02 \cdot 10^{-14}$
7.	11,5	3	$1,28 \cdot 10^6$	$1,33 \cdot 10^2$	$17,58 \cdot 10^{-4}$	$4,75 \cdot 10^{-14}$

d = diameter of the airborne particle

T = Transparency measured on the airborne particle, in relative unities.

D = radioactive decay

p = airborne particle

Tab. 3

Measured activities of P^{32} airborne particles

Nr. crt.	Partic. diam. um	Transparency	decay/mm^2	decay/part.	Activity Decay/s/part.	Ci/partic.
1.	6,9	8	$5,6.10^5$	20,93	$2,77.10^{-4}$	$0,75.10^{-14}$
2.	6,9	7	$5,8.10^5$	21,67	$2,87.10^{-4}$	$0,77.10^{-14}$
3.	6,44	15	$4,6.10^5$	14,67	$1,98.10^{-4}$	$0,54.10^{-14}$
4.	20,01	5	$6,0.10^5$	188,59	$24,94.10^{-4}$	$6,74.10^{-14}$
5.	33,35	3	$6,4.10^5$	558,78	$73,91.10^{-4}$	$19,98.10^{-14}$
6.	9,20	4	$6,2.10^5$	41,19	$5,45.10^{-4}$	$1,47.10^{-14}$
7.	1,38	8	$5,6.10^5$	30,14	$3,99.10^{-4}$	$1,08.10^{-14}$
8.	6,9	9	$5,4.10^5$	20,18	$2,67.10^{-4}$	$0,72.10^{-14}$
9.	18,4	6	$5,9.10^5$	156,8	$20,74.10^{-4}$	$6,6.10^{-14}$
lo.	36,8	2	$6,5.10^5$	891,0	$91,40.10^{-4}$	$24,70.10^{-14}$
11.	6,9	10	$5,3.10^5$	19,81	$2,62.10^{-4}$	$0,71.10^{-14}$
12.	9,66	7	$5,8.10^5$	42,49	$5,62.10^{-4}$	$1,52.10^{-14}$
13.	11,27	8	$5,6.10^5$	55,83	$7,38.10^{-4}$	$1,99.10^{-14}$
14.	8,05	8	$5,6.10^5$	28,49	$3,77.10^{-4}$	$1,02.10^{-14}$

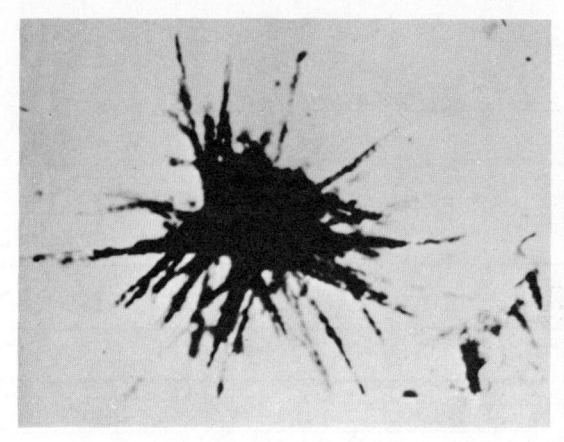

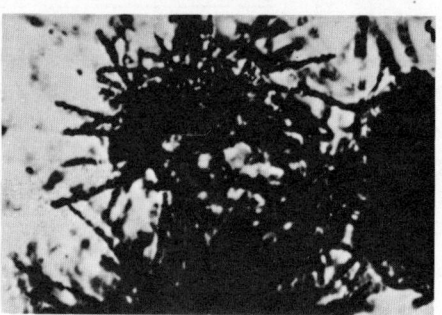

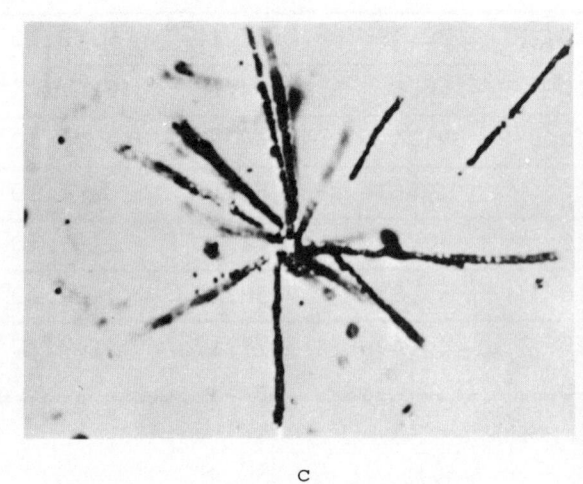

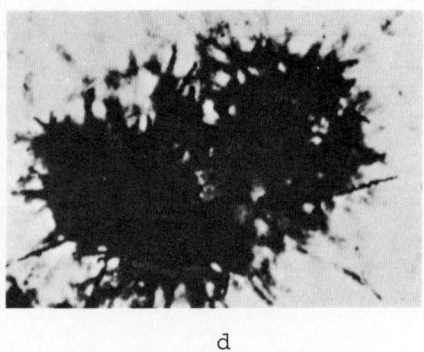

Fig. 1.

(a-d). Microscopical particles of radioactive mineral observed on some patterns (x 1700).

Fig.2. Outside view of the IFA - SDN - 8 spectrodosimeter for neutrons.

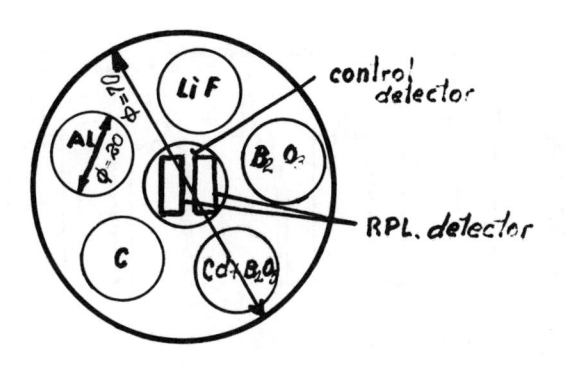

Fig.3. α-n targets used in the spectrodosimeter for neutrons.

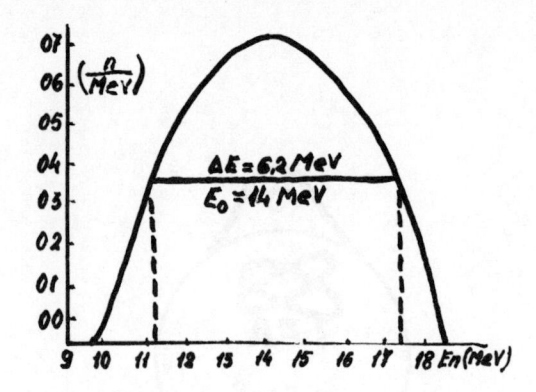

Fig.4. Experimental energy spectrum of a 14 MeV neutron beam.

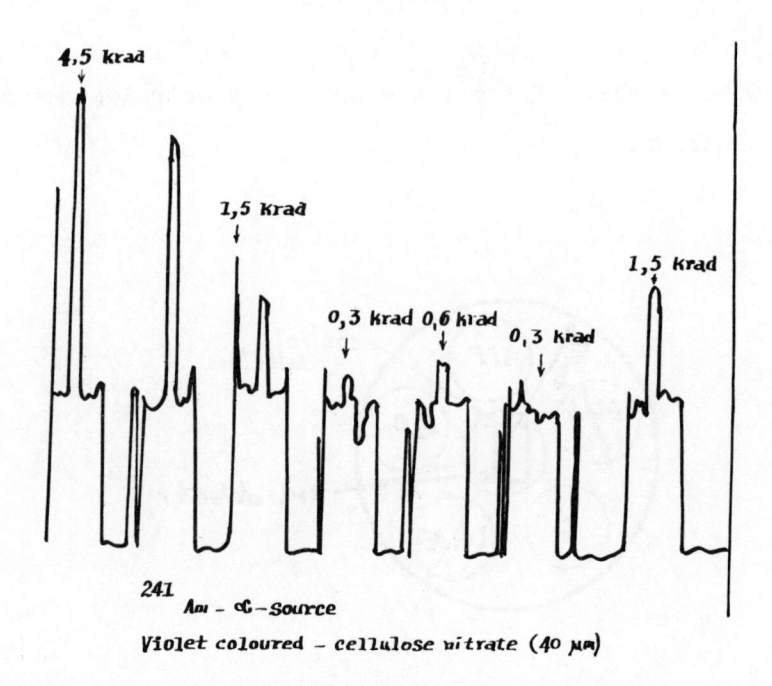

Fig.5. Curves of optical transparency measured on radiocromatic plastic sheets exposed by α - particles of different doses, etched after exposure.

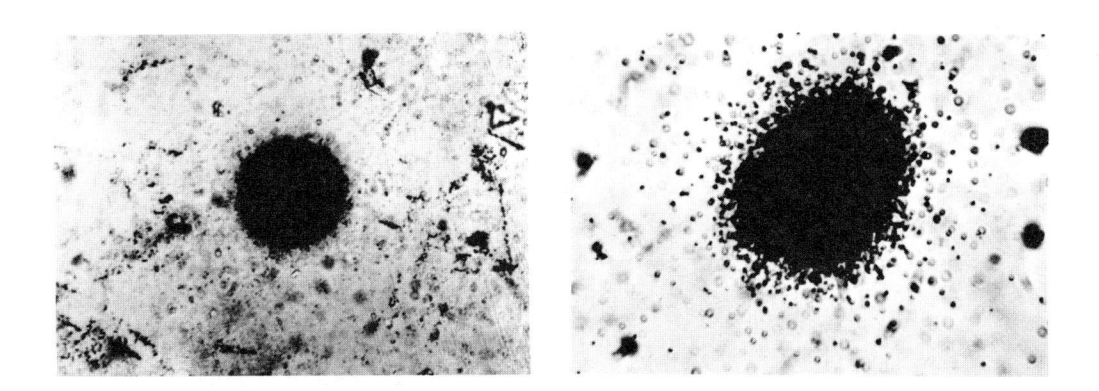

Fig.6. Microautoradiographic images of B radioactive airborne particles of P^{32}.

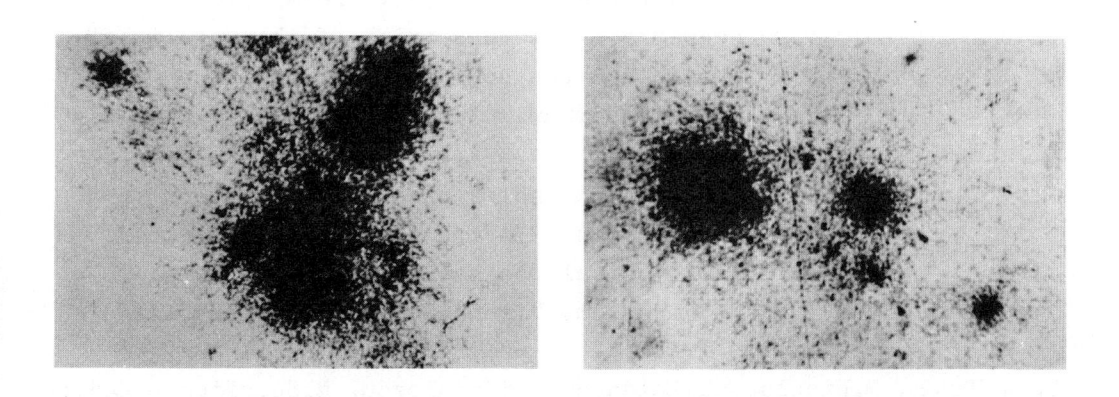

Fig.7. Microautoradiographic images of B radioactive airborne particles of Au^{198}.

REFERENCES

1. M. Nicolae, C. Păun – Determination of the internal α-radioactive contamination on respiratory tractus by spute autoradiography, Rev. Roum. Phys. under press.
2. K. Beker – Health Phys., 16, 113 (1969).
3. J.W.N. Tuyn – Rad. Effects, 5, 75 (1970).
4. F. Spurny – Tracks of charged particles formed in polimers foils by fast neutrons; implications to fast neutron dosimetry, Proc. 8-th Int. Conf. Nucl. Photography and Solid State Track Detectors, Bucharest 1972, p. 353.
5. A. Dragu, M. Nicolae – Application du nitrate de cellulose à la dosimetrie des neutrons rapides, Proc. 8-th Int. Conf. Nucl. Photography and Solid State Track Detectors, Bucharest 1972, p.362.
6. A. Dragu, D. Haşegan, M. Nicolae, Neitronîi spectrometrî dlea smeshnovo polea izlucenia, Problemî obespecenia radiationoi bezopasnosti pri exploataţii atomnîh electrostanţii; Sbornik docladii naucino tehniceskoi konferenţii SEV, Usti nad Labem, Sept. 1975, Praga 1976.
7. B. Arcipiani, M. Marsequerra – Interpretation of the results of some neutron spectrum elaboration procedures, Nucl. Instrum. Methods, 108, (1973) p. 301.
8. W.L. Mc. Laughlin, M. Rosenstein, H. Levine – Bone and Musche Equivalent solid chemical Dosimeters for Photon and Electron dose above one krad, Biomedical Dosimetry, Proceeding of a Symposium, Vienna 10-14 March 1975, I.A.E.A. Vienna, 1975, p. 267.
9. A. Ugron, M. Nicolae, C. Matache, T. Săndulescu, M. Spiridon – Radiocromic sistem for α and γ dose measurement in medical dosimetry, Rev. Roum. Phys., under press.
10. M. Nicolae, R. Pop – Microdosimetry of B-radioactive airborne particles by autoradiography, Rev. Roum. Phys., under press.

STATUS REPORT: DIRECT TRACK DETECTOR DOSIMETRY IN NEGATIVE PION BEAMS*

H. B. Knowles*, F. H. Ruddy*, G. E. Tripard*,
G.M. West**, and M. M. Kligerman***

*Washington State University, Pullman, Washington 99163, U.S.A.
**Cancer Research and Treatment Center, University of New Mexico
Albuquerque, New Mexico 82131, U.S.A.
***Los Alamos Scientific Laboratory, Los Alamos,
New Mexico 87545, U.S.A.

Abstract

Cellulose nitrate and Lexan polycarbonate track detectors have been exposed in a water phantom to 350 rads in a stopping negative pion beam. The dose is measured at the "peak" or star region. The vertical beam from the Biomedical Channel at the Los Alamos Medium-Energy Physics Facility (LAMPF) was used. Heavy ions from negative pion interactions in the "peak" region and "plateau," or entrance, region were determined by the step etch technique. The ions, and thus the Linear Energy Transfer spectrum from pion interactions with oxygen, have been tentatively identified. The cellulose nitrate track detectors have also been used to measure superficial dose contours of the high LET portion of the peak pion beam in the first human biology trials. These results will be discussed.

---

*This work was supported in part by NIH Contract CA-14052 from the NCI Division of Research Resources and Centers to the Cancer Research and Treatment Center, University of New Mexico, Albuquerque, New Mexico, USA.

**Present Address: Lackland Air Force Base, San Antonio, Texas, USA.

I. Introduction

The track detector work by the physicists at Washington State University was initially started because of our interest in the negative pion therapy problems, both of macrodosimetry and microdosimetry. The products from the pion "stars" which arise from the capture of a negative pion by a nucleus in tissue, are typically of very low energy ($\leq$ 1 MeV/ atomic mass unit) and it is extremely difficult to measure these by conventional electronic detectors. There is, moreover, likely to be a high background of lightly ionizing radiations in the pion beam--the pions themselves, electron and muon contamination of the pion beam, together with protons which are energetic products of stars and also are produced by collisions with fast neutrons, which are the most prolific product particles from pion stars. However, it is the "high linear energy transfer" (LET) particles from the pion stars that are of special interest to the negative pion radiotherapy preclinical trials, and the track detectors offer a method of measuring these particles in the presence of many particles of "low LET." Thus, trials of both Lexan polycarbonate and cellulose nitrate were made in the negative pion beam at Los Alamos Scientific Laboratory, and subsequently in the actual preclinical trials themselves, for patient dosimetry.

II. Depth-Dose Curves and Procedure

The negative pion beam from LAMPF is a vertical beam, as shown in Figure 1. This design was for the comfort of the patient, but proved to be invaluable in our experimental problem. Pions from the 800 MeV proton beam are collected and focused by the upper three quadrupoles, achiomatically focused on a wedge absorber, bent by a second magnet and finally refocused by an array of five quadrupoles. Probably because of the wedge absorber (which has less effect on muons than on pions of the same momentum), the pion beam at the treatment-Table level is extremely pure, as shown by Figure 2. Here the time of flight spectrum is shown with "time reversal" so that the large peak is that of the pions, with about 7% of muon and 7% of electron contamination indicated. When the beam is stopped in a water phantom, a conventional dosimeter measurement recorded a pronounced dose increase at 18.7 cm depth, the Bragg peak of the pions. Some of this dose maximum is carried by the high LET fragments from the pion stars. These data, which were taken by J. F. Dicello, are indicated in Figure 3.[1] Particles of all LETs are of course included in this measurement.

Both Lexan polycarbonate and cellulose nitrate of the Kodak Pathé CA 80-15 type were employed in this experiment. In an approximate way, it may be stated that the cellulose nitrate has its registration threshold at such a level that it records only particles of high LET while Lexan records those of very high LET. The latter are in fact not very useful in radiation biology because the effective relative biological effectiveness (RBE) tends to drop below unity as a result of "overkill."[2]

Pairs of both cellulose nitrate and Lexan polycarbonate films were made up in "sandwiches," 4.0 cm x 4.5 cm in size and the halves of each sandwich sealed together against water intrusion with Silastic, a commercial silicone-rubber sealant used for caulking bath tubs. Only one type of plastic was used in each sandwich. It was initially expected that detector to detector interface would yield a background for correction of the water to detector interface tracks. The sandwiches, shown to the left in Figure 4, were mounted in pairs (one cellulose nitrate and one Lexan) at different variable depths in a plastic frame, being held in position by nylon monofilament and rubber bonds. The detectors were grouped near the dose peak of the pions but were extended in both directions so that the entrance or "plateau" properties of the pion beam could be observed, as can be seen in Figure 5.

The framework was set into a battery jar and filled with distilled water, and then set under the pion beam. The photograph in Figure 6 is not of the exposure being reported here (a brass collimator, designed for the actual preclinical trials, is seen protruding into the water). However all photographs of the uncollimated depth dose exposure were lost, and Figure 6 gives an indication of the geometry.

Following an exposure of a nominal 350 rads as measured at the pion peak by Dicello, the sandwiches were removed from the phantom and returned to Pullman for etching and microscopic examination.[2]

b. <u>Results</u>

The cellulose nitrate films received a 2 hour etch in 6.25 normal NaOH at 30°C while the Lexan films were etched for 2 hours at 60°C in Lexan-saturated NaOH of the same normality, to which a wetting agent had been added. Experience suggested that the Lexan tracks would be fully developed. The Lexan results appear in Figure 7.

The track densities at the water-Lexan interface are everywhere about twice as large as those on the Lexan-Lexan interface. Since Lexan can normally record slow boron ions and those heavier, it is not surprising that there is this discrepancy, because the products from a pion star can have a maximum atomic number of one less than the original capturing nucleus. Elementary calculations suggest that, if only the oxygen atoms in both Lexan and water are assumed to be the origin of the heavy fragments observed, the track density ratios can be explained. The fragments produced are probably carbon or nitrogen.[3]

It is interesting that there are no tracks observed in Lexan at any distance from the pion Bragg peak, so that it appears that Lexan, although it is so insensitive that its track density is statistically insignificant, could be used to localize the peak of stopping negative pions in clinical dosimetry.

The cellulose nitrate data are shown in Figure 8. Initially, when the sandwich layers were separated and etched, the side that had been wetted and dried (the cellulose nitrate to water interface) showed a track density about half that of the dry side (the cellulose nitrate to cellulose nitrate interface) after the same etch time. This surprising result has now been explained as a result of more recent work, as reported elsewhere in this conference.[4] Before it was explained, a cure was found for the problem--the cellulose nitrate film is soaked for two hours in water before initiating the etch and the "equivalence" of both sides is restored. Basically, the etch induction time for ions of the same energy can be forced to become the same by presoaking the exposed films.

However, it may not be possible to obtain exact equivalence between the two sides of the cellulose nitrate detectors because of an intrinsic difference in the yield of detectable particles from pion stars in carbon and oxygen. Cellulose nitrate is unusually rich in oxygen for a plastic-- its approximate chemical composition may be given as $C_{12}H_{16}N_4O_{18}$--but there is now other evidence that suggests that about half of the negative pions may capture on carbon atoms.[5] In consequence, if pion stars on carbon produce a higher yield of detectable heavy ions than do pion stars on oxygen, the track density at the cellulose nitrate to cellulose nitrate interface may be somewhat greater than that at the cellulose nitrate to water interface. This is an effect that seems to occur at the peak of the dose curve in Figure 8.

It is also noted that there is an effect of heavy ion production in the plateau--presumably by energetic negative pions which can interact "in flight" with both oxygen and carbon nuclei.

A comparison of cellulose nitrate-cellulose nitrate and Lexan-Lexan track densities is shown in Figure 9. It is clear, both in the "plateau" and "peak" region of the pion dose curve from the LAMPF biomedical channel, that high LET fragments exceed very high LET fragments by a factor of at least 40, at least in the (approximately) tissue equivalent compounds in the detectors.

III. Spatial Distributions of Pion Doses

a. Equipment and Procedures

Before the preclinical trials, it was suggested that cellulose nitrate detectors might serve as useful *in situ* dosimeters during the preclinical trials. To achieve this, ten pairs of 4.0 cm x 4.5 cm cellulose nitrate films were sealed together into "sandwiches" by transparent tape and shipped to Los Alamos Scientific Laboratory. They were laid over a large flat malignant melanoma on the skin of a patient who was being treated with negative pions for relief of several metastatic skin tumors. To have the pion and dose peak at the surface of the patient's body, it was necessary to degrade the pion beam in a block of polyethylene which was placed inside the collimator which is visible in Figure 6: the thickness of the polyethylene was approximately equivalent to that of 18.5 cm of water so the dose appeared at the surface. The axes of the track detectors were aligned with the beam axes approximately as shown in Figure 10. After elevating the patient into contact with the collimator-degrader, the track detector sandwiches were securely lodged between patient's skin and the polyethylene. On each of the ten exposures, a maximum <u>nominal</u> pion dose of 116 rads was delivered. The cellulose nitrate sandwiches were stored under refrigeration until air reshipment to Pullman, where they received the conventional 2 hour etch in 6.25 normal NaOH at 30°C.

b. <u>Results</u>

Typical profiles of the over track densities are shown in Figures 11 and 12. In Figure 11, the actual total number of tracks in four fields of view are shown: these were always taken off the upper layer of the sandwich to avoid complications in the etch process that might have resulted from previous contact with the skin of the patient. The count of 125 tracks per four fields of view suggests a slightly higher dose than the 114 nominal rads received by the patient, but only about 20% higher. This calibration was done from the data shown in Figure 9 and must be corrected for the effects of collimation.

Nine of the ten detectors sandwiches showed high LET dose patterns similar to those in Figures 11 and 12. One did not, exhibiting track densities which were about twice as large as those shown. It was later ascertained, from a search of the records, that this detector sandwich has received two independent irradiations by inadvertance. The striking clinical results on the tumor in this patient have been reported elsewhere.[6,7]

IV. Conclusions

Cellulose nitrate track detectors can be used to measure the high LET dose profiles from negative pion beams; in both the longitudinal and lateral directions. When properly calibrated, they could provide an excellent permanent record of the total dose of high LET particles received by the patient. Lexan track detectors could be used to identify the stopping region quite precisely but yield much lower track densities than does cellulose nitrate. As dose integrators, they have many attractive features, of which the principal one is their small size, so that they do not interfere with the therapy process.

References

(1) J. F. Dicello, H. I. Amols, T. F. Lane, A. Lundy, J. D. Doss, H. B. Knowles, and J. E. Barnes, "Dosimetry for Pion Beams at the Los Alamos Meson Physics Facility," Rad. Res. $\underline{62}$, 562 (1975).

(2) P. Todd, "Heavy-ion Irradiations of Cultured Human Cells," Rad. Res. Suppl. $\underline{7}$, 196 (1967).

(3) H. B. Knowles, G. E. Tripard, and F. H. Ruddy, "Plastic Track Detectors as Negative Pion Macro- and Microdosimeters," Med. Phys. $\underline{2}$, 163 (1975).

(4) F. H. Ruddy, H. B. Knowles, S. C. Luckstead, and G. E. Tripard, "Etch Induction Time in Cellulose Nitrate: A New Particle Identification Parameter," proceedings of the present conference.

(5) H. B. Knowles, F. H. Ruddy, G. W. West, and M. M. Kligerman, "High-LET Contours under the LAMPF Biomedical Channel by Plastic Track Detector Scanning," in preparation for submission to J. Radiology.

(6) M. M. Kligerman, G. W. West, J. F. Dicello, C. J. Sternhagen, J. E. Garnes, K. Loeffler, F. Dobrowolski, H. T. DAvis, J. N. Bradbury, T. F. Lane, D. F. Petersen, and E. A. Knapp, "Initial Comparative Response of Metastatic Superficial and Surrounding and Unerlying Normal Tissue to Peak Pions and X-Rays." Presented at American Radium Society Meeting, May 4, 1975.

(7) M. M. Kligerman, "Meson Radiobiology and Therapy," <u>Proc. Seventh Inter. Conf. on Cyclotrons and Their Applications</u> (Zurich, Switzerland, 1975), p. 419.

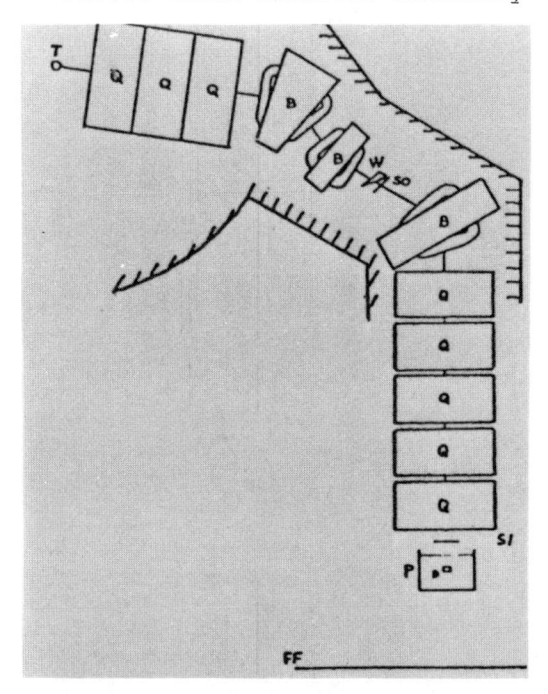

Figure 1. Vertical Diagram of the LAMPF Biomedical Channel

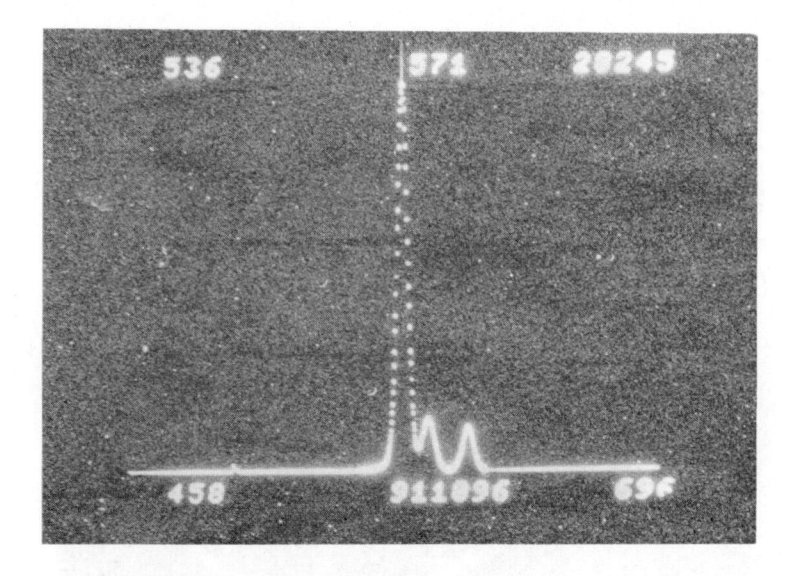

Figure 2. Time of Flight Analysis of the Pion Beam from the LAMPF Biomedical Channel

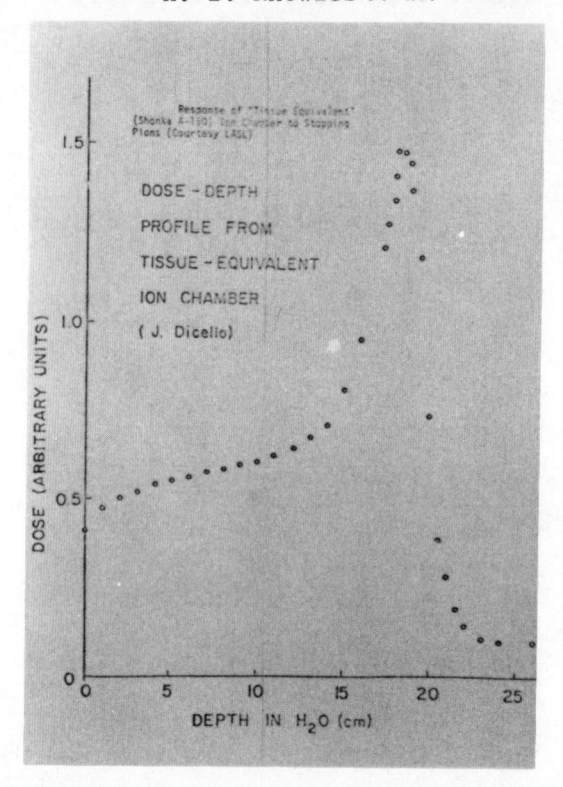

Figure 3. Depth-Dose Curve in a Water Phantom

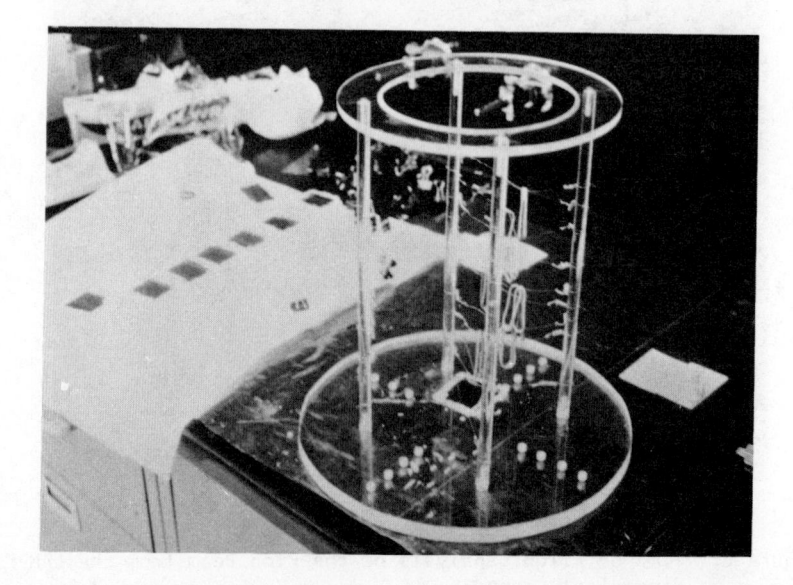

Figure 4. Assembly of Track Detector Phantom for High LET Depth Dose Curve

Figure 5. Assembled Track Detector Phantom

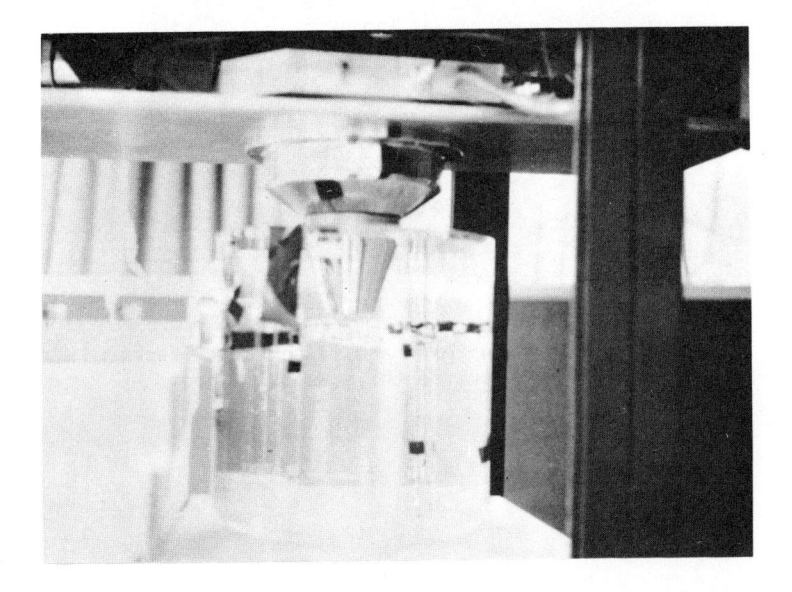

Figure 6. Representative View of Phantom under the Biomedical Channel (Collimator in Place)

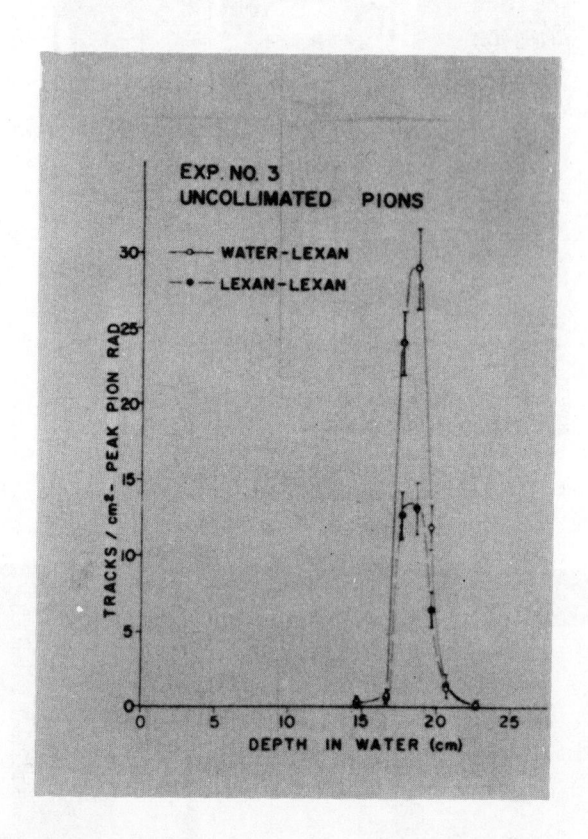

Figure 7. Lexan-Lexan and Lexan-Water Track Density versus Depth in Water Phantom

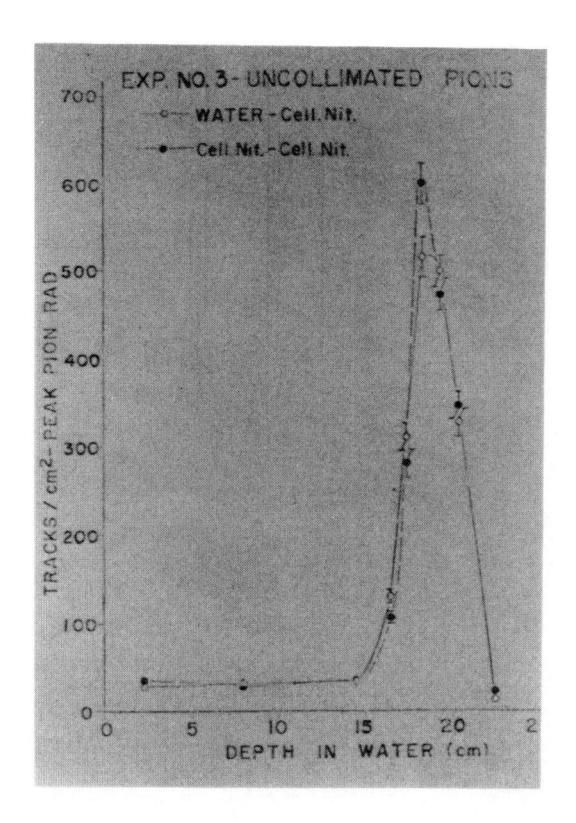

Figure 8. Cellulose Nitrate-Cellulose Nitrate and Cellulose Nitrate-Water Track Densities versus Depth in Water Phantom

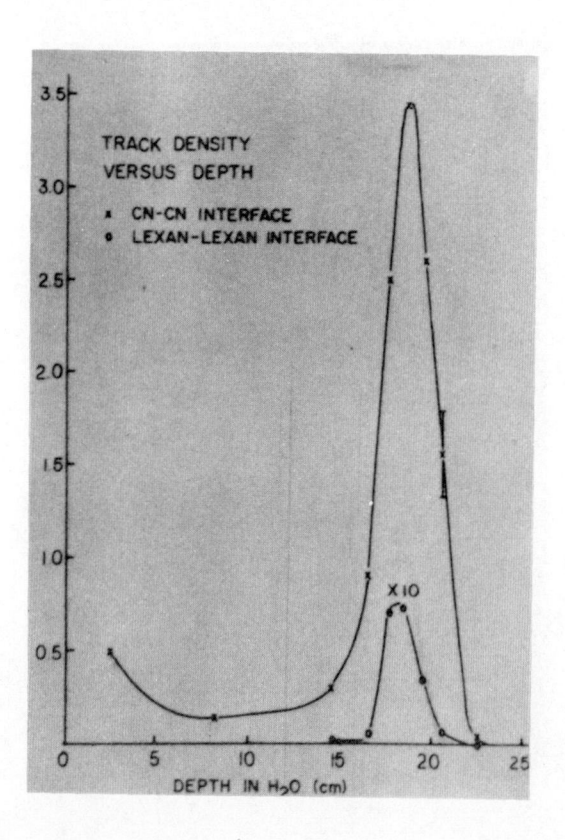

Figure 9. Cellulose-Nitrate-Cellulose Nitrate and Lexan-Lexan Track Densities Compared

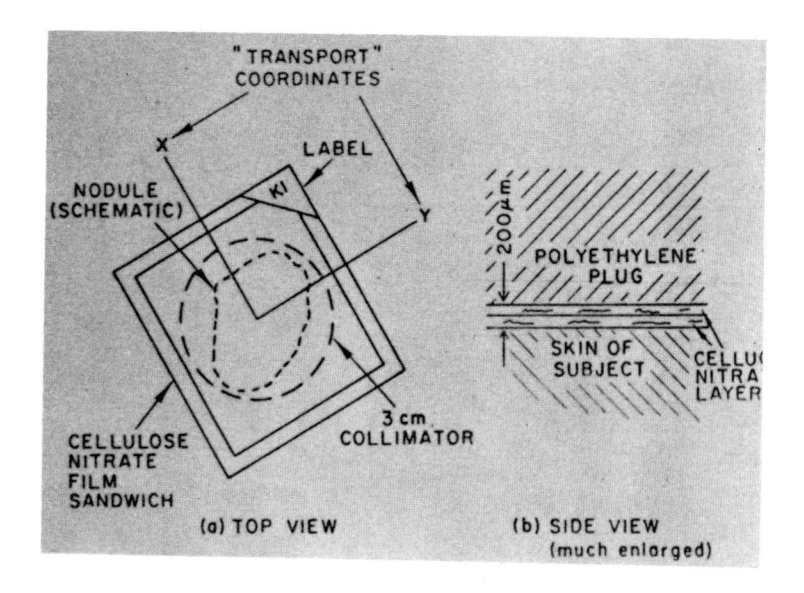

Figure 10. Assembly and Orientation of Cellulose Nitrate Track Detector "Sandwiches" for Preclinical Dosimetry

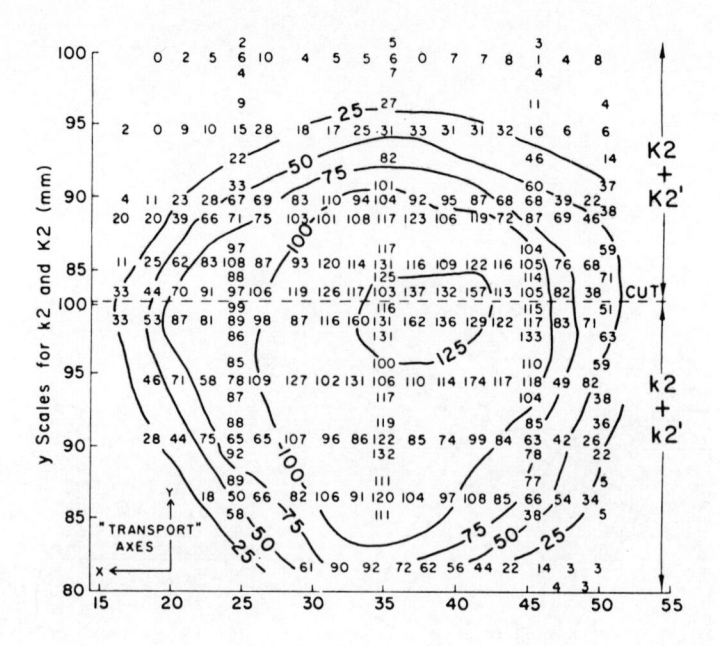

Figure 11. Track Densities for One Preclinical Exposure of 114 Rads--Track Counts Shown

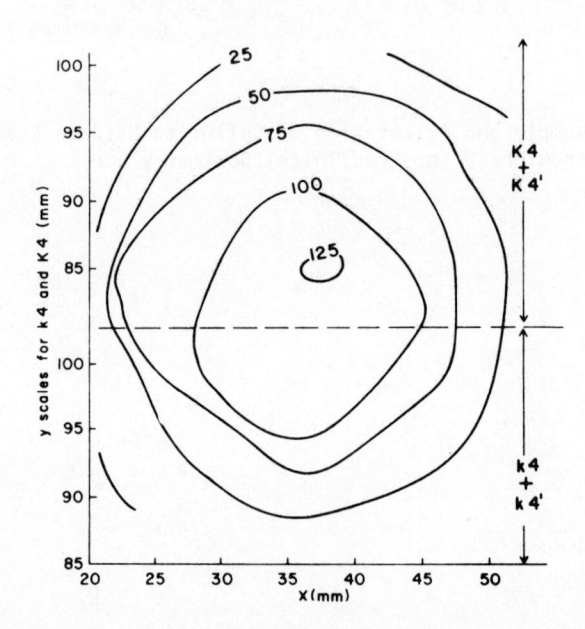

Figure 12. Track Density Contours for 114 Rad Exposure.

THE MEASUREMENT OF RADON BY ALPHA-SENSITIVE PLASTIC TRACK DETECTORS FOR USE IN URANIUM EXPLORATION

H. A. Khan and R. A. Akber

Nuclear Engineering Division, Pakistan Institute of Nuclear Science & Technology (PINSTECH), Nilore, Rawalpindi, Pakistan

ABSTRACT

Experiments have been carried out to establish the technique of radon detection by Solid State Nuclear Track Detectors (SSNTD) for use in uranium exploration. A series of studies have been made for the development and standardization of this new 'Nuclear Geological' technique. The actual use of the method in the field has indicated some new 'Hot Points'.

1. INTRODUCTION

The experience shows that the gamma surveys are mainly useful for surface or near surface mineralizations.[1-3] The prospection of ore bodies lying deeper than a few meters needs methods, like deep drilling followed by gamma logging.[4] This latter technique though probably the most direct one, is extremely expensive. Therefore, the need for some alternative approach had been long felt.

Here, we describe our efforts made to establish the technique of radon detection by Solid State Nuclear Track Detectors[5,6] for Uranium exploration. With the latest knowledge of the technique, it has been applied for uranium exploration in some areas of Pakistan.

2. THE EXPERIMENTAL ARRANGEMENTS AND THE RESULTS

2.1. ALPHA SENSITIVE PLASTIC TRACK DETECTORS

Generally speaking, alpha sensitive track detectors are materials like cellulose acetate, cellulose nitrate and cellulose acetate butyrate so that when one alpha particle falls on it and passes through, a physical damage inside the body is produced. This latent damage can be seen with an electron microscope under a magnification of about 20,000 or so. An electron microscope is quite a costly instrument, so it cannot be made available in an average laboratory. To overcome this difficulty, these latent damage trails are enlarged by immersing the plastics in a suitable chemical reagent. The chemical 'Etches Out' the damaged region and produce an 'Enlarged Version' of the latent damage trail, called an 'Etched Track' or simply a 'Track'. A careful analysis of the shape, size, and the orientation of the track can yield important information concerning the charge, mass, energy and direction of the alpha particle.

2.2. THE METHOD OF RADON DETECTION AND ITS USE URANIUM EXPLORATION

Let us imagine an 'ore-body' lying at a certain depth from the top surface. Over the geological time period, the uranium present in the ore body decayed and thus produced a chain of daughter products. Among these daughter products, ^{222}Rn is a gas and decays through alpha emission with a half life of $T_{\frac{1}{2}} = 3.825$ days. A part of the radon gas so produced comes out of the ore body and diffuses out in all the directions. Let us concentrate on the amount of gas reaching the top surface. It is quite obvious that the point situated just on the top of the ore body would receive more radon than those lying away from it. If, therefore, some radon detection systems are employed over a large area, the point lying on or near the top of the ore body would show a peak in the radon concentration distribution.

2.2.1. RADON DETECTION BY AMMANOMETERS

One of the methods consist of collecting radon from a certain area in a chamber and then counting the decaying atoms of radon by scintillators or other alpha sensitive detectors. This sort of system is called an Ammanometer.

The system has been found to have the following drawbacks :

i) The procedure of radon extraction is not satisfactory. Varying amounts of air is introduced in the counting system and thus non uniform air-gas mixtures are counted, which make the reproducibility of results very poor.

ii) If one instrument is used to cover a large number of points (one after the other) the error introduced by the varying amounts of radon concentration in air would be significant.

iii) On the other hand, if a large number of such sets are employed. The cost of investigation per point would go very high.

iv) Experience has shown that with prolonged use, a constant source of background goes on building up due to the radon daughter products which stick to the walls and the surface of the counting chamber. This factor reduces the sensitivity of the system.

2.2.2. THE MEASUREMENT OF RADON BY ALPHA SENSITIVE PLASTIC SOLID STATE NUCLEAR TRACK DETECTORS

On the other hand, the alpha sensitive Solid State Nuclear Track Detectors can be made to measure the radon concentration distribution in a large area without the above mentioned disadvantages of the ammanometers.

The plastic detectors are at first contained in small tubes and then burried at shallow depths for a duration of about three to four weeks. The radon gas reaching the detector tubes remains there are appreciable time. Some of the radon atoms decay by emitting alpha particles. A fraction of these alpha particles fall on an alpha sensitive solid state nuclear track detector and produce damage trails in it. This sort of long exposure of detectors (burried at shallow depths) made over a large area averages out the already mentioned background effects besides reducing them to a marked extent. The detectors being extremely cheap reduce the cost per point of investigation to such an extent that the method of radon detection by plastic track detectors becomes the most inexpensive among the presently available techniques.

2.2.2.1. THE RADON MEASUREMENTS IN SIMULATION EXPERIMENTS

The measurement of radon by track detectors was carried out in a number of simulation experiments. The following investigations were made :

A. Studies of the effects of the number, the strength, the dimensions, and the position of the 'ore body' on the accumulated tracks for a fixed geometry of the detector tube;

B. The studies of the effects of the size and the geometry of the detector tube, and the position of the detector in the tube on the track accumulation rates for a fixed source;

C. The effects of the exposure time and the environmental conditions on the track accumulation rates;

D. The determination of the diffusion rates of radon through various media of geological interest.
E. The differentiation between uranium and thorium deposits.
F. The background problems and the optimization of the design of the detector-tube system.

After performing the above mentioned experiments the detector-tube systems were first calibrated in a known area of mineralization and then applied in the field for uranium prospection and exploration in some areas of Pakistan.

A. THE STUDIES CONCERNING THE EFFECTS OF THE NUMBER, POSITION AND DEPTH OF THE ORE BODIES ON TRACK DENSITY DISTRIBUTION

In this set of experiments a lump of uranium ore was burried in sand at a certain depth, D, and then plastic track detectors were imbedded near to the top surface. The detectors were contained in small plastic cylinderical tubes with their open ends facing down. The depth, D, of the source was varied from 5 cm to 15 cm in three steps. After an exposure of about two months, the detectors were taken out, etched and scanned for track densities. The results are shown in Fig. 1a. It is quite interesting

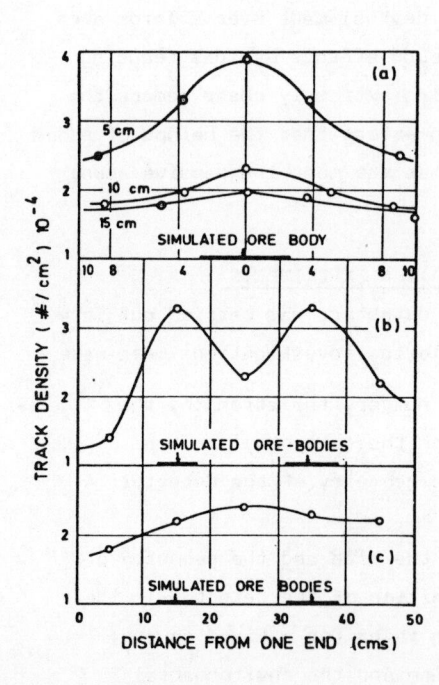

Figure 1
The simulation experiments employing, (a) a single source, (b) two sources lying at the same depth, and (c) two sources lying at different depths.

to see a peak just on the top of the burried 'ore body'. A systematic study
concerning the variation of the peak width with the size and the depth of
the ore-body is currently being made. It is expected to the helpful in
estimating the depth of the underground mineralization. It may be mentioned
at this juncture that more pronounced peaks would be obtained if the experiments are performed with soils which have attained equilibrium in radon
concentration (as is usually the case in the field).

Experiments similar to those described above were performed by
using two 'ore-bodies' lying at varying depths. Two results of the two
set ups worth mentioning here are shown in Figs. 1b and 1c. It is quite
obvious from the figures that for the simple arrangements of two sources
(Fig. 1b), the track density distribution shows two well resolved peaks.
On the other hand, the arrangement shown in Fig. 1c is complicated - the
weaker source lying at a shallower depth while the stronger source
lying at a greater depth. The track density distribution is quite
complicated and warns us about one of the drawbacks of the
'SSNTD - Technique'.

B. THE STUDIES CONCERNING THE EFFECTS OF THE POSITION OF
THE PLASTIC TRACK DETECTOR IN THE DETECTOR TUBE

It is quite obvious that the track density accumulated on an
SSNTD (while placed in the detector-tube) is a function of the solid
angle subtended by the effective 'air-gas' column at the detector.
Therefore, the position of the detector in the detector tube is bound
to play an important role in the alpha track accumulation. The following
experiments were carried out in this connection.

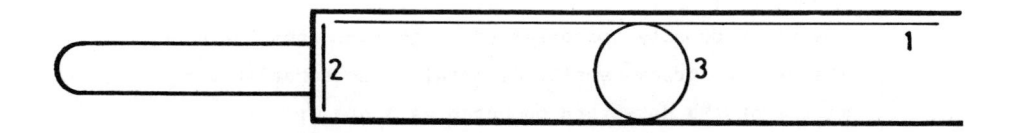

DETECTOR TUBE

Figure 2
An sketch of the detector tube along with the respective
positions of the plastic detectors employed for studying
the effects of the position of the detector on track
accumulation rates.

Plastic films were positioned in a detector tube as shown in Fig. 2. The tube was kept in radon atmosphere for a few weeks and then the detectors were analysed. Fig. 3 shows the results of track density distribution of the detectors numbered as 1 and 2.

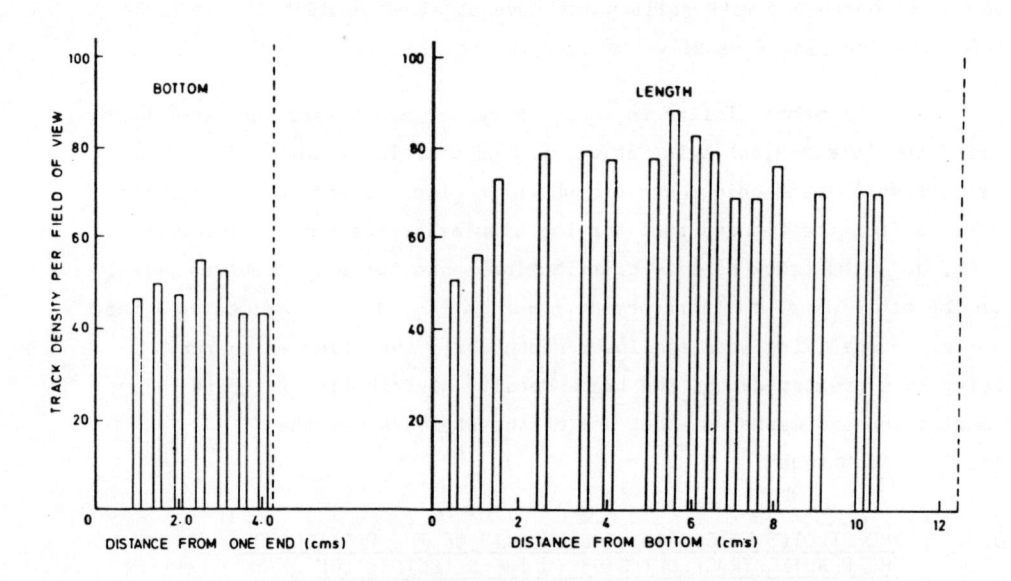

Figure 3
Track density distribution obtained on various parts of (a) the detector # 1 (on the right) and (b) the detector # 2 (on the left) when kept in radon atmosphere as shown in figure 2.

The results can be summarized as follows :

i) The track density is the least on the corners of the closed end;

ii) The track density increases as we go away from the ends;

iii) The maximum track density is obtained somewhat at the middle of the detectors numbered as 1 and 2;

iv) The maximum track density obtained on the detector number 1 is almost double of the maximum track density obtained on the detector number 2.

The analysis of the detector number 3 yielded the results as shown in Fig. 4.

The results show the following :

i) A uniform track density is obtained on the inner side of the loop;
ii) A non-uniform track density is obtained on the outer surface of the loop;

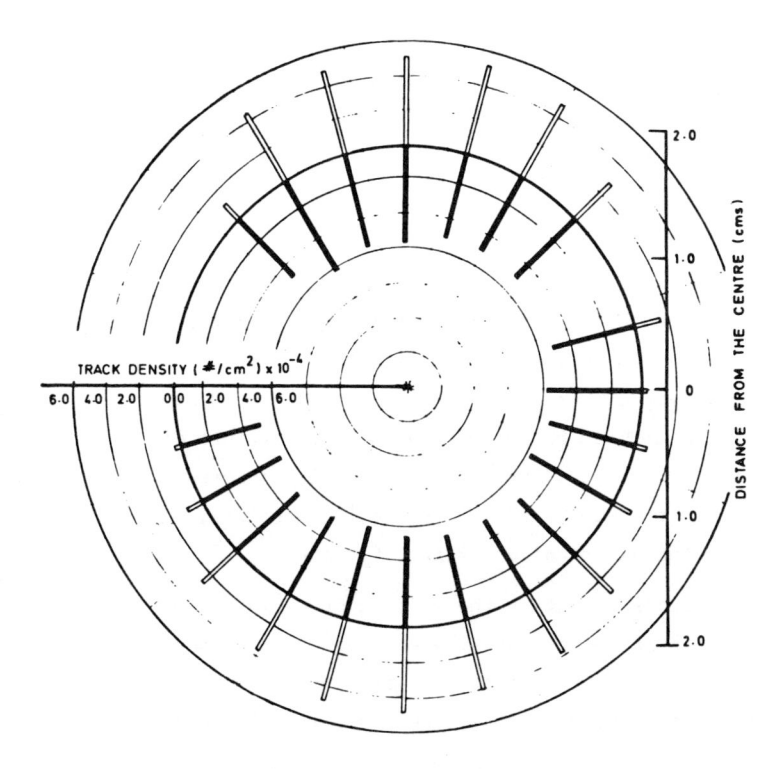

Figure 4
The track density distribution as obtained on the two sides of the loop detector (detector # 3) when exposed in the configuration shown in figure 2.

iii) For 2.5 cm diameter tube, the maximum track density on the outer surface is comparable to the track density on the inner side of the loop;
iv) The maximum track density obtained on the loop was smaller than the maximum track density obtained in the middle of detector number 1.

The maximum values so observed were found to increase with the tube size upto the tube diameter of about 3.5 cm.

C. THE STUDIES CONCERNING THE EFFECTS OF THE EXPOSURE TIME AND THE ENVIRONMENTAL CONDITIONS ON THE TRACK ACCUMULATION RATES

It was important to measure the track accumulation rate as a function of the exposure time in order to find out the time interval required for producing a saturation stage in the exposure step. Both, plastic track detectors and a sillicon surface barrier detector were employed for these studies.

The following conclusions were made from the results thus obtained :

i) The accumulation rate vs. exposure time curve follows an exponential type of pattern;
ii) The saturated values increase with increasing tube size upto about 3.5 cm tube diameter;
iii) Normally, a saturation value is obtained in about 2-3 weeks.

The 'detector - tube' system is usually left in the field for 3-4 weeks under adverse conditions of heat and humidity. Experiments conducted to study this effect indicate that the cellulose nitrate detectors can retain alpha particle tracks even if they are heated at $70-80^{\circ}C$ for a few weeks. It has been estimated that the temperature of the detectors lying at the depths at which they are normally burried does not exceed $30^{\circ}C$ even in the months of intense heat. Therefore, they are fairly safe as regards the danger of annealing of tracks.

The experiments conducted to study the effects of humidity on the track registration properties of plastic track detectors show that a slight enhancement in efficiency of the detector is observed. The exact contribution of humidity on detectors lying in the field is being currently evaluated.

D. THE MEASUREMENT OF DIFFUSION PARAMETERS OF RADON IN VARIOUS MEDIA OF INTEREST BY USING PLASTIC TRACK DETECTORS

In these measurements a container having a lump of uranium ore lying on the base were filled with sand, earth, gravel, and their mixtures. Plastic SSNTDs were imbedded in the body at different depths from the ore-body. The detectors were retrieved after about four months and scanned for track densities. Similar arrangement was realized for a big piece of sandstone. It was observed that among these materials sandstone offers the maximum resistance to the flow of radon. Other materials were found to have similar diffusion coefficients. More work on these lines is in progress.

E. **THE PROPERTY OF PLASTIC TRACK DETECTORS USED FOR DIFFERENTIATION BETWEEN URANIUM AND THORIUM DEPOSITS**

Similar to the production of radon in uranium decay, thoron ($T_{\frac{1}{2}}$ = 55 seconds) is given off by thorium deposits. We conducted experiments to differentiate between thoron and radon from the differences in tracks formed by alpha particles in their respective decays. It has been observed that radon alpha tracks (E_α = 5.49 Mev) are first to etch out than the thoron alpha tracks (E_α = 6.29 Mev). These differences in track development properties have been found to be helpful in differentiating between uranium and thorium deposits.

F. **THE BACKGROUND PROBLEMS IN RADON MEASUREMENTS FOR URANIUM EXPLORATION**

It has been found that the biggest source of background is the presence of radon in the atmosphere. To overcome this problem, the detector tubes are burried at some depth instead of keeping them on the surface. But making a pit and the retrieving of the tube not only increases the cost per point of investigation but also it causes a lot of inconvenience and the wastage of time. To reduce this factor, a number of experiments were performed in order to optimize the depth of burrial. It has been found that an optimum depth of about 15 cm is needed. The experience shows that a pit of this depth can be made in about a minute by using a portable drilling machine and that the retriving step is very convenient and can be accomplished by using very little labour.

The above mentioned experiments indicated that for the best results, the following points regarding the tube dimensions and its design are to be kept in mind;

i) The tube diameter and the length should be more than 3 cm and 6 cm respectively;

ii) There should be some arrangement of opening the tube in the middle for ease in the positioning of the detector.

Keeping these points in view and also considering the available drilling machines and the ease in handling the tube during the installation stages, a detector tube having dimensions as shown in Fig. 5 was designed, constructed, and used in the field.

Figure 5
A photograph of the 'optimum geometry' detector tube employed in the present studies.

3. THE APPLICATIONS OF THE TECHNIQUE FOR URANIUM EXPLORATION

3.1. THE CALIBRATION

After getting some confidence from the simulation experiments and obtaining the necessary basic information, it was decided to apply the technique in the field.

To start with, it was planned to test the technique in an area known for uranium mineralization. The 'detector-tube' systems were installed in such an area in Pakistan. Fig. 6 shows the results of 'SSNTD - Method' and that of direct drilling and 'Gamma Logging technique'. The two sets of the results have quite a good agreement inspite of many assumptions in their calculations.

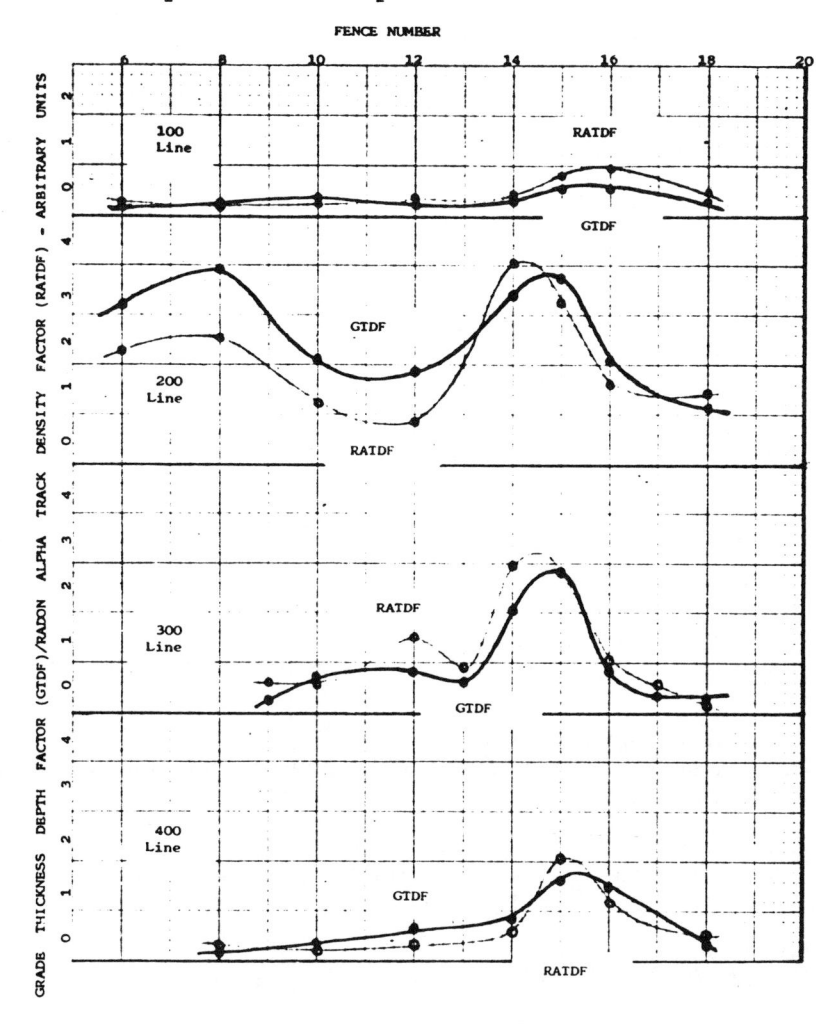

Figure 6

The results obtained from the calibration of tube system against gamma measurements. Here, Radon Alpha Track Density Factors (RATOF) have been compared with Grade Thickness Depth Factors (GTDF) in a known area of mineralization.

3.2. URANIUM EXPLORATION

Uranium Exploration was attempted in some prospective areas of Pakistan.

The work carried out uptill now has indicated the following :

i) A couple of hot points in 'Area A';

ii) A couple of hot points near the Camp Site of 'Area B';

iii) Barren area in a terrace near to the 'Area B';
iv) A hot region, in 'Area A'. This hot region is expected to throw some light on the missing part of a channel in the area.

More exploratory and prospection work has been planned for other prospective areas of Pakistan.

REFERENCES

1. S.H.U. Bowie, M. Davis and D. Ostle
    Proceedings NATO-Sponsored Adv. Study Institute on
    Methods of Prospecting for Uranium Minerals, London 1971,
    Inst. Min. Metall., London (1972).

2. C. Weiss
    Proceedings Third Int. Geochemical Exploration Symp.
    Toronto, 1970. Can. Inst. Min. Metall., Montreal,
    Special Vol. 11 (1971) 502.

3. W.H. Little and A.Y. Smith
    Proceedings of the IAEA Meeting on 'Uranium Exploration Methods',
    Vienna, 1972, page 131.

4. W.K. Hawkins and M. Gearhart
    Proceedings of the IAEA Meeting on 'Uranium Exploration Methods',
    Vienna, 1972, page 213.

5. R.L. Fleischer, P.B. Price and R.M. Walker
    Nuclear Tracks in Solids : Principles and Applications
    University of California Press, 1975.

6. H.A. Khan
    The Nucleus 8 (1971) 63.

THE USE OF CA80-15 AND LR-115 CELLULOSE NITRATE TRACK DETECTORS FOR DISCRIMINATION BETWEEN RADON AND THORON

H.A. Khan*, R.A. Akber*, A. Waheed*, M. Afzal**,
P. Chaudhary** S. Mubarakmand** and F.I. Nagi***

*Nuclear Engineering Division, PINSTECH, Nilore,
Rawalpindi, Pakistan
** Nuclear Research Laboratory, Government College,
Lahore, Pakistan
***Pakistan Atomic Energy Minerals Centre, Lahore, Pakistan

ABSTRACT

Attempts have been made to employ CA80-15 and LR-115 cellulose nitrate track detectors (obtained from Kodak Pathé, France) for discriminating between radon and thoron gases. The variation of the etched track parameters with the energy of the incoming alpha particle has been exploited in these experiments. The results so obtained seem to be particularly useful in uranium/thorium exploration and in the design of personnel dosimeters.

1. INTRODUCTION

Recent advances in the method of uranium/thorium exploration by detecting radon emanations, and radon/thoron personnel dosimetry in mines[1,2] resulted in the development of new types of alpha detection arrangements. Perhaps, the most important among these are the detection systems called Solid State Nuclear Track Detectors[3-5] (SSNTD). In this respect, cellulose nitrate plastic detectors have been found extremely useful.

Here, we describe our efforts to exploit the difference in the track development properties existing in the latent damage produced by varying energy alpha particles for discrimination between radon and thoron. CA80-15 and LR-115 detectors have been used in these investigations.

2. THE EXPERIMENTAL ARRANGEMENTS

Uranium ores having known concentrations and standard thorium nitrate samples were used as radon and thoron sources. The detectors were kept at large distances from the main sources so that alpha particles from the decay of uranium and thorium series could not reach them directly. Afterwards, the sample containers were covered with perspex lids.. Thus, the arrangements were such that the detectors could receive alpha particles only from the decays of radon and thoron, which escaped from the ore and diffused up in the air in the sample containers. Two different geometries were tried for the exposure of the detectors. Firstly, the detectors were kept at a distance of 6.3 cm from the top of the radon and thoron sources so that the thickness of the gas (air + radon/thoron) column effectively contributing alpha tracks to the detector extends up to their range (5.5 cm) in the gas mixture (hereafter termed as 'Extended Geometry'). Secondly, another plastic detector was introduced between the source and the main detector so that the distance between the top detector (the main detector) and the second detector (the shield) was 3 mm (hereafter termed as 'Limited Geometry'). In the second case, the main detector was receiving alpha particles mostly from those atoms which were decaying in the air-gas column lying between the two detectors. By doing so, we selected two different effective thicknesses of the air column. In the first case, the particles had a wide energy distribution, while in the second case the energies of the particles had a very little spread and were nearly equal to 5.49 Mev for radon and 6.29 Mev for thoron alpha decay.

The detectors were etched in 10% aqueous solution of NaOH kept at $52 \pm 1^{\circ}C$ for time intervals ranging between 10 and 400 min. The detector exposure time to the radioactive gases was about three days.

The thermal stability of the latent damage trails was studied by carrying out the annealing of these detectors at various points in the temperature range of $50^{\circ}C$ to $150^{\circ}C$.

3. RESULTS AND DISCUSSION

3.1. ETCHING PROPERTIES OF LATENT DAMAGE TRAILS

The study of the etching properties of the latent damage trails produced by alpha particles by alpha particles from radon/thoron decays yielded the following information :

1) All the curves showed at first, a rapid increase in the track density and then a levelling off, followed by a fall.

2) The slope for the curve obtained from the thoron source was higher than the slope of the curve obtained from the radon source (uranium). Also, it was found that the etching time required to just reveal the tracks increases with the increasing thorium to uranium ratio.

An improvement in the resolution (measured by the separation between the peak values of the track density vs. etching time curves) is obtained when only a limited volume of the source (Limited Geometry) is used, instead of utilizing (Extended Geometry) whole of the available sensitive volume. The reason behind this improvement is that in the case of the 'Limited Geometry' exposure, the difference between the two alpha groups is much greater (due to less energy spread) than the difference produced when whole of the sensitive volume (Extended Geometry) is utilized. This is because in the Extended Geometry case a lot of energy spread in the two energy groups is caused when they pass through thick column of air/gas mixtures.

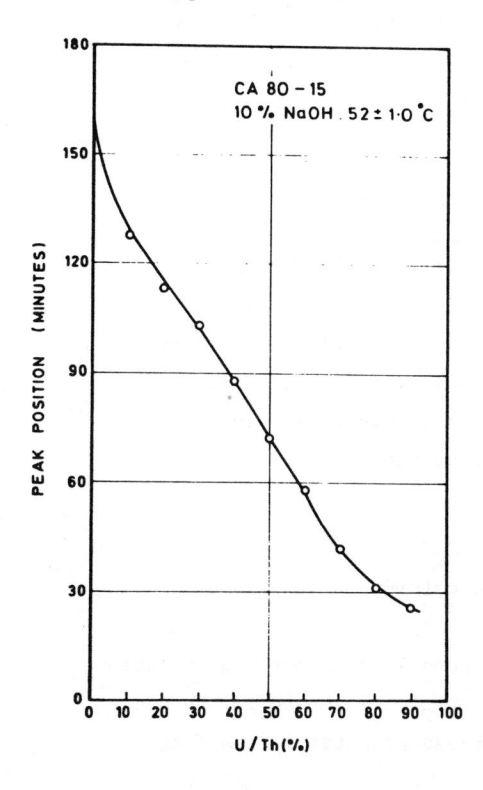

Figure 1
The variation of the peak position (on time scale) as a function of uranium to thorium ratio. The peak positions were obtained from track density vs. etching times curves of CA80-15 track detectors exposed to samples containing different mixtures of uranium and thorium. The detectors were etched in 10% NaOH kept at $52 \pm 1°C$.

Besides using the above mentioned uranium/thorium samples,

different mixtures of uranium and thorium were also employed. The exposures were carried out in 'Extended Geometry' and after etching the detectors under the conditions already mentioned, the variation of track density as a function of etching time and uranium to thorium ratio was obtained. From these curves, the etching times corresponding to the highest track densities as a function of uranium to thorium ratios. These final results are shown in Fig. 1. It is worth noting that the peak position (on time axis) decreases with increasing uranium to thorium ratio.

The distribution and variation of etched track length as a function of the etching time was also studied. As expected it was observed that both the average track length, l, and the maximum etched track length, l_{max} at first increase with the etching time and then become more or less constant. It is also seen that l_{max} for radon alphas saturates earlier than the l_{max} for thoron alphas.

One plausible explanation of the above trends is that the latent damage trails due to higher energy alphas from thoron lie deeper in the detector body and take longer time to develop fully than those damage trails due to radon alpha particles.

3.2. ANNEALING PROPERTIES OF LATENT DAMAGE TRAILS PRODUCED BY ALPHAS FROM RADON AND THORON

The annealing properties of the latent trails due to radon and thoron alpha groups was also studied. The results showed that under the presently employed etching condition, the l_{max} values are higher for radon than for thoron. This probably indicates that the shrinkage of the latent damage trails is more severe from the higher energy ends, which is closer to the upper surface.

4. CONCLUSIONS

One might draw the following conclusions from the above results :

1. Both, the development and the annealing properties of latent damage trails due to alpha particles produced in the decay of radon are different from those produced in the decay of thoron.

2. CA80-15 and LR-115 Cellulose Nitrate Detectors can be effectively employed in discriminating between radon and thoron in their mixed atmospheres. This can be used in separate estimations of radon and thoron contributions in mines.

As the above mentioned detectors are capable of separating the radon/thoron contributions, these materials buried in cups in the prospective areas can be made to yield valuable information about the relative abundance of uranium and thorium deposits in the region. Initial experiments indicate that for an ore body having 0.05% uranium and buried at a depth of about 130 m, an exposure time of four weeks would result in reasonable track density.

The accuracy obtained in the above mentioned characteristic differences in the tracks produced by radon and thoron alpha particles up till now can yield only rough estimation of the relative abundances of the two types of minerals. For more accurate determinations more work in the perfection of the technique is needed.

ACKNOWLEDGEMENTS

We are grateful to our colleagues from the three centres for many fruitful discussions. One of us (H.A.K) is grateful to Pakistan Atomic Energy Commission for allowing him to participate in this collaborative programme.

REFERENCES

1. Proc. IAEA Conf. 'Radon in Uranium Mining', No.IAEA-PL-565/8(1974).

2. Thompkins R.W. Can. Min. J. **91**, 103 (1970).

3. Fleischer R.L., Price P.B. and Walker R.M. In Nuclear Tracks in Solids – Principels and Applications. University of California Press, Berkeley (1975).

4. Khan H.A. Nucl. Instr. Meth. **113**, 55 (1973).

5. Khan H.A. Nucleus (Karachi), **8**, 63 (1971).

RESULTS OF AN INTERNATIONAL ALPHA PARTICLE REGISTRATION INTERCOMPARISON WITH SOLID STATE NUCLEAR TRACK DETECTORS

Herwig G. Paretzke

Institut für Strahlenschutz der GSF, D-8042 Neuherberg, F.G.R.

Abstract

In view of the widespread interest in alpha particle registration with solid state nuclear track detectors, an international intercomparison of such measurements has been arranged. Sixteen sets of fourteen detectors each were sent to GSF-Neuherberg, irradiated there carefully with various alpha particle fields, and then returned for evaluation. Fourteen irradiation runs were made for each set simulating seven different irradiation situations commonly encountered in practical applications.

The results of this intercomparison reported in this paper are based on the data of eight sets. They show good agreement with respect to the determination of track densities in the case of vertical incident alpha particles. Also the results obtained for determination of particle energies and angle of incidence in most cases were rather accurate.

However, apparently it is still rather difficult to determine accurately and precisely the specific activity of alpha emitters on a thick filter positioned at some distance from the detectors, i.e. for the case of 2π-incidence and a broad energy spectrum.

1. Introduction

Solid state nuclear track detectors (SSNTDs) are often used as integrating registration devices for alpha particles, e.g. in Radon dosimetry, Uranium prospection, biological and technical autoradiography, radiochemistry, and in nuclear and solid state physics. Further on, alpha particles from radioactive sources or accelerators are also frequently employed for investigations on detection properties of plastic materials for tracks of nuclear particles.

This broad interest in alpha particle registration with SSNTDs made an intercomparison of results obtained in alpha particle measurements by different research groups eventually employing quite different methods and techniques highly desirable. Therefore, an International Alpha Particle Registration Intercomparison with Solid State Nuclear Track Detectors (ALRIT) was organized in the GSF-Institut für Strahlenschutz at the occasion of the 9th International Conference on Solid State Nuclear Track Detectors (9th ICNTD) taking place from Sept. 30 to Oct.6, 1976, at the Gesellschaft für Strahlen- und Umweltforschung mbH (GSF) Neuherberg/München.[1] The aim of this intercomparison has been to achieve estimates on the precision and accuracy of alpha particle measurements with SSNTDs.

The participants were informed of the irradiation geometries, particle spectra, order of magnitude of the track densities, etc. in April 1976. Sixteen sets of plastic track detectors were submitted by twelve scientists from 7 countries. They were irradiated by the end of June and returned immediately for evaluation. This report is based on those eight results which were received in time. The results and conclusions were discussed among the participants during the 9th ICNTD. They came to the consensus to have another intercomparison performed by the end of 1977.[2]

2. Simulated Irradiation Situations

Seven measurement situations commonly encountered were simulated:

1) determination of track density for vertical incident and mono-energetic particles
2) determination of track density for vertical incidence and various energies or a broad energy spectrum
3) determination of track density for 2π-incidence and mono-energetic particles (encountered e.g. in thin layer auto-radiography)

---

[1] The results of an International Neutron Fluence Measurement Intercomparison with SSNTDs performed in the same institute at the same occasion are reported in (1).

[2] Scientists interested to participate in ALRIT II are kindly invited to contact the author.

4) determination of track density for 2π-incidence and various energies or broad energy spectrum (encountered e.g. in thick layer autoradiography and in Radon dosimetry),
5) determination of track density for irradiation from an extended source at a short distance emitting various energies (encountered e.g. in Radon-dosimetry using filtering),
6) determination of angle of incidence for monoenergetic unidirectional particles,
7) determination of energy of vertical incident particles.

For situation No.1, 2, 3 and 5, several particle fluence values and for situation No.7 two energies were provided, thus 14 irradiation runs (Table I) were performed per set of detectors:

Table I

Run No.	Situation No.	Alpha source	Geometry	Air pressure	Track density [cm^{-2}]	Irrad. time [min]	Quantity to be determined
1	1	Am-241	90°	<0.01 Torr	10^2	1/6	track density
2					10^3	1	
3					10^4	7	
4	2	Am+Cm+Pu	90°	<0.01 Torr	10^2	1	track density
5					10^3	15	
6	3	Pu-238	2π, direct contact	726 Torr	10^3	1/3	track density
7					10^4	3	
8	4	Am+Cm+Pu	2π, direct contact	726 Torr	10^3	1/6	track density
9	5	Radon-daughters	2π, 2cm distance	726 Torr 26°	10^2	10	track density
10					10^2	30	
11					10^3	125	
12	6	Am-241	30°-60°	<0.01 Torr	10^3	1	angle on incidence
13	7	degraded Am-241	90°	<0.01 Torr	10^3	1/6	particle energy
14						1	

The track densities given in Table II indicate the order of magnitude only. The actual densities were measured either directly with semiconductor- or proportional counter-devices or only the density ratios were determined by irradiation time measurement.

3. Performance of Irradiations

The irradiations were performed with thin alpha sources (about 7 mm diameter) of Am-241, Cm-244, Pu-238 and Pu-239, and with Radon daughter products deposited on a fibre filter (17 cm diameter) by intake of environmental air. The alpha energies and their relative abundancies are given in Table II:

Table II

Nuclide	Energies MeV
Americium-241	5.48 (85%), 5.44 (13%)
Curium-244	5.80 (77%), 5.76 (23%)
Plutonium-239	5.15 (72%), 5.13 (17%), 5.10 (11%)
Plutonium-238	5.50 (72%), 5.46 (28%)
Polonium-218	6.00 (100%)
Polonium-214	7.69 (100%)

The irradiations for situations 1, 2, 6 and 7 were performed in a vacuum chamber at pressures less than 0.01 Torr. Typical residence times of detectors in this chamber before irradiation were between 10 and 60 minutes. For situation 1, 2 (vertical incidence, density) and 7 (angle), the source-to-detector distance was 134 ± 0.5 mm ensuring that all angles of incidence were within ± 5° from vertical incidence. The irradiation times were determined with a photographic shutter to ± 0.5 sec.

During irradiations for situation 6 (angle), the distance was 121 ± 2 mm at an angle of 55.8° ± 1.5° to the detector surface. Both source and detectors were covered except for a small central area of 1 mm respective about 6 mm diameter.

The Am-241 particle energies were degraded by energy absorber foils for situation 7. The transmitted spectra were measured with a semiconductor detector set-up with 30 keV resolution.

For irradiation in situation 3 and 4 (2π incidence), the sources were brought into direct contact with the detector unit applying slight compression.

In situation 5, the detectors were mounted on a spinning turntable 20 ± 1 mm above a fibre filter (Fa.Schleicher und Schüll, No. 8, 17 cm diameter). The Radon daughter products on this filter were supposed to be in radioactive equilibrium after 4 hours of air intake at a rate of about 83 m^3/h. The out-door radon concentration was monitored and did not change during the irradiations by more then 10% (about 150 pCi/m^3). Temperature and pressure during exposure were 26 C and 726 Torr. The specific alpha activity of Radon daughter products on the filter as measured afterwards with a flow-through proportional counter was 43.9 dpm/cm^2 and decreased after some minutes with a half value time of about 43 min.

4. Results and Discussion

The data reported and the materials and methods used by those eight ALRIT-participants who returned their results in time are compiled in (2). Mainly, Kodak Special Films LR 115 and CA 80-15 were employed but also Makrofol G and E (Bayer AG) and custom made cellulose nitrate foils were used. Four detector units had no energy absorber foils at all, the residual sets had typically 20 μm thick polycarbonate absorber foils.

The detectors were evaluated with optical microscopes at magnifications of 200 - 600 x typically.

Concerning situations 1 and 2, i.e. in the case of vertical incident alpha particles, it can be seen in figs. 1 and 2 that many results agree reasonably well with the nominal values.

This agreement is getting worse (fig.3) for applications in which an alpha emitter has to be brought into direct contact with a detector, as it usually is the case in autoradiography applications (situations 3 and 4). The counting efficiency here can be

considerably smaller than one, and it depends on the development
and evaluation techniques. In fig.3 it becomes apparent that
not all efficiency factors were accurately known. The precision
of measurements can be checked by comparing the track density
ratios between runs 6 and 7 with the correct value. In a few cases
good agreement was found for monoenergetic particles, wheras
for polyenergetic particles (run 8) more data were necessary
for significant conclusions.

However, the tendency showing up already in the results for situations 3 and 4 is confirmed by the results reported for the three
irradiations with alpha particles from Radon daughter products
deposited on a fibre filter (fig.4). The overall efficiencies
differ up to a factor of ten among the detector sets. The track
density ratios for the 3 runs deviate strongly from each other
and from the correct values. Apparently, most of the measurement
methods have to be improved to obtain higher accuracy and precision in Radon dosimetry.

Astonishingly good results were obtained in situations 6 and 7,
i.e. in determinations of unknown alpha particle energies and
of unknown angles of incidence (fig.5). Especially the results
for run 13 (3.95 MeV particle energy) show only one result outside the correct region. For run 14 (2.45 MeV), four values are
outside this region indicating some energy resolution problems
at lower energies. Several measurement techniques were empolyed
to determine the particle energies, e.g. etching from the rear
side and measurement of track lengths or track diameters. The
angles of incidence were obtained by measurement of horizontal
projections and depths of tracks, or of the diameters of the
track ellipse.

5. Conclusions

Because of the brevity of this report and of the rather limited
information based on only 8 out of 16 sets, it is difficult to
draw general conclusions as to the accuracy and precision of
alpha particle measurements with plastic track detectors. However, there is some evidence that track densities for vertical

incident particles, particle energies and angles of incidence can be determined with higher accuracy than track densities for 2π-incident particles. The methods employed in the latter irradiation situations will have to be refined since especially these situations are encountered in such important applications as autoradiography and radon dosimetry.

Acknowledgement

I should like to express my sincere gratitude to my colleagues Drs. R.Winkler and H.Hötzl for valuable advice and help with the Radon measurements.

References

(1) H.Schraube and H.G.Paretzke, "Neutron Fluence Measurements with Track Detectors - Results of INFIT", paper presented at the 9th ICNTD, Neuherberg, 1976

(2) H.G.Paretzke "Alpha Particle Registration with SSNTDs - Full Results of ALRIT I", GSF-Report-S 402, 1976.

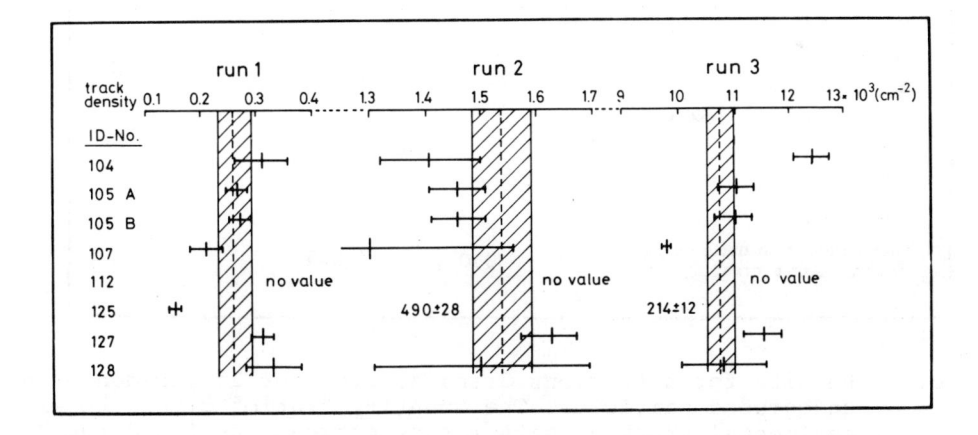

Fig.1   Results for situation 1, i.e. for vertically incident undegraded Am-241 particles. Dashed line indicates the nominal mean value, dashed area indicates region within which 67% of the experimental results have been expected.

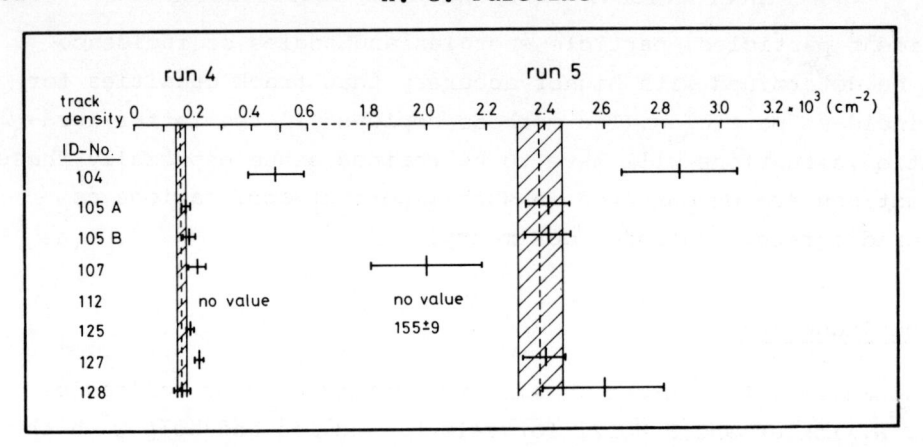

Fig.2   Results for situation 2, i.e. for vertically incident undegraded Am-241, Cm-244 and Pu-239 particles. Dashed line indicates nominal mean value, dashed area indicates region within which 67% of the experimental results have been expected.

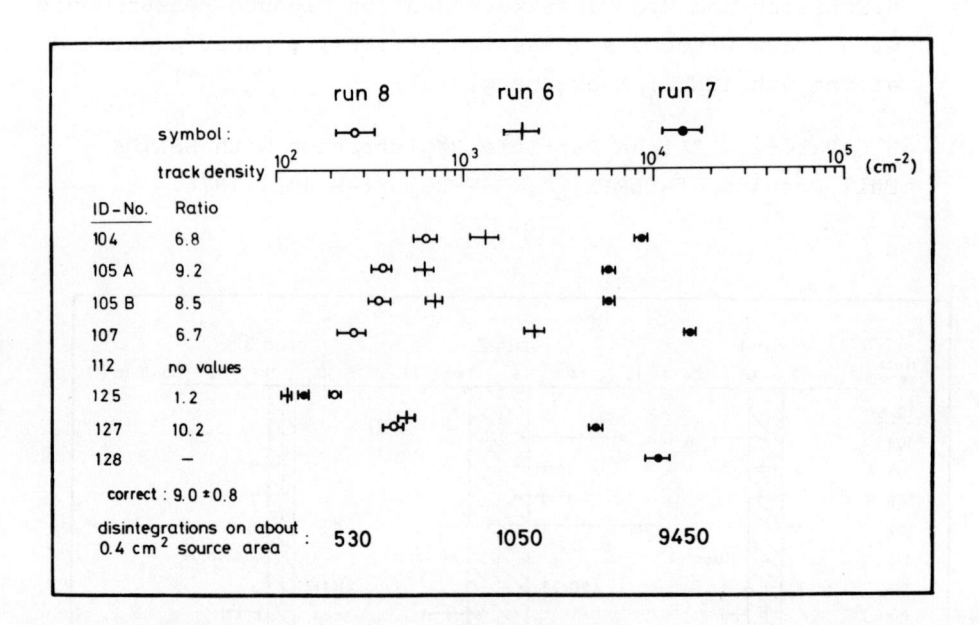

Fig.3   Results for situations 3 and 4, i.e. for 2π-incidence and undegraded particles. The quantity "ratio" gives the experimental track density ratios between run 6 and run 7.

International alpha particle registration   829

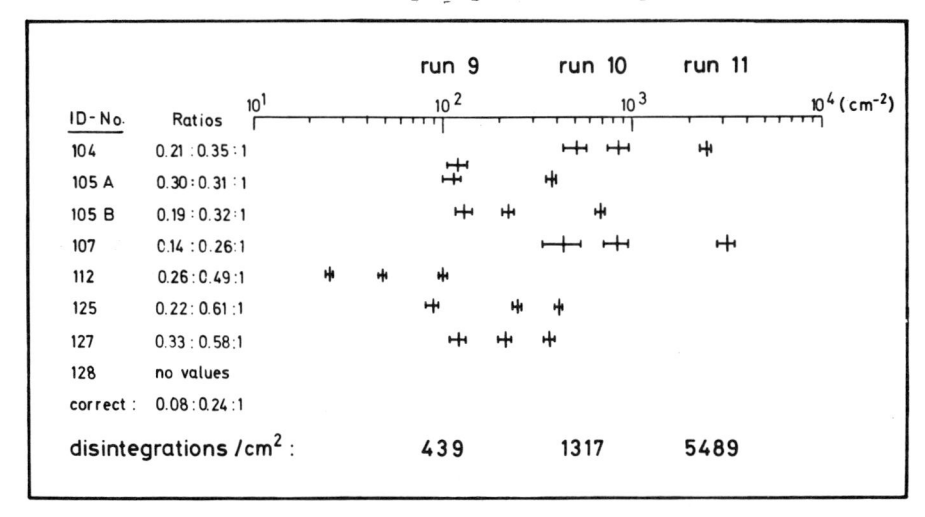

Fig.4   Results for irradiation situation 5, i.e. for irradiation from radon daughter products on a filter. The ratios between the measured track densities between run 9/run 11 and run 10/run 11 respectively, are also given to eliminate the influence of different counting efficiency factors.

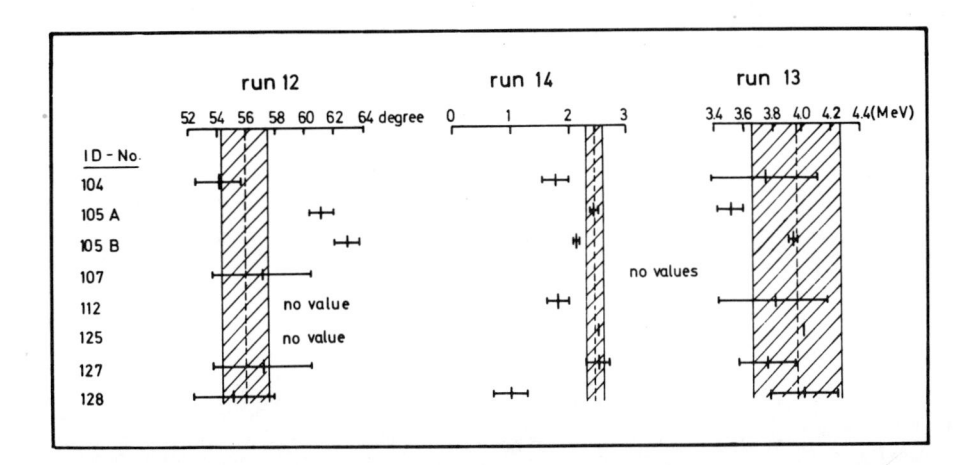

ID-No.	Ratios	run 9	run 10	run 11
104	0.21 : 0.35 : 1		⊢⊣ ⊢⊣	⊢⊣
105 A	0.30 : 0.31 : 1	⊢⊣	⊢⊣	
105 B	0.19 : 0.32 : 1	⊢⊣ ⊢⊣	⊢⊣	
107	0.14 : 0.26 : 1		⊢⊣ ⊢⊣	⊢⊣
112	0.26 : 0.49 : 1	⊢⊣ ⊢⊣ ⊢⊣		
125	0.22 : 0.61 : 1	⊢⊣	⊢⊣ ⊢⊣	
127	0.33 : 0.58 : 1	⊢⊣	⊢⊣ ⊢⊣	
128	no values			
correct:	0.08 : 0.24 : 1			
disintegrations/cm² :		439	1317	5489

Fig.5   Results for situations 6 and 7, i.e. for determination of unknown angle of incidence and of unknown particle energy. Dashed region indicates mean angle and largest deviation geometrically possible for run 12, and mean energy and fwhm of the energy specta for runs 13 and 14.

FACTORS AFFECTING THE SENSITIVITY OF PLASTIC DETECTORS TO IONIZING α-PARTICLE REGISTRATION

R. Antanasijević, Z. Todorović, A. Stamatović, and D. Miocinović

Institute of Physics, 11001 Belgrade, SFR, Yugoslavia

Abstract

The detection characteristics of cellulose nitrates of different physico-chemical properties have been investigated. Their densitometric characteristics, detection efficiency and the possibility of an automatic counting of individual α-particle tracks have been determined.

Introduction

Cellulose nitrates represent the most sensitive detectors in the class of solid state track detectors. Therefore the study of various physico-chemical factors affecting their detection characteristics is of interest. In the present work, the detection properties of different kinds of cellulose nitrates have been studied with the view of their dosimetric application.

Experimental

Samples of ten kinds of commercial cellulose nitrates of different physico-chemical properties (%N from 11.9 to 13.2, and η from 1.0 to 20 cP) were prepared by Benton's method [1].

Maximum surface areas of dried films of a thickness of 50-60 µm were about 200 cm^2. The films were exposed to α-particles from ^{241}Am. The distance between the source and detector was 2 cm. All the samples were exposed under the same conditions for 1-1000 sec. Etching was made in a 10% NaOH solution at 40°C for the time needed to dissolve a top layer of a thickness of 1 µm of the nonexposed layer.

The densitometric characteristics were determined using a "Zeiss" microphotometer. The possibility of a semiautomatic counting of individual tracks was examined on an adapted "Leitz" microscope whose schematic view is given in Fig. 1.

Preliminary investigations have shown that the nitro-cellulose under study may be classified according to detection characteristics into three groups of similar characteristics, hence a sample of each kind has been analyzed in detail.

Results and Discussion

The characteristics of the cellulose nitrates used are given in Table 1. Bulk etch rate (r_B) and the diameter of track decrease with increasing % N and η as has earlier been observed also in the case of nonplastified films [2]. Track form also depends on physico-chemical properties. In samples 1, 2 and 3, tracks

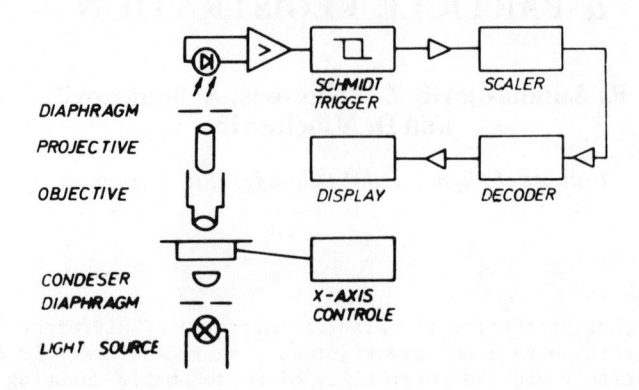

Fig. 1. Scheme of the system for semiautomatic track counting.

appear only in pit form, in both pit and conical forms, and only in conical form, respectively. The magnitude of diameter and the form of α-particle track are dependent on the physico-chemical properties of the cellulose nitrate specimen. As % N and η increase a regular, conical track form is obtained.

Commercial name	$\eta(C_p)$	% N	r_B (μm/h)	d (μm)	efficiency reg. (%)
NC-75	0,8	11,97	1,0	3,5	∼ 75
NC-650	5,0	12,25	0,5	1,8	∼ 80
NC-1800	18,4	12,35	0,3	0,5	− 45

Table 1. Detection characteristics and physico-chemical properties of the cellulose nitrates used.

The densitometric characteristics of the CN under study are shown in Fig. 2.

In the figure it may be seen that they are directly dependent on track form. The magnitude of diameter of a track and its pit form in sample 1 influences the high densitometric sensitivity at low exposures (up to 100 sec), whereupon a saturation and a decrease in density occur.

The sensitivity of sample 2 is lower and its densitometric curve is shifted to longer exposure times, while sample 3 exhibits a densitometric sensitivity only at very high exposures (exceeding 50 sec).

The efficiency of automatic track counting is also dependent on track form. For samples 1, 2 and 3 the upper limit of the number density of tracks at which they may still be counted is ∼10^5, ∼10^6 and ∼2×10^6 respectively.

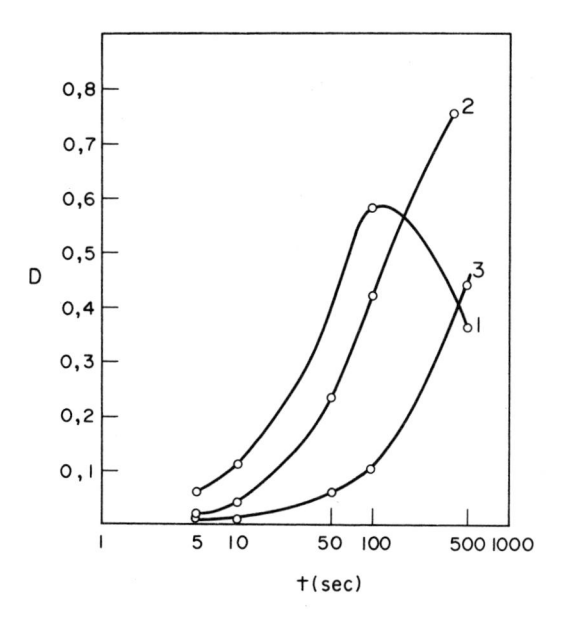

Fig. 2. Optical density of blackness as a function of exposure to ^{241}Am.

References

1. E. V. Benton: A Study of Charged particle track in cellulose nitrate, Ph.D. Thesis, 1968.
2. R. Antanasijević, M. Jurić: Proc. VIII Int. Conf. Nuclear. Phot. and Solid State Track Detectors, Bucharest, 232, 1972.

ALPHA AUTORADIOGRAPHY BY CELLULOSE NITRATE LAYER

J. Simonović, J. Vuković and R. Antanasijecić

Institute of Physics, Medical School, University of Belgrade, SFR, Yugoslavia

Abstract

From domestic cellulose nitrate bulk material thin layers for α-particle autoradiography were prepared. An artificial test specimen of a uniformly alpha labelled grid source was used. The efficiency of autoradiographs by cellulose nitrate was calculated comparing with data from an Ilford K2 nuclear emulsion exposed under the same conditions as the cellulose nitrate film. The resolution was determined as the distance from grid pitch edge at which the track density fell considerably.

Thin plastic layers of a thickness of about 50 μm were prepared for autoradiography from domestic sorts of cellulose nitrates produced at the "Milan Blagojević" plant in Luchani (Yugoslavia). They were made of cellulose nitrate bulk material in such a way that they were first dissolved in acetone and precipitated with water in order to remove impurities, and then dried at a temperature of 100°C. The samples thus prepared were there upon dissolved in appropriate solvents with addition of a plasticizer and finally dried according to Benton[1]. Of several kinds of thin plastic cellulose nitrate films only two types, NC-900 and NC-1800, were selected as the most suitable for alpha autoradiography. They have the lowest bulk etch rate (0,20 μm/h and 0,15 μm/h) and a relatively high alpha-ray detection efficiency.

The plastic layers so chosen were used as detectors for alpha autoradiography in a broad sense, special attention being paid to their application in future in the autoradiography for histological investigations.

In order to obtain a radiographic alpha source of most fully known characteristics in the sense of well defined alpha active and inactive areas, which would simulate an alpha radiographic object, use was made of a special electron - microscope grid of type H-12 with a regular arrangement of bars 25 μm and 12.5 μm in width and 100 μm in length, of square aperture (Fig. 1). A 300 Å layer

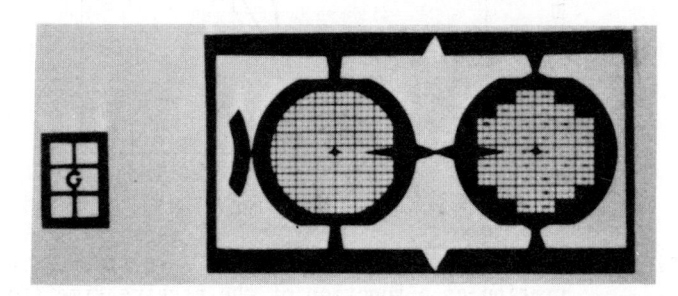

Fig. 1. Electron-microscope grid, type H-12, as an alpha radiographic source (thickness of natural uranium on the grid is about 300 Å).

of natural uranium was vacuum evaporated onto one of the surfaces of such a grid, providing a radiographic source of α-rays of an energy of 4 MeV. The activated grids were placed in contact with the prepared plastic cellulose nitrate layers for 30 days. To ensure close contact they were kept in an evacuated polyethylene bag.

In order to determine the detection efficiency and other characteristics of interest for alpha autoradiography, an Ilford K2 nuclear emulsion was exposed to the same grid specimen, in the same way and at the same time as in the case of the plastic foils.

On the basis of grid patch geometry of the given bar of 25 μm and counting the α-particle tracks in the emulsion it has been established that the number of tracks at a distance larger than 10 μm from the bar edge falls to one half, hence the radiographic image in the emulsion has a width of 45 μm instead of 25 μm. The situation is similar also with the number of tracks at the site of the narrower bar, where width increases from 12.5 to 30 μm. This increase in dimension of radiographic picture is due to the isotropic emission of α-particles from the edges of the given grid, and to the ranges of the α-particles.

The plastic films were etched in 10% NaOH water solution at 25°C. After each hour of etching the density of tracks on a given surface has been measured. Etching was stopped when the number of tracks on the surface ceased to increase with the prolongation of etching (6 hours). During this time of etching the efficiency of the registration of alpha particles of natural uranium has been determined and it was found to be 33% and 20% for NC-900 and NC-1800, respectively (Fig. 2). For these conditions of etching the tracks of alpha particles show both conical and pit shapes with a small aperture diameter - 1.0 and 0.8 μm, respectively. For the above conditions of etching the change of transparency was not found to be of importance, but this is not the case for higher temperatures of etching.

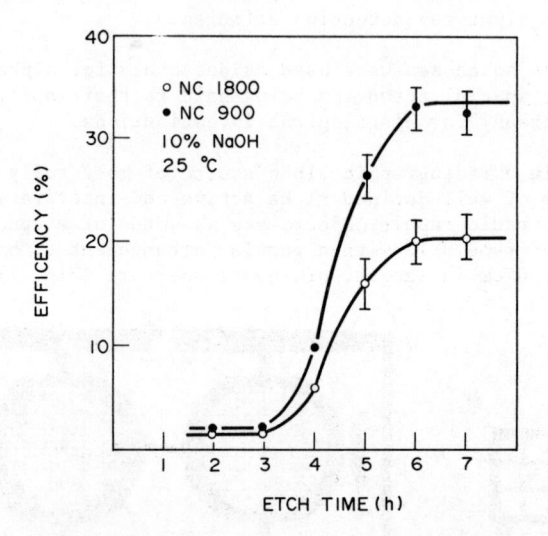

Fig. 2. The efficiency of 4 MeV alpha particles registration as a function of the etching time (for NC-900 and NC-1800).

The dimensions of radiographic source (bars) are compared with dimensions obtained by measuring radiographic picture in plastic detectors as in emulsion. Radiographic picture disagree by 6 μm from real dimensions.

Available results show that such characteristics of α-ray relocalization may satisfy the conditions for radiography of geological specimens and, at higher radiological concentrations, those of histological preparations. Thus radiography by cellulose nitrates is closer to contract radiography that to that based on detected tracks, especially if the active specimens to be investigated are of a size of the order of 1 mm.

Reference

1. E. V. Benton: A Study of charged particle tracks in cellulose nitrate, Ph.D. Thesis, 1968.

NEUTRON DOSIMETRY BY MEANS OF SOLID STATE NUCLEAR TRACK DETECTORS

F. Spurný and K. Turek

Laboratory for Radiological Dosimetry, Czechoslovak Academy of Sciences, Praha, Czechoslovakia

ABSTRACT:

A review of the neutron dosimetry applications of solid state nuclear track detectors is presented.

Firstly, the present state and the principal dosimetric properties of neutron dosimeters based on SSNTD´s are discussed, their advantages with the comparison to other possible detectors are given. Particular attention is devoted to personal neutron dosimetry applications, here the author´s original works also are discussed.

The possibilities of other future applications of SSNTD´s in neutron dosimetry are discussed in the second part of work. Both, the technical improvements as well as the new fundamental approaches are analyzed.

I. INTRODUCTION

The goal of dosimetry, in general, is to characterize both quantitatively and qualitatively the effect of ionizing radiation on a matter; the most usual "reference matter" is a human body, the effects to be characterized biological effects. The methods used in dosimetry are however up to now practically exclusively physical ones. If one take into account the deep differences between the physical and biological phenomena, one can easily imagine how much and how difficult problems the dosimetry must or should resolve.

It is particularly true if one discuss the problems of neutron dosimetry. In spite of a great recent progress in the field, it exists here still a large number of problems which are not yet resolved in a satisfactory way. The growing number and importance of nuclear installations accentuate at the same time the necessity of fast resolution at least of some of them.

That is the reason for a great interest of neutron dosimetrist´s in the dosimetry applications of solid state nuclear track detectors; I would like to analyze firstly the present state of SSNTD´s applications in neutron dosimetry.

II. PRESENT STATE

The fundamental unity which should characterize quantitatively an effect of ionizing radiation is the absorbed dose. It is the unity which is measurable directly by proper chosen physical method (calorimetry etc.). However, if the ionizing radiation in the matter are neutrons and one would measure the absorbed dose in the human body, serious difficulties arise.

Firstly, the influence of a human body on usual neutron field is very profound, the dose absorbed in elementary volume of human tissue in a body differs very seriously of the dose absorbed in the same volume isolated; beside that, it varies very significantly with the position in the body.

Secondly, with the regard to the complexity of neutron interactions and the character of the neutron energy deposition processes it is rather difficult to realize a device which would measure directly the absorbed dose.

And, last but not least, it is known that the relative biological effectiveness of neutron radiation is higher than, for exemple, of gamma radiation, and depend on the neutron energy and other factors.

From all these reasons the practice of particularly neutron personal dosimetry is based:
- on the model calculations of neutron energy deposition in the human body; and
- the utilization of neutron dosimeters which response is given by one or several properly chosen neutron interactions; generally quite different of the neutron interactions in the human tissue.

Solid state nuclear track detectors (SSNTD's) represent in the most of theirs present day applications, very important contribution and extension of this last type of neutron dosimeters.

They exist two basic types of them /1-3/. The first one make use of fission fragment registration, the second one of alpha and light recoil particles detection. Advantages of them with comparison to other types of comparable neutron dosimeters can be summarized in the following manner:
- they are practically insensitive to beta and gamma radia-

tion;
- they are nearly free of disturbing environmental influences at usual conditions, certain of them (glasses, minerals) can be used even at extreme conditions (like in the reactor core);
- they are very versatile, very easy to handle;
- they are generally more sensitive, they cover a very wide dose range (more than 6 orders of magnitude) independently on the dose rate; it permit to use them in a large variety of conditions, from natural background measurements up to in-core reactor experiments;
- theirs development and evaluation is not too complicated, can be automated; however, it is useful to mention that the fission fragments tracks are more simple to recognize, theirs counting is more easy and the possibilities of automation more promisive.

The neutrons are for the dosimetric purposes usually divided to four basic energetical regions:
- thermal neutrons with the energy inferior to Cd cut-off (0.4 eV);
- intermediate neutrons with the energy between 0.4 eV and about 100 keV;
- fast neutrons with the energy between 100 keV and 20 MeV; and
- high energy (relativistic) neutrons with the energy superior to 20 MeV.

SSNTD´s can be used in all these neutron energy regions, the typical possibilities are given in the table 1.

As concerns the <u>thermal neutrons</u>, one make use of both types of dosimeters.

Practically all types of SSNTD´s can be used for fission fragment registration, the most frequent are mica /4-6/, different glasses /7-9/, as well as organic polymers /10-14/. Typical sensitivities are for pure ^{235}U in free-air irradiation geometry about 3.10^{-3} tracks per 1 thermal neutron /7,13/.

Only some of the organic polymers (cellulose derivatives, polycarbonates) are sufficiently sensitive to detect alpha and light recoil particles; therefore, only these can be used for thermal neutron dosimetry using (n,α) reactions. Typical sensitivities in free-air irradiation geometry are about 6.10^{-5} tracks per 1 neutron /15,16/.

If one attach a dosimeter with a thermal neutron detector to a human body, the sensitivity arises due to the albedo neutrons. Nagarajan et al. /17/ found out, for exemple, that it is equal to $2.6.10^{-3}$ tracks per 1 neutron using B_2O_3 radiator.

Taking into account:
- the conversion factors between the thermal neutron fluence and absorbed dose in a human body /18/;
- the sensitivities mentioned; and
- the reasonable lowest limits of detection (track densities 1 cm^{-2}, resp. 10^2 cm^{-2} for spark; resp. visual counting as the most typical possibilities of evaluation),

one can appreciate the lowest "measurable" neutron absorbed dose. They are given also in the table 1.

Thermal neutron detectors based on SSNTD's can be also used in the Bonner spheres. Theirs properties permit very long term measurements without any problem of device stability or of the time factors lying to the activation detectors. "Rem-meters" based on this principle can measure the dose equivalents as low as 10^{-6} $J.kg^{-1}$ (10^{-4} rem) /19/.

One has, up to now, no entirely satisfactory and simple methods of the intermediate neutron's dosimetry, especially for low dose measurements. Practically only methods having acceptable energetical dependence are based on the albedo princip (20-24/. The thermal neutron detectors using a SSNTD'S mentioned above /4-17/, represent a very important contribution also for the dosimetry in this neutron energy region. They have very low detection threshold (see table 1), with the comparison to common ^{6}LiF -^{7}LiF thermoluminescent couple detectors are entirely free of the disturbing effect of

Table 1: Neutron dosimetry by means of different solid state track detectors

Neutron energy region	Neutron interactions used	SSNTD	Detection threshold (free-air) for	
			spark counting	visual counting
thermal	$^{235}U(n,f)$	all	10^{-8} Gy	10^{-6} Gy
	$^{10}B(n,\alpha)$ $^{6}Li(n,\alpha)$	some organic polymers	$> 10^{-6}$ Gy	10^{-4} Gy
intermediate	$^{235}U(n,f)$	all	10^{-6} Gy	10^{-4} Gy
	$^{10}B(n,\alpha)$ $^{6}Li(n,\alpha)$	some organic polymers	$> 10^{-4}$ Gy	10^{-2} Gy
fast	$^{232}Th(n,f)$ $^{237}Np(n,f)$	all	10^{-4} Gy 10^{-5} Gy	10^{-2} Gy 10^{-3} Gy
	(n,n) (n,α) (n,n')	some organic polymers	?	10^{-2}–10^{-3} Gy
high energy	$U(n,f)$ $Bi(n,f)$ $Au(n,f)$ $Ta(n,f)$ spallation	all	$\sim 10^{-5}$ Gy	$\sim 10^{-3}$ Gy
	(n,n) (n,α) (n,n') and others	some organic polymers	?	$\sim 10^{-2}$ Gy

gamma radiation.

Both mentioned types of neutron dosimeters based on SSNTD´s can be used also in the fast neutron dosimetry.

Fission fragment registration by SSNTD's and its application in fast neutron dosimetry drew attention from the early years of SSNTD's history /1/, it is a lot of works accomplished in this direction /4-7, 11-14, 25-32 and others/. One know that the fission fragments track density is proportional to the fission cross section /7, 11-13/, the most convenient fissionable material for personal neutron dosimetry is from this point of view the ^{237}Np /3, 25, 32/, it is also the most sensitive. Nevertheless, it is the ^{232}Th which is more frequently used, the reasons are generally other than physical ones. The lowest detectable absorbed doses for both materials are given in the table 1, one can see that, using spark counting, both materials cover the dose range necessary even for personal neutron monitoring, they are still introduce in several laboratories in routine practice /11, 25, 27, 29, 30/.

However, the presence of a fissionable material in a personal dosimeter represents a certain number disadvantages, like the risk of loss, risk of contamination, and radiation hazard. A number of authors tried therefore to develop a a fast neutron dosimeter based on the alpha and light recoil particles registration /32-50 and others/. It was shown that the sensitivities of cellulose derivatives and/or polycarbonates vary only a little with the energy of neutrons above about 1 MeV /33, 36, 37, 48-50/; for the fission neutrons they are of the order of 10^{-5} tracks per 1 neutron. It corresponds to the absorbed dose detection threshold in tissue about 10^{-2} to 10^{-3} Gy in the case of visual counting (see table 1), the spark counting of these tracks has not yet given sufficiently satisfactory results /51, 52/. For avoiding rather tedious counting of alpha and recoil particles tracks in usual polymer foils (see figure 1A), several authors tried to use for neutron dosimetry film KODAK LR 115

/44, 45/. It consists of thin (~10 μm) layer of intensively red coloured cellulose nitrate deposited on the colourless polyester base. Some of the neutron induced tracks can be etched through the cellulose nitrate layer and appear as light spots on a rather dark background (see figure 1p). KODAK LR 115 is practically free of background throughetched tracks (density of the order 1 cm^{-2} /53/), the high contrast permit even microdensitometric evaluation /45/. Unfortunately, its sensitivity drops very rapidly with the neutron energy /44/, the detection thresholds for fission neutrons seems to be the same as for other polymer foils.

In our laboratory we tried to analyze more deeply the dosimetric properties of some fast neutron dosimeters based on SSNTD´s, I would like to present you some of the results acquired. The dosimeters studied were chosen in this way:
- microscopic glass in the contact with ^{232}Th as a representative of fission fragment detectors;
- polycarbonate Makrofol E as a representative of common alpha-sensitive polymers; and
- KODAK LR 115 as a representative of a detector where only a part of neutron induced tracks (throughetched) is counted.

Etching conditions adopted are given in the table 2.
All detectors were evaluated by visual counting, basic properties studied were theirs energetical and angular dependences.

As it was stated, the energetical dependences of fission fragments track neutron dosimeters are given by the energetical evolution of fission cross sections, we have found out the same behaviour in the case of our detector, the mean efficiency of fission fragment registration in our case was 0.42 /54-55/. In the case of polymer materials and alpha or recoil particles detection the situation is much more complex. That´s why we tried to analyze theirs energetical dependences more profoundly, both theoretically as experimentally /48-50,56/.

The theoretical calculations were carried out supposing three different groups of energetical restrictions of develo-

Table 2: Etching conditions adopted for studied fast neutron solid state nuclear track detectors

Detector	Etching agent	Etching temperature	Removed layer /μm/
Microscopic glass	39% HF	25°C	~60
Makrofol E	15 g KOH + 40 g C_2H_5OH in 45 g H_2O	70°C	6.5
KODAK LR 115	2.5 N NaOH	40°C	5

pability of a track, the groups are presented in the table 3. The results of calculations, the theoretical sensitivities, are shown in the figures 2 and 3, one give there the theoretical numbers of tracks created in 1 μm-layer of a polymer. Experimental sensitivities $C(Source)_{exp}$ were established for several neutron sources (Am F; ^{252}Cf; D/D reaction; Am Be; T/D reaction), they are compared with the theoretical values for the same sources /49, 50, 56/ in the figures 4 and 5. Generally, the ratio of experimental and theoretical sensitivities must be lower than removed layer and should depend only a slightly on the neutron energy. Coming out from this statement one can appreciate from the figures 4 and 5 which group of energetical restrictions for the developability of a track is the most reasonable. It is seen that it is the group I if one count all tracks, the group III if one count only through-etched tracks in KODAK LR 115. Taking into account the energies given in the table 3 it is possible to conclude that neutron detection threshold should be about $E_n \sim 0.7$ MeV if one count all tracks, about 3 MeV for throughetched tracks in KODAK.

Comparison of energetical dependences of studied neutron dosimeters, expressed in the track densities for 1 Gy absorbed dose in the element 57 of Snyder's human phantom /18/, is given in the figure 6. On can see, in agreement with the analysis given above, that:
- it is the Makrofol E which is the most sensitive; its energetical dependence being slightly better than for ^{232}Th in the contact with glass;
- the track density in KODAK LR 115 depend too much on the neutron energy; its using should be limited only to rather well known neutron spectra.

Angular dependences, i.e. the dependences of counted track densities on the fast neutron incidence angle are given in the figure 7. It is the fission fragment detector which angular dependence is the least important, one observe also a certain influence of neutron energy.

Table 3: Energetical restrictions chosen for developability of a track

Group	Particle type	Track of a particle is developable if its energy is higher than
I	all ($Z \geq 2$)	0.2 MeV
II	alpha	0.2 MeV
	recoils ($Z \geq 4$)	0.6 MeV
III	alpha	0.5 MeV
	recoils ($Z \geq 4$)	1.0 MeV

To summarize:

Our studies showed that all three types studied dosimeters based on SSNTD's are suitable for fast neutron dosimetry. Every of them has some advantages, the choice will depend on the actual conditions of their application (neutron spectrum, dose and dose rate, personal or field monitoring etc.).

Also <u>high energy neutrons</u> can be detected by means of SSNTD's. The most of experience is with the fission fragment registration, one use U, Bi, Au, Ag, or Ta as radiator materials /57-61/. The lowest absorbed doses measurable are again given in the table 1, one can see that they are sufficiently low for neutron personal monitoring. One has, as far as the authors of this review are informed, a very little experience with the high energy neutron detection by means of organic polymer materials alone. The author's own experiments in the neutron beam of synchrocyclotron at Dubna ($E_{max} \sim$ 680 MeV) showed /62/, that their sensitivity is about 10^{-5} tracks per 1 neutron, it corresponds to the lowest detectable tissue absorbed dose (by visual counting) 10^{-2} Gy. Of course, much more experiments are needed for establishing the sensitivities to high energy neutrons more accurately (influence of gamma background, proton's component etc.).

III. OUTLOOK

In spite of a vast variety of neutron dosimetry applications of SSNTD's it is safe to predict that theirs possibilities in the field are still far from being exhausted.

Firstly, with the regard to very important recent development, the practical, especially routine, applications of SSNTD's are not so wide-spread as possible or as one can expect.

In the case of fission fragment detectors, the reasons for it were mentioned above, i.e. mainly their radioactivity and, also, their cost. The detectors based on the direct interaction of fast neutrons in sensitive polymer foils are free of disadvantages connected with the fissionable materials; their counting is, on the other hand, rather tedious and one look still for a simple and cheap method of their automatic and well reproducible evaluation.

Perhaps, further technical development can resolve this problem. One of the most interesting improvements in this direction is represented by electrochemical etching/63/. By this technique, tracks are enlarged to visibility with the bare eye, it simplifies to a great extent their counting. Sohrabi et al. /64,65/ found out that the lowest detectable fast neutron dose for polycarbonate as SSNTD is, using this technique and usual visual counting, of the order of 10^{-4} Gy, it means up to two orders of magnitude lower than in the case of classical etching. Moreover, the evaluation should be rather easy to automate, for exemple by a microdensitometric method.

The electrochemical etching is not only possible improvement in etching technique. Although the search of new etching agents progress very rapidly, one can still expect the development of new, better solutions; beside that, the possibilities of quite different track amplification principles are also not excluded /3/. Very important influence on optimizing the detector sensitivity, can have also the different methods of foil's treatment, like UV-irradiation, oxidation effects etc. /77-81/.

All these improvements lead and should lead to still better distinguishability of tracks, to better reproducibility of theirs paramaters; the necessary conditions for any method of automatic evaluation. It exist a large variety of them /2,66/, it is impossible to analyze all in this paper. Nevertheless, the exclusive importance of spark counting, developped originally by Cross and Tommasino /67, 70/, particularly for low track density and/ or large area evaluation must be emphasized. As far as the high track density automatic evaluation method is concerned, the situation is more complex. There are a several competitive possibilities here; some of them are based on the automatic analysis of microscopic image, others on some global optical phenomena (scattering, absorption etc.). Generally speaking, they are not so satisfactory as it is the case of spark counting. The improvements in track amplification methods mentioned above can, however, bring a very important contribution in this connection; one can imagine also a further principles of automatic evaluation (using laser light, ionizing particle penetration etc.), utilizable for fission fragment as well as alpha and recoil particle tracks registration.

It is still one very important problem which could be resolved with the aid of new more sensitive materials and/or new improved methods of track amplification, i.e. easy and reproducible registration of proton's tracks. A great progress was registrated in this field /71-74/, Lück was even able to prepare a cellulose nitrate which permit the proton spectrometry in the energy region from 300 keV to 1 MeV /75/. Unfortunately, a polymer of comparable properties is not commercially available; the practical, especially routine, application of the method in neutron dosimetry rests still a great task for a future /82/.

New detectors and/or new methods of track amplification and counting are not only ways for further extension of neutron dosimetry applications of SSNTD's. Certainly, one can imagine still further particles to be detected, still new problems to be resolved. But, the extension is also possible taking into account quite different aspects of neutron radiation action.

The applications of SSNTD's discussed up to now were generally designated to characterize a neutron field from the classical macrodosimetric point of view, more quantitatively than qualitatively. And, some of the properties of SSNTD's, as evoked François at the last conference at Bucharest /76/, are very attractive also from the point of view of microdosimetry (dependence of track parameters on the local energy deposition density etc.). Beside that, the tissue like composition of some of them would permit to simulate different, even biological effects of neutron radiation in the tissue. SSNTD´s could bring in this manner a grand contribution to the basic task of dosimetry in general, i.e. to characterize an ionizing radiation action both quantitatively and qualitatively.

It is outside of any discussion that the SSNTD´s still acquired a very important and, one can say, nearly irreplacable position between the methods of neutron dosimetry. Moreover, theirs properties and their flexibility permit to believe that SSNTD´s will contribute to the resolution of many other neutron dosimetry problems. Looking backward, one can notice that an intimate contact of SSNTD´s with the neutron dosimetry had an important influence not only on the resolution of neutron dosimetry problems. The stimulating effect of neutron dosimetry on the development of SSNTD´s is also outside of any doubt. It can be remarked in the technique of SSNTD´s (automatic evaluation etc.) as well as in their theory. It is hopeful to believe that this contact will be mutually even more beneficial and fruitful in the future.

Acknowledgements:

The authors are much obliged to Dr. Z. Spurný for helpful and stimulating discussions during the preparation of this work.

REFERENCES

/ 1/ R.M.Walker, P.B.Price, R.L.Fleischer: Appl.Phys.Lett. $\underline{3}$, (1963), p.28

/ 2/ R.L.Fleischer, P.B.Price, R.M.Walker: Ann.Rev.Nucl.Sci. $\underline{15}$, (1965), p.1

/ 3/ K.Becker: in "Topics in Radiation Dosimetry"; Suppl.1.: red.: F.H.Attix et al., Academic Press, New York 1972, p.79

/ 4/ C.O.Widell: in "Neutron Monitoring"; IAEA, Vienna 1967; s.417

/ 5/ M.A.Gomaa, A.M.Eiol; A.M.Sayed: in "Neut.Monit.Rad.Prot. Purp."; IAEA, Vienna 1973, vol. II; p.219

/ 6/ W.Stolz, B.Dörschel: Kernenergie $\underline{12}$, (1969), p.244

/ 7/ K.Becker: Health Physics $\underline{12}$, (1966), p.769

/ 8/ F.Běhounek, J.Novotný, Z.Spurný: Czech.J.Phys. $\underline{B18}$, (1968), p.743

/ 9/ J.W.H.Schreurs, A.M.Friedman, O.J.Rokop, M.W.Hair, R.M. Walker: Rad.Effects $\underline{7}$, (1971), p.231

/10/ E.T.Agard, R.E.Jervis, K.G.McNeill: Health Phys, $\underline{21}$, (1971), p.625

/11/ S.B.Prêtre: in "Neut.Monit.Rad.Prot.Purp."; IAEA, Vienna 1973; vol.II; p.99

/12/ S.B.Prêtre: Rad. Effects $\underline{5}$, (1970), p.103

/13/ C.M.Unruh, W.V.Baumgartner, L.F.Kocher, L.W.Brackenbush, G.W.R.Endress: in "Neutron Monitoring"; IAEA, Vienna 1967, p.433

/14/ W.V.Baumgartner, L.W.Brackenbush: BNWL-332; 1960

/15/ K.Becker: Health Physics $\underline{16}$, (1969), p.113

/16/ B.J.Tymons, J.W.N.Tuyn, J.Baarli: in "Neutron Monit.Rad. Prot.Purp.", IAEA, Vienna 1973, vol.II, p.63

/17/ P.S.Nagarajan, D.Krishnan: Health Physics $\underline{17}$, (1969), p. 323

/18/ J.A.Auxier, W.S.Snyder, T.D.Jones: in "Radiation Dosimetry", 2nd edition, eds.: F.H.Attix, W.C.Roesch, E.Tochiliń; vol. I, Academic Press, New York 1968, p.275

/19/ P.F.Rago, R.C.Barrall, T.G.Carter: Health Phys. 26, (1974), p.102

/20/ J.A.Dennis, J.W.Smith, S.J.Boot: in "Neutron Monitoring", IAEA, Vienna 1967, p.537

/21/ D.F.Hankins: LA-4832, LASL, 1972

/22/ A.Korba, J.E.Hoy: Health Physics 19, (1970), p.331

/23/ E.Piesch, B.Burgkhardt: in "Neutr.Monit.Rad.Prot.Purp.", IAEA, Vienna 1973, vol.II, p.31

/24/ R.G.Alsmiller,Jr., J.Barish: Health Phys. 26, (1974), p.13

/25/ I.B.Keirim-Markus et al.: Atomnaya Energiya 34, (1972), p.11

/26/ M.Sohrabi, K.Becker: ORNL-TM-3605, ORNL, Oak Ridge 1971

/27/ J.Trousil, J.Singer, M.Maršál: in "Neut.Monit.Rad.Prot. Purp.", IAEA, Vienna 1973, vol.II.; p.99

/28/ M.Heinzelmann, H.Schürer: in "Proc. 1st Symp. Neut.Dos. Med.Biol.", EURATOM, EUR 4896 d.e.f, (1972), vol.II, p.315

/29/ C.H.Distenfeld,J.R.Klemish,Jr.: BNL-17452, BNL, New York 1973

/30/ K.Buijs, J.P.Vaane, B.Burgkhardt, E.Piesch: in "Neut. Monit.Rad.Prot.Purp.",IAEA, Vienna 1973, vol.II,p.159

/31/ M.A.Gomaa: Atomkernénergie 23, (1974), p.161

/32/ W.G.Cross, H.Ing: Health Physics 28, (1975), p.511

/33/ K.Becker: Health Physics 16, (1969), p.113

/34/ L.Medweczky, G.Somogyi: in "Proc.2nd Symp.Health Phys.", Pecz 1966, vol.I., p.60

/35/ J.W.Tuyn: Trans.Amer.Nucl.Soc. 13, (1970), p.523

/36/ K.Jozefowicz: in "Neut.Monit.Rad.Prot.Purp.", IAEA, Vienna 1973, vol.II.,p.183

/37/ E.Piesch: in Advances Phys.Biol.Rad.Det.", IAEA, Vienna 1971, p.399

/38/ V.Nishiwaki, T.Tsuruta, K.Yamazaki: J.Nucl.Sci.Techn. 8, (1971), p.162

/39/ B.J.Tymons, J.W.N.Tuyn, J.Baarli: in"Neut.Monit.Rad. Prot.Purp.", IAEA, Vienna 1973, vol.II., p.63

/40/ M.Heinzelmann, W.Haschke: Jül-787-ST, Jülich 1971
/41/ H.A.Khan: Nucl.I$_n$str.Meth. 113, (1973), p.55
/42/ A.L.Frank, E.V.Benton: Rad. Effects 3, (1970),p.33
/43/ A.Dragu, M.Nicolae: Radioprotection 7, (1972),p.87
/44/ G.M.Hassib, L.Medweczky: in "Proc.2nd Symp.Neutr.Dos. Biol.Med.", Neuherberg, 1974, EURATOM, EUR 5273-d-e-f, 1975, vol.I., p.535
/45/ J.Tripier, R.Oppel, G.Remy, M.Debeauvais: in "Proc.2nd Symp.Neutr.Dos.Biol.Med.", Neuherberg 1974, EURATOM, EUR 5273 d-e-f, 1975, vol.I., p.509
/46/ G.Fängewisch, A.Scharmann: Kerntechnik 16, (1974),p.13
/47/ M.A.Gomaa: Atomkernenergie 23, (1974), p.161
/48/ F.Spurný, J.Lochmanová: Jaderná energie 20, (1974),p.233
/49/ F.Spurný, J.Lochmanová: Jaderná energie 20, (1974),p.306
/50/ F.Spurný, J.Lochmanová, K.Turek: Radioprotection 9 (1974), p.307
/51/ K.Becker, M.Abd-el-Razek: ORNL-TM-4460, ORNL 1974
/52/ K.Becker, M.Abd-el Razek: Nucl.Instr.Meth. 124, (1975), p.557
/53/ F.Spurný, K.Turek: Czech.J.Phys. B26, (1976), p.235
/54/ K.Turek, F.Spurný: Výzkumná zpráva LRD ČSAV 15/1975, Praha 1975
/55/ F.Spurný, K.Turek: Výzkumná zpráva LRD ČSAV 18/1976, Praha 1976
/56/ F.Spurný, K.Turek: paper presented at the 9th Inter. Conf.Solid State Nucl.Track Det., Neuherberg 1976
/57/ J.Hudis, S.Katcoff: BNL-13124, Brookhaven 1969
/58/ H.A.Wollenberg, A.R.Smith: UCRL-19364, University of California 1969
/59/ M.Heinzelmann, H.Schüren: Jül-670-ST, Jülich 1970,p.147
/60/ G.K.Svensson: in "Proc.2nd Int.Cong. IRPA", Brighton, 1970
/61/ M.Debeauvais, R.Stein, J.Ralarosy, P.Cüer: Nucl.Phys. A90, (1967), p.186
/62/ F.Spurný: to be published
/63/ L.Tommasino, C.Armellini: Rad.Effects 20, (1973),p.253
/64/ M.Sohrabi: Health Phys. 27, (1974),p.598

/65/ M.Sohrabi, R.Z.Morgan: in "3rd Eur.Cong.IRPA", Amsterdam, May 1975

/66/ W.Abmayr, G.Burger, P.Gais, H.Paretzke: in "Proc.8th Int.Conf.Nucl.Phot.Solid State Track Det.", Bucharest 1972, IFA, 1972, vol.II, p.425

/67/ W.G.Cross, L.Tömmasino: Rad. Effects 5, (1970),p.85

/68/ L.Tommasino, W.G.Cross: Health Phys. 23, (1972),p.403

/69/ W.G.Cross, L.Tommasino: in "Proc. 1st Symp.Neut.Dos. Biol.Med.", EURATOM, EUR 4896-d-e-f, vol.II, p.283

/70/ L.Tommasino, W.G.Cross: in "Proc. 8th I_{nt}.Conf.Nucl. Phot.Solid State Track Det.", Bucharest 1972, IFA 1972, vol.II, p.440

/71/ M.Várnagy, S.Szegedi, S.Nagy: Nucl.Instr.Meth. 89,(1970), p.27

/72/ M.Várnagy, J.Csikai, J.Szabó, S.Szegedi, J.Bánhalmi: Nucl.Instrum.Meth. 119, (1974), p.451

/73/ B.S.Carpenter, P.D.Lafleur: Int.J.Appl.Rad.Isot. 23, (1972), p.157

/74/ H.B.Lück: Nucl.Instr.Meth. 116, (1974), p.613

/75/ H.B.Lück: Nucl.Instr.Meth. 119, (1974), p.403

/76/ H.François: in "Proc.8th Int.Conf.Nucl.Phot.Sólid State Nucl.Track Det.", Bucharest 1972, IFA, 1972, vol.II., p.333

/77/ R.P.Henke, E.V.Benton, H.H.Heckmann: Rad.Effects 3, (1970), p.43

/78/ W. DeSorbo, J.S.Humphrey: Nature 220, (1968), p.1313

/79/ W.Enge, H.O.Schmitt, K.P.Bartholomä, R.Beaujean: paper presented at 12th Int.Conf.on Cosmic Rays, Hobart, Tasmania, August 1971

/80/ G.Siegman, K.P.Bartholomä, W.Enge: IFKKI 75/1, (1975)

/81) G.Somogyi: Rad.Effects 16, (1972), p.233

/82/ R.Griffith: in "4th AEC Workshop on Personnel Neutron Dosimetry"; eds. E.J.Vallario, D.E.Hankins, C.M.Unruh, BNWL-1777, (1973), p.35.

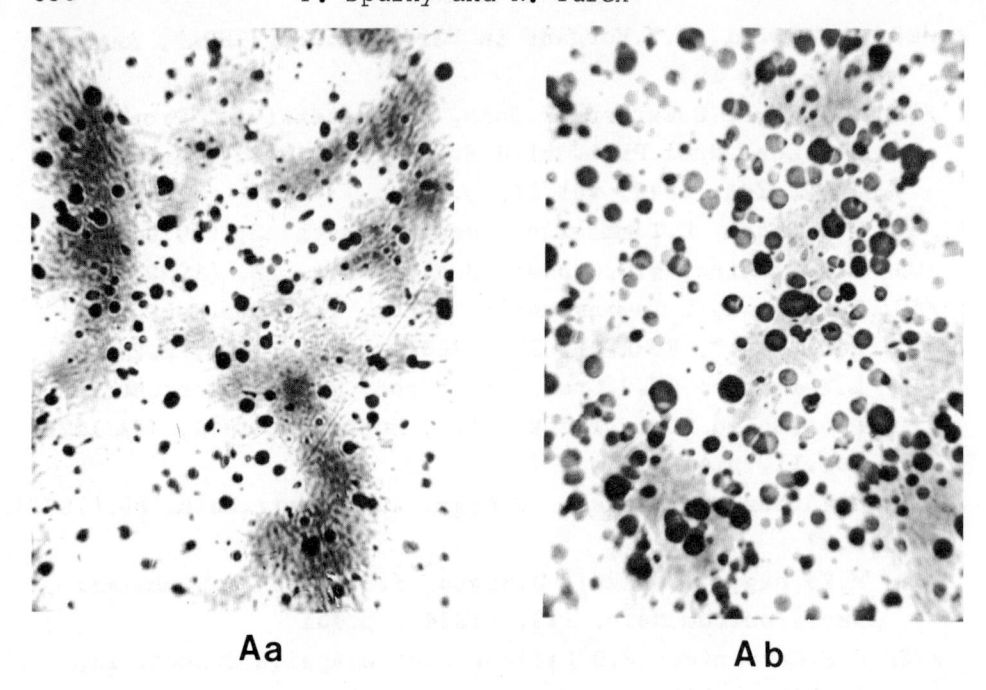

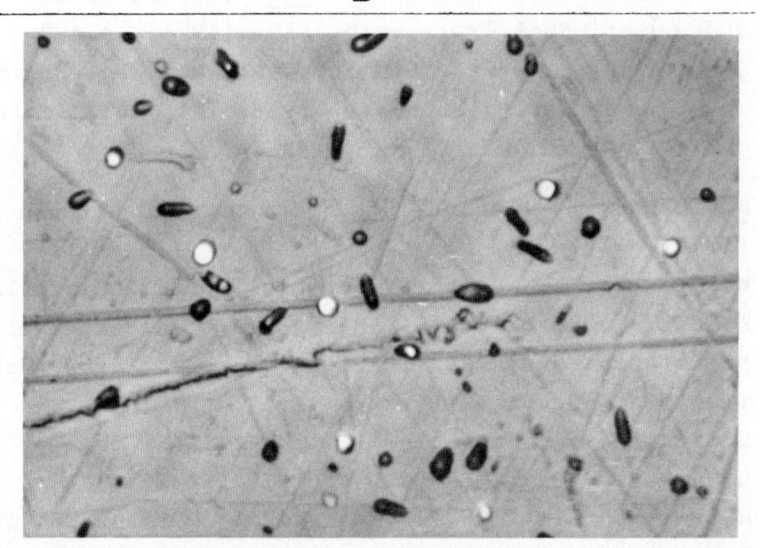

Figure 1: Neutron induced alpha and recoil particle tracks in polymers.
A. Makrofol E: E_n = 14.7 MeV a) etched 20 min
   b) etched 40 min
B. KODAK LR 115: E_n = 14.7 MeV; etched 8 hours

Neutron dosimetry 859

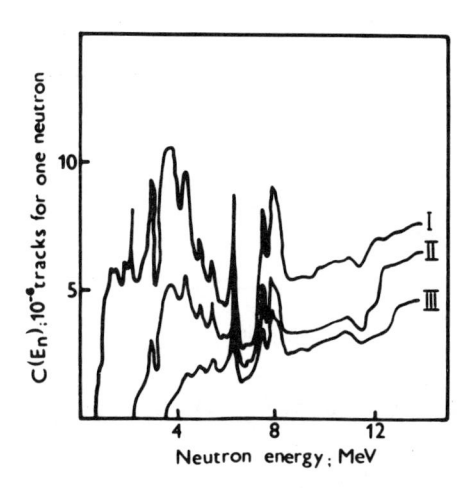

Figure 2: Theoretical sensitivity of polycarbonate

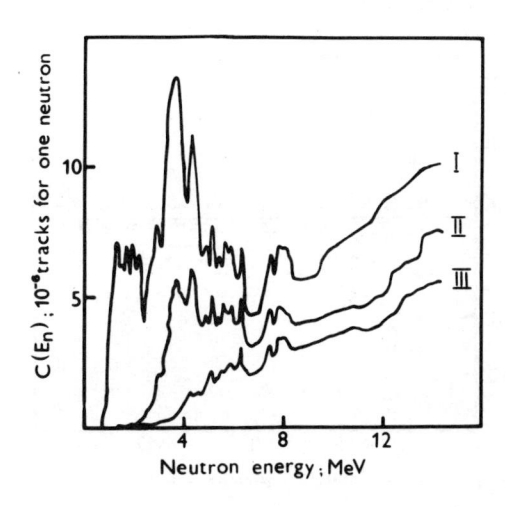

Figure 3: Theoretical sensitivity of cellulose nitrate

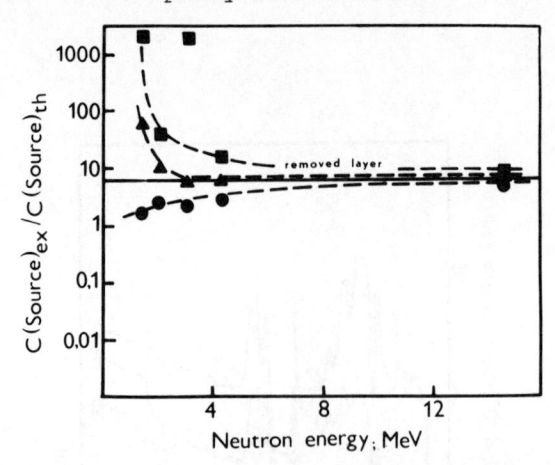

Figure 4: The ratios $C(Source)_{exp}/C(Source)_{th}$ for polycarbonate:
- ● - taken $C(Source)_{th}$ for the group of energetical restrictions I
- ▲ - taken $C(Source)_{th}$ for the group of energetical restrictions II
- ■ - taken $C(Source)_{th}$ for the group of energetical restrictions III

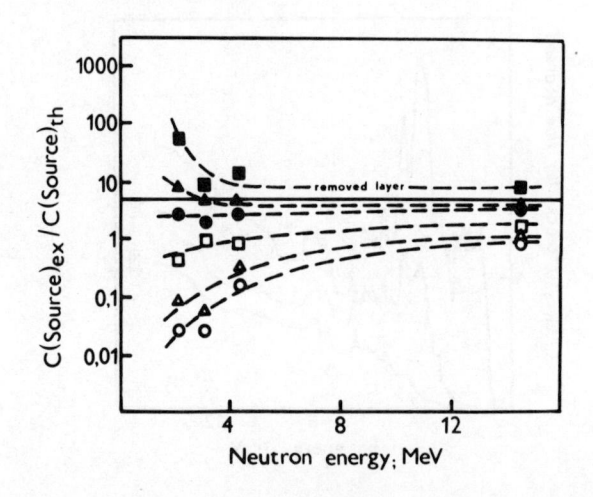

Figure 5: The ratios $C(Source)_{exp}/C(Source)_{th}$ for cellulose nitrate; the description the same as in figure 4; full signs - all tracks counted; open signs - only throughetched tracks counted

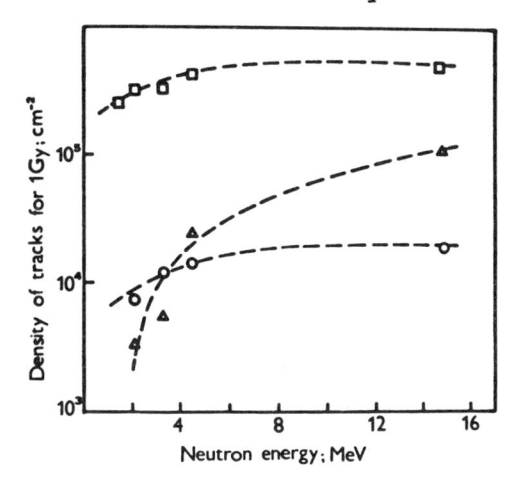

Figure 6: Energetical dependences of studied fast neutron dosimeters based on SSNTD s:
○ - glass with ^{232}Th
□ - Makrofol E
△ - KODAK LR 115

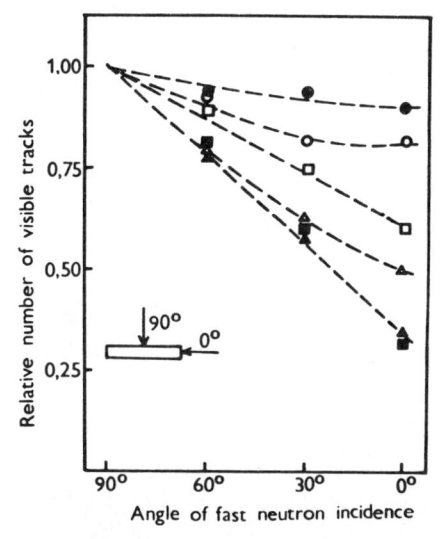

Figure 7: Angular dependence of studied fast neutron dosimeters:
○ - glass with ^{232}Th
□ - Makrofol E
△ - KODAK LR 115
full signs - IBR 30 reactor neutrons; open signs -

ON THE ENERGETICAL DEPENDENCE OF POLYMER SOLID STATE NUCLEAR TRACK DETECTORS AS FAST NEUTRON DOSIMETERS

F. Spurný and K. Turek

Laboratory for Radiological Dosimetry, Czechoslovak Academy of Sciences, 180 86 Praha 8, Na Truhlárce 39/2a, Czechoslovakia

ABSTRACT

The energetical dependences of some polymer solid state nuclear track detectors as fast neutron dosimeters were studied both theoretically and experimentally.

The numbers of heavy charged particles ($Z \geqslant 2$) formed during the fast neutron interactions in plastics (polycarbonates, cellulose derivatives), and the tracks of them should be etchable, were calculated for several groups of assumptions as regard the developability of tracks.

The experimental studies were carried out with several plastic materials and for a number of different neutron sources, the detectors were evaluated by visual counting in a microscope.

The experimentally obtained sensitivities for Makrofol E polycarbonate and KODAK LR 115 cellulose nitrate are compared with the theoretical data. The comparison permit to appreciate the conditions necessary for the developability of a track as well as the neutron detection energy threshold for the detectors and the method of evaluation used.

I. INTRODUCTION

Neutron interactions in some polymer solid state nuclear track detectors (SSNTD's) lead to charged particles which tracks are developable by usual etching technique. The most known polymers of that type are cellulose derivatives and polycarbonates. Foils of these types of polymers could therefore represent a very cheap neutron dosimeter, very easy to handle. That's why the dosimetric properties of these materials were studied by a certain number of authors /1-10/.

One of the most important property of any neutron dosimeter is the dependence of its response on the neutron energy. One knew, that the neutron registration threshold of the detectors mentioned should be about 1 MeV or a little below /5,11,12/, the information on full energy scale were generally rather limited.

That's why we decided to study these dependences more profoundly, both theoretically and experimentally /13/. This paper gives a survey of results obtained in this direction.

II. Theoretical Analysis

Let N_j is the number of nuclei of j_{th} element per 1 cm^3 of a polymer material. The number $C(E_n)$ of heavy charged particles ($Z \geqslant 2$) created during the interactions of one neutron of the energy E_n in the layer of the thickness ΔX, and theirs tracks are developable, is given by the equation:

$$C(E_n) = \sum_i \sum_j \zeta_{ij}(E_n) \, \sigma_{ij}(E_n) \, N_j \, \Delta X \quad , \tag{1}$$

where

$\zeta_{ij}(E_n)$ is a factor which express what fraction of char-

Table 1: Energetical restrictions chosen for developability of a track.

Group	Particle type	The track of a particle is developable if its energy is higher than
I	all ($Z \geq 2$)	0.2 MeV
II	alpha	0.2 MeV
	recoils ($Z \geq 4$)	0.6 MeV
III	alpha	0.5 MeV
	recoils ($Z \geq 4$)	1.0 MeV

ged particles created during i-th neutron interaction with nucleus of j-th element form the developable tracks and

$\sigma_{ij}(E_n)$ is the cross section of i,j-th neutron interaction.

The neutron interactions taken into account were published elsewhere /14,15/, the factors $\zeta_{ij}(E_n)$ were calculated for three different groups of energetical restrictions (see table 1). If more than one heavy charged particle are created during a neutron interaction, we supposed that they form only one track. The calculation were carried out for the thickness of a polymer material $\Delta X = 1$ μm and for the monoenergetic neutrons of the energy between 0.1 and 15 MeV. The typical results obtained for polycarbonate and cellulose nitrate are given in the figures 1 and 2 [+]. One can see there that the values $C(E_n)$ are for both materials of the order of 10^{-6} tracks for one neutron they are a little higher for the cellulose nitrate, especially for the energies superior to about 10 MeV.

Using the values for monoenergetic neutrons we have calculated also the theoretical sensitivities for some common neutron sources $C(Source)_{th}$. They are given for both mentioned materials and all three groups of energetical restrictions in the table 2, the spectra for polyenergetic sources were taken from the recent works /16,17/.

III. EXPERIMENTAL RESULTS

The experimental studies were carried out with several polymer materials /13-15,18/, we present here only the results for Makrofol E and KODAK LR 115; the behaviour of other materials being very similar to that of Makrofol E /14,15,18/.

Makrofol E was etched by the solution of 15 g KOH + 40 g C_2H_5OH + 45 g H_2O at 70°C /20/. The results discussed in this work were obtained for etching time 20 min, it corresponds to

-----
[+] The results for other cellulose derivatives are comparable /14/.

Table 2: Theoretical number of tracks $C(Source)_{th}$ developable after the interaction of one neutron in the polymer layer $1 \mu m$.

Neutron Source	Group of Energetical Restrictions	The Values $C(Source)_{th}$ for	
		Polycarbonate	Cellulose Nitrate
Am F $E_n = 1.5$ MeV	I	$4.65 \cdot 10^{-6}$	$4.78 \cdot 10^{-6}$
	II	$1.49 \cdot 10^{-7}$	$1.21 \cdot 10^{-7}$
	III	$3.95 \cdot 10^{-9}$	$2.81 \cdot 10^{-8}$
^{252}Cf $E_n = 2.1$ MeV	I	$4.90 \cdot 10^{-6}$	$5.15 \cdot 10^{-6}$
	II	$1.31 \cdot 10^{-6}$	$1.35 \cdot 10^{-6}$
	III	$2.72 \cdot 10^{-7}$	$2.96 \cdot 10^{-7}$
$D(d,n)^3He$ $E_n = 3.3$ MeV	I	$8.56 \cdot 10^{-6}$	$10.78 \cdot 10^{-6}$
	II	$3.61 \cdot 10^{-6}$	$3.67 \cdot 10^{-6}$
	III	0.00	$2.48 \cdot 10^{-7}$
Am Be $E_n = 4.4$ MeV	I	$6.59 \cdot 10^{-6}$	$7.41 \cdot 10^{-6}$
	II	$3.34 \cdot 10^{-6}$	$3.54 \cdot 10^{-6}$
	III	$1.33 \cdot 10^{-6}$	$1.37 \cdot 10^{-6}$
$T(d,n)^4He$ $E_n = 14.7$ MeV	I	$7.70 \cdot 10^{-6}$	$10.20 \cdot 10^{-6}$
	II	$6.57 \cdot 10^{-6}$	$7.65 \cdot 10^{-6}$
	III	$4.73 \cdot 10^{-6}$	$5.58 \cdot 10^{-6}$

removed layer about 6.5 µm.

KODAK LR 115 was etched by 2.5 N NaOH at 40°C, the etching time was 8 hours, it corresponds to removed layer about 5 µm.

Both detectors were evaluated by visual counting, the magnification varied from 400x to 720x. For the evaluation of KODAK LR 115 we applied a green optical filter.

We used following neutron sources:
- Am F, resp. Am Be standard radionuclide sources; theirs neutron emission rates were $4.6 \cdot 10^5$; resp. $2.09 \cdot 10^6$ s^{-1};
- ^{252}Cf neutron source with the neutron emission rate $1.6 \cdot 10^6$ s^{-1};
- neutron generator SAMES, it produced the neutrons of the energy 3.3 MeV[+]; and
- neutron generator NA2; it produced the neutrons of the energy 14.7 MeV.

Generator produced neutrons were calibrated using activation detectors, neutron fluence for other sources were calculated from theirs emission rates taking into account the irradiation geometry.

Number of tracks produced for one neutron from used sources, $C(Source)_{exp}$, are given in the table 3; one can see there that, if one count all visible tracks, the detection efficiency are for both materials very similar. Nevertheless, if one count in KODAK LR 115 only throughetched tracks, the detection efficiency decreases with the neutron energy very rapidly.

IV. DISCUSSION

The comparison of theoretical and experimental efficiencies (see table 2 and 3) show that, if one count all visible tracks, the experimental values are generally higher. It is

---
[+] These experiments were carried out at the Nuclear Research Center at Fontenay-aux-Roses; the authors are much obliged to MM. Portal, Médioni and Chériot for their cooperation in these irradiations.

Table 3: Detection efficiency of some polymer solid state nuclear track detectors in different neutron spectra

Neutron Source	Energy /MeV/	Number of tracks counted for 1 perpendicularly incident neutron in		
		Makrofol E	KODAK LR 115	
			all tracks	throughetched
Am F	1.5	$(8.0\pm0.8)\cdot10^{-6}$	-	-
^{252}Cf	2.1	$(1.3\pm0.2)\cdot10^{-5}$	$(1.2\pm0.1)\cdot10^{-5}$	$(1.2\pm0.2)\cdot10^{-7}$
D(d,n)^{3}He	3.3	$(1.8\pm0.3)\cdot10^{-5}$	$(1.8\pm0.3)\cdot10^{-5}$	$(2.4\pm0,4)\cdot10^{-7}$
Am Be	4.4	$(2.0\pm0.2)\cdot10^{-5}$	-	$(1.0\pm0.2)\cdot10^{-6}$
T(d,n)^{4}He	14,7	$(4.0\pm0.4)\cdot10^{-5}$	$(3.5\pm0.3)\cdot10^{-5}$	$(8.8\pm1.0)\cdot10^{-6}$

quite understandable if one take into account that the theoretical values are for the thickness $\Delta X = 1\,\mu m$, while the removed layer are 6.5 or 5 μm, the counted tracks originate therefore from thicker layer of a material.

The ratio $C(Source)_{exp}/C(Source)_{th}$ could in this manner give some information on the thickness of a material from which the tracks are in our experimental conditions counted; such comparison permit also to appreciate which group of energetical restrictions is the most reasonable. The ratios obtained are given in the figures 3 and 4.

Generally, the values of the ratios should not be higher than the removed layers. Beside that, theirs changements with the energy of neutrons should not be too steep, nevertheless, with the tendency to increase with the energy of neutrons /15, 18/.

One can see in the figures 3 and 4 that such behaviour represent for Makrofol and KODAK (all tracks) the ratios for the energetical restriction group I. One can stated therefore, that in these cases one count in our experimental conditions the tracks of all heavy charged particles of the energies higher than about 0.2 MeV.

As far as is concerned the throughetched tracks in KODAK LR 115, one see in the figure 4 that it is the energetical restriction group III which gives the most reasonable values of ratios $C(Source)_{exp}/C(Source)_{th}$. Of course, the values are generally lower than in other cases, it is caused by the fact that minimal removed layer for throughetching of a track is, for exemple for alpha particles, about 4 μm /19/.[+]

V. CONCLUSIONS

The results of our studies show that the heavy charged particles form the developable tracks in polycarbonates and

---

[+] We must mention that the sensitivity for throughetched tracks depends strongly both on the thickness of removed layer and residual thickness. Both must, therefore, be very thoroughly checked. If no, the accuracy of evaluation deteriorates.

cellulose derivatives if their energy overstep about 0.2 MeV. With the regard to the fact, that only neutron interaction which is capable to lead to heavy charged particles for the neutrons below 1 MeV is the elastic scattering, one can appreciate the energy threshold for the neutron detection. It should be about 0.7 MeV, the energy at which the recoiled carbon nuclei have the energy 0.2 MeV. If one count only through-etched tracks is KODAK LR 115, the energy threshold is of course higher, about 3 MeV.

REFERENCES

/ 1/ K.Becker: Health Physics 16, (1969), p.113
/ 2/ J.W.Tuyn: Trans.Amer.Nucl.Soc. 13, (1970), p. 523
/ 3/ Y.Nishiwaki, T.Tsuruba: J.Nucl.Sci.Technol. 8, (1971), p. 162
/ 4/ E.Piesch: in "Advances in physical and biological radiation detectors", IAEA, Vienna 1971, p.399
/ 5/ K.Jozefowicz: in "Neutron Monitoring for Radiation Protection Purposes"; IAEA, Vienna 1973, vol.II.,p.139
/ 6/ A.Dragu, M.Niedae: Radioprotection 7, (1972), p.87
/ 7/ H.A.Khan: Nucl.Instr.Methods 113, (1973), p.55
/ 8/ G.M.Hassib, L.Medweczky: in "Proc. 2nd Symp.Neutr. Dos. Biol.Med.", Neuherberg 1974, EURATOM, EUR 5273 d-e-f, 1975, vol.I., p.535
/ 9/ J.Tripier, R.Oppel, G.Remy, M.Debeauvais: in "Proc.2nd Symp.Neutr.Dos.Biol.Med.", Neuherberg 1974, EURATOM,EUR 5273 d-e-f, 1975, vol.I., p.509
/10/ G.L.Fängewisch, A.Scharmann: Kerntechnik 16, (1974),p.13
/11/ K.Becker: in "Topics in Radiation Dosimetry"; Suppl.1.: red.: F.H.Attix et al., Academic Press, New York 1972, p.79
/12/ F.Spurný: Jaderná energie 22, (1976), p.49
/13/ F.Spurný: in "Proc. 8th Inter.Conf.Nucl.Phot.Solid State Nucl.Track Det.", Bucharest 1972, vol.II., p.355
/14/ F.Spurný, J.Lochmanová: Jaderná energie 20, (1974),p.233
/15/ F.Spurný, J.Lochmanová, K.Turek: Radioprotection 9, (1974), p.307
/16/ G.Burger, G.Eckl, H.Gredel: in "Advances in physical and biological radiation detectors", IAEA, Vienna 1971, p.467
/17/ "Prompt Fission Neutron Spectra"; IAEA,Vienna, 1972
/18/ F.Spurný, J.Lochmanová: Jaderná energie 20, (1974), p.306
/19/ F.Spurný, K.Turek: Czech.J.Phys. B26, (1976), p.235
/20/ G.Somogyi, J.Gulyás: Radioizotopy 13, (1972), p.549

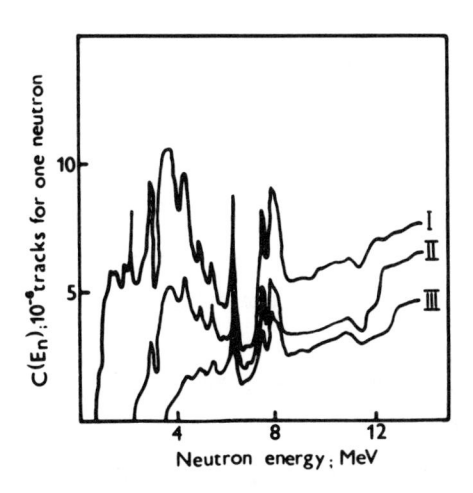

Figure 1: Theoretical sensitivities of polycarbonate as fast neutron detector

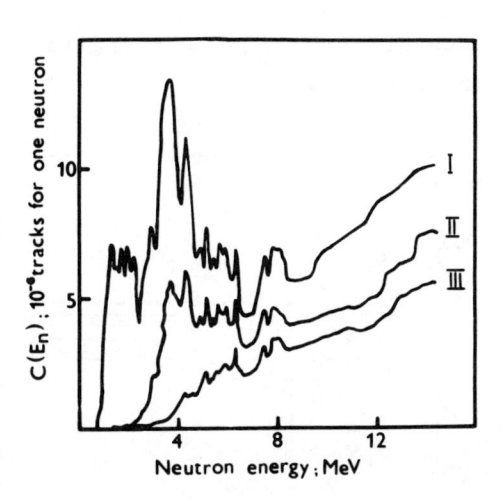

Figure 2: Theoretical sensitivities of cellulose nitrate as fast neutron detector

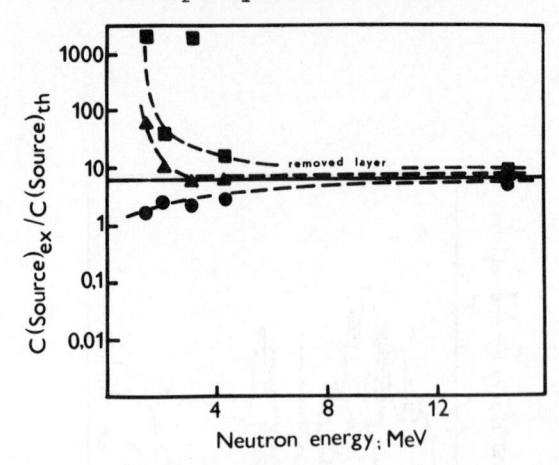

Figure 3: The ratios of $C(Source)_{exp}/C(Source)_{th}$ for Makrofol E:

- ●- taken $C(Source)_{th}$ for the energetical restriction group I
- ▲- taken $C(Source)_{th}$ for energetical restriction group II
- ■- taken $C(Source)_{th}$ for energetical restriction group III

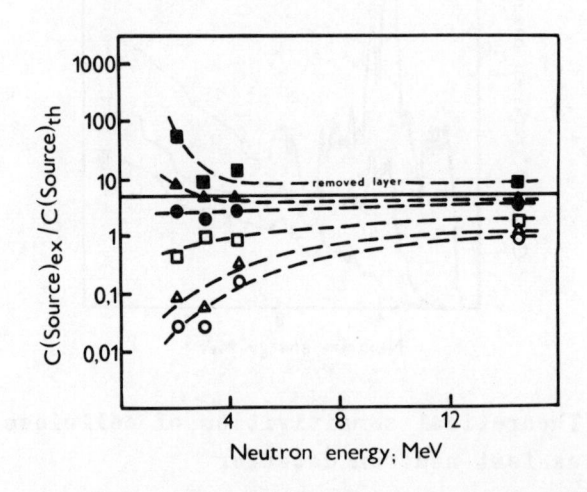

Figure 4: The ratios of $C(Source)_{exp}/C(Source)_{th}$ for KODAL LR 115; signs as in figure 3; open ones for the throughetched tracks

FAST NEUTRON DOSIMETRY BY TRACK DETECTION WITH CELLULOSE NITRATE FILMS

U. Lotz, E. Pitt, A. Scharmann and B. Vitt

I. Physikalisches Institut, Universität Giessen, Heinrich-Buff-Ring 16, D 3600 Giessen, West Germany

Abstract   Track etching has become a common method for the indication and the dosimetry of neutrons, charged particles and fission fragments. At the present cellulose nitrate foils are the most sensitive detectors. Their response to γ-rays is much less than to neutrons. Thus they are usable as a selective neutron dosimeter.

When irradiated by neutrons, latent tracks are created, which are visible after etching. The etching behavior of the Kodak LR 115 film and the physical properties such as the sensitivity after irradiation with neutrons of different energies were investigated by an optical image analyzing system.

Our investigations show, that the sensitivity of the LR 115 film can be increased. By using the spark counting methode of CROSS and TOMASSINO we have a selective personal neutron dosimeter of high sensitivity, easy operation, and low costs.

I. Introduction

At present cellulose nitrates are the most sensitive known track detectors (1). Our investigations are concentrated on Kodak LR 115 film, which consists of a 12 µm thick cellulose nitrate layer, coated on a 100 µm thick polyestar base.

The high natural response of this material to fast neutrons allows to avoid fission converter screens if using it in fast neutron dosimetry. The intensitivity to γ-rays in common dose ranges enables the use of this detector material as a selective fast neutron dosimeter.

To investigate neutron irradiated and etched films we used an optical-electronic image analyzing system "Leitz-Classimat" and constructed a portable spark counter according to CROSS and TOMASSINO.

The computer operated TV analyzing system is counting the tracks and measuring the area of these tracks (2). Under optimum optical conditions this device needs 15 minutes to investigate an area of 1 mm^2.

Image analyzing systems are expensive, very complicated, they are operating rather slow and so they are not suitable for the evaluation of a large number of dosimeters.

An alternative evaluation was done by the meanwhile well known spark counting method. We constructed a portable spark counter (3,4), which also works by battery power supply. The evaluation area is 2.5 cm^2 and it takes five minutes to evaluate one detector foil three times including presparking. We used the stripped cellulose nitrate layer of LR 115 film as detector for spark counting. The grounded electrode was aluminized polycarbonate of Bayer AG, Leverkusen, with an aluminium layer thickness of 0.1 µm.

II. Experiments

Optimum etching conditions were given by 2.5 n NaOH of 60°C as etchant and an etching time of 90 min. The etchant was stirred mechanically. Under these conditions the bulk etching rate was 4 µm/h. This value was calculated from transmission spectra measurements.

Fast neutron irradiations were carried out by a neutron generator and an Am-Be-source. The response of the detectors to monoenergetic generator-neutrons of 3.3 and 14 MeV was determined to $5 \cdot 10^{-5}$ tracks/n and $3.3 \cdot 10^{-5}$ tracks/n for Am-Be-neutrons by the Classimat. The lower number of tracks for Am-Be-neutrons is caused by the spectrum of the source. The energy of 20% of its neutrons is below 1.5 MeV and therefore not sufficient to give rise to the number of etchable tracks. But there is a remarkable difference in the track diameters, which can be seen in fig. 1.

This results from the fact, that 14 MeV neutrons are creating recoils of higher energy than 3.3 MeV ones.

Fig. 1

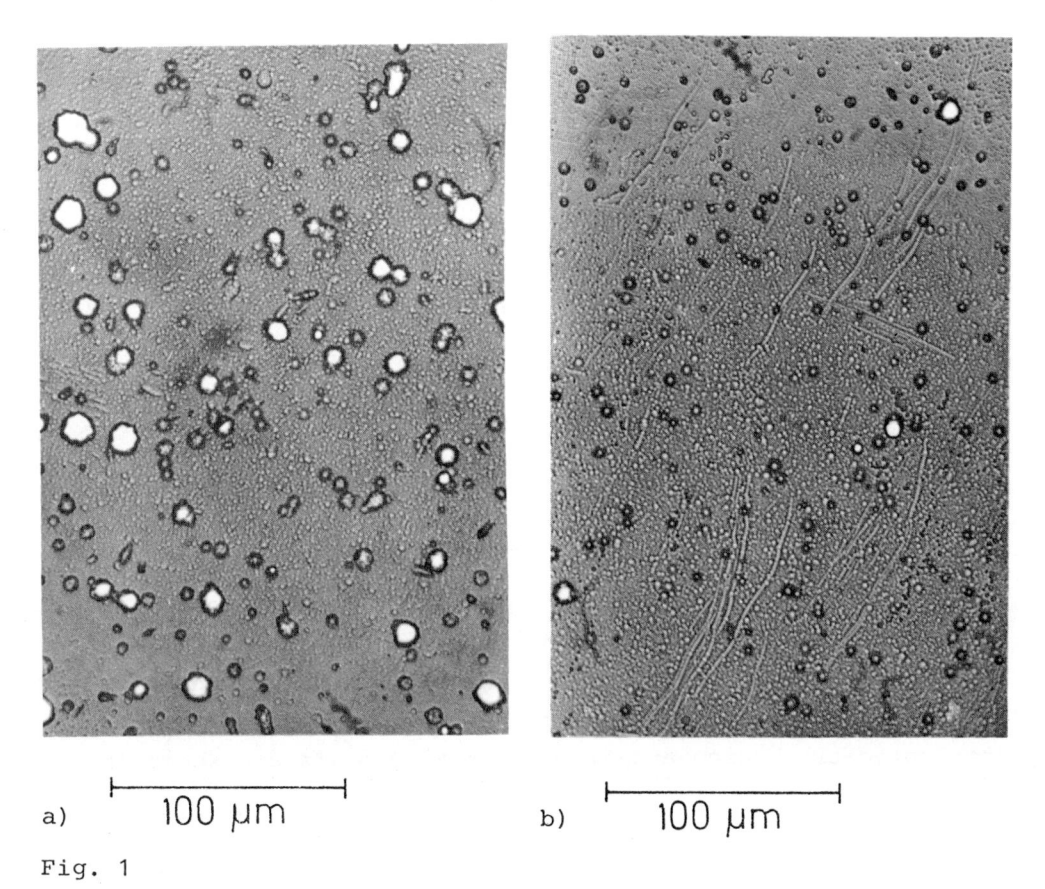

a) tracks of 14 MeV neutrons in LR 115
b) tracks of 3.3 MeV neutrons in LR 115   dose: $5 \cdot 10^9$ n/cm^2

Furthermore at 14 MeV neutron energy nuclear reaktions like ^{12}C(n,n')3α are possible but the probability is very low (1).

Results are given in table 1. The 6% of big tracks caused by Am-Be-neutrons are due to the 5.5 and 10 MeV peaks of the Am-Be-source's spectrum (5). The quantity of big tracks caused by 3.3 MeV neutrons is too low to give a proper explanation.

Table 1: Mean track diameters in LR 115

neutron energy	tracks < 5 μm		tracks > 5 μm	
	quantity %	mean diameter μm	quantity %	mean diameter μm
14 MeV	77	3.2	23	8.0
3.3 MeV	98	1.7	(2)	(8.7)
Am-Be	94	1.9	6	7.0

To investigate the energy dependence of the track diameters from neutron, respectively recoil energy, we irradiated some foils with α-particles from an ^{241}Am-source (E_α = 5.47 MeV). Energy was varied by changing the distance between source and detector in air. Detection efficiency and mean track diameter were determined by the Classimat. As all tracks were nearly circular the mean track diameters could be computed from the integrated track area and track number. Results are shown in fig. 2.

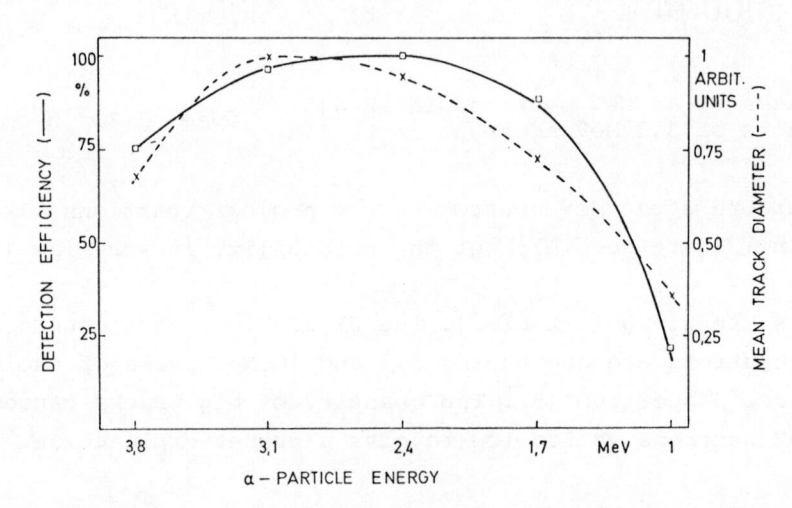

Fig. 2
Correlation between track diameter and detection efficiency

Detection efficiency and track diameter are of similiar energy dependence, both exhibiting maxima between 2 and 3 MeV.

This can be explained by the following mechanisms: α-particles of energies higher than 3 MeV are reaching their critical value of energy loss in some distance beyond the surface of the detector, so at first the etchant has to remove a certain layer of material with normal bulk etching rate before it is reaching the preferred etching region of the latent track. Therefore the etchant has less time to attack the walls of the track and to increase its diameter. α-particles between 2 and 3 MeV have an energy loss about the critical value in or near the surface, these are attacked immediately. Particles of lower energy are creating only small or not etchable tracks. This explains the lower number and smaller diameter of the tracks in the low energy range of fig. 2. This also corresponds with the fact, that in 3.3 MeV neutron irradiated foils there are nearly no big tracks, whereas Am-Be-neutrons cause an amount of 6% and 14 MeV neutrons of 23% tracks with diameters greater than 5 μm. These measurements were performed by the Classimat.

In the following dosimetric properties of the spark counter are described. While the Classimat is also counting tracks, which are not perfectly etched, the spark counter is only registrating totally or nearly perfectly etched ones, which can be penetrated by the pre-sparking voltage. Thus the counting efficiency of the spark counter for neutron tracks is much lower. This disadvantage is compensated by larger evaluation areas, shorter counting times and lower background. We presparked the etched foils with 900 V three times and counted them several times with voltages between 400 and 700 V afterwards.

We got reproducible results at a voltage of 400 V. This voltage caused the smallest burned areas in the aluminium electrode and was therefore of best efficiency.

We didn't use high pressure air or inertial gas atmospheres to keep the technical equipment of the counter as simple as possible. The number of tracks counted by the spark counter was controlled by optical counting.

After several sparking processes the foils were cleaned from evaporated aluminium particles, which lead to double sparkings.

The counting efficiency is 10^{-7} tracks/n for Am-Be and $2 \cdot 10^{-6}$ tracks/n for 14 MeV neutrons. The difference is explained by the larger diameters of 14 MeV neutron tracks, which could be penetrated by the presparking voltage with higher probability. Despite the lower efficiency, the spark counter is able to indicate a dose of 1 rem for Am-Be-neutrons and of less than 200 mrem for 14 MeV neutrons. This is caused by the detection area of 2.5 cm^2 compared to the detection area of 1 mm^2 of the Classimat and a very low background.

First investigations of the chemical properties of the latent track by EPR measurements showed a correlation between the signal and the irradiation dose. Further results of films of another production pointed out, that this signal might be highly influenced by manufacturing conditions. Final results will be published when the comparison of a greater number of detectors is finished.

Summary

The response of Kodak LR 115 film to α-particles and fast neutrons was investigated by both an image analyzing system and a spark counter.

The expensive image analyzing microscope has excellent conditions in basic research, not at least because it is computer operated, but is too slow for the evaluation of a larger number of dosimeters and has a lower detection limit of 5 rem. The detection limit of the quick and cheap spark counter is 200 mrem for 14 MeV and 1 rem for Am-Be-neutrons.

Our further efforts will be to get under controll the energy dependence and to enhance the response by modified etching processes and a new kind of detector foils.

References

(1) K. Becker, A. Scharmann
    Introduction into Solid-State Dosimetry
    Thiemig Verlag, München 1974

(2) G.-L. Fängewisch, A. Scharmann
    Proceed.Int.Symp.Neutron Dosimetry, EUR-4896, 1972

(3) W.G. Cross, L. Tomassino
    rad. eff. $\underline{5}$ (1970) 85

(4) S. Prêtre, K. Heusi
    Eidgen. Institut f. Reaktorforschung Würenlingen Juni 1972

(5) M.N. Thompson, J.M. Taylor
    Nucl. Inst. Meth. $\underline{37}$ (1965) 305

A NEUTRON DOSEMETER USING CELLULOSE NITRATE FILM

I. Y. Khadduri and I.K. Al-Haddad

Nuclear Research Institute, Tuwaitha, Baghdad, Iraq

A neutron dosemeter using cellulose nitrate film coated on both sides with lithium tetraborate (Kodak CA 80-15 type B film) has been prepared for the survaillance of personal exposed to thermal and epithermal neutron doses. A calibration of the track density in the film versus absolute thermal and epithermal neutron fluences has been obtained. The response of the dosemeter is found to be $1.06\pm0.2\times10^{-3}$ tracks per thermal neutron and $3.65\pm0.5\times10^{-4}$ tracks per epithermal neutron for a K/E neutron energy spectrum.

Introduction:

The use of cellulose nitrate plastic track detectors in dosemetry has found a number of applications [1,2], including fast neutron dosemetry [3]. Their insensitivity to gamma and beta radiations renders them useful for the detection of alpha particles or neutrons (via a suitable converter) in mixed fields of radiation. Furthermore, their use is relatively simple and safe.

The sensitivity of the cellulose nitrate film Kodak CA 80-15 type B to thermal and epithermal neutrons posed the possibility of their use for dosemetry purposes for personal working in neutron fields around nuclear reactors and other neutron sources. A film badge for carrying the cellulose nitrate film and allowing for the measurement of thermal and epithermal neutron fluences simultaneously as well as the absolute calibration of the film for dosemetry purposes is described below.

The Neutron Dosemeter and Principles of Detection:

The dosemeter consists of a 1 cm x 5 cm film of the specially treated cellulose nitrate CA 80-15 type B [4] which is sandwiched between two plastic holders that contain a 2 cm x 0.5 cm window at one end and a 1 cm diameter x 0.5 mm thick cadmium disc at the other end, see figure 1. The window allows for the thermal and epithermal neutrons to impinge freely on the film while the cadmium stops the thermal neutrons from reaching the film. The film itself consists of a 100 μm thick lightly rose-tinted cellulose nitrate which is light insensitive. It is primarily intended for recording alpha particles or fission fragments. The film is coated on both sides with lithium (natural) tetraborate which is dispensed in a water soluble binder that is easily washed off with ordinary water. The lithium tetraborate serves as the neutron converter with the following reactions, ^{6}Li(n,α)T and ^{10}B(n,α) Li. The tracks of the alpha particles that penetrate the cellulose nitrate are greatly enhanced into pitted holes upon etching in a strong alkaline solution and can be easily seen under an optical microscope.

Upon exposing the badge to a given neutron fluençe, the lithium tetraborate coating is removed by allowing running water to slowly dissolve the coating without any rubbing. The film is then etched in a 10% KOH solution at 40±0,1°C in a constant temperaure water bath for 3 hours. It is then dipped into a 2% acetic acid solution, rinsed in water for one hour, and dried.

The etched film is then placed in its original position on the plastic holder and viewed under an optical microscope with an objective x25, see figure 2. The areas under the the cadmium disc and the window were respectively scanned and the average number of etched tracks per field of view of area 0.264 mm^2 is determined. The average diameter of the etched track is about 7 μm. The scanner was asked to reject all those tracks whose diameter appeared to be less than one half the average diameter.

The background of etched tracks in the film was on the average four tracks per field of view. The background is mostly due to a deterioration in the quality of the cellulose nitrate film because of corrosion due to a chemical reaction between the lithium tetraborate and the film in a humid and warm atmosphere (5). This deterioration can be avoided by storing the film in a refrigerator.

The Response of the Dosemeter:

The cellulose nitrate film detects the alpha tracks emitted by the boron 10 and the lithium 6 in the coating and the sensitivity of the dosemeter to thermal and epithermal neutrons depends on the absorption cross section of these reactions. The response of the dosemeter to thermal neutrons, expressed as the number of etched tracks per unit area, can be written as

$$R(th) = C [A \sigma_B(th) + B \sigma_{Li}(th)] \phi(th) \qquad (1)$$
$$= D \phi(th)$$

where the $\sigma_B(th)$ and $\sigma_{Li}(th)$ are the absorption cross sections for boron 10 and lithium 6 at thermal energy and $\phi(th)$ is the thermal neutron fluence. The coefficients A and B represent the concentration of boron 10 and lithium 6 atoms respectively in the converter. The coefficient C takes into account the distance in the converter at which the alpha particle is emitted and the efficiency of detection of the alpha paricle in the cellulose nitrate. The coefficient D is determined experimentally by exposing the dosemeter to a known thermal neutron fluence and counting the number of etched tracks per unit area. It also depends, evidently, on the etching conditions and the scanning procedure.

The response of the dosemeter to epithermal neutrons can be written as

$$R(epi) = A \int_{0.4eV} \sigma_B(E) \phi(E) dE + B \int_{0.4eV} \sigma_{Li}(E) \phi(E) dE$$

where the lower energy limit corresponds to the cadmium cut off for thermal neutrons. The upper energy limit depends on the cross sections of the boron 10 and lithium 6 reactions, see figure 3. Of more importance, however, is that the response of the dosemeter is a function of the epithermal neutron energy spectrum that is to be monitered. In our case, the epithermal neutron spectrum is approximatly a K/E spectrum.

Using the same physical set-up as that outlined below, Chapuis (6) obtained neutron fluences with two other energy spectra, and the theoretical calculation of the above equation revealed that the response to neutrons with energies between 1 KeV and 10 KeV is very small in comparison to those with energies between 0.4 eV and 1 KeV.

Calibration of the Dosemeter:

Several dosemeters were irradiated for various known fluences at the reactor of the Laue-Langenvin Institute at Grenoble, France. The geometry of irradiation is a cavity in the graphite pile, 5 cm diameter by 3 cm length. The ratio between the thermal flux and the epithermal flux per lethargy unit is 26. The epithermal neutron energy spectrum is approximatly K/E, see figure 3. Specified irradiations were performed of bare dosemeters for the thermal fluences and for dosemeters under cadmium for the epithermal fluences. The monitoring was made using bare and under cadmium cover indium foils.

The thermal neutron fluence is related to the difference in the track densities in the area under the cadmium which registers the epithermal neutrons only from that under the window which registers both the thermal and epithermal neutrons. For our conditions, the value of the D coefficient in equation (1) was found to be $1.06 \pm 0.2 \times 10^{-3}$ tracks per thermal neutron which is close to that found by Chapuis (6). In order to find the thermal neutron dose, one applies the following relationship :

Thermal neutron dose (mRem) =

(track density under the window - density under the cadmium)

$$X \frac{100 \ (mm^2/cm^2)}{\text{area of field of view } (mm^2)}$$

X (1.06 x10^{-3} tracks/neutron)$^{-1}$ X (9.6 x10^5 neutron/cm^2.mRem)^{1}.

The epithermal neutron fluence is related to the difference in the track densities of the background from that under the cadmium. For our conditions, the response of the dosemeter to epithermal neutrons for a K/E spectrum and for neutrons between 0.4 eV and 1 KeV is 3.65 ±0.5x10^{-4} tracks per epithermal neutron (in comparison to 6.5 x10^{-5} tracks per epithermal neutron reported by Chapuis (6) for reasons not yet ascertained). The epithermal neutron dose can be found from the following relationship:

Epithermal neutron dose (mRem) =

(track density under the cadmium - background track density)

$$X \frac{100 \ (mm^2/cm^2)}{\text{area of field of view } (mm^2)}$$

X (3.65 x10^{-4} tracks/neutron)$^{-1}$ X (9.6 x10^5 neutron/cm^2.mRem)$^{-1}$.

The minimum detectable dose is limited by the background and the number of scanned fields of view. The upper range of the dosemeter is limited by the overlapping of etched tracks. In our situation, the measurable dose range is the following:

    1 mRem to 1 Rem for thermal neutrons,

    5 mRem to 4 Rems for epithermal neutons.

Conclusion:

A neutron dosemeter utilizing the (n,α) reactions of lithium 6 and boron 10 in conjunction with a cellulose nitrate film has been described. It allows for the detection of thermal neutrons and epithermal neutrons in the enrgy range 0.4 eV to 1 KeV with minimum detectable doses of 1 mRem and 5 mRem, respectively. Its response to epithermal neutrons is, however, dependant on the epithermal neutron energy spectrum it is being exposed to.

The dosemeter's sensitivity might be enhanced by using lithium enriched in lithium 6 in the lithium tetraborate coating which contains at present natural lithium. Due to the expense of such enriched lithium and in order to avoid washing it away after each exposure, a permanent converter might be attached to both plastic holders of the dosemeter along with cellulose nitrate film CA 80-15, and not type B.

The authors acknowledge with gratitude the interest of A.M.Chapuis in this work and her assistance and that of M. Bricka in exposing the dosemeters to standard fluences. Thanks is also extended to J. Matloub for the scanning of the films.

References:
1- Chapuis,A.M.,Gerard,N.,Goudain,G., Bull. Soc. France Radioprotection, (June, 1971)
2- Barbier,J., Journal Phot. Sci., 19(1971)p.108.
3- Tripier,J.,Uppel,R.,Remy,G.,Debeauvais,M., Nucl. Inst. and Meth., 125(1975)487.
4- Barbier,J., Trans. Am. Nucl. Soc., 13 (1975) 530.
5- Barbier,J., private communication, October 1975.
6- Chapuis,A.M., "La Dosimetrie des Neutrons par Detecteurs a Ionogravure", presented at the IXeme Congress sur le Controle des Rayonnements Ioisants, Grenoble, April 1975.

Figure 1. The neutron dosimeter with the Film in between the two plastic holders.

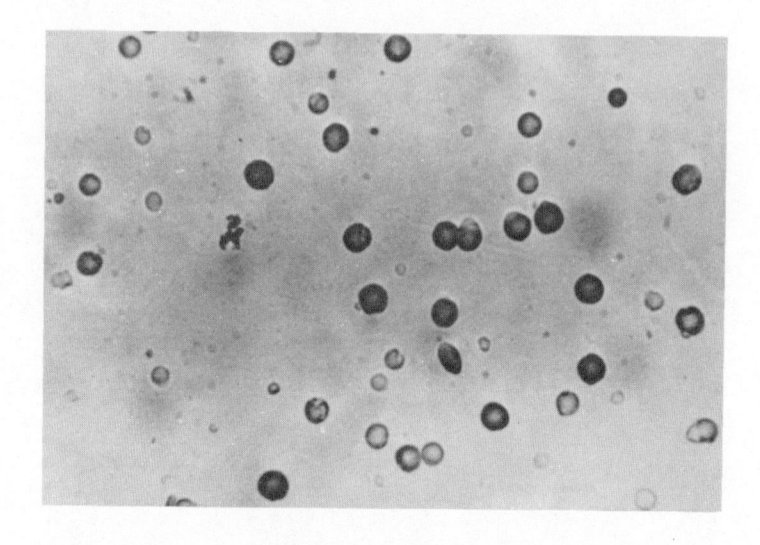

Figure 2. A photograph of the etched alpha tracks in the cellulose nitrate film (x 540).

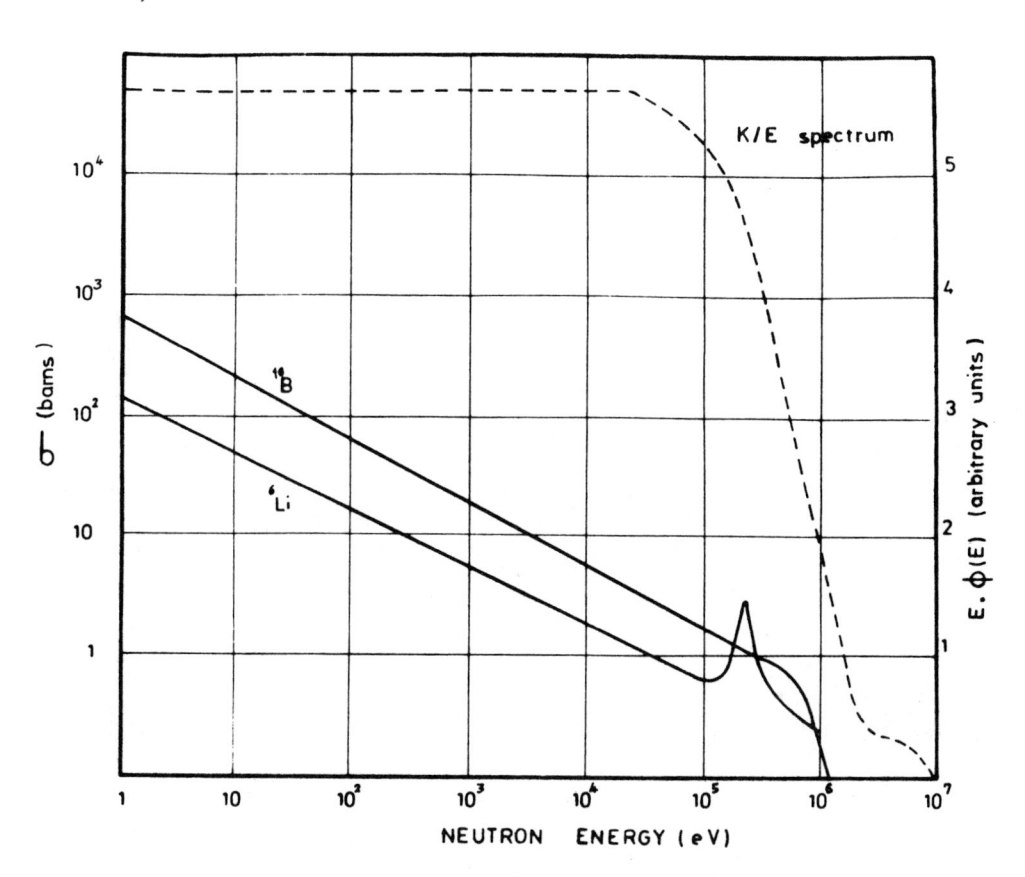

Figure 3- The K/E neutron energy Spectrum and the ^{6}Li(n,a) and ^{10}B(n,a) cross-sections versus neutron energy.

STATUS REPORT: DIRECT TRACK DETECTOR DOSIMETRY IN FAST NEUTRON BEAMS

H. B. Knowles[*], F.H. Ruddy[*], G. E. Tripard[*],
H. Bichsel[**], J. Eenmaa[**] and J.E Smathers[***]

[*]Washington State University, Pullman, Washington 99163, U.S.A.
[**]University of Washington, Seattle, Washington, 99815, U.S.A.
[***]Texas A and M University, College Station, Texas 77843, U.S.A.

ABSTRACT

Cylindrical phantoms of both water and polyethylene 25 cm diameter by 25 cm high, have been exposed to fast neutrons from the therapeutic neutron beams at the University of Washington (where the neutrons are produced by 21 MeV deuterons striking a beryllium target) and also at the TAMVEC cyclotron at Texas A and M University (where the incident deuteron energy is 50 MeV). Both phantoms were loaded before irradiation with cellulose nitrate and Lexan polycarbonate track detectors, arranged in a regular pattern. The detectors are in direct contact with the water or polyethylene, thus permitting registration of the heavy ions produced by scattering or reaction in oxygen and carbon. These detectors have been partially analyzed and spatial macrodose distributions have also been measured. Additionally, the differential yields of various heavy ions from oxygen and carbon from the two different neutron spectra are presented and discussed.

I. Introduction

Following some preliminary track detector work with negative pion beams as therapeutic modalities (to be reported later) it occurred to us that the same objection to conventional dosimetry could be raised in the case of fast neutrons that has been made in the case of negative pions: specifically, that the materials used for construction of both macro- and microdosimeters, were designed to be "tissue equivalent" only for low energy neutrons and gamma-rays. This was the specific goal of Shonka and his associates, whose "A-150 plastic" is now widely used in dosimetry.[1] This material consists principally of carbon, for which the fast neutron nonelastic cross sections are only moderately well-known, whereas real tissue is about 80% oxygen by atomic fraction, and very little has been reported on the nonelastic cross sections of oxygen. Thus it is not at all certain that either macrodose distributions or microdose spectra (such as the spectra of lined energy, y, as used by Rossi[2] would be properly reproduced in a fast neutron beam. Because plastic track detectors are small and thus relatively non-interfering recorders for heavy ion spectra, it was postulated that something could be learned from their use in fast neutron beams used for therapy. It was possible to arrange to make exposures with our colleagues at the University of Washington cyclotron in Seattle, Washington and also at the TAMVEC facility at Texas A and M University in College Station, Texas (in the latter case, by shipping preloaded phantoms to and from College Station) so that two of the existing American facilities for fast neutron cancer therapy have participated in these preliminary tests.

II. Equipment and Procedure

An "oxygen" and a "carbon" phantom were constructed. The first consisted of a circular battery jar 25 cm in diameter by 25 cm high, to be filled with distilled water before irradiation. An exiguous structure of plastic and rubber bands supported an array of 11 track detector "sandwiches" inside the jar, in the geometry shown in Figure 1. Six of the sandwiches were made of two sheets of Kodak Pathé CA 80-15 cellulose nitrate, 100 μm thick and five of the sandwiches of sheets of Lexan polycarbonate, approximately 125 μm in thickness. The edges of the sandwiches were sealed with Silastic, a commercial adhesive we have found to be effective for sealing these plastics against liquid intrusion between the two sheets but which at the same time does not appear to react chemically with either plastic. As shown in Figure 1, the odd-numbered sandwiches are cellulose nitrate and the even-numbered sandwiches are Lexan polycarbonate. The sandwiches were 4.5 cm by 4 cm in size and were spread 2.25 cm apart in the direction of the neutron beam. The carbon phantom was of the same size and shape but was built up of slabs of polyethylene, each 2.5 cm thick, and sandwiches of cellulose nitrate and Lexan polycarbonate (sealed at the edges by transparent tape in this phantom) were placed in corresponding positions, as seen in Figure 2. The difference in spacing between the layers was designed to provide an approximate compensation between the two phantoms for the ratio of the densities of polyethylene to water. The loaded components of the carbon phantom were assembled on two threaded steel rods and bolted together.

The phantoms were taken to Seattle by automobile, where the distilled water was added to the oxygen phantom just before the irradiation. Each phantom was placed in the approximate position of a patient in front of the neutron collimator, and received a 200 rad exposure (as measured at the surface), as shown in Figures 3 and 4. They were then returned to Pullman for removal of the track detector sandwiches which were processed at Washington State University, by a method to be described below. Subsequently, the phantoms were loaded and shipped to Texas A and M University, again exposed to a dose of about 200 rads (measured approximately at the surface) as illustrated in Figures 5 and 6. They were then shipped to Pullman at about the time the carrier was on the point of going out of business and thus became lost for about 10 days, finally being found in the airport at Lewiston, Idaho. This is not simply a complaint about shipment problems but may be pertinent because the delay could have contributed to some unexpected features found in our initial data.

III. Results

All cellulose nitrate films were stepetched for two hours in 6.25 normal NaOH at 30°C. Some Lexan films have also been etched at a higher temperature: the track densities appear to be low and the Lexan data has thus received little attention. Microphotographs of the resulting track densities on the films on the sandwich surfaces facing the neutron beam appear in Figures 7, 8, 9, and 10. For comparison purposes, those from sandwich No. 1 are shown in each case. A magnification of 400X was used: the fine grid divisions are 1.25 μm, to give a scale. Figure 7 shows the results from the carbon phantom, and Figure 8 those from the oxygen phantom, exposed to the "low energy" neutron spectrum of the University of Washington cyclotron. Here, the average neutron energy is about 6 MeV, being produced by 21 MeV deuterons on a beryllium target. Figures 9 and 10, respectively, are from the carbon and oxygen phantoms set in the higher energy TAMVEC neutron beam, which is produced by 50 MeV deuterons on beryllium and thus has an average energy of 22 MeV. Some results of the track density versus depth from the cellulose nitrate track detectors appear in Figure 11. Only part of the data from the TAMVEC runs are shown. The individual sets of data are self-consistent, revealing a relaxation length of about 15 cm, which is in general agreement with other data.

IV. Discussion

The track detectors reveal only what radiation biologists call the high-LET portion of the dose. This would consist of alpha-particles produced by inelastic reactions and the fast heavy ions produced by elastic scattering. At the higher energies from TAMVEC, it is not surprising that there are more of these than from the corresponding lower neutron energies available at the University of Washington. A perusal of Figures 9 and 10 shows many long tracks and also a large number of tracks that have entered the detector surface almost at a normal angle--that is, elastic heavy ion recoils which have appeared at 0°. (These are in fact difficult to see in the photograph.) By contrast, in Figures 7 and 8 there are only a few short tracks of the type that might be made by an alpha particle from inelastic scattering-induced breakup and the heavy ion recoil tracks are, in general, larger (having entered at low energies).

What is surprising is that oxygen seems to yield a higher track density than does carbon, with the low energy-neutron beam, as illustrated in Figure 11. Although the neutron spectrum from the University of Washington cyclotron is very broad, it is expected that few neutrons will have sufficient energy to produce the excitation necessary to break up ^{16}O (the threshold for this reaction exceeds 12 MeV) in contrast to the low threshold needed to break up ^{12}C, which is about 4.6 MeV. The additional tracks on the oxygen film may result from elastic scattering, although there is not very much difference between the elastic scattering cross sections of the two nuclei.

Environmental effects on the detectors may also play a part in the track densities. In a previous paper, it was noted that cellulose nitrate films that had been exposed to high temperatures showed lower etch induction times than comparable films that had been refrigerated.[3] After a 2-hour etch, the thermally damaged films might show more tracks, and because the phantom shipment from College Station was delayed--and in rather warm weather--the high track densities for the TAMVEC exposure may be accounted for in this manner. Thus the matter is not as yet completely resolved. Thermal effects could not be used to explain the oxygen-carbon ratio for the University of Washington exposures, because the two phantoms were kept together and transported during cool weather. It was noted in our study of etch induction time that track formation in cellulose nitrate might be delayed by wetting and then drying the exposed film. However, this might tend to reduce the track densities in the wetted (oxygen) films but would have no effect on the dry (carbon) films. It is thus fairly certain that these low energy neutrons produce a larger concentration of high-LET particles in oxygen than in carbon, although it is impossible at this time to make a more detailed statement. There is however good reason to be skeptical of the fast neutron microdosimetry that is done using Shonka A-150 plastic.

References

(1) F. R. Shonka, J. E. Rose and G. Failla, "Conducting Plastic Equivalent to Tissue, Air and Polystyrene," Proc. 2nd U.N. Conference on Peaceful Uses of Atomic Energy 21, 184 (1958).

(2) H. H. Rossi, "Specification of Radiation Quality," Rad. Res. 10, 522 (1959).

(3) F. H. Ruddy, H. B. Knowles, S. C. Luckstead, and G. E. Tripard, "Etch Induction Time in Cellulose Nitrate: A New Particle Identification Parameter," proceedings of the present conference.

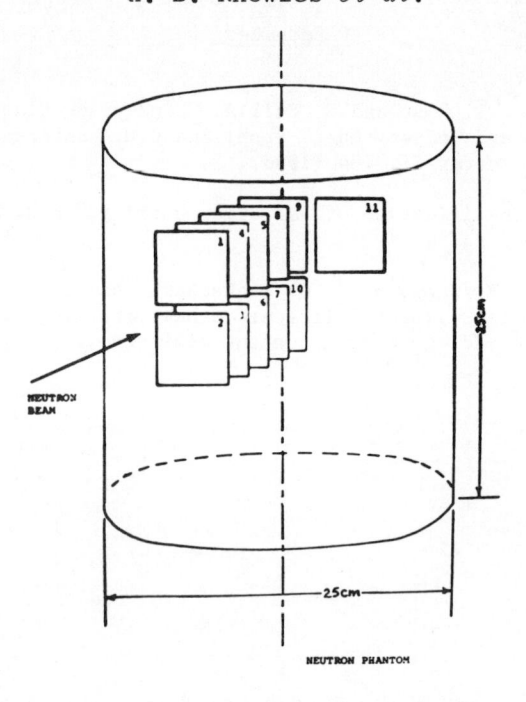

Figure 1. Schematic Diagram of Track Detector Array in Phantoms

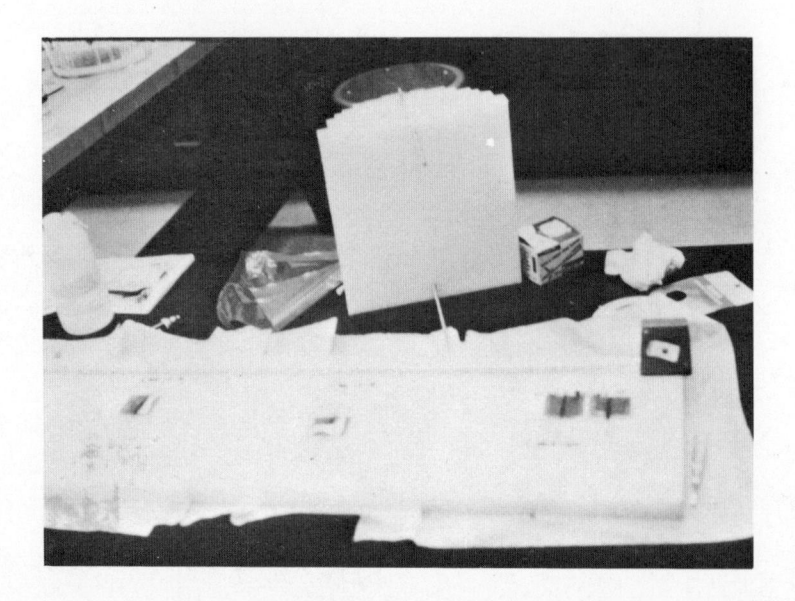

Figure 2. Assembly of the Carbon Phantom

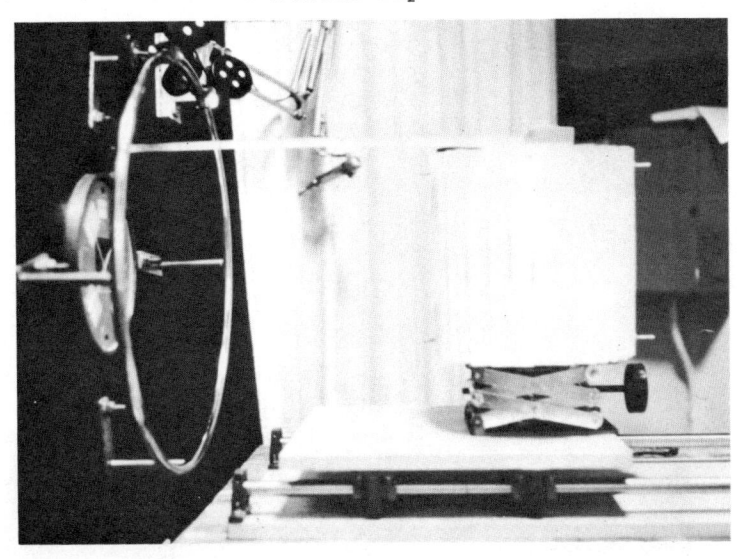

Figure 3. Exposure of Carbon Phantom in U. Washington Neutron Beam

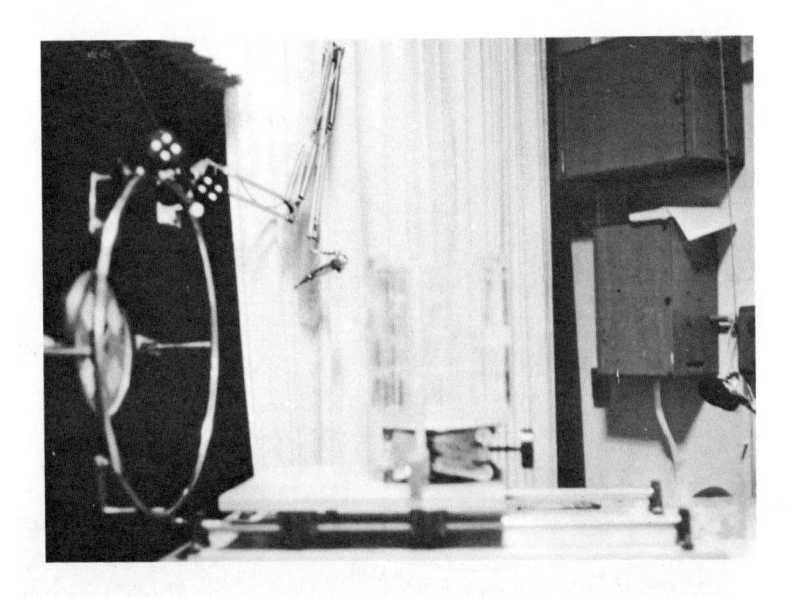

Figure 4. Exposure of Oxygen Phantom in U. Washington Neutron Beam

Figure 5. Exposure of Carbon Phantom in TAMVEC Neutron Beam

Figure 6. Exposure of Oxygen Phantom in TAMVEC Neutron Beam

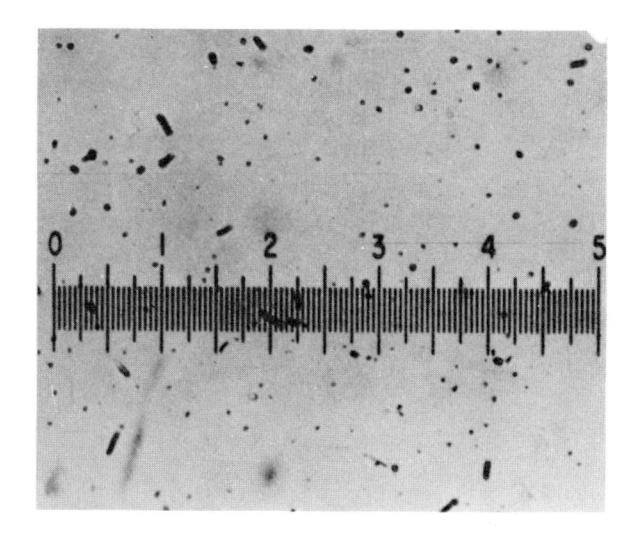

Figure 7. Track Density on Carbon-Detector Interface, U.W. Neutron Beam

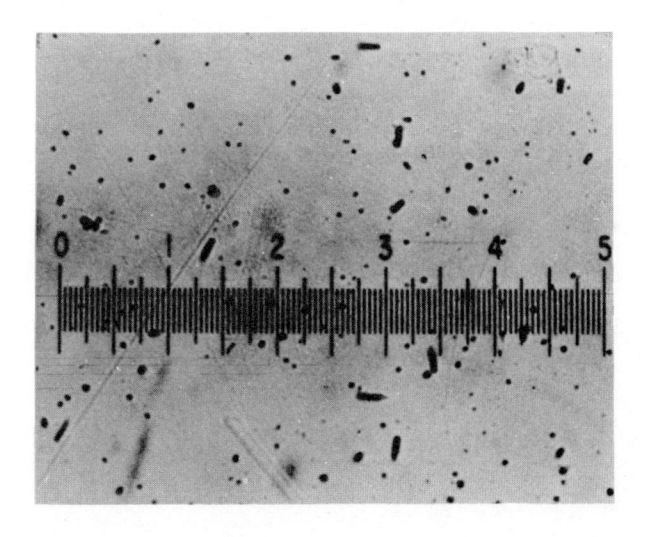

Figure 8. Track Density on Oxygen-Detector Interface, U.W. Neutron Beam

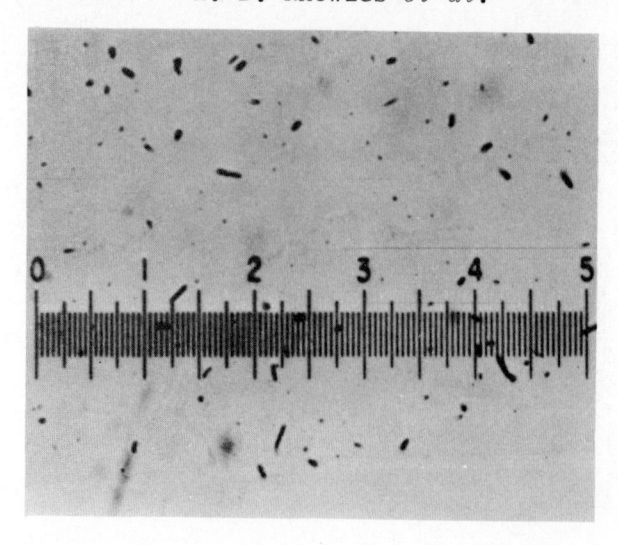

Figure 9. Track Density on Carbon-Detector Interface, TAMVEC Neutron Beam

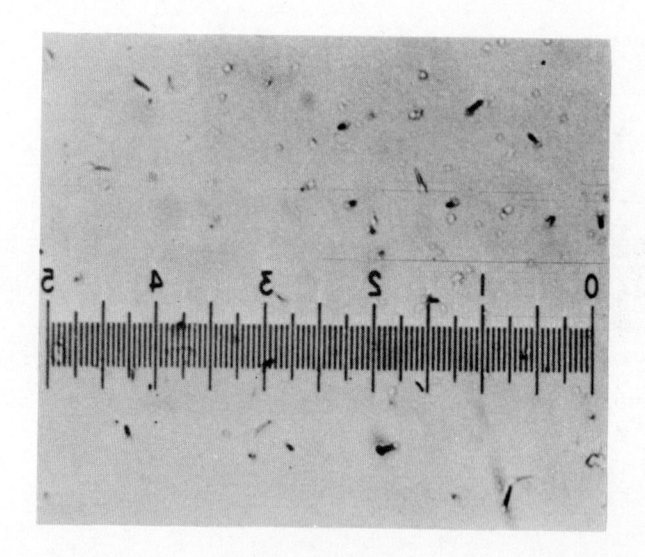

Figure 10. Track Density on Oxygen-Detector Interface, TAMVEC Neutron Beam

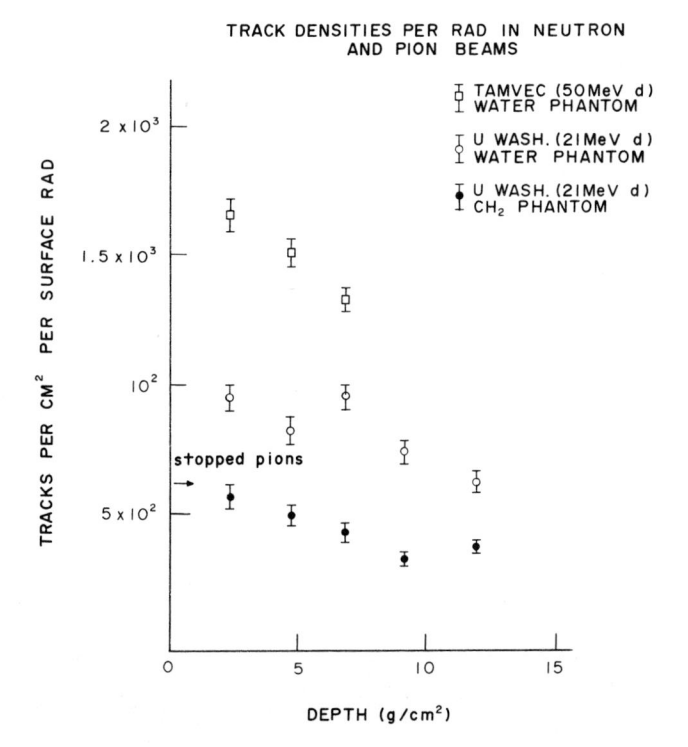

Figure 11. Track Density versus Depth in Phantom

ON THE ELECTROCHEMICAL ETCHING OF NEUTRON-INDUCED TRACKS IN PLASTICS AND ITS APPLICATION TO PERSONNEL NEUTRON DOSIMETRY

G.M. Hassib*, J.W.N. Tuyn and J. Dutrannois

CERN, Geneva, Switzerland

ABSTRACT

The recently developed electrochemical etching method seems to be a promising technique to reveal neutron-induced recoil tracks in plastics. However, very little work has been done and the data presented are preliminary.

The purpose of the work presented here was to investigate in some detail the most important parameters affecting the electrochemical etching kinetics of recoil tracks in Makrofol polycarbonate (~ 500 μ) and to find the optimum conditions for applying this technique to personnel neutron dosimetry.

The electric track diameter and the neutron sensitivity were studied as a function of frequency, applied voltage, etching condition, neutron dose, neutron spectrum, and the directional dependence.

Special cells were designed in such a way that only one surface of the plastic foils was etched electrochemically while the other surface was made conductive and then connected with the earth terminal of a high voltage function generator. With this technique the electric tracks appear as dark stars on a quite clear background, which permits a more reliable evaluation of the track density, using any automatic or semi-automatic scanning equipment.

---

* Present address : Atomic Energy Establishment, Cairo, Egypt.

INTRODUCTION

During the last few years extensive efforts have been devoted to the search for a personnel neutron dosimeter with better characteristics than those of the conventional photographic nuclear track emulsions. Dielectric track detectors have been proposed as a valuable alternative method because of several advantages. Combinations of fissile materials and thin plastic foils to be counted by a spark counter have been used in routine and accident personnel neutron dosimetry in different laboratories.

However, radiation hazard and contamination risk inherent in fissile materials have not encouraged a large-scale application of this method. The detection of fast neutrons by registration of induced recoil nuclei and/or alpha particles produced in (n,α) reactions in plastic track detectors may overcome most of these limitations. The application of this technique in practice has encountered some problems due to the fact that the resulting tracks are characterized by a wide spectrum of diameters and ranges which make the visual counting difficult, unreliable and time-consuming.

The recently developed electrochemical etching technique introduced by Tommasino[1] makes it possible to overcome the difficulties of using non-fissile radiators. Employing the electrochemical etching method the recoil nuclei and alpha particle tracks can be enlarged to such a size that they can be seen by the naked eye or displayed on a screen by using a slide projector (Fig. 1).

The method is based on the fact that the particle damage trails in plastics are characterized by a high conductivity compared with that of the undamaged region. On applying a pulsed high voltage during the etching process an electrical breakdown is induced at the end of the damage trails in the bulk of the plastic foils.

In this technique the plastic foil to be etched electrochemically is placed between two insulated containers filled with the etchant in such a way that a good contact between the plastic detector and the containers maintains the electrical insulation of the two solutions. Platinum electrodes in the containers are connected with a high-voltage function generator.

In preliminary experiments using this technique (2), the application of the electrochemical etching method for dosimetry purposes proved to have a number of advantages over other photographic or non-photographic nuclear track registration methods.

To use this technique in the optimal way it is essential to have a well established fundamental knowledge of the influence of several parameters of the method. Among the essential parameters are the type of detector and the thickness of the foil, the applied voltage and frequency, the nature, temperature and concentration of the etchant, the neutron energy and cell geometry. So far these parameters have never been studied systematically.

It is the aim of this work to investigate in detail the influence of the most important parameters affecting the registration of neutron-induced tracks in polycarbonate plastic detector. The study was restricted to the nature and concentration of the etchant, etching time, frequency, applied voltage, neutron energy, and directional dependence. Special cells were designed in such a way that only one surface of the plastic foils was etched electrochemically while the other surface was made conductive and then connected with the earth terminal of a high-voltage function generator. With this technique the electric tracks appear only on one surface of the plastic foil while the other surface remains clear without any disturbing background. All measurements were done using 500 μ thick Makrofol* E sheets after being irradiated with 14 MeV and PuBe neutrons.

---

*   Trade name Bayer, Germany.

CONCENTRATION AND NATURE OF ETCHANT

An extensive effort was made to study the effect of concentration and nature of etching solution on the track diameter and registration efficiency of Makrofol E. For these experiments a special cell with six separate chambers and a common electrode was designed such that six foils could be etched simultaneously using different etchant concentrations or different etching solutions.

KOH was chosen to study the effect of etchant concentration on the track diameter in a given electrochemical etching condition. Figure 2 shows the results of these measurements for solutions containing 15% - 40% KOH. In this figure the electric track diameter increases with increasing etchant concentration to a maximum at about 25% and then decreases to a minimum at about 35% KOH, after which the diameter increases again. This tendency follows approximately the electrical conductivity curve at the region between 15% and 35% as it may be expected from electrochemical kinetics. It means that for a given etchant there is an optimum concentration at which the electrical conductivity reaches its maximum value. In the case of KOH the optimum concentration was found to be around 25% at room temperature.

Studies have shown that the addition of ethyl alcohol to the etching solution increases the bulk etching rate of the plastic detectors (3). In this work we have tested different mixtures of ethyl alcohol and 25% KOH solution. The highest track diameter was found at a mixture containing 1:1 volume ratio of alcohol and 25% KOH solution as shown in Fig. 3. At a given etching condition the rate of track diameter growth using alcoholic solution was found to be more than three times higher than that for 25% KOH.

EFFECT OF ETCHING TIME

Compared with the conventional etching method, the electrochemical etching technique is controlled by additional parameters affecting the rate of track growth. Figure 4 shows the electric track diameter as a function of etching time at different time intervals. In this figure the rate of track growth increases with increasing time interval. This tendency may be attributed to the continuous generation of heat due to the electric current in the etching solution. The induced heating accelerates the rate of track diameter growth up to a point where the rate of heat generation is equal to the rate of heat loss by radiation. In this condition the growth of the tracks behaves in the same way as in the case of conventional etching.

ELECTRICAL PARAMETERS

The efficiency of the electrochemical etching is strongly dependent on the frequency and the applied voltage. Figure 5 shows the effect of the frequency on the maximum track diameter and the neutron sensitivity at frequencies between 1 and 10 kHz. The track diameter increases with increasing frequency up to a point where it levels off, while the change in the neutron sensitivity is negligible within ± 20%.

Figure 6 shows the increase in the neutron sensitivity as a function of applied voltage using two etching solutions with different etching rates. From these experiments one can conclude that there exists a threshold value of the applied voltage for the appearance of the electric tracks. Above this threshold and for a given neutron dose the number of electric tracks increase with increasing applied voltage up to about 3 kV, where the track density levels off. Moreover, one can see from Fig. 6 how the nature of the etching solution is strongly affecting the registration efficiency of the detector. It shows that the higher the etching rate of the solution, the higher the registration efficiency. The alcoholic solution gives an increase of a factor of two for the neutron sensitivity compared to 25% KOH in a given condition.

These results show that for maximum efficiency and reproducible quantitative measurements using 500 μ thick foils, one should carry out the electrochemical etching process using alcoholic solutions at more than 3 kV and more than 5 kHz at room temperature. Of course, optimum etching conditions have to be reestablished whenever a different foil type or thickness, etchant or etching temperature is used.

NEUTRON ENERGY AND DIRECTIONAL DEPENDENCE

For the application to neutron dosimetry of the electrochemical etching technique three main points have been briefly investigated. The first was the energy dependence on the neutron sensitivity, the second was the directional dependence, and the third was the lower detectable limit in relation to the background tracks. Foils were irradiated with PuBe and 14 MeV neutrons at different angles. The foils were etched electrochemically under identical conditions. The neutron doses were derived from measurements with a Rem counter.

Table 1 shows the results of these measurements. It can be seen that the neutron sensitivity varies by a factor of about 3 between PuBe and 14 MeV neutrons for normal incidence. This factor increases as the angle of neutron incidence decreases. These results agree well with those of previous studies[4] using visual hole counting of cellulose nitrate LR-115.

It can be seen also from Table 1 that the detector sensitivity is strongly dependent on the direction of neutron incidence. It is of interest to point out that, unlike conventional etching, the directional dependence is less pronounced with 14 MeV neutrons than with PuBe neutrons. This tendency may be due to the differences in the critical particle detection angle.

BACKGROUND TRACKS AND LOWER DETECTION LIMIT

One of the important problems we have faced is the background tracks in the plastic foils. The background track density depends on the age of the plastic foils, the condition of storage, and the etching condition and determines the reliability of the method for measurement of low neutron doses. After several attempts to solve this problem we reached the following conclusion.

The plastic foils have to be etched electrochemically to reveal the background tracks using the same etching condition which will be applied later for etching the exposed foils. After irradiation, a short conventional etching before applying the high voltage gives the background tracks a quite different configuration than the newly induced neutron tracks as shown in Fig. 7. By using this process the lower detection limit, within ± 10% reproducibility, becomes lower than 500 mrem in the present condition. This limit could be lowered by using new thinner plastic foils.

REFERENCES

1. L. Tommasino, CNEN Report RT/PROT (71) 1 (1970).

2. M. Sohrabi, Health Physics, 27, 598 (1974).

3. G. Somogyi and D.S. Srivastava, Proc. 7th Int. Colloq. Nucl. Photogr. and SSTD, Barcelona, 1970, p. 711.

4. G.M. Hassib and L. Medveczky, ATOMKI, 16, 315 (1974) and Proc. 2nd Symp. Neutron Dosimetry, Neuherberg/Munich, 1975, Vol. I, p. 535.

Table 1

Directional dependence of the registration efficiency
of Makrofol E at different neutron energies

Direction	PuBe spectrum		14 MeV	
	t/cm^2/rem	%	t/cm^2/rem	%
90°	50	100	135	100
45°	30	60	100	74
0°	15	30	60	44

Etching condition : 2 kV, 3 kHz, 25% KOH + C_2H_5OH, ~ 3 hours.

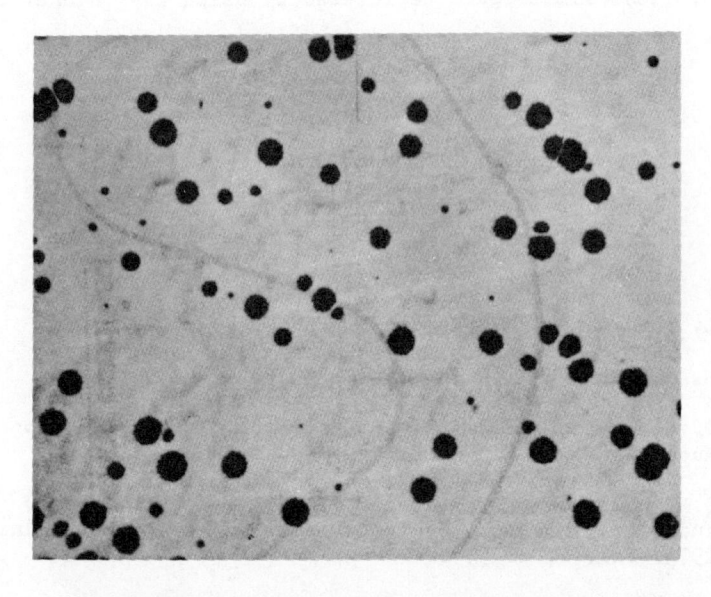

Fig. 1. Micrograph of recoil particle tracks in 500 μ thick polycarbonate etched 2 h in alcoholic solution at room temperature, 3 kV, 3 kHz (x 50).

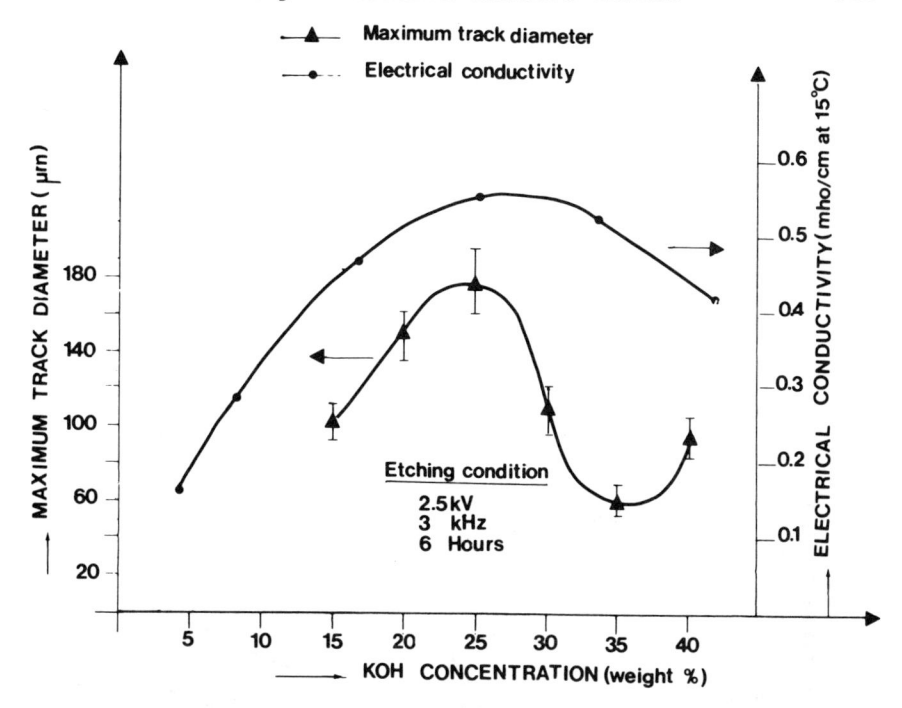

Fig. 2. Effect of the concentration of KOH on the electrochemical etching of 500 µ thick Makrofol E.

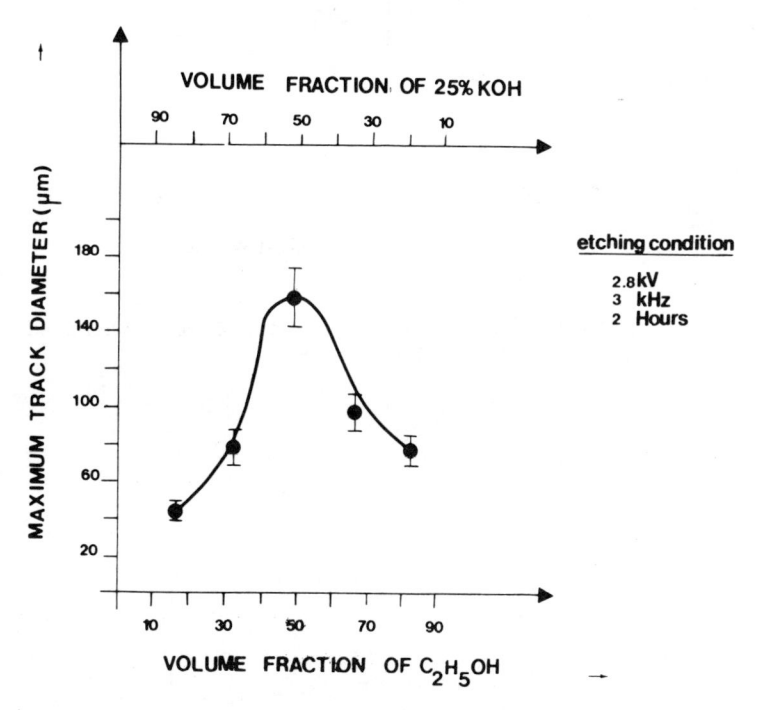

Fig. 3. Effect of mixing 25% KOH with ethyl alcohol on the electrochemical etching of 500 µ thick Makrofol E.

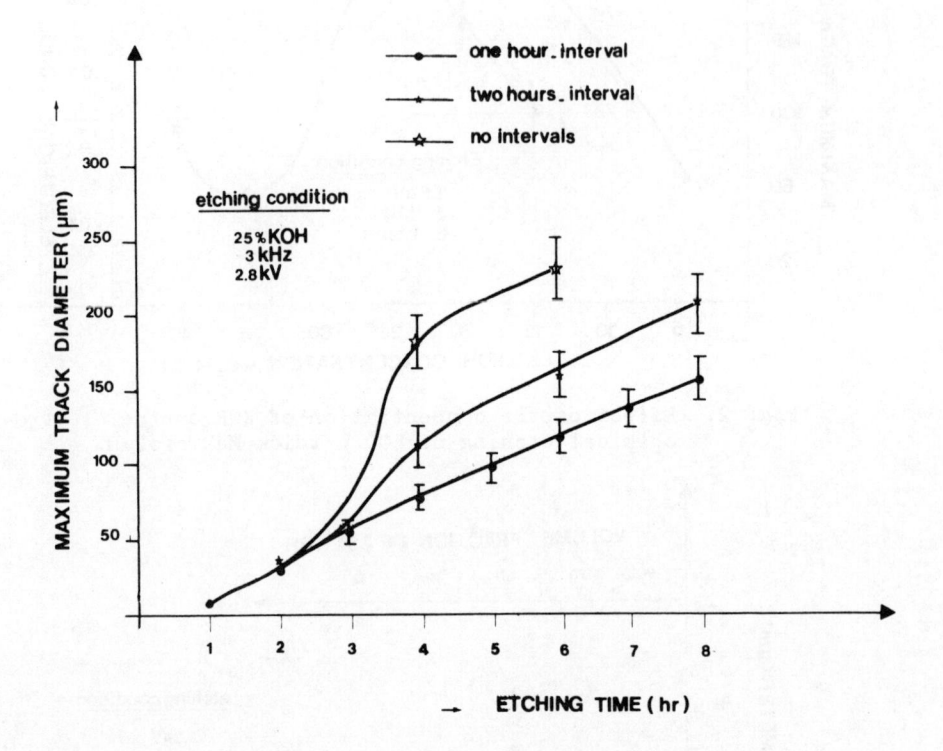

Fig. 4. Effect of etching time on the electrochemical etching of 500 μ thick Makrofol E.

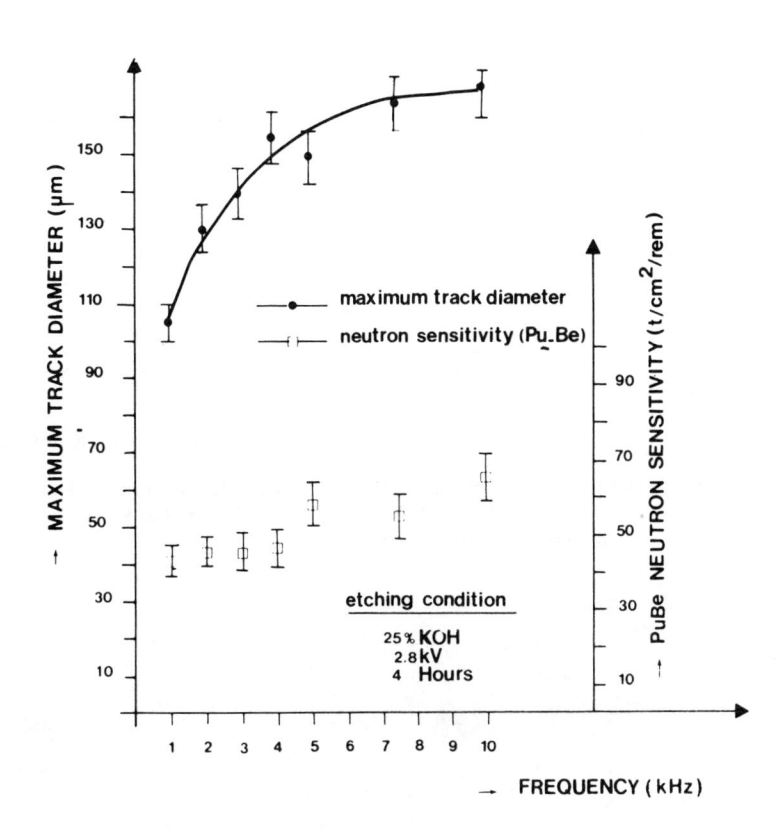

Fig. 5. Effect of frequency on the electrochemical etching of 500 μ thick Makrofol E.

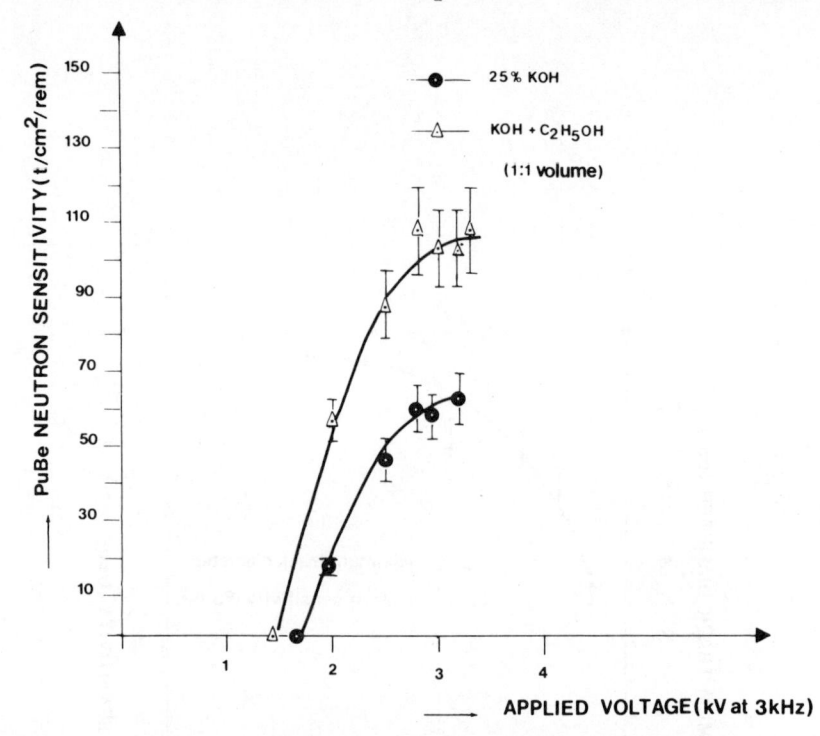

Fig. 6. Effect of applied voltage on the neutron sensitivity of 500 μ thick Makrofol E irradiated with fast neutrons from a PuBe source.

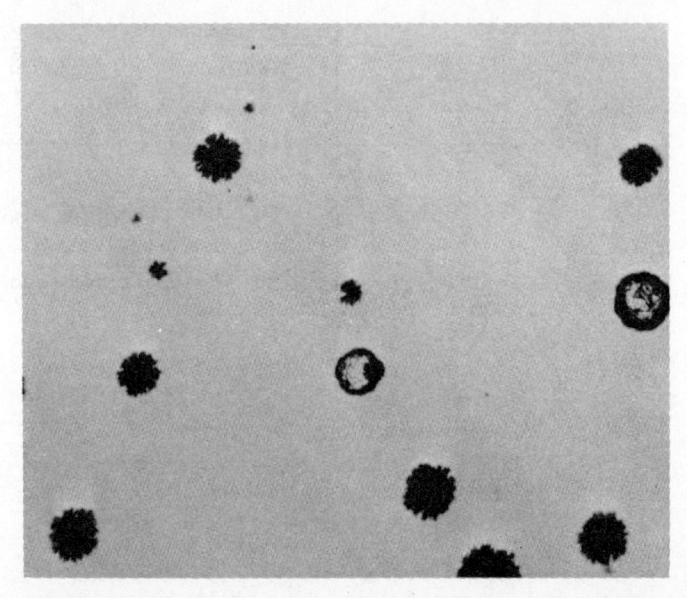

Fig. 7. Micrograph showing the difference between the background tracks (pale) and the neutron induced tracks (black) by applying the process described here (x 50).

NEUTRON FLUENCE MEASUREMENTS WITH SOLID STATE NUCLEAR TRACK DETECTORS — RESULTS OF AN INTERNATIONAL INTERCOMPARISON

H. Schraube and H. G. Paretzke

Institut für Strahlenschutz der GSF, D-8042 Neuherberg, F.G.R.

Abstract

An intercomparison of solid state nuclear track detectors, as they are used in neutron dosimetry and monitoring, has been performed at the GSF neutron calibration facility. The detectors of 20 participants were exposed in low scattering environment to monoenergetic neutron fields of 0.57, 5.23 and 15.1 MeV respectively, and to a Cf-252 fission neutron field. For every participant and for each neutron energy three fluences were applied, the magnitudes of which were reasonable for radiation protection interests. Neutron fields were monitored by means of ionization and moderator detectors. A comparison of neutron fluence data measured with SSNTDs with absolute monitor data is made on the basis of the results of 12 participating laboratories.

Especially at low energies and low doses the deviations of the reported values from the true neutron fluences and from each other are considerable. This fact should lead to further studies on the systems themselves as well as on evaluation and calibration techniques.

1. Introduction

Solid state nuclear track detectors (SSNTDs) have increasingly been used in neutron monitoring, personnal and criticality dosimetry. Therefore it appeared meaningful to perform an intercomparison programme at GSF Neuherberg at the occasion of the 9th International Conference on Solid State Nuclear Track Detectors (9th ICNTD) taking place from Sept.30 - Oct.6, 1976, at the same location.

The dosemeters of 20 participating laboratories were irradiated with 4 different neutron spectra and three doses each. The doses were chosen to be reasonable for personal dosimetry purposes. The participants were informed of the neutron energies and the order of magnitude of the neutron fluences to be applied to the individual dosemeters. Immediately after irradiation the detectors were returned. The participants were asked to report neutron fluence values, as well as to give a detailed description of their systems. This paper gives a description of the experimental procedure and discusses the main results of this intercomparison.

2. Experimental Arrangement

The irradiations were performed in the GSF-neutron calibration hall under low scattering conditions (1). The target for neutron production is located in the centre of the hall at the end of the beam line of a 3 MeV-Van-De-Graaff-Generator (Fig.1).
The neutron beam was monitored by a tissue equivalent parallel plate ionisation chamber and a moderator detector with a response proportional to the Rem-response. Other instruments for absolute and for spectral monitoring served as an internal control.

Californium-252 irradiations were performed with a source of about 1 mg Cf. Part of the handling system is shown in Fig.1, namely the transport-pipe to the external storage, the internal storage in the basement of the pit and the pipe to the ground floor.

The SSNTDs were arranged in circular order and fixed to a fabric, which was stretched in an aluminum frame. This arrangement was set up at 1.5 m distance from the target for accelerator neutrons and at 1.2 m distance for Californium irradiations. The alignment was checked axially and vertically by means of telescopes.

Fig.2 shows the dosemeter setup during californium irradiation. The movement of the source is done with accurate mechanics. Therefore, monitoring is not necessary, and the TE-chamber is removed. The Cf-positioning tube is removable for accelerator neutron irradiations.

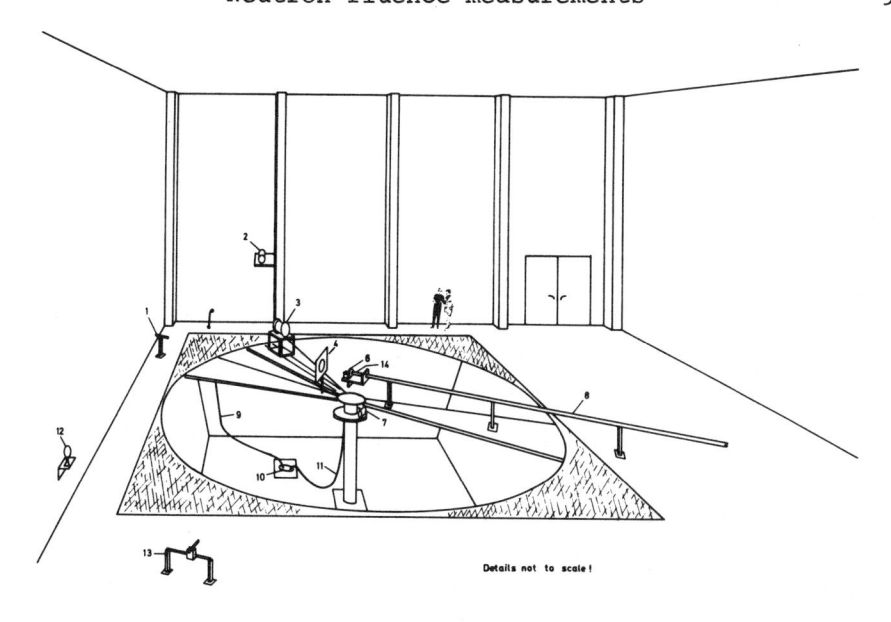

Fig.1　View of the GSF neutron calibration hall with beam-line (8), target (14), TE-monitor chamber (6), Precision-Long-Counter (3), Moderator detectors (2, 7), SSNTD holding frame (4), Cf-252 system (9, 10, 11) and aligning equipment (1, 12, 13).

Fig.2　View of the irradiation arrangement. In contrast to Fig.1, the TE-monitor chamber is removed, and the Cf-positioning tube (11) is prepared for irradiation.

3. Irradiation Conditions

The irradiation geometry is shown in Fig.3. The maximum angle between the beam axis and the dosemeters is 6.5°, the minimum angle 4.2° and the mean 5.3°. The solid angle is small enough, that all neutrons hitting the dosemeters had to penetrate the TE-chamber. The angles were chosen such that a reasonable homogeniety was achieved.

The relative neutron fluence distribution in the plane perpendicular to the beam axis at 1.5 m distance from the target can be deduced from Fig.4. The highest inhomogeniety occurs for the $D(d,n)He^3$-Reaction with ± 2% at 2.3 MeV particle energy and 5.23 MeV neutron energy. The inhomogeniety reduces to ± 1% for the $T(p,n)He^3$-Reaction with 1.5 MeV particle energy and 0.57 MeV resulting mean neutron energy. For the $T(d,n)He^4$-Reaction the fluence variation is less than ± 0.5%. These inhomogenieties are due to geometrical distance effects as well as to the kinematics of the neutron producing reactions. For the Californium runs the inhomogenieties are due only to geometrical effects and are in the same order of magnitude as for the $T(d,n)He^4$-irradiations.

4. Monitor Data

The monitor and fluence values for the 12 irradiations are listed in table 2. All detectors of the participants for each dose were irradiated at the same time. The third column of the table gives the TE monitor reading, which was used for calculation of neutron fluences at the circular reference line (see Fig.4). The ratio of the two monitor readings, as listed in the fourth column, serves to check the irradiation conditions. Except for the medium dose 15.1 MeV irradiation (an aluminum cap for calibration was left at the target by mistake) no significant changes are to be observed. No attempt was made to correct this one value, because of the much larger spread of the SSNTD-results. The absolute values have been deduced from the European Neutron Dosimetry Intercomparison (2), which was recently performed in the same laboratory. In that intercomparison several laboratories

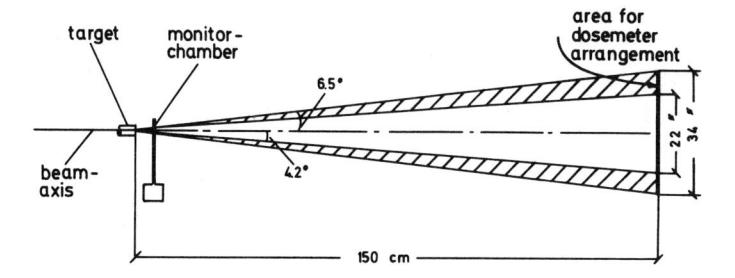

Fig.3   Irradiation geometry at the accelerator target

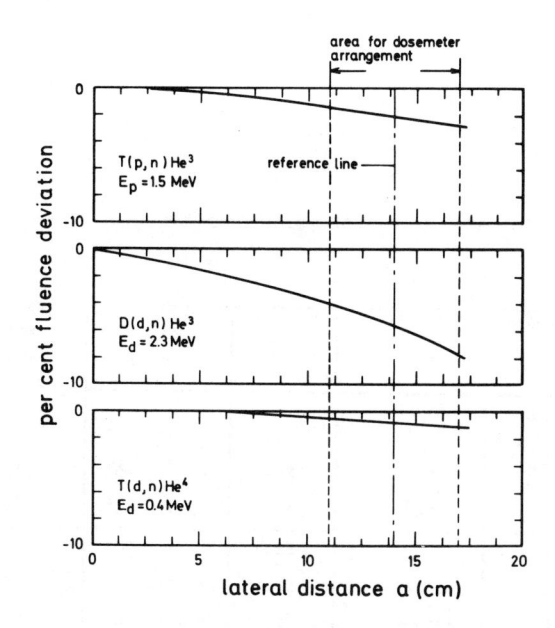

Fig.4   Relative neutron fluence distribution at 150 cm distance from the target versus the lateral distance "a" from the beam axis. (Irradiation geometry as shown in Fig.3).

Table 1 Neutron field specifications for the intercomparison of nuclear track detectors
1) Fluence-kerma (standard-man) conversion factor (3)
2) Fluence-dose equivalent conversion factor (4)

reaction	particle energy (MeV)	mean neutron energy at $\vartheta_L = 5.3°$ (MeV)	energy spread (± %)	d_k 1) (rad cm^2)	d_e 2) (rem cm^2)	fluence spread (± %)
T(p,n)He^3	1.5	0.57	25	$1.7 \cdot 10^{-9}$	$2.6 \cdot 10^{-8}$	≤1
D(d,n)He^3	2.3	5.23	5	$4.3 \cdot 10^{-9}$	$3.8 \cdot 10^{-8}$	2
T(d,n)He^4	0.4	15.1	5	$7.1 \cdot 10^{-9}$	$5.4 \cdot 10^{-8}$	≤0.5
Cf-252	-	2.1 (Maxwell. distr.)	-	$2.95 \cdot 10^{-9}$	$3.6 \cdot 10^{-8}$	0

Table 2 Monitor and neutron fluence values

Neutron Energy E_n (MeV)	Irradiation run No.	TE-Monitor reading M_1	Monitor ratio M_2/M_1	Fluence ϕ_o (1/cm^2)	Standard error $\Delta\phi_o/\phi_o$
0.57	1	117	329	$2.85 \cdot 10^6$	± 3%
	2	914	328	$2.22 \cdot 10^7$	
	3	7767	329	$1.89 \cdot 10^8$	
5.23	4	583	52.7	$2.54 \cdot 10^6$	± 2%
	5	5696	52.1	$2.49 \cdot 10^7$	
	6	45529	52.0	$1.99 \cdot 10^8$	
15.1	7	689	53.9	$1.75 \cdot 10^6$	± 2%
	8	4544	57.7	$1.16 \cdot 10^7$	
	9	68219	53.5	$1.73 \cdot 10^8$	
Cf-252	10	-	-	$3.00 \cdot 10^6$	± 5%
	11	-	-	$1.26 \cdot 10^7$	
	12	-	-	$8.70 \cdot 10^7$	

Table 3 Review of general properties of the nuclear track detector systems

	ID-number:	39	33	25	12	13	4	5 (1,2)	15	16A (2)	16A (3)	16A (4)	16B	18	1
Detector	Makrofol	x	x			x	x				x		x		
	Cellulose nitrate			x	x		x	x		x					
	Mylar													x	x
	Soda glass								x						
External converter	Uranium	x				x					x				
	Thorium							x	x		x		x		
	Neptunium							x						x	
	(n,α)-converter					x									
Evaluation method	Microscope	x	x	x	x		x			x	x	x			
	Microscope-proj.				x				x						
	Spark counter							x					x	x	x
Type of calibration	Theoretical							x							
	Use of published response functions	x				x			x						
	Am-Be, Pu-Be-source									x	x		x	x	x
	Cf-source							x		x	x	x			
	Target-reactions		x	x	x					x	x	x	x		
	Reactor													x	

Table 4 Processing methods and conditions for Makrofol (Ma) and Mylar (My) detectors. (US = additional ultrasonic treatment during rinsing).

ID-NO	Detector thickness (μm) and material		Etchant		Etching		Rinsing		Drying	
					temp.	time	medium	time	temp.	time
1	8	My	25%	KOH	40°	105 min	H_2O	10 min	20°	6 h
5(1)	10	Ma	33%	KOH	60°	30 min	H_2O	5 min (US)	25°	24 h
5(2)	12	Ma	33%	KOH	60°	55 min	H_2O	5 min (US)	25°	24 h
13	200	Ma	26%	KOH	60°	80 min	H_2O	2-3 min	20°	(filter paper)
16A(4)	200	Ma	15% KOH 40% C_2H_5OH		70°	20-40min	H_2O	20 min	25°	120 min
16B	8	My	25%	KOH	50°	110 min	H_2O	60 min	25°	24 h
18	10	Ma	28%	KOH	60°	30 min	$HCl+H_2O$	2 min	40°	30 min
33	200	Ma	15% KOH 40% C_2H_5OH		60°	80-90min	H_2O	2 min	20°	5 min
39	200	Ma	17%	NaOH	60°	100 min	H_2O	200 min	50°	24 h

measured the Kerma-rate at a certain distance from the target by means of absolute ionization chamber techniques. From those values the fluences have been calculated by using reaction kinematics and inverse distance squared law. The error given in the last column is the unweighted standard error of the mean of these intercomparison results. Additional errors, e.g. for conversion factors, are not taken into account.

5. Irradiated Track Detector Systems

In table 3, a summary of general properties of those 14 nuclear track detector systems is shown, the results of which were returned to our laboratory. Tracks were registrated in 6 detectors of Makrofol, 5 of cellulose nitrate, 2 of Mylar foils, and 1 of soda glass. It is seen that 8 systems were used with an external fission converter, while 6 did not use any converter.

The etch pits were counted by microscope-techniques (10x), by microscope-projectors (2x) and by spark-counters (4x). With the spark-counters one or two pre-sparkings at 900 or 1000 V were applied, and counting took place at 550 V or 600 V. Four groups used theoretical considerations or published response functions for converting track densities to neutron fluence. Ten groups calibrated their systems experimentally by using accelerator neutrons or standard radionuclide sources.

The processing methods and etching conditions employed by the participants are listed in Tables 4 and 5.

6. Results and Discussion

Figs. 5-8 give the results reported by the participants. The abscissa show the neutron fluence, $\emptyset_o$ being the absolute monitor results. The ordinate gives the identification number. The total fluence values increase by about a factor of ten among the three irradiations for each neutron field. For the 15.1 MeV irradiation, for instance, about 60% of the results are within a factor of 2 for the low dose irradiation. For the medium dose, 60% are within a factor of two and 40% are better than a factor of

Neutron fluence measurements

Table 5 Processing methods and conditions used for cellulose nitrate (CN) and microscopic soda glass (SG) detectors. (US=additional ultrasonic treatment during rinsing)

ID-No	Detector thickness (μm) and material		Etchant		Etching		Rinsing		Drying	
					temp.	time	medium	time	temp.	time
4	8	CN	17%	KOH	50°	120 min	10% acetic acid	10 min	20°	30 min
12	100	CN	25%	KOH	60°	20 min	H_2O	2 min (US)	20°	5 min
15	8	CN	10%	NaOH	60°	80 min	H_2O	20 min	40°	10 min
16A(2)	1300	SG	37%	HF	25°	20 sec	H_2O	20 min	25°	60 min
16A(3)	10	CN	9.1%	NaOH	40°	8 h	H_2O	20 min	25°	120 min
25	90	CN	19%	NaOH	50°	30 min	H_2O	30 min	20°	60 min

Table 6 Comparison of the neutron fluence ϕ_0 deduced from monitor reading with $\bar{\phi}_1$, $\bar{\phi}_2$ calculated from the participants' results.

Neutron Energy E_n (MeV)	Fluence ϕ_0 (1/cm²)	$\bar{\phi}_1$ (1/cm²)	$\bar{\phi}_2$ (1/cm²)
0.57	$1.89 \cdot 10^8$	$(4.4 \pm 4) \cdot 10^8$	as $\bar{\phi}_1$
5.23	$1.99 \cdot 10^8$	$(3.2 \pm 3.7) \cdot 10^8$	$(1.6_9 \pm 0.8_3) \cdot 10^8$
15.1	$1.73 \cdot 10^8$	$(1.4_7 \pm 0.5_2) \cdot 10^8$	as $\bar{\phi}_1$
Cf-252	$8.70 \cdot 10^7$	$(1.0_0 \pm 0.6_2) \cdot 10^8$	as $\bar{\phi}_1$

$E_n = 0.57$ MeV

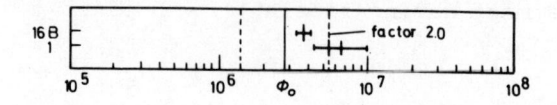

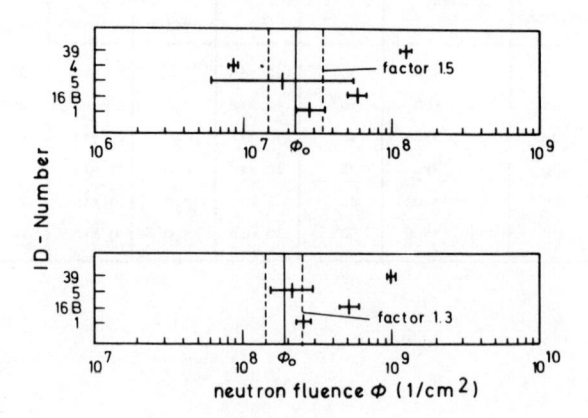

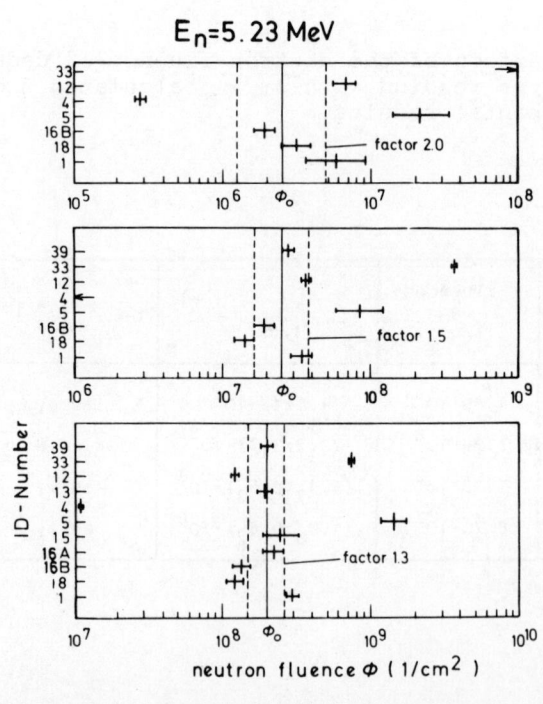

$E_n = 5.23$ MeV

Fig. 5, 6 Neutron fluence results reported by the participants for 0.57 and 5.23 MeV neutron energy. Factors of 1.3, 1.5 and 2.0 off the monitor values ϕ_0 are drawn to guide the eye.

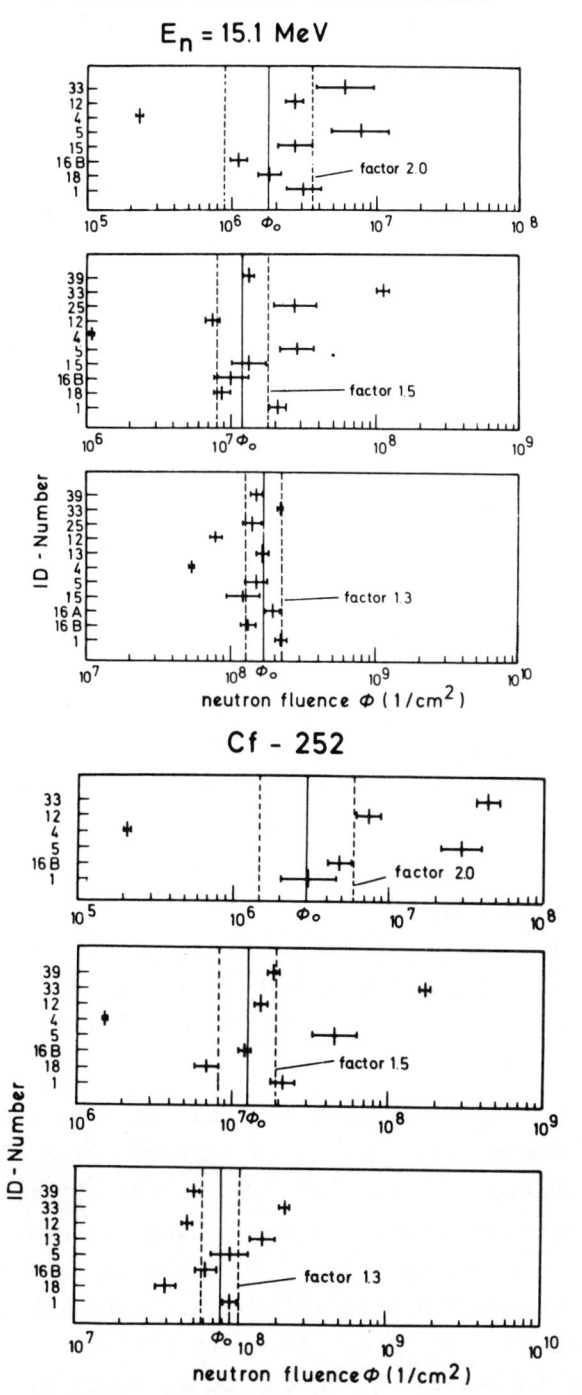

Fig.7, 8 Neutron fluence results reported by the participants for 15.1 MeV neutron energy and Cf-252 neutrons. Factors of 1.3, 1.5 and 2.0 off the monitor values ϕ_o are drawn to guide the eye.

1.5. For the high dose, 90% are within a factor of 1.5, including 80% better than a factor of 1.3. The results of the other irradiations are similar, but tend to less accurate distributions. In addition, only few results were reported for low energies and doses.

In Table 6, unweighted means $\bar{\phi}_1$ of the values reported by the participants are compared with the monitor values ϕ_o for the respective high dose irradiations. An attempt was made to test the reported values for outliers. It was assumed, that the data follow approximately a normal distribution. Then the tables of Dixon (5) were used on a level of significance of 10%. This led to the withdrawal of the two highest values of the 5.23 MeV irradiation, and a new mean value $\bar{\phi}_2$ was determined. It should be clearly pointed out, that this method was used tentatively only since the underlying statistical distribution is unknown and certainly is influenced by systematic errors.

7. Conclusion

The following conclusions can be drawn from the results of this intercomparison:

i) The fluence values reported by the participants have high variances and sometimes differ from each other by a factor of hundred and more. The mean values agree fairly well within the statistical errors with those values, which were calculated from the monitor data and which were assumed to be correct.

ii) It would have been of high interest to find some general hints on the optimum technique to be employed with SSNTDs in neutron dosimetry with respect to detector types, to processing, to evaluation methods, and to calibration procedures. But no definite conclusion could be drawn in this direction because of the poor statistical significance. However, it became clear that it is difficult to get good results with SSNTDs at low energies and low doses especially without using an external convertor.

iii) There is a neccessity for reliable neutron dosemeters in
neutron operational dosimetry. This is a dose range, where
doses in the range of about 50 mRem - 5 Rem are predomi-
nating, and which can be classified between personal and
accidental dosimetry. Before SSNTDs can successfully be
used in this field, further investigations will be neccessary
to decrease the lower limits of SSNTD-systems with respect
to neutron energy and dose[1].

Acknowledgement

We are grateful to Dipl.-Phys.A.Morhart for his help during the
experiments and to Dr.F.Schulz and his group for running the
accelerator.

References

(1) Schraube H., F.Grünauer and G.Burger, "Calibration problems with neutron moderator detectors" Neutron monitoring for radiation protection purposes, IAEA STI/PUB/318 Vienna (1973) p.453

(2) Broerse, J.J., G.Burger and M.Coppola, "Preliminary results of the European neutron dosimetry intercomparison project (ENDIP)" Basic physical data for neutron dosimetry, EUR 5629, Luxembourg (1976), p.257

(3) Ritts, J.J., M.Solomito and P.N.Stevens, "Calculation of neutron fluence-to-kerma factors for the human body" Nucl.Appl. Techn. 7 (1969) 89

(4) Nachtigall, D. and G.Burger "Dose equivalent determinations in nuetron fields by means of moderator techniques" F.H.Attix (ed.), Topics in radiation dosimetry, Suppl.1, (1972) p.385

(5) Dixon, W.J., "Processing data for outliers" Biometrics 9 (1953) 74

---

[1] Scientific laboratorys, which plan to test and calibrate their dosemeter systems with fast neutrons are invited to contact for irradiations Dr.G.Burger of this laboratory or the authors.

THE DEVELOPMENT AND APPLICATIONS OF PLASTIC TRACK DETECTORS FOR NEUTRON FLUX AND GAMMA DOSE MEASUREMENTS

H.A. Khan, R. A. Akber and G. Hussain

Nuclear Engineering Division, Pakistan Institute of Nuclear Science and Technology (PINSTECH) Nilore, Rawalpindi, Pakistan

ABSTRACT

Plastic track detectors have been developed for the measurement of neutron fluences and high gamma doses. The calibration of a CA80-15 Cellulose Nitrate Track Detector coated with 'Lithium Tetra Borate' for thermal and epi-thermal neutrons has been done. The results have been compared with those obtained by using other methods. Various plastics have been tested and applied for the measurement of high gamma doses in a reactor core. Finally, the results about the thermal neutron fluences, and gamma doses in the thermal column and the fuel elements respectively of Pakistan Research Reactor (PARR) have been given.

1. INTRODUCTION

Plastic Track Detectors find their extensive application for light and heavy charged particles registration over a wide energy range. The possible use of these detectors for thermal and fast neutron detection by placing a convertor foil against the detector or doping some fissile material into it has already been reported.[1-4] Kodak Pathe', France has recently produced cellulose nitrate plastic films which are thinly coated with Lithium Tetra Borate on both of its sides. In the first part of this paper we describe the calibration and the use of this detector as a thermal and epithermal neutron detection system. The neutron detection efficiency of these films has been calculated on the basis of a simple model. Later the model has been tested by comparing the results with those obtained by using the gold foil activation technique. The results of the flux measurements in the thermal column of a research reactor are also reported.

It is seen that for the detectors exposed to high gamma doses ~ several Mr (Mega rads) the general velocity of etching and the track width is enhanced[5]. This technique is used for the high gamma dosimetry after shut down and at low power operation of Pakistan Research Reactor (PARR). The gamma doses in the spent fuel elements are also determined.

2. CALIBRATION OF THE DETECTORS

2.1. THERMAL AND EPITHERMAL NEUTRON FLUX MEASUREMENTS

The plastic sheets of 100 micron (μm) thick cellulose nitrate $(C_{12}H_6O_{18}N_4)_n$ coated with white natural lithium tetraborate ($Li_2B_4O_7 - 5\ H_2O$) layer of 40 to 50 microns thickness on each side are used for neutron flux measurements in the present studies. Density of the layer material as provided by the manufacturer is 2.0 gms/cc. Table one provides the ways in which a thermal neutron incident on the detector can produce an (n, α) interaction.

The development of plastic track detectors

TABLE I : THE POSSIBLE (n,α) INTERACTIONS IN LITHIUM TETRA BORATE

REACTION	GROUP	ENERGY (Mev)	RELATIVE PROBABILITY	CROSS-SECTION (barns)
$^6Li(n,\alpha)^3H$	1	2.05	1.00	940 ± 4
$^{10}B(n,\alpha)^7Li$	2	1.78	.064	3837 ± 9
$^{10}B(n,\alpha)^7Li^*$	3	1.50	.936	

A certain percentage of the charged particles produced as a result of any of the above mentioned (n,α) reactions reaches the cellulose nitrate film to produce the etchable damage trails. For the ith energy group of alpha particles this fraction N_i per cm^2 is

$$N_i = (F\sigma_i)(x_i f_i)(\tfrac{1}{2} R_i) \mu\mu' \cos^2\theta_c \qquad (1)$$

where,

F = The thermal neutron fluence in #/cm^2 (2)
  = ϕt

σ_i = The microscopic interaction cross-section for (n,α) reaction for the ith isotope in cm^2

x_i = The number density of the ith isotope in #/cm^3

f_i = The relative probability for an (n,α) reaction

R_i = The 'effective thickness' in cms of the convertor material for the alpha particles of the ith energy group. The 'effective thickness' is half the range of the particles.

The range (= 2 R_i cms) can be related to the range in air by Bragg Kleeman Rule^8 as

Range in convertor (cms) = $3.2 \times 10^{-4} \frac{\sqrt{a_{eff}}}{d}$ Range in air (cms) (3)

Here a_{eff} and d are the effective atomic weight and the density of the convertor material.

It is quite evident that only those alpha particles reach the plastic detector which are ejected within a cone of side equal to the range of the particles and its basal plane on the detector convertor interface. This fact has been taken into account by a factor of 0.5 in equation (1). Correction factors for the flux depression inside the convertor above and within the effective thickness, respectively are $\mu\mu'$. Their commulative effect equals 0.941.

The effect of critical angle of etching (θ_c) upon the latent damage trails reduces the number of etchable tracks by a factor of $\cos^2 \theta_c$. Our unpublished work for ≤ 2.0 Mev alpha particles and using 30% NaOH as etchant shows that θ_c varies as $\leq 7°$. This makes $\cos^2 \theta_c \simeq 0.99$. This factor being close to unity has been ignored.

The over all track density N, which is the summation of the track densities due to individual groups as obtained by equation 1, can thus be related to the neutron fluence F as,

$$F = \phi t = 7.42 \times 10^2 \, N \qquad (4)$$

While deriving equation (4) it has been assumed that the triton with 1.78 Mev and ^{7}Li with 1.02 Mev and 0.86 Mev which are also ejected as (n,α) reaction products do not produce any etchable damage trails.

The unetchability of 1.78 Mev energy tritons which are equivalent to about 0.6 Mev protons is found in the literature[10,11]. No work about ^{7}Li of this energy is, however available. A second model which takes into account the effect of ^{7}Li as well is therefore developed, but experimental results are seen to be closer to equation (4).

The validity of equation (4) was tested by comparing the thermal flux as measured by

a) Lithium Tetra Borate Coated Cellulose Nitrate Films by using equation (4) and
b) Gold Foils by using activation technique.

The films and the foils were exposed at the same points on the face of the thermal column of a research reactor. The comparison indicates that the thermal neutron flux as measured by the gold foils is invariably greater by an average deviation of about 10%.

If the activation technique is assumed to be a standard then the two most probable sources of this error in the model.a) About 9.8% over estimation of the range of alpha particles by Bragg-Kleeman Rule or b) The existance of an energy limit at about 300 Kev below which no alpha particles produce etchable damage trails, or otherwise mixed contribution of these two sources of error.

In any case, taking gold foil activation technique as a standard method for the thermal neutron flux measurement the neutron fluence by the present detector can fairly be approximated as,

$$F = 800\ N \qquad (5)$$

For epi-thermal neutron flux measurement for which an average $\bar{E}$ of the neutron energy spectrum can be selected

$$F = 4.42 \times 10^3\ \sqrt{\bar{E}}\ N \qquad (6)$$

gives the one group approximation of the neutron fluence at energy $\bar{E}$ (eV).

N is the alpha track density ($\#/cm^2$) on the detector. It is assumed that within the neutron energy limits for equation (6), ^7Li and ^{10}B are perfect $1/v$ absorbers.

2.2. FAST NEUTRON FLUX MEASUREMENTS

It is observed that the Lithium Tetra Borate Coated Cellulose Nitrate Films are not fit for fast neutron flux measurements (neutron energy $\sim$ Mev) mainly due to

    a) a rapid fall in the (n,α) interaction cross-section
    b) the background produced by the recoil tracks.

Figure (1) gives the results of one of the attempts made to record the fast neutrons by the (n,α) interaction on the film under study. The neutron source was an Am-Be source and the experimental set up was as in the inset of the figure. The graph indicates that the signal to noise ratio is very significant.

2.3. HIGH GAMMA DOSE MEASUREMENTS

The exposure of plastic track detectors to high gamma doses changes the etching characteristics. The extent of the change can be employed as an index for the gamma dose. Little work has been done on the use of Plastic Track Detectors for the high gamma dosimetry. One

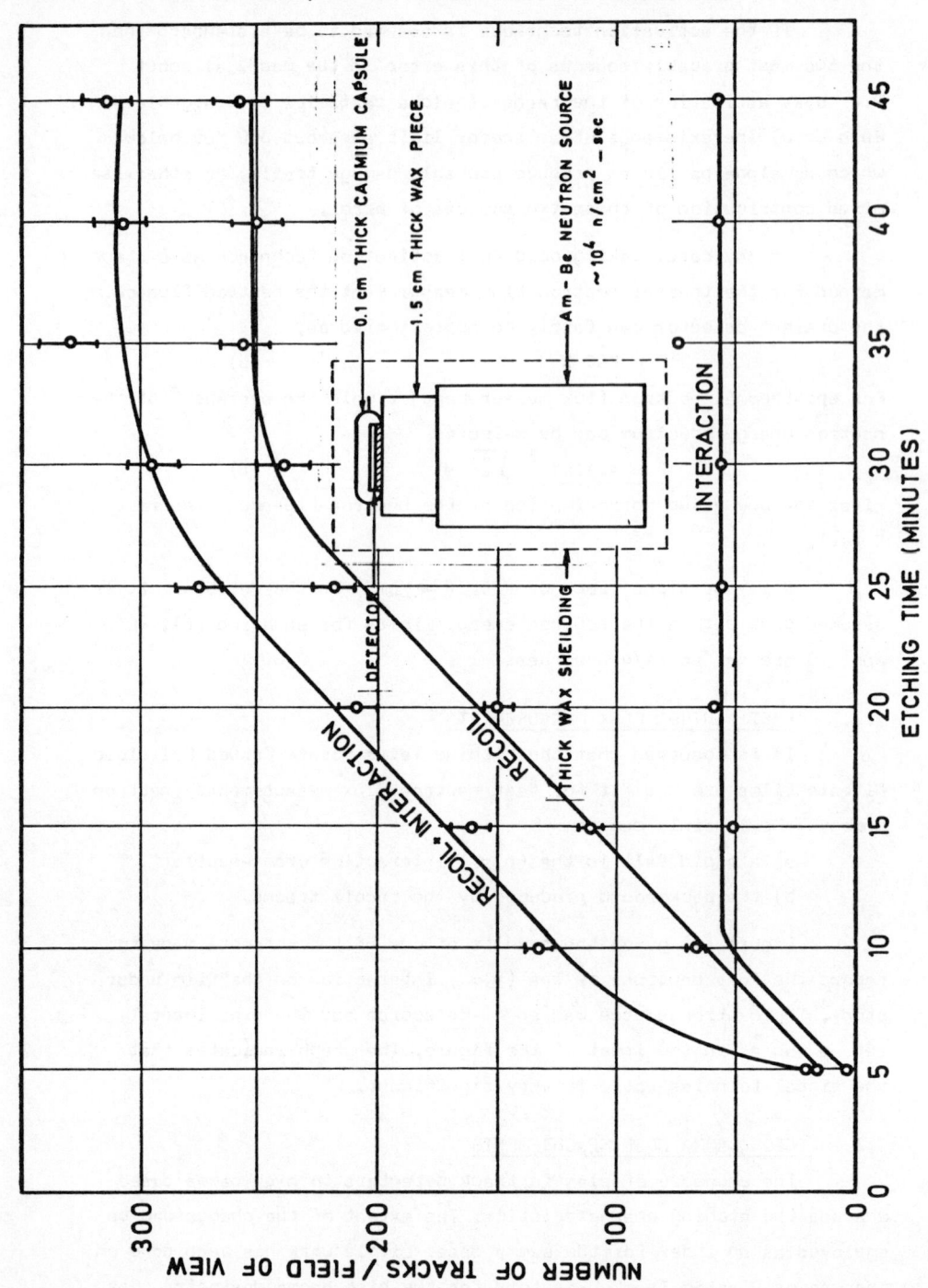

Figure 1 : The variation in the neutron interaction + recoil and recoil track densities as recorded on a Lithium Tetra Borate coated and bare CA80-15 respectively. The insert shows the experimental set up.

The development of plastic track detectors

paper reports the feasibility of the use of plastic SSNTDs and provides the necessary calibration curves depending upon the following characteristics:

a) The fractional rise in the average (maximum) track width with gamma dose 'D' megarads. The tracks being produced by a ^{252}Cf fission fragment source before exposing to the high gamma dose from a large ^{60}Co source. The etchant used 30% NaOH at $50°C$, Figure 2.

b) The fractional change in the general velocity of etching with the gamma dose.

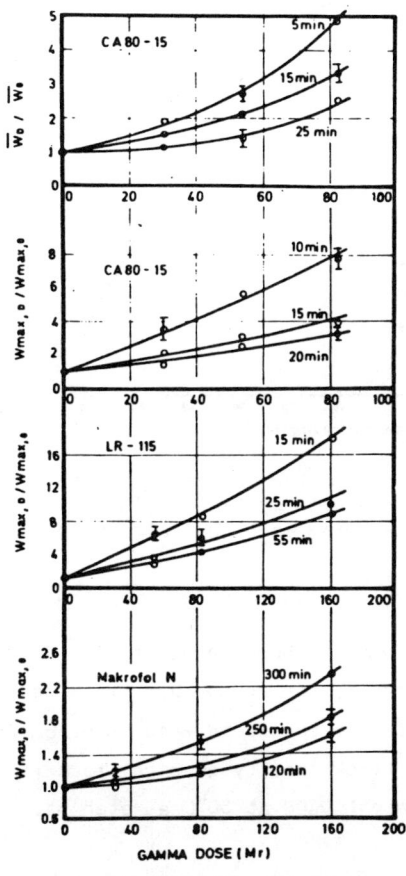

Figure 2
A set of curves showing the variation of W_o/W_o and $W_{max,D}/W_{max,o}$ for CA80-15, LR-115, and Makrofol N plastic track detectors, as a function of gamma dose. Both the ratios increase with the increasing gamma dose. It is interesting to note that cellulose nitrate (CA80-15 and LR-115) show a decrease, while the polycarbonate (Makrofol N) shows an increase in the ratio with the prolongation of the etching process.

3. **EXPERIMENTS AROUND A REACTOR**

3.1. **FLUX MEASUREMENT IN THE THERMAL COLUMN**

The thermal and epi-thermal neutron flux on the face of the thermal column of the Pakistan Research Reactor (PARR) was measured. Figure (3) shows the thermal flux distribution pattern along the horizontal axis of the thermal column. The dotted lines show the flux as mapped by the gold foils and are drawn for the sake of comparison.

Figure (4) gives the relative epithermal neutron flux above the cadmium cut at 0.4 ev. The epithermal flux as mapped by the lithium tetra borate coated cellulose nitrate film 'necks out' at the centre. The dotted line shows a corresponding fall in the recoil track density due to energetic neutrons.

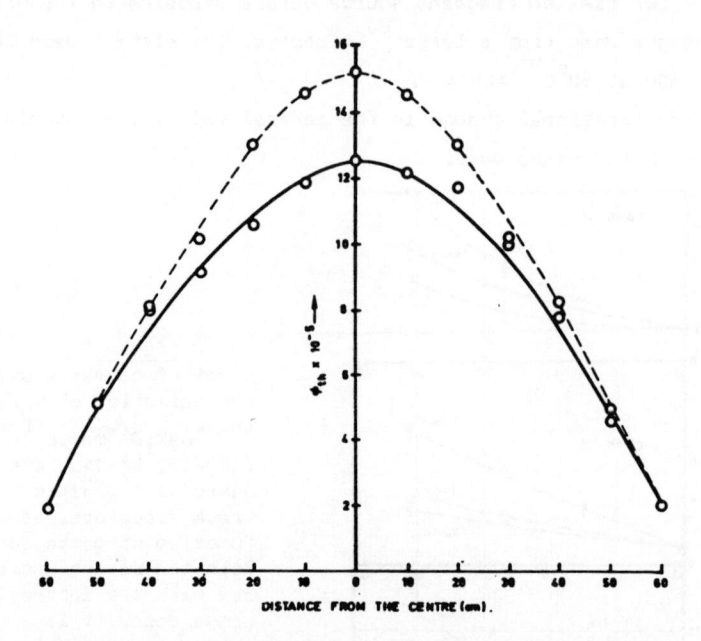

Figure 3 : Thermal flux distribution along the horizontal axis of the thermal column as measured by SSNTDs. The dotted line shows the flux measurement by gold foils.

An important precaution in the use of these detectors is to avoid the over lapping of the etched tracks. For reasonable track density the total fluence experienced by the detector should be $\sim 10^8$ neutrons/cm^2. The suitable etching time at which the track density goes to its maximum is 20 minutes provided we use 30% by weight NaOH at 50°C as etchant.

3.2. GAMMA DOSE MEASUREMENT IN 'PARR' SPENT FUEL ELEMENTS

The Pakistan Research Reactor (PARR) is a swimming pool type reactor with MTR type parallel plate fuel elements. The spent fuel elements rest in a rack about 25 feet deep in water. Special stainless

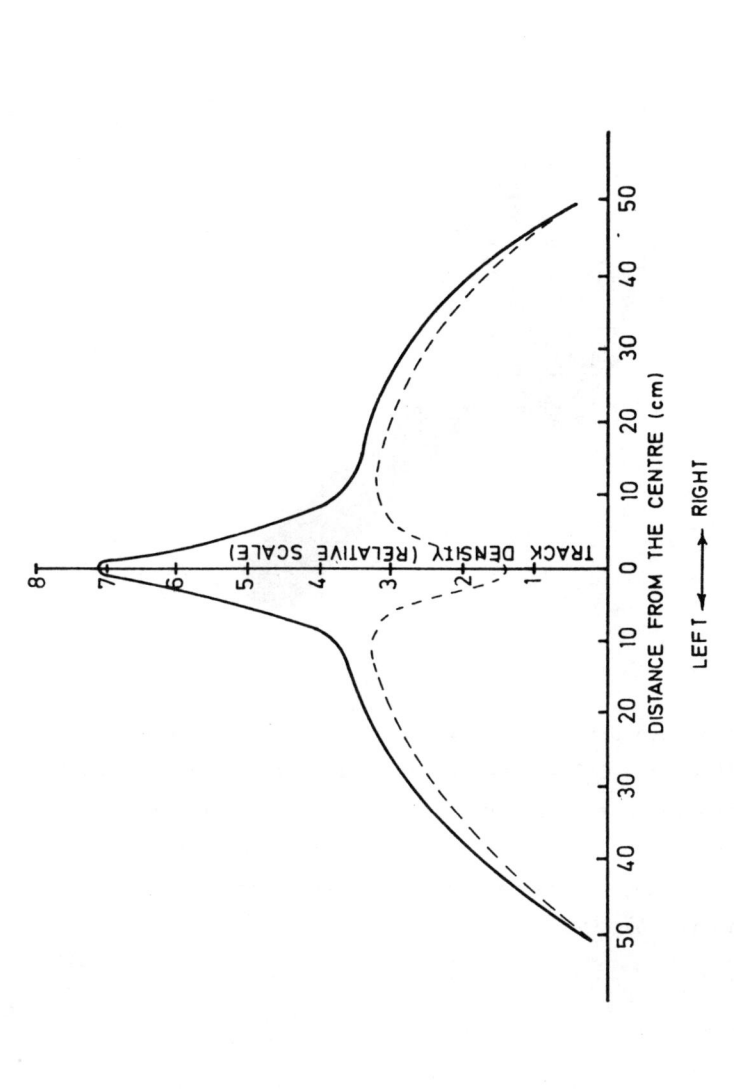

Figure 4 : Epithermal flux distribution along the horizontal axis of the thermal column. The flux necks out in the centre while the corresponding fast neutron flux falls. The fast neutron flux distribution is shown by the dotted line.

steel stringers were designed which went into the fuel plate spacing (≃ 0.3 cms) in the spent fuel elements. The stringers contain pockets in them in which the detectors can be placed (Figure 5).

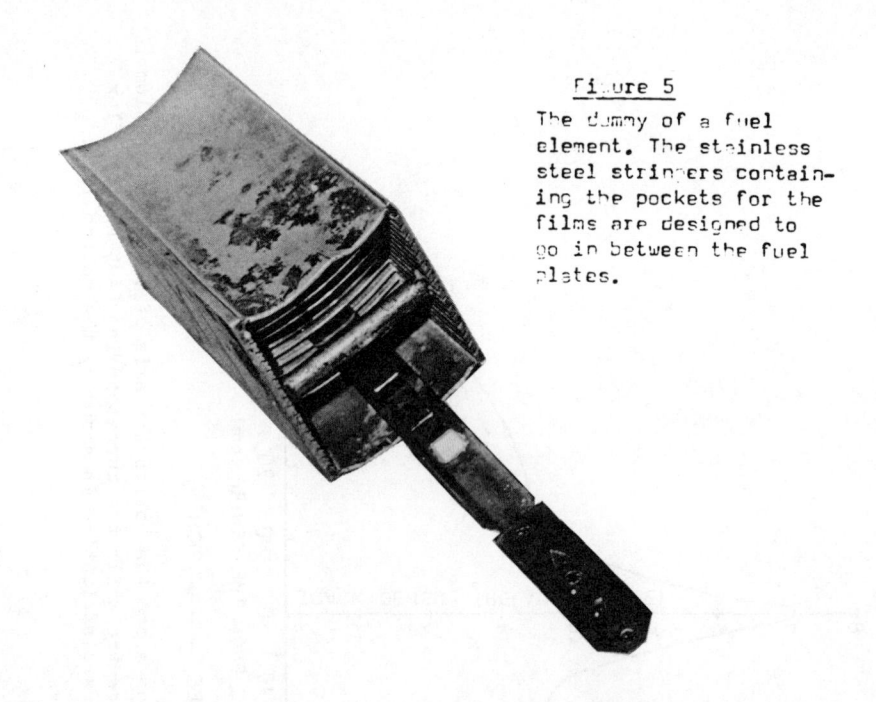

Figure 5
The dummy of a fuel element. The stainless steel stringers containing the pockets for the films are designed to go in between the fuel plates.

The gamma dose distribution is found by using the average track width method in a CA80-15 cellulose nitrate detector and by using the calibration curve as given in Figure (?). The detectors were exposed for 72 hours and the dose rate in R/hr was found by assuming a little change in gamma dose during this time. The results are shown in Figure (6).

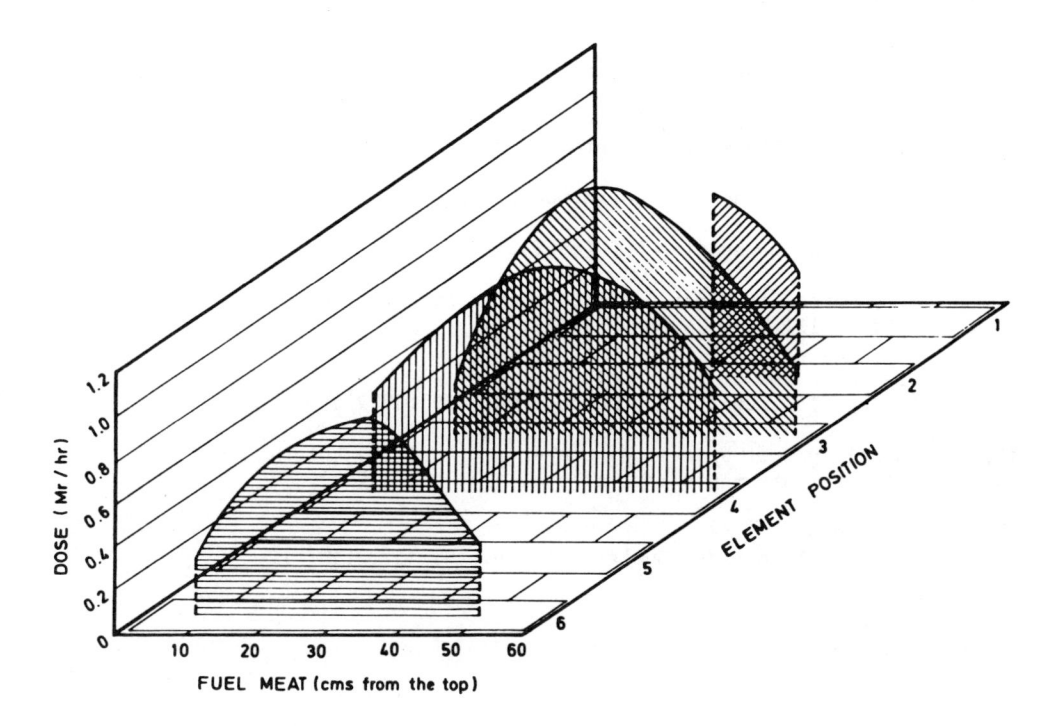

Figure 6 : The gamma dose distribution in the spent fuel elements lying in the rack.

REFERENCES

1. Krystyna Jozefowiez
   1332/IXA/PR Institute of Nuclear Research, Wassa.

2. Young Soo Yoo and Seung Gy Ro
   Journal of Korea Physical Society
   $\underline{5}$ No. 1, 22 (1972).

3. Paul F. Rago, R.C. Barral, T.G. Carter,
   Health Physics, $\underline{26}$, 102 (1974).

4. Klaus Becker and J.S. Jun,
   An Oak Ridge National Laboratory Report.

5. H.A. Khan, M.A. Atta, Shaukat $Y_a$meen, M.R. Haroon & Athar Hussain,
   Nuclear Instruments and Method, $\underline{127}$, 105 (1975).

6. Kodak Pathe, Personal Correspondence.

7. Neutron Cross-section BNL (1955).

8. W.J. Price, 'Nuclear Radiation Detection' 2nd Edition
   McGraw Hill Book Company (1967).

9. H.A. Khan and S.A. Durrani
   Nuclear Instruments and Methods $\underline{98}$, 229 (1972).

10. Paul B. Hahn, Margaret A. Wecker and Adolf F. Voigt,
    Nuclear Instruments and Methods, $\underline{123}$, 111 (1975).

11. H.A. Khan
    Journal of Physical Society, Japan, $\underline{39}$, 1159 (1975).

DETECTION OF FAST AND SLOW NEUTRONS BY ETCH PIT METHOD OF NUCLEAR TRACK REGISTRATION IN PLASTICS

M. A. Kenawy, M. El-Fiki, S. El-Konsol, M. A. Fadel and A. M. Basha

Ain Shams University, NIS, AEA, Cairo University, Egypt

Abstract

A plastic, cellulose acetate, was irradiated with neutrons, from Ra-Be source and ET-RR-I, to investigate its usefulness as flux detectors. Elements such as aluminium, copper, graphite, teflon and gold were used as radiators in case of fast neutron detection, while lithium has been used in the case of thermal neutrons. Since these elements undergo (n,α) reactions with neutrons of different energies, they can in principle be used for neutron dosimetry via registration of alpha particles and other reaction products. After etching with suitable chemical reagents, the number of etch pits were counted by ordinary optical microscope. To avoid track counting, a direct read-out method by the optical density measurements was carried out using a micro-densitometer. A comparative study between the electric conductivity of the plastic with both track scanning and optical density methods has been carried out.

Introduction

When plastics are irradiated with neutrons, the neutrons may interact with some of the atoms in material producing charged particles which should cause radiation damage to the substance (1, 2). Through suitable chemical etching reagents, the damage sustained by certain types of irradiated plastics, can be made visible by ordinary optical microscope. The fact that the number of etch pits produced per unit area is proportional to neutron fluence supports the possibility of utilizing nuclear track registration in plastics as a means of measuring fast and slow neutrons(3).

Since the range of carbon and oxygen recoils from fast neutron interaction in cellulose acetate is very short, tracks of α-particles produced by (n,α) threshold reactions are mainly registered in the plastic (4,5). To increase number of (n,α) threshold reactions different materials, aluminium, copper, graphite, teflon and gold, could be used as external radiators. Lithium or boron external radiators could be used in case of thermal neutron detection. The detection in the latter case will be via the registration of alpha particles and triton as reaction products from li (n,α) T reaction. The sensitivity of such detector depends on (n,α) cross-section of the external radiators, concentrations of the radiator and range of the alpha particles in plastics. Consequently, the sensitivity of the α-particles registration detector is less than that of fission fragments detectors in which fissionable materials are used as a external radiator.

Higher as the sensitivity of the fission fragments detectors is the α-particles registration detector (as compared to that of fission fragments detector) has several advantages for personal dosimetry in particular. Since, no expensive fissionable materials with its inherent safety problems are required and neither the skin near the dosimeter nor the other gamma radiation sensitive detectors in the dosimeter are exposed to the gamma radiation emitted from fissionable materials. In addition, no back-ground increase due to spontaneous fission is present. Spontaneous fission may, for instance in uranium containing detectors, lead to a considerable "predose" over extended periods of use.

The aim of this work, therefore, was to investigate the influence of the different external radiators in comparison with that in case of using fissionable materials. Furthermore, different parameters affecting the sensitivity of these nuclear track detectors have been studied. The track counting method, induced change in the optical density of the plastic material by irradiation (6,7) as well as the method which is based on the induced change in electric conductivity of the plastic (8) have been studied.

A comparative study between the electric conductivity of cellulose acetate with both track scanning and optical density methods have been carried out.

Experiment and Irradiation Facility:

Cellulose acetate samples used in the present work have the thickness 80 micron were supplied by VEB-Wolfen fabric. Lithium has been used as external radiator for slow neutron detection, while for fast neutron detection the external radiators were aluminium, copper, teflon, graphite and gold as well as uranium nitrate. The external radiator was sandwiched by plastic samples. External radiators of uniform thickness, greater than the alpha-particle range in its materials, were chosen to insure that all alpha-particles produced will be registered. To prevent any scattering of alpha-particles or decrease in its energy, cellulose acetate was tightly packed together with the external radiator.

ET-RR-1 (2 MW research reactor), was used for slow neutron detection. Samples were irradiated in a vertical channel over the thermal column at various distances.

For fast neutron detection, the Ra-Be source was embedded in a paraffin block of dimensions $100 \times 100 \times 90$ cm. The samples were exposed to neutrons at a fixed position 2 cm away from the source.

After irradiation plastics were etched for different time ranging from 15-240 minutes in a 28% KoH solution at 60°C. An ordinary optical microscope was used to count the number of etch pit as well as a micro-densitimeter for measuring the optical density. The electric conductivity of the sample was measured before etching since it is affected by the temperature and the etching solution (8).

A correction for the back-ground was necessary especially in case of uranium nitrate as external radiator. This was done by setting an unirradiated sample in contact with uranium nitrate for the same time of irradiation of the other sample. The unirradiated sample was etched under the same conditions and a considerable number of back-ground etchpits were observed. For other external radiators a plastic sample was also etched under the same condition, to determine the back ground. A similar correction using the unirradiated plastic sample was also necessary in case of both optical density and electrical conductivity measurements.

Results and Discussion:

The variation of the number of etchpits per unit area with etching time for cellulose acetate with and without external radiators, irradiated by fast and thermal neutrons, is shown in Fig. 1 (a, b). It is clear that the external increase the number of tracks registered. The increase in the sensitivity of the detector is due to its proportionality with the (n,α) reaction cross section.

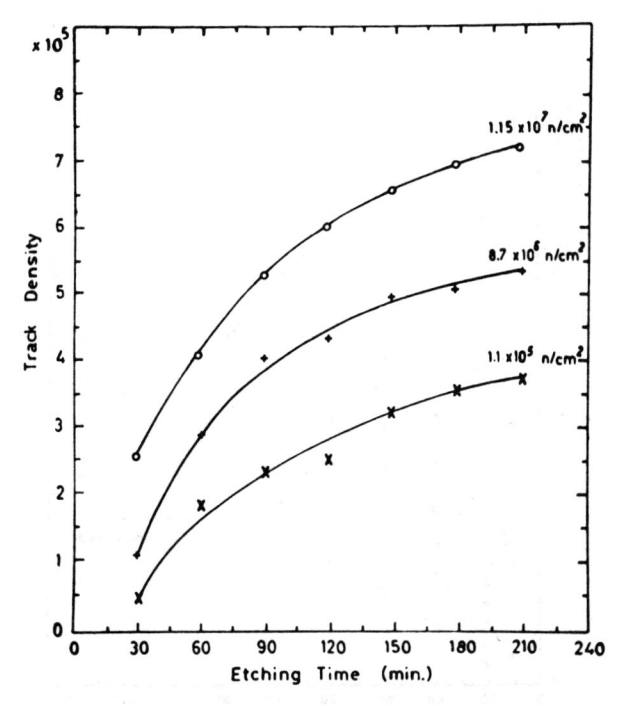

Fig. 1a. Relation between track density and etching time for thermal neutron using C.A. with Lithium as external radiator.

The number of etch pits per unit area was found to increase linearly especially in the early period of etching. But beyond certain limits in etching time, the variation starts to decrease. This could be attributed to the fact that when plastics are irradiated with fast neutrons the radiation damage sustained is not only on the surface but also in the deep layer. As the surface layers are removed by etching, the interior damages are removed etch pits which add to the number of etch pits. It could be seen that aluminium as external radiator is most sensitive compared to the other radiators, but its sensitivity is less than that when uranium is used. Fig. 2 shows the change of the diameter of etch pits with etching time for different external radiators. It is clear that the diameter due to fission fragments is much more greater than any other. Furthermore the diameter due to alpha particles from different external radiators vary according to the energy of the produced α-particle. In Fig. 3 A and B - the optical density is given as a function of etching time for both fast and thermal neutrons respectively. The optical density changes linearly with time to a certain limit of about 150 min. The nonlinearity above this might

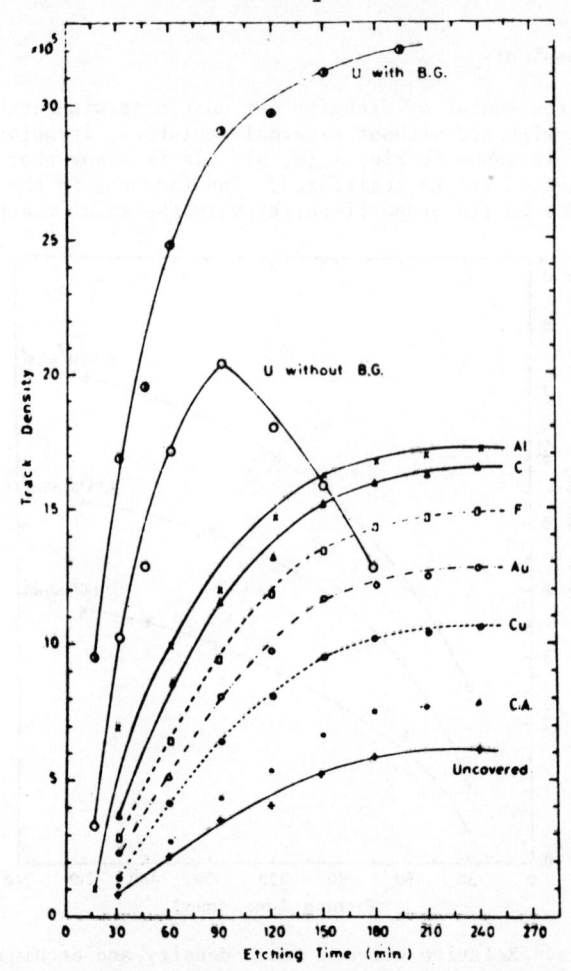

Fig. 1b. Relation between track density and etching time for fast neutron using C.A. with different external radiators (fluence 10^9 n/cm^2).

be due to track overlapping. A comparative relation between optical density and track density at different etching times is drawn in Fig. 4 for samples irradiated by 10^9 n/cm^2 and different external radiators. A flattening could be seen at high track densities. This might be attributed to the difficulty in counting the tracks due to surface background roughening as to make it difficult to avoid counting loss. The nonlinearly above this time might be due to track overlapping.

Fig. 5 shows the relation between the neutron track density as a function of neutron fluence, in case of using aluminium as external radiator with etching conditions (3 hr. time in 28% KoH solution at 60°C). This linear relationship between the track density and the neutron fluence can be used as a calibration curve for neutron fluence ranging from 10^7 to 10^{11} n/cm^2. Below this range one may not recommend using the method of track counting for neutron detection

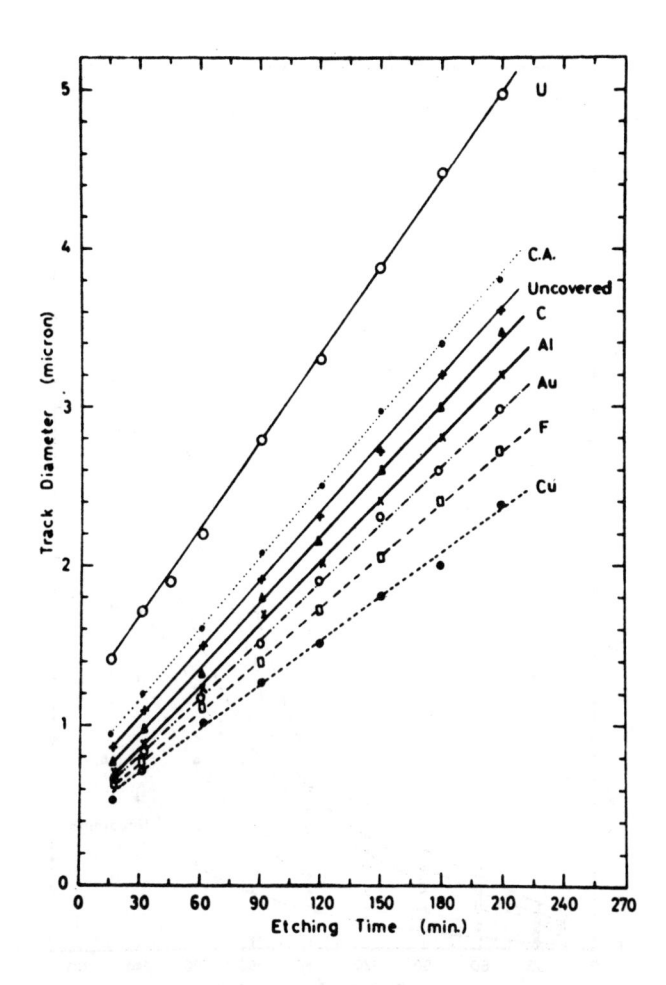

Fig. 2. Relation between track diameter and etching time.

especially when cellulose acetate is used. This disadvantage is due to the low sensitivity, although aluminium has been used as external radiator. Above this range the track density measurement is lower than the actual value. This is due to an overlapping effect of the tracks and roughening of the back-ground as well.

Similar calibration curve has been presented under the same condition between the optical density and neutron fluence as shown in Fig. 6. It is seen that the linearity is extended over a wide neutron fluence range from 10^8 up to 10^{13} n/cm^2. So one might recommend the use of optical density measurements for neutron fluence in the range from 10^8 - 10^{13} n/cm^2.

The bulk electrical conductivity of cellulose acetate as a reference as well as, of cellulose acetate sample with aluminium as external radiator exposed to different neutron fluences have been measured. The results are represented in both figures 7 and 8.

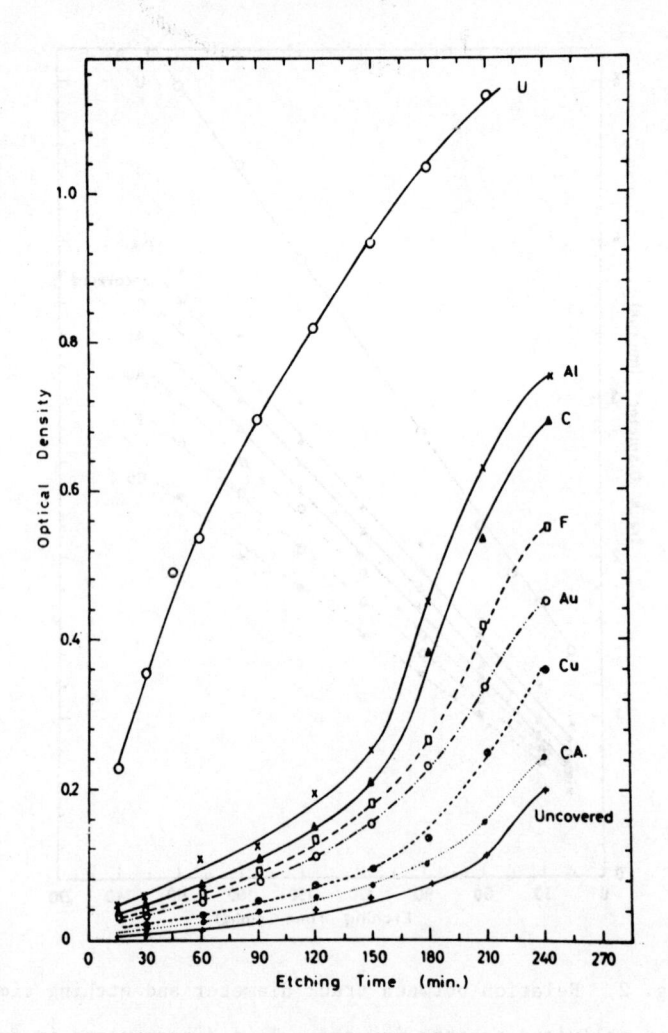

Fig. 3A. Relation between optical density and etching time for fast neutron using C.A. covered with different external radiators.

The data shows that the use of aluminium as a radiator increases the damage subtended by neutrons, and also increases the conductivity of the materials under considerations.

Conclusion:

It might be concluded that aluminium as external radiator has the higher sensitivity in comparison with the fissionable materials as external radiator in case of fast neutrons. Consequently, one can replace the use of α-particle registration detector instead of fissionable fragment detector, moreover, the

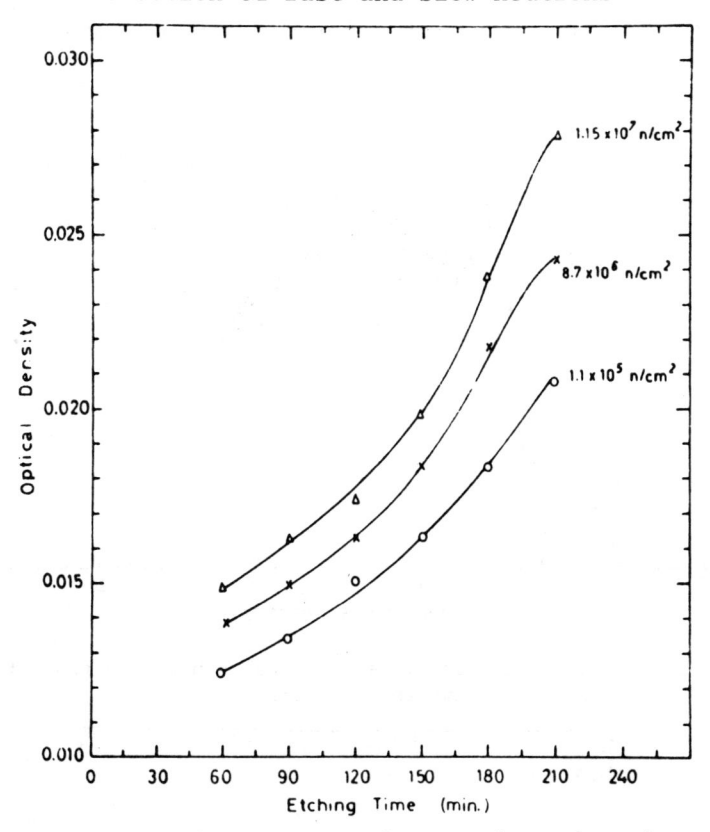

Fig. 3B. Relation between optical density and etching time for thermal neutron using C.A. and Lithium.

change in electric conductivity with neutron fluence could be used over a wide range in neutron dosimetry, particularly at low fluence, while the track scanning and optical density measurements could be used at high neutron fluence up to 10^{13} n/cm², respectively.

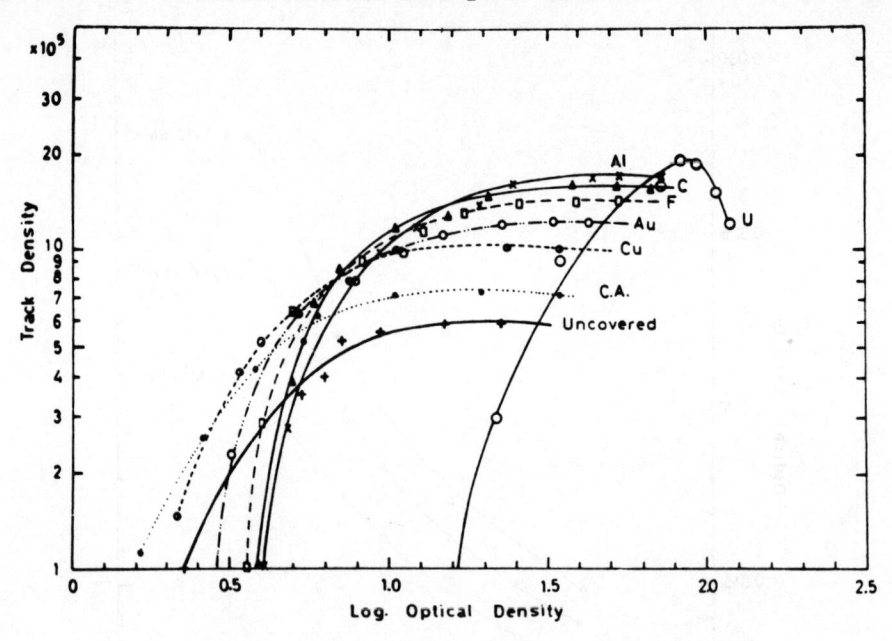

Fig. 4. Track density versus optical density in case of different external radiators.

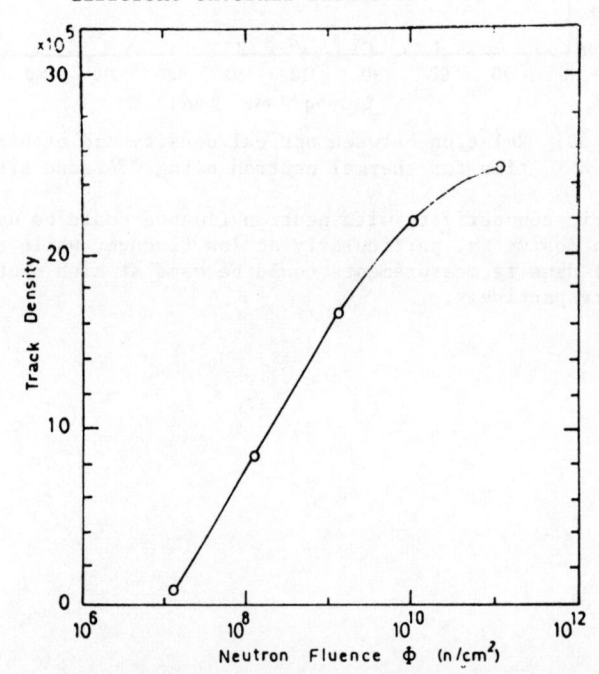

Fig. 5. The relation between the track density and the fast neutron fluence, aluminium as external radiator.

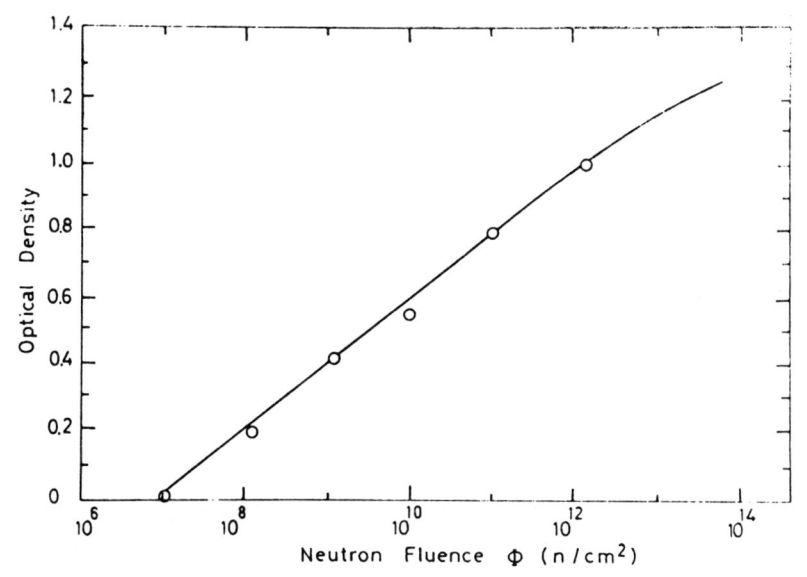

Fig. 6. The optical density versus neutron fluence, aluminium as external radiator.

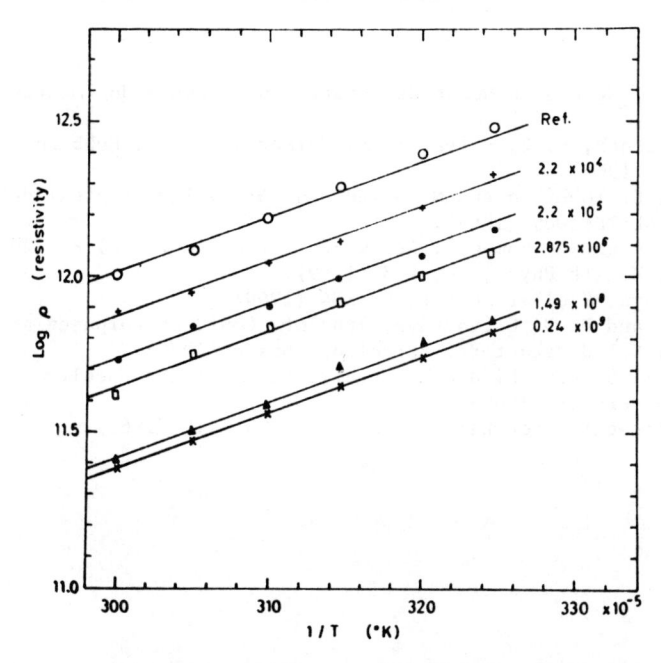

Fig. 7. Relation between logarithm of the resistivity and the reciprocal of temperature for C.A. with external radiator aluminium exposed to different fluence.

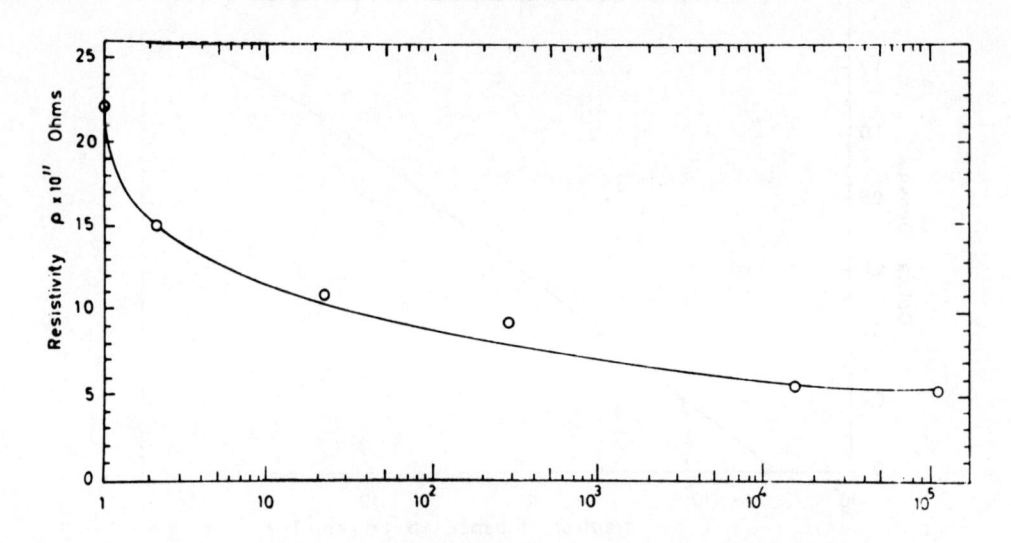

Fig. 8. The Resistivity (ohm) versus the neutron fluence for aluminium as external radiator at (t = 4pC).

References:

(1) Kerntechnik, Neuere Methoden der Neutronen personen dosimetrie, 7, 253 (1965).
(2) R. L. Fleischer, P. B. Price, R. M. Walker and E. L. Hubbard, Phys. Rev., 133A, 1443 (1964).
(3) S. Pretre, E. Tochilin and N. Goldstein, Naval Radiological defence, Report UNSDL-TR-1089 (1966).
(4) L. Medveczky and G. Somogyi, Atomki-Bulletin (Debrecen) 8, 226 (1966).
(5) K. Becker, Health Phys., 16, 113 (1969).
(6) J. W. N. Tuyn, Kjeller Report, Kr-134 (1969).
(7) G. Somogyi and D. S. Srivastava, 7th Int. Conf. on corpuscular Photograph and Visual Solid detectors, Barcelona, Spain (1970).
(8) M. A. Fadel, S. El-Fiki and M. Kenary, Bull. of the Faculty of Science of Baghdad University (1969).
(9) Tablice Wartosci Q reackji Jasdroych, Nr. 4, 110 (1963).

APPLICATION OF SOLID STATE NUCLEAR TRACK DETECTORS FOR PERSONNEL MONITORING AROUND HIGH ENERGY ACCELERATORS

J. Dutrannois and J. W. N. Tuyn

CERN, Geneva, Switzerland

ABSTRACT

In the framework of the design of a better system for personnel neutron monitoring, an extensive systematic study has been carried out to evaluate the feasibility of different nuclear track detector-radiator combinations.

This paper is concentrated on the use of LR115 cellulose nitrate and Makrofol polycarbonate foil with different etching and counting techniques. For LR115, classical etching was applied and automatic track counting and analysis were performed with a Quantimet. For the Makrofol, in addition to the standard spark counting method, electrochemical etching was used. Because of the wide neutron energy spectrum to be covered it was necessary to use a selection of radiators. It appears that an empirical formula for the neutron dose can be derived from the readings of the track density under the chosen radiators consisting of plastic, boron-loaded material, and bismuth. A comparison of the results obtained with this method and the dose recorded by standard radiation survey instrumentation has been performed.

INTRODUCTION

In the radiation environment of high energy proton accelerators the problem of personnel neutron monitoring is impossible to solve satisfactorily by the use of either nuclear emulsions or simple albedo dosimeters because of the strong variation of the neutron spectrum in time and space, the more so since the neutron spectrum covers an extremely wide energy range. An approach to solve this problem at CERN by the use of solid state nuclear track detectors in conjunction with different radiators has been reported elsewhere (1,2) whereby for example ^{6}LiF thermoluminescence detectors (TLD) were used as TL detectors as well as radiators for cellulose nitrate. The present paper describes the results of a further study of the feasibility of different nuclear track detector-radiator combinations for personnel neutron monitoring. Basically two different detector materials were used: Makrofol polycarbonate and LR115 cellulose nitrate foil with different etching and counting techniques. The wide neutron energy spectra to be covered necessitated the use of at least two different radiators for an acceptable neutron dose estimation, as will be shown by results of field tests obtained at different realistic positions around the CERN high energy proton accelerators compared with the standard radiation survey instrumentation used at CERN. Results obtained with fissile radiators are included for comparison only, since it is felt that such radiators owing to their radiotoxicity are not recommended for personnel dosimetry but should be reserved for example for neutron fluence measurements.

EXPERIMENTAL PROCEDURES

The series of investigations were concentrated on two different detector materials:

- LR115 foil consisting of an 8 μm thick layer of cellulose nitrate incorporating a red dye attached to a transparent polyester backing.
- Makrofol E polycarbonate foil in two different thicknesses: 10 μm for use with the spark counting technique developed by Cross and Tommasino (3) and 500 μm for use with the technique of electrochemical etching developed by Tommasino (4) and described in detail in another paper at this conference (5).

The LR115 foil was used with three different radiators: an 8 μm layer of natural boron on glass backing for detection of low energy neutrons through the $^{10}B(n,\alpha)^{7}Li$ reaction, polythene for detection of fast neutrons mainly through C recoils arising from neutron interactions in radiator as well as detector foil, and finally Bi for additional contributions of reaction products from spallation and fission reactions induced by high energy particles. The LR115 foils were etched for 90 minutes in a 2.5 N NaOH solution at 60°C, while the induced tracks which appear under a microscope as light spots on a dark ground were counted by using a Quantimet, an image analysing computer, which not only counts the track density but evaluates the hole size distribution as well. This is shown in Figs. 1 and 2, where the size distribution as a function of hole diameter is given for the areas under the different radiators. As can be seen, the reaction products of the $^{10}B(n,\alpha)^{7}Li$ reaction cause a significantly different distribution compared to the two other radiators, where holes caused by heavier particles are dominating. For routine use all holes exceeding a threshold diameter of 7.4 µm are counted to discrimate against unwanted background. A total surface of 0.2 cm^2 is scanned routinely.

The spark counting technique was used for the 10 µm Makrofol foils in contact either with Makrofol or with Th and Bi foils as fissile radiators. In the first case neutron induced C and O recoils from the radiator or the detector foil itself are detected by etching the foils for 72 hours in a 6.25 N KOH solution at room temperature, after which only 2 µm of the foil is left, necessary because of the short range of the recoils. For fissile radiators 22 hours of etching is sufficient. The tracks were after etching enlarged at 900 V and counted at 550 V. Typical recoil induced sparks versus voltage graphs are shown in Fig. 3.

Finally the electrochemical etching technique with 500 µm thick Makrofol foil was used according to the optimal conditions mentioned elsewhere (5): 2.8 kV at 3 kHz in a 1:1 mixture of 25% KOH and C_2H_5OH at room temperature for 3 hours.

RESULTS AND DISCUSSION

Irradiations of foils were performed at different locations outside the shielding of the CERN 28 GeV proton synchrotron at which the neutron spectrum was known to vary strongly. The results obtained were compared with the dose equivalent obtained from the standard CERN survey instrumentation consisting of a Rem Ion Chamber (RIC) for neutrons up to ~ 20 MeV and ^{11}C activation detectors for the detection of hadrons above 20 MeV. The RIC moderator was used as a phantom as well. The ratio of the dose equivalent measured by these two devices is a sensitive neutron spectrum index. In a previous study an empirical relationship for the neutron dose equivalent derived from the hole densities in LR115 behind a ^{6}LiF teflon thermoluminescent detector disc and a polythene cover was found to give an estimation within ~ 20% of the neutron dose (6). For practical reasons the new CERN proposed personnel monitor uses the ^{6}LiF TL detector only as a neutron indicator and separately boron as a radiator. Taking into account the higher hole density under the boron radiator the revised relationship becomes:

$$H \text{ (mrem)} = 5.2 \left(\frac{\text{holes/cm}^2 \text{ under B}}{600} + \text{holes/cm}^2 \text{ under polythene}\right) \quad (1)$$

The hole density under the Bi radiator has not been included in eq. (1) in spite of the higher hole density compared to polythene because of its higher background due to the presence of natural alpha emitters in Bi.

The results obtained for three locations with the detectors attached to front (F) and back (B) of the moderator in strongly varying neutron spectra as can be seen from the $H_{HEP}/H_{neutron}$ ratio are presented in table I. Relation (1) approximates the dose for the hard spectrum with the accuracy as found in the past. For the medium and soft spectra an overestimation is made, partly owing to the statistical error in the number of holes under the polythene, partly to the high and locally strongly varying contribution of low energy neutrons.

The neutron sensitivity of Makrofol after electrochemical etching is somewhat lower for fast neutrons, while its sensitivity for low energy neutrons underneath the boron radiator is higher than LR115. Therefore, already at low doses, overlapping of tracks is a problem since the tracks are enlarged considerably by electrochemical etching, as can be seen from the size distribution in Fig. 4, where the distribution for foils with and without overlapping tracks are shown. The maximum in the distribution is found around a diameter of 0.5 mm, which is considerably larger than for LR115 (see Figs. 1 and 2) with 0.02 mm. Therefore scanning of Makrofol can be done at a lower magnification than LR115. Background elimination in LR115 is however simpler with the Quantimet than in Makrofol. In the latter, background is reduced by electrochemical and conventional etching before irradiation (5) and the distinction between background and newly induced tracks still requires some human judgement. For both methods background subtraction is the main limitation. We found that LR115 background varies from one badge to the next and even storage at different temperatures may considerably influence the background. For the LR115 samples used in this study a background of 56 ± 24 holes/cm^2 was found. The pre-etched background of Makrofol varied between 50 and 120 tracks/cm^2 for the foils used. The background to be subtracted after pre-etching is much lower, ~ 15 tracks/cm^2. The lower background combined with the lower magnification and consequently larger surface scanned is reflected in the higher statistical accuracy for Makrofol compared to LR115 in Table I.

The use of the spark counting method was tested under different conditions with PuBe neutrons, neutrons produced by 600 MeV protons on a Be target, and outside the shielding of the PS West Hall. The results are given in table II in comparison with some measurements with 500 μm thick Makrofol. A few conclusions can be drawn: the directional dependence with sparks due to recoil particles is much stronger than with the electrochemically etched Makrofol for PuBe neutrons, and the neutron energy dependence is strong for the spark counting method regardless of whether Th or Makrofol is used as radiator. It has been shown in the past that the Bi/Th ratio could be used as a spectrum index to estimate the sparks/rem for Th covered foils (2). However, the sensitivity of Bi is low. In spite of its stronger energy and directional dependence, Makrofol without fissile radiators should be preferred for personnel neutron dosimetry to avoid the spread of radioactive material. Unfortunately the fact that for recoil-induced sparks the foils have to be etched down to a thickness of 2 μm, close to electrical breakdown, makes the method extremely sensitive to variations in etching conditions or foil thickness. The sensitivity of the method is sufficient for personnel neutron monitoring around accelerators since a surface of 20 cm^2 is easily sparkable and would produce ~ 800 sparks/rem in addition to ~ 10 sparks of background. Another application was found in measuring the depth distribution of high LET particles in a polythene absorber in front of the 600 MeV beam. The preliminary result is shown in Fig. 5 compared with the depth-dose distribution measured with a tissue-equivalent ionization chamber. The interpretation of this distribution might be difficult because of its possible dependence on etching conditions as shown by Becker et al. (7).

CONCLUSIONS

The results obtained with the three different techniques described in this paper lead to the following conclusions.

The spark counting method for recoils has sufficient sensitivity and an attractive lower detection limit for personnel neutron monitoring around high energy proton accelerators because of its low background. The strong energy and directional dependence combined with the possible errors in the evaluation procedure are reasons to prefer the use of LR115 or Makrofol electrochemically etched. The final choice for use at CERN to replace the nuclear emulsion for personnel neutron monitoring will strongly depend upon the way in which the background of the foils can be further reduced by better preparation techniques for LR115 or an improved pre-etching procedure for Makrofol. The routine large scale electrochemical etching of Makrofol may create additional practical problems.

ACKNOWLEDGEMENTS

The contributions of G.M. Hassib, S. Hertzman and M. Höfert to this study is gratefully acknowledged.

REFERENCES

1. B.J. Tymons, J.W.N. Tuyn and J. Baarli, Proc. Symp. Neutron Monitoring for Protection Purposes, IAEA, Vienna, 1972, Vol. II, p. 63.

2. B.J. Tymons, J. Dutrannois and J.W.N. Tuyn, Working Party on Space Biophysics, Geneva, 1973, CERN Report DI/HP/168.

3. N.G. Cross and L. Tommasino, Proc. Int. Topical Conf. on Nuclear Track Registration in Insulating Solids and Applications, 1969, Vol. 1, p. 73.

4. L. Tommasino, CNEN Report RT/PROT/71)1 (1970).

5. G.M. Hassib, J.W.N. Tuyn and J. Dutrannois, On the electrochemical etching of neutron induced tracks in plastics and its application to personnel neutron dosimetry, this conference.

6. B.J. Tymons and J.W.N. Tuyn, Personnel neutron dosimetry by means of cellulose nitrate film combined with LiF as both radiator and TLD, Health Physics (in press).

7. K. Becker and M. Abd-el-Raseh, ORNL-TM-4460 (1974).

Table I

Comparison between LR115 and Makrofol (electrochemically etched) in different neutron spectra

Exposure condition	Radiation survey result mrem		LR115 tracks cm^{-2} rem^{-1}			H_{LR115} mrem	Makrofol E tracks cm^{-2} rem^{-1}	
			Boron	Radiator CH$_2$	Bismuth		Boron	Radiator Makrofol
PS East Hall top shield	$H_n = 1066$ $H_{HEP} = 753$ $H_{n+HEP} = 1819$ $\frac{H_{HEP}}{H_n} = 0.706$	F	2.72×10^4	192 ± 28	242 ± 30	2276	3.70×10^4	129 ± 12
		B	1.25×10^4	133 ± 25	340 ± 31	1473	2.32×10^4	142 ± 14
PS East Hall entrance door	$H_n = 402$ $H_{HEP} = 29$ $H_{n+HEP} = 431$ $\frac{H_{HEP}}{H_n} = 0.06$	F	1.27×10^5	134 ± 68	374 ± 77	816	2.09×10^5	90 ± 19
		B	2.13×10^4	330 ± 71	259 ± 73	822	6.95×10^4	74 ± 19
PS Linac	$H_n = 492$ $H_{HEP} = 0$ $\frac{H_{HEP}}{H_n} = 0$	F	3.69×10^4	170 ± 82	339 ± 86	756		
		B	4.22×10^4	236 ± 84	536 ± 92	783		

Table II

Comparison of neutron sensitivity and directional dependence of various detectors

Exposure condition	Makrofol (electrochemically etched) tracks cm^{-2} rem^{-1}	Makrofol (spark counting) sparks cm^{-2} rem^{-1}		
		Makrofol	Radiator Thorium	Bismuth
PuBe neutrons:				
Normal incidence	50	4.2	18	
45° incidence	30	2.1		
Parallel incidence	15	0.42		
PS West Hall $\frac{H_{HEP}}{H_n} = 0.56$		43	73	5.5
SC 600 MeV neutron beam		10	109	11

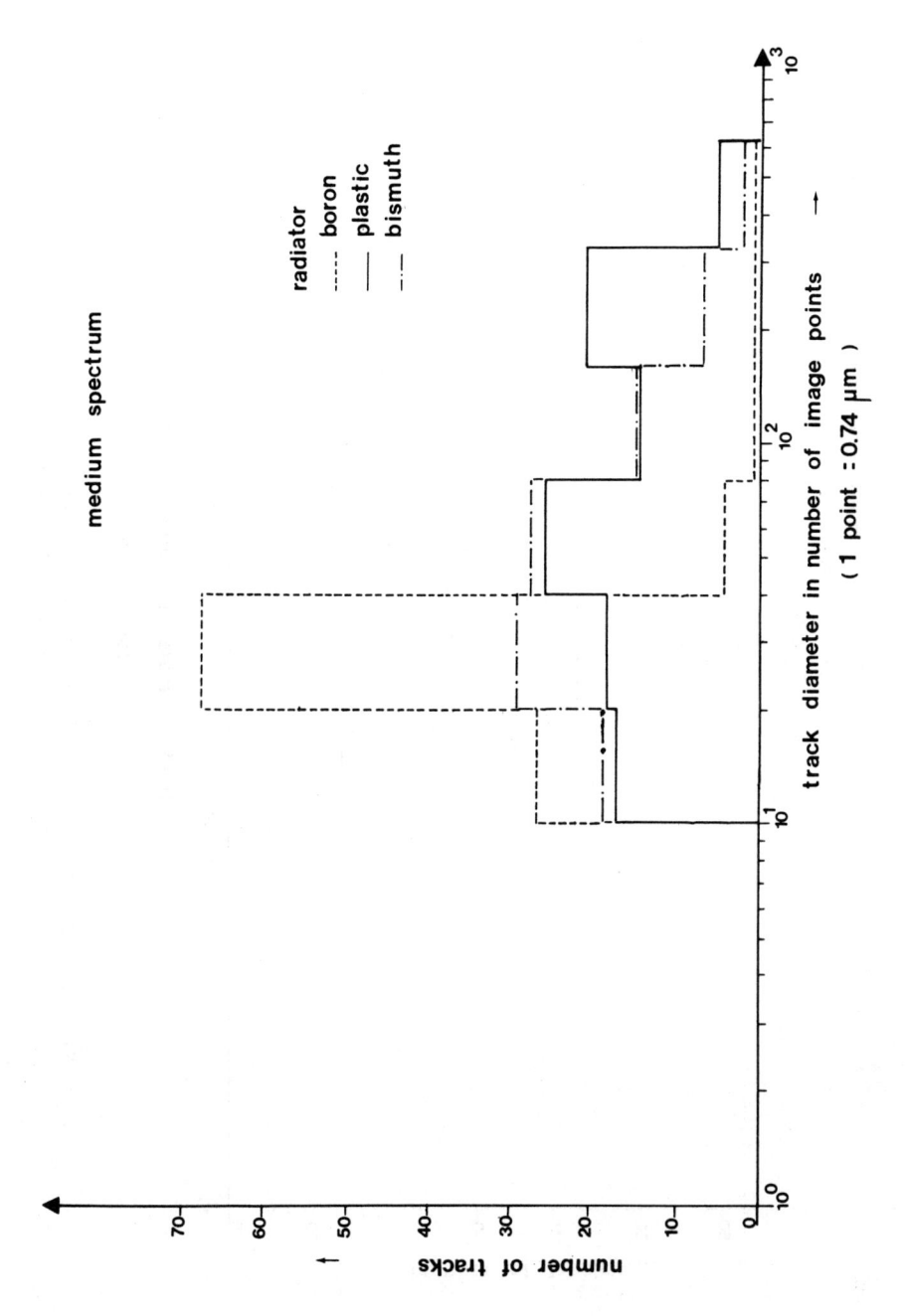

Fig. 1. Track diameter distribution in LR115 behind various radiators for a "medium" neutron spectrum.

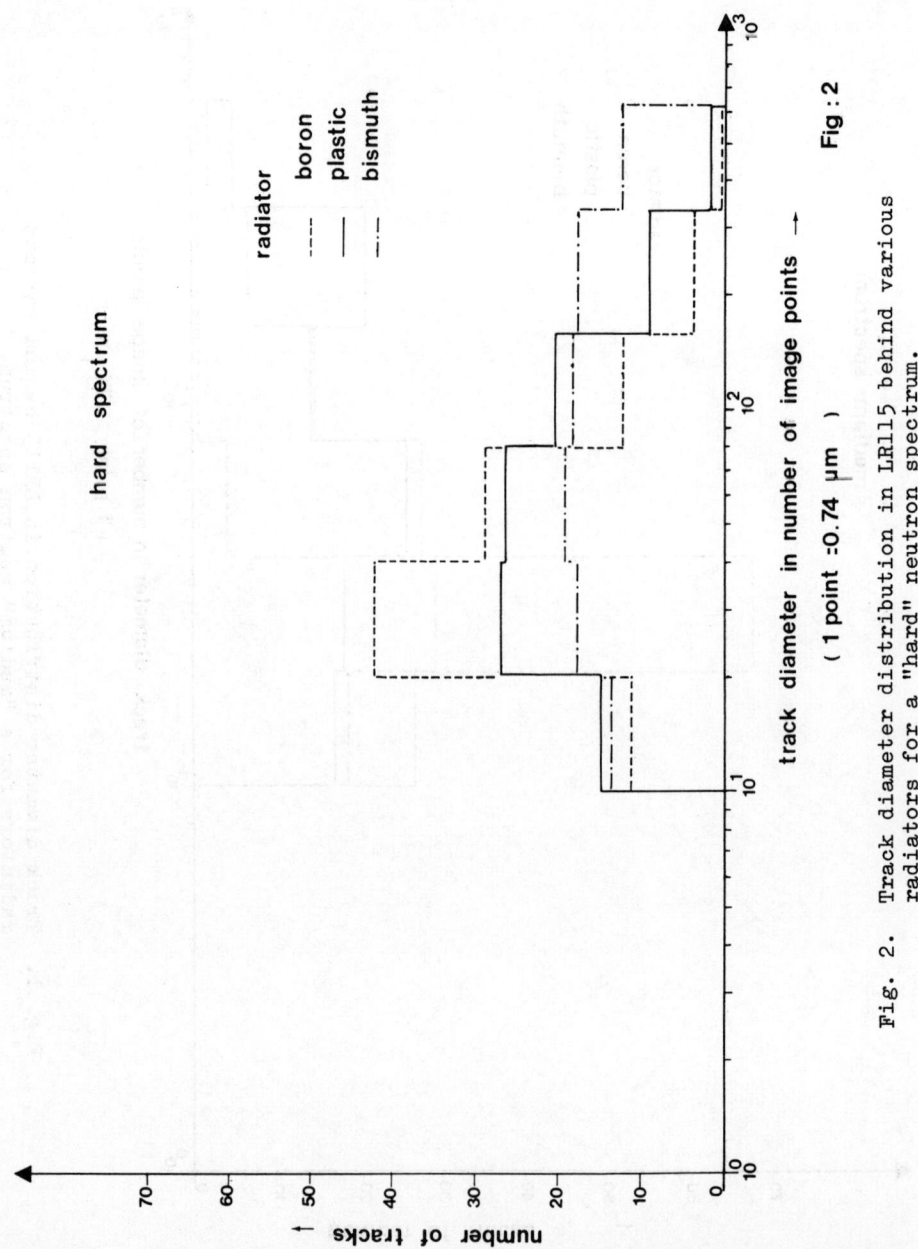

Fig. 2. Track diameter distribution in LR115 behind various radiators for a "hard" neutron spectrum.

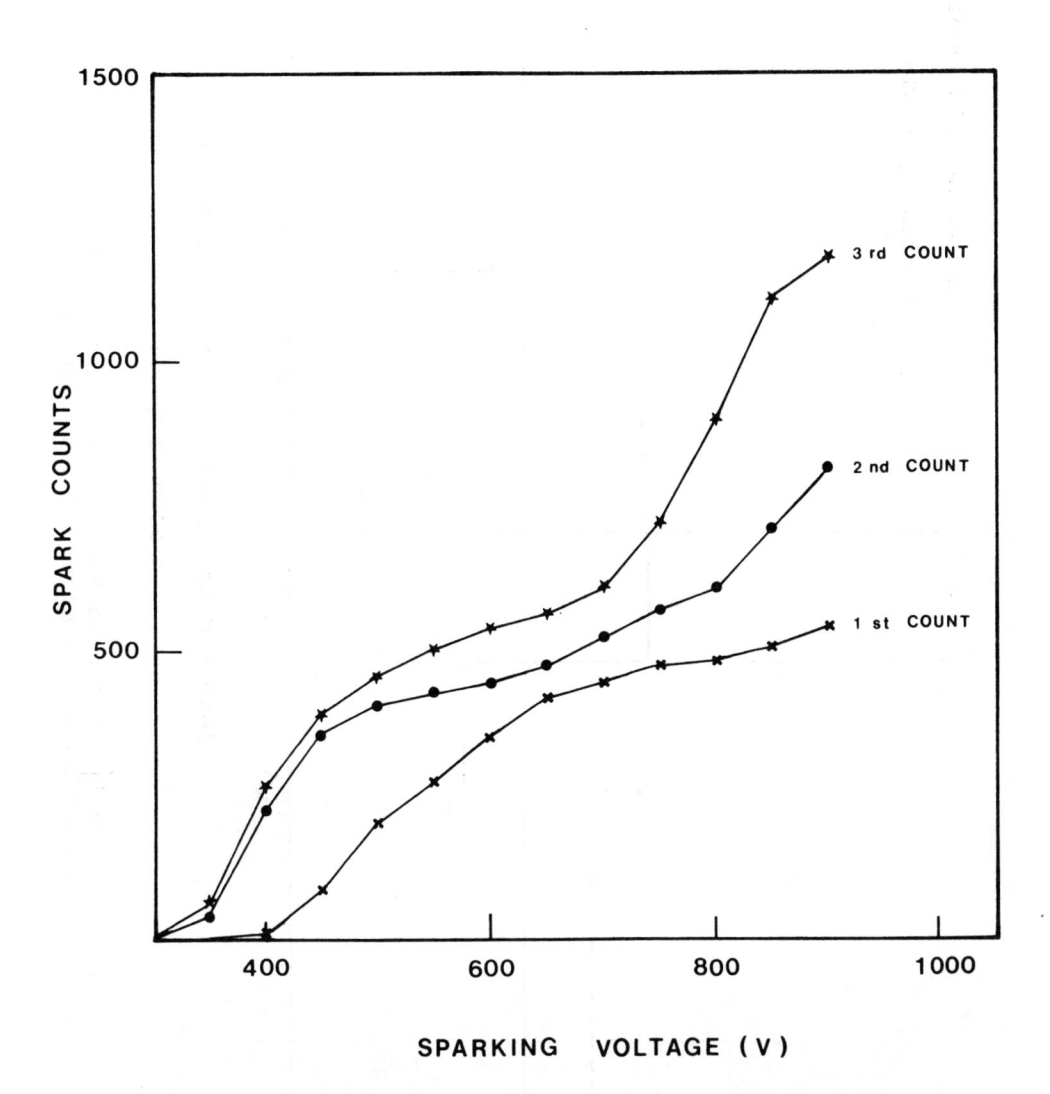

Fig. 3. Spark counts as a function of sparking voltage after 72 hours etching.

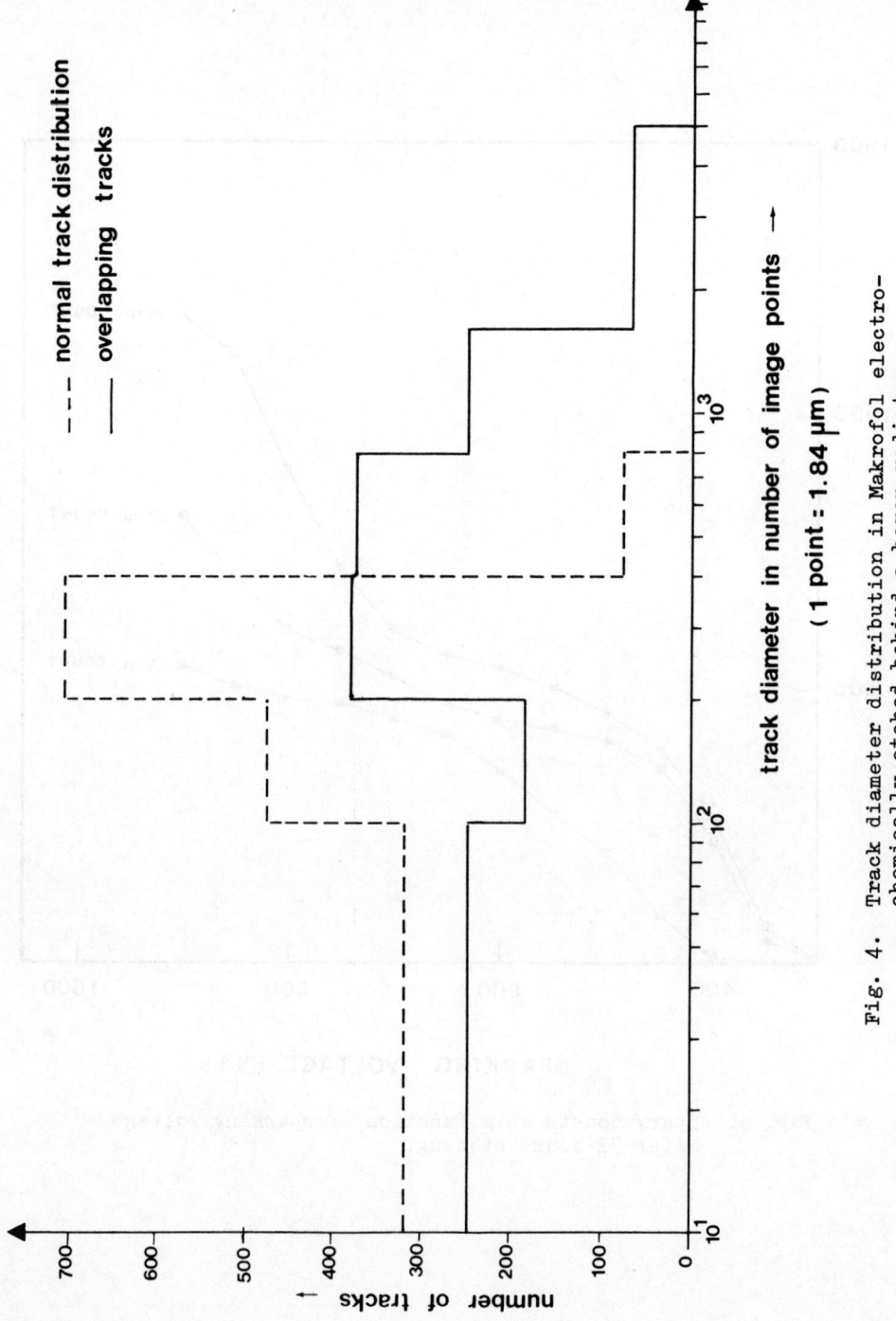

Fig. 4. Track diameter distribution in Makrofol electro-chemically etched behind a boron radiator.

Fig. 5. Depth-dose distribution in a 600 MeV neutron beam compared with a recoil spark distribution.

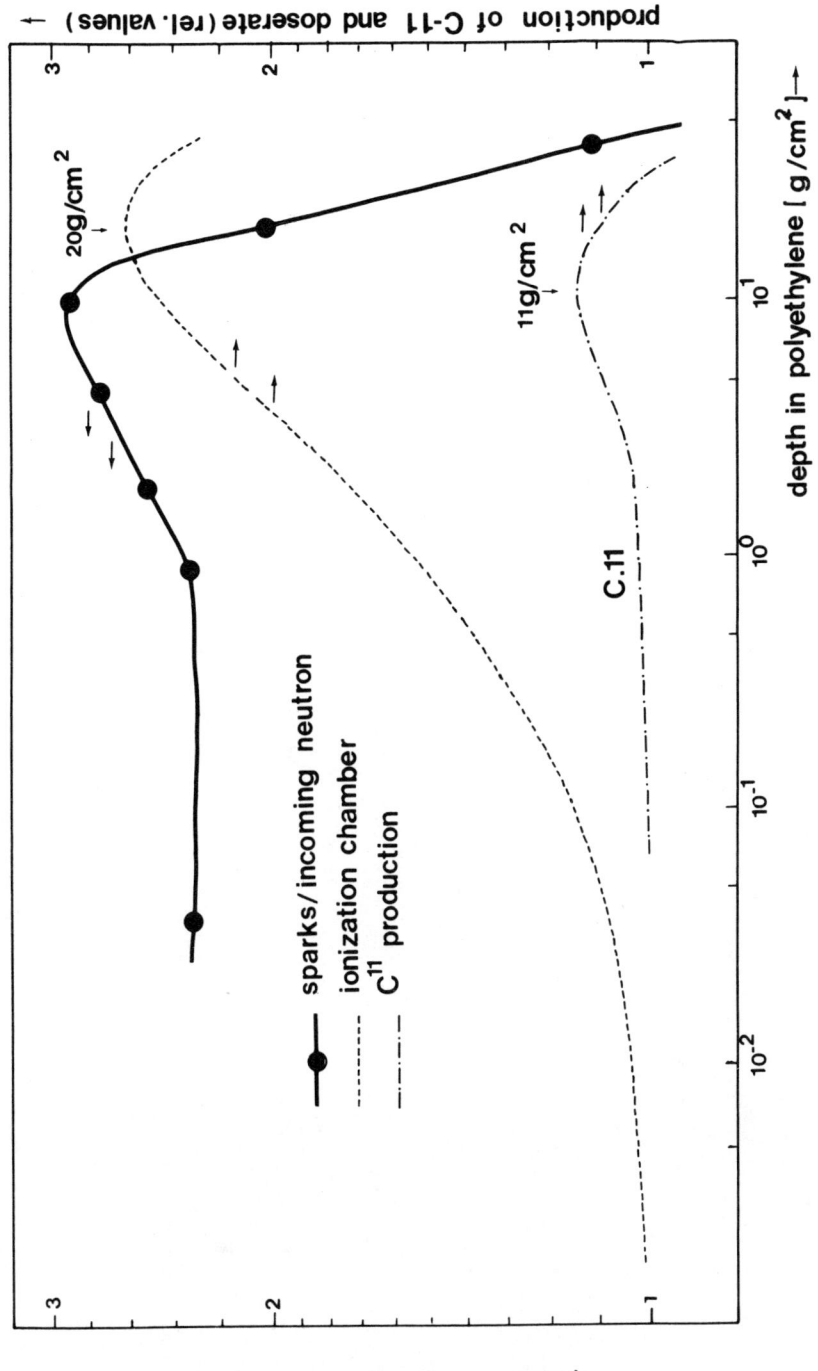

PERSONAL DOSIMETER FOR HIGH ENERGY CORPUSCULAR RADIATION

D. Hasegan, A. Dragu, M. Nicolae and A. Apostol

Institute for Atomic Physics, Bucharest, Romania

Abstract

The personal dosimeter for high energy corpuscular radiation presented in this paper consist from a plastic badge of rectangular form with combined dosimetric system fixed in. For protons, α particles and heavy ions is using as detector a combined stack of cellulose nitrate, cellulose acetate and nuclear emulsion, sheets of 200 μm thick and 5 × 5 cm large.

For neutrons of different energy, a solid state track detector spectrometric system is used. It contains four elements of circular form with ∅ = 2 cm, each of them consisting from a n-α target and a cellulose nitrate foil as detector of the α-particles generated by the neutrons in the target. Using a method of model spectra and the density of nuclear tracks recorded in each detector foil, the approximative spectra and the neutron dose can be determined. The gamma ray dose is measured by two small containers of thermoluminescent LiF powder.

Some experimental data obtained by this kind of dosimeter exposed on supersonic aircrafts at the altitude of 16 - 20 Km and some practical conclusions concerning their use in routine are presented.

1. Introduction

Supersonic Aircrafts operate at altitudes much higher than subsonic transport. Consequently, the intensity of cosmic radiation is higher and in broader charge and energy spectra. The cosmic rays has been studied by investigations with balloons, rockets and satellites and thus the basic features of cosmic radiation reaching the top of the earth atmosphere are well-known (1 - 10). Nevertheless, investigations have been carried out for a better knowledge of cosmic radiation at the SST level, especially from the dosimetric point of view (11 - 15). Although the results of such investigations have shown that no real danger is expected to arise for passengers, a personal monitoring for crew members should be used.

We present a personal dosimeter developed for this purpose at the Institute for Atomic Physics - Bucharest.

2. Personal dosimeter

The personal dosimeter to be used on supersonic aircrafts consists in a plastic holder with combined dosimetric systems fixed in:
- nuclear emulsion and plastics - for heavy ions and stars;
- a modified variant of the I.F.A. SDN-5 spectrodosimeter for neutrons;
- LIF thermoluminescent detectors - for gamma radiation.

Figure 1 represents schematically the dosimeter.

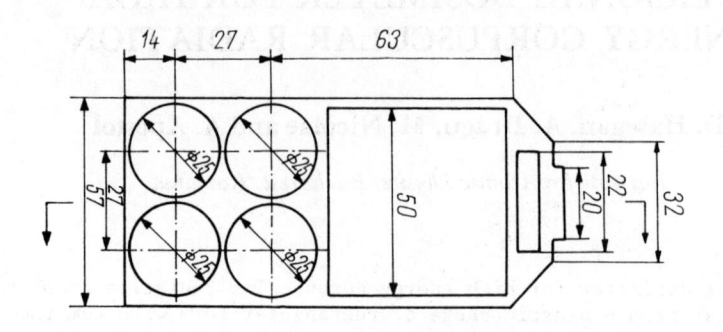

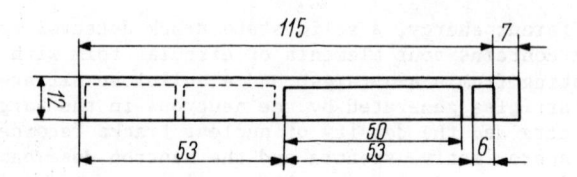

Fig. 1. The scheme of the personal dosimeter

2.1. Heavy ions

For the heavy ion detection a stack of six Kodak cellulose nitrate sheets, 30 × 35 × 0,2 mm³ each, is used. After the exposure of the personal dosimeter the plastic sheets are processed in the following etching conditions: NaOH 10% solution, 40°C and 12 hours. After the etching, in order to neutralize the natrium hydroxide remained inside the pits, the sheets are immersed into an acid bath for 5 min. Under these conditions the bulk etch rate is:

$$V_B = 0.014 \text{ μm/min.}$$

Considering that the energy loss spectrum is, in first approximation, enough for the dosimetric purposes, we used so far the cone diameter method (15, 16, 17), i.e. the minor and major - axis are measured, from which the etching rate, V_T, is computed. Using a calibration curve $V_T = V_T (dE/dx)$, one can obtain the energy loss spectrum.

Figure 2 shows the differential energy loss spectrum obtained between March 19 and July 4, 1974.

In order to estimate the contribution of heavy ions to the total dose, a weighted mean value of the absorbed dose is calculated. The results are presented in Table 1. For comparison we give the figure obtained by using the intensity of stopping nuclei as given in (18) and dE/dx values from (19).

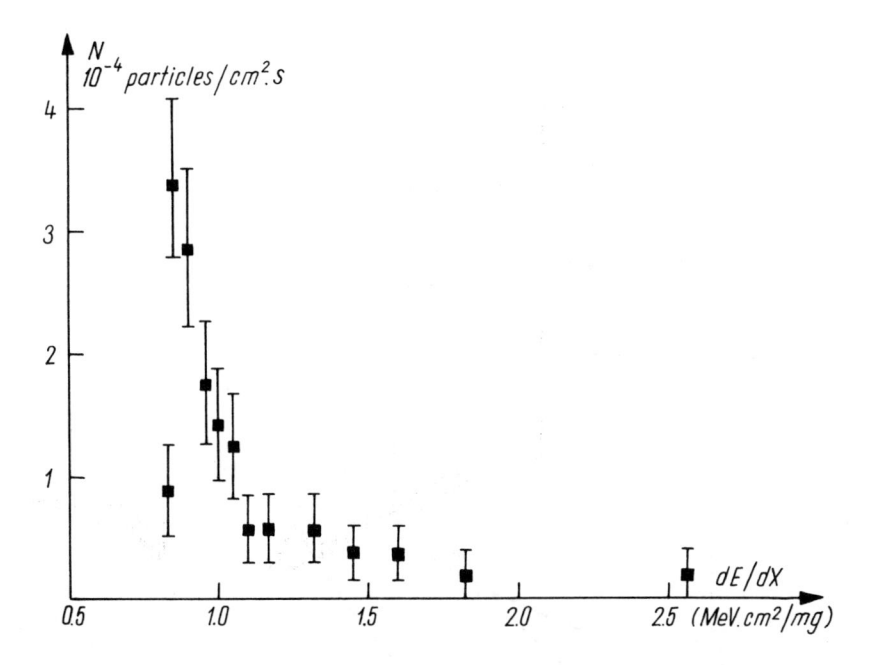

Fig. 2. The differential energy loss spectrum of the heavy ions.

Table 1

Absorbed dose due to the heavy ions

Altitude	Exposure		Dose	
(Km)	period	time	mrad	mrad/h
	March-July 1974	20	$1.67 \cdot 10^{-3}$	$8.35 \cdot 10^{-5}$
16	Oct. 1974-March 1975	2	$6.04 \cdot 10^{-4}$	$3.02 \cdot 10^{-4}$
30	Oct. 1974-March 1975	-	$4.9 \cdot 10^{-4}$	-
20	calculated (enders only)	-	-	$1.3 \cdot 10^{-5}$

For more precise radiobiological analyses, one has to take into account the characteristic features of heavy ions: a very high amount of energy deposited inside a very narrow cylinder around the trajectory as well as the spread of energy due to electrons, i.e. the spatial distribution of the energy deposited around the ion path, which can be calculated using different models (20 -23). A very suitable method for visualizing this distribution is photometrical measurements on heavy ion track in nuclear emulsion (24, 25). Usually these measurements were carried out on tracks of fast or very fast heavy ions. We performed an experiment in the Bragg peak region (26), namely on the last 100 µm of tracks of Ne and Ar. Measurement were carried out on a MPV Leitz Photometric Microscope with a slit of 9.66 × 2.53 µm object. Figure 3 shows the opacity variation as a function of the ion velocity. It also contains the

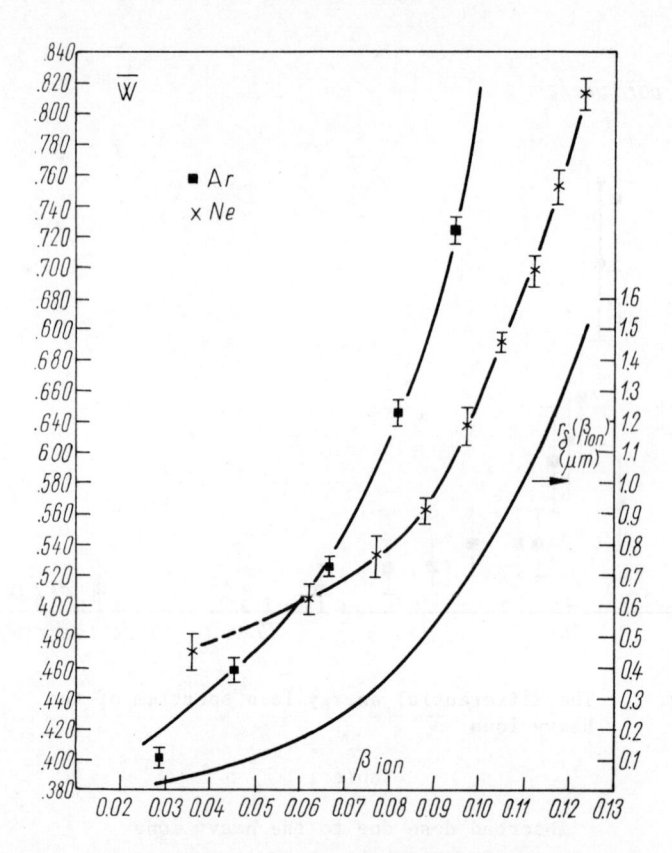

Fig. 3. The track opacity vs. ion velocity.

curve $r_\delta = r_\delta(\beta_{ion})$, r_δ being the characteristic thickness (20) of δ electrons. As can be seen, when the track width becomes greater than the diameter of one grain ($r_\delta > 0.11$ μm) the opacity of Ar tracks is higher than that of Ne tracks. Figure 4 shows the variation of the ratio W_{Ar}/W_{Ne} against the ratio of the square effective charge. There is a good agreement between these results and the expectation since the differential cross sections of δ electron production dn/dω is proportional with the square effective charge. Using Katz's theory we calculated the width of tracks as the double of the distance from the trajectory where the mean energy deposited in a grain equals the threshold of energy for a grain to become developable (for I.F.A. EN3 nuclear emulsion $E_{th} = 5.7 \times 10^5$ erg/cm^3). This was compared with the experimental track width, which equals the product between the slit width and the opacity. The result is shown in figure 5.

2.2. Disintegration stars

Physical characteristics of the stars produced by high energy particles on emulsion nuclei are rather well known (27,28). Yagoda et al. (24) have reported the utilisation of the disintegration stars originating in the gelatin matrix for measuring the tissue dose that would be produced in an equal volume of tissue replacing the emulsion. H. J. Schaefer et al. (29) used nuclear

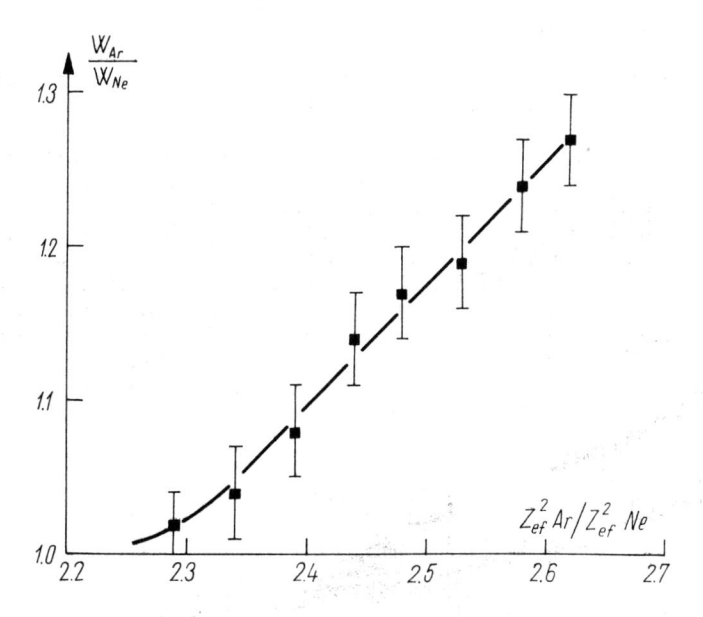

Fig. 4. The variation of W_{Ar}/W_{Ne} ratio with the ratio of the square effective charges.

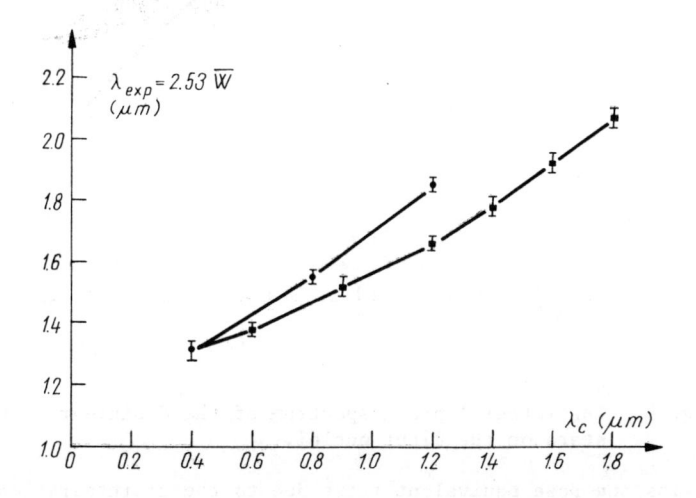

$\lambda_{exp} = 2.53\ \overline{W}$
(μm)

$\lambda_c\ (\mu m)$

Fig. 5. The comparison between the experimental and calculated track widths. Each pair of values is obtained at the same ion velocity.

emulsion for estimating the dose contribution from tissue disintegration stars to the total dose of Apollo XI mission.

Personal dosimeter presented here is provided with a stack of nuclear emulsion sheets for the estimation of the dose due to the star phenomenon.

From the dosimetric point of view the interest lies in the stars originating in the gelatin, i.e. the disintegration stars on the light nuclei. The estimated fraction of stars due to collisions with heavy nuclei is (28): $\alpha_H = 0.727 \pm 0.016$. Consequently from the total sample, those originating in the light nuclei from the nuclear emulsion is $N_1 = N_{tot}(1 - \alpha_H)$. Since the exposure is usually under uncontrolled temperature and humidity conditions, the fading is expected to be rather high. Therefore the stars are analysed according to the frequency distribution given in (28), as it was done in (15) for the dosimeter exposed in March - July 1974. Figure 6 represents the integral prong spectrum. As in (29, 30) a mean energy per star prong of (14 MeV and a mean QF of 6.5) were used.

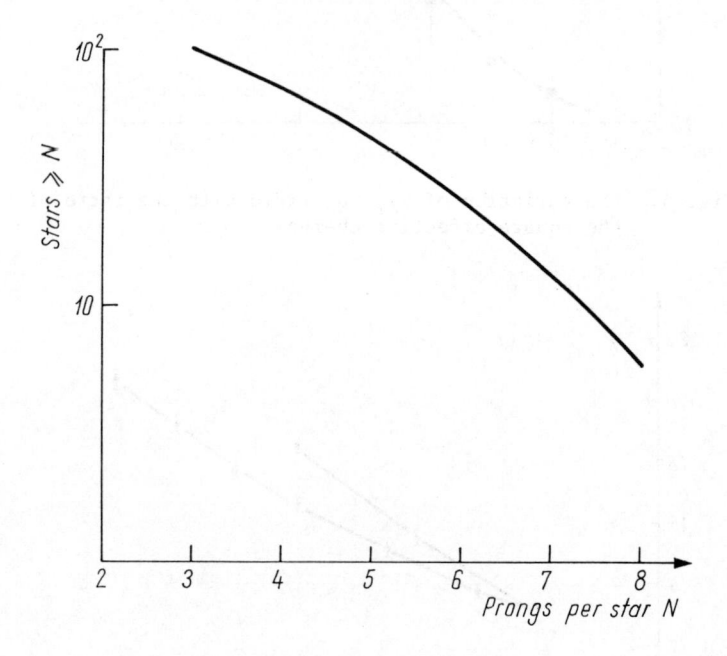

Fig. 6. The integral prong spectrum of the disintegration stars on the light nuclei.

Table 2 contains the dose equivalent rates due to the disintegration stars at SST level and in the cosmic space.

2.3. Neutrons

The neutrons encountered at the SST level have a rather wide energy spectrum. Therefore it is not enough to obtain the dose due to fast neutrons. The personal dosimeter uses a spectrometric system (32) which is able to estimate the neutron energy spectrum in the energy range between 0.1 eV up to 20 MeV Its basic phenomenon is the (n, α) reactions taking place in several unfission-

Table 2

Dose equivalent rate due to the disintegration stars

Experiment	Altitude (Km)	Star density ($g^{-1}.h^{-1}$)	Dose equivalent rate (mrem/h)
Davison (30)	20.4	-	0.35
Fuller and Clarke (31)	18.25-19.75	70	0.5
Present work	16	57	0.24
Apollo XI (29)	Cosmic space	139	0.48

able targets. Alpha particles are recorded in cellulose nitrate plastic detectors. After processing, their surface density under each target is measured by track counting. The energy spectrum is obtained by applying the model spectra method, namely the unknown spectrum is considered to be a linear combination of several given spectra. The coefficients of this combination have to be calculated by taking into account the experimental results, the atom density of the target materials and the cross sections of (n,α) reactions.

As targets we use Li, B, C and Al.

Table 3 contains the track density obtained by a personal dosimeter exposed at the SST level between March - July 1974. The witness was exposed in the same dosimeter without any target, therefore the track number found contains the α particles from (n, α) reactions on the elements from the cellulose nitrate detector, as well as the recoils inside the detector. Figure 7 represents the integral neutron spectrum. For calculating the dose due to neutrons the values of the factor η ($Rem/n/cm^2$) as given in (33) were used. The dose rate obtained is $D = (8.25^{+0.53}_{-0.22})$ mrem/h. Table 4 contains a comparison between this result and others given in literature.

Table 3

Neutron spectrodosimeter

Target	Track density (cm^{-2})	
	Uncorrected	Corrected
LiF	1494.2	686.8 ± 35.0
B_2O_3	1852.4	865.9 ± 28.1
Al	1743.2	811.3 ± 54.0
C	2867.7	1373.5 ± 50.8
Witness	120.6	-

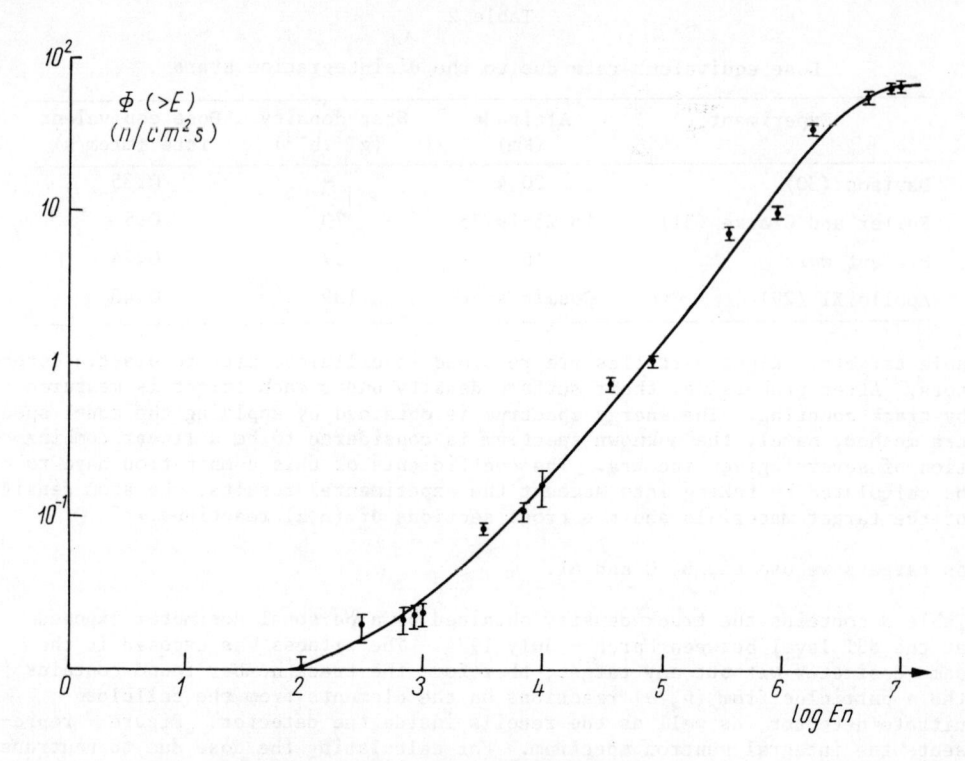

Fig. 7. The integral neutron spectrum.

Table 4

Equivalent dose rate due to the neutrons

Author	dose rate (mrem/h)
Davison (30)	0.4
Fuller and Clarke (31)	0.6
Kaiser (13)	0.44
Allkofer (34)	0.65 - 8.7
	100 - 300
Present work	8.25 $^{+\ 0.53}_{-\ 0.22}$

3. Conclusions

- The personal dosimeter presented is a compact cassette having dimensions of 122 × 57 × 12 mm.

- It is a complex dosimeter which is suitable for dosimetric monitoring of practically all kinds of radiations encountered at the SST level:
  - heavy ions by plastic sheet stack;
  - protons by nuclear emulsion;
  - neutrons by spectrodosimeter;
  - γ radiation by thermoluminescent dosimeters;
  - disintegration stars by nuclear emulsion.

- All dosimetric system contained inside the personal dosimeter are integrator and the fading is practically nonexistent, except for nuclear emulsion.

- The spectrodosimeter uses unfissionable elements as targets.

- Due to its small dimensions, the personal dosimeter can be easily worn on the flying suit, being changed at every two months.

References

1. M. M. Shappiro, R. Silberg, Ch. Tsao. Contribution to the George Ganow Memorial volume, ed. by F. Reines, April 1970.
2. M. M. Shappiro, R. Silberg, Ch. Tsao. Proc. Int. Conf. Cosmic Rays, Jaipur 1963.
3. M. M. Shappiro, R. Silberg, Ch. Tsao. Proc. Int. Conf. Cosmic Rays, London 1965.
4. W. R. Webber. Handbuck der Physik, Springer Verlag, vol. XLVI/2 1967, 181.
5. E. Schopper. Handbuck der Physik, Springer Verlag, vol. XLVI/2 1967, 372.
6. O. C. Allkofer, M. Simon. Atompraxis 16, 1, 1970.
7. J. Engelmann. Int. Congress on Protection against Accelerator and Space Radiation, CERN, Geneva 1971.
8. O. C. Allkofer. 8th Int. Conf. on Nuclear Photography and SSTD, Bucharest, 1972.
9. C. J. Waddington. Proc. Nucl. Phys. 8, 1, 1960.
10. H. J. Schaefer, J. J. Sullivan. Aerospace Med. 38, 1, 1967.
11. H. J. Schaefer, J. J. Sullivan. Int. Congress on Protection Against Accelerator and Space Radiation, CERN, Geneva, April 1971.
12. H. J. Schaeffer, J. J. Sullivan. 8th Int. Conf. on Nuclear Photography and SSTD, Bucharest, 1972.
13. Reunion du Groupe de Recherches sur les Effets Biologiques du Rayonnements Cosmique à l'altitude du vol du trasport supersonique, Strasbourg, 1972.
14. K. Fukui, P. S. Young, Y. K. Lim. Nuovo Cimento X, 61, 210, 1969.
15. D. Hasegan. Rev. Roum. Phys. 21, 2, 177, 1976.
16. R. P. Henke, E. V. Benton. Nucl. Instr. Meth. 97, 483, 1971.
17. D. Haçegan. Rev. Roum. Phys. 17, 9, 1023, 1972.
18. O. C. Allkofer. Groupe de Recherchers sur les Effets Biologiques du Rayonnement Cosmique à l'altitude du vol du transport supersonique, Strasbourg 21 Dec. 1972.
19. J. Tripier, G. Remy, J. Ralarosy, M. Debeauvais, R. Stein, D. Huss. Nucl. Instr. Meth. 115, 26, 1974.
20. E. J. Kobetich, R. Katz. Nucl. Instr. Meth. 71, 226, 1969.
21. M. J. Berger. 4th Symposium on Microdosimetry, Verbania-Pallanza 24 − 28 Sept. 1973.

22. J. Fain, M. Monin, M. Montret. 8th Int. Conf. on Nuclear Photography and SSTD, Bucharest, 1972.
23. H. G. Paretzke. 4th Symposium on Microdosimetry, Verbania-Pallanza, 24-28 Sept. 1973.
24. M. Jensen, L. Larsson, O. Mathiesen, R. Rosander. LUIP-CR-73-06, 1973.
25. P. H. Fowler, V. M. Clapham, V. G. Cowen, J. M. Kidd, R. T. Moses. Proc. Roy. Soc. London A, 318, 1-43, 1970.
26. D. Haçegan, A. Apostol. to be published.
27. C. G. Powell, P. H. Fowler, D. H. Perkins. "The study of elementary particles by the photographic method". Pergamon Press, New York, 1959.
28. E. M. Friedländer, A. Friedman. Nuovo Cim. 52 A, 912, 1967.
29. H. J. Schafer, E. V. Benton, R. P. Henke, J. J. Sullivan. Rad. Res. 49, 245, 1972.
30. P. J. N. Davison. Thesis, University Bristol, 1967.
31. E. W. Fuler, M. T. Clarke. AWRE Report No. 064/68, 1968.
32. A Dragu, D. Hasegan, M. Nicolae. CAER Conference, Prague, 8-12 Sept. 1975.
33. P. S. Nagarajan. Proceedings of a regional seminar, Bombay, 9-19 Dec. 1968.
34. O. C. Allkofer. Réunion du group de travail de biophysique spatial, Strasbourg, 11-12 June 1970.

DOSIMETRY WITH ACTIVATED EMULSION

C. Heilmann, H. Francois and C. Jacquot

SADVI, Centre de Recherches Nucleaires, 67037 Strasbourg Cedex

ABSTRACT

The quantity of silver which is contained in a nuclear emulsion after its chemical development is easely measured with good accuracy by the analysis of the gamma spectrum given by this emulsion after being activated in a neutron source. Using simultaneously different emulsion sensitivities it is possible to give with the total dose, the composition spectrum in ionisation power of this dose. Applications are discussed in medical physics and space dosimetry. Charged and neutral particles are concerned.

The aim of this new technique is to measure the LET spectrum of an arbitrarily mixed particle field. This flux may be composed of many types of particles of different energies. It is particularly suited for instance for cosmic radiations, initial or degenerated accelerator beams, or any sources.

In a first step a stack made of nuclear emulsions of different sensitivities (K5, K2, K0, K-2) is exposed to the field. Each emulsion records the track of a particle according to its LET and the emulsion sensitivity.

In a second step these emulsions are developped using the well known "temperature technique". This chemical treatment transforms the AgBr crystals into Ag-metallic grains along the particle path. The residual AgBr is dissolved by the chemical fixing procedure. The Ag mass contained in the plates after this chemical treatment depends on the initial flux, the sensitivities and the LET of the particles.

In the third step the plates are irradiated with a well calibrated neutron flux in order ot activate the remaining Ag grains.

In a fourth step the plates are analyzed with a multi-channel spectrometer in order to measure with high precision and sensitivity the quantity of Ag left inside the emulsion plates. Measurements are being performed on longlife isotope Ag^{110m} (270 days). The results of this measurement are compared to that obtained with a second stack irradiated with a well calibrated accelerator beam, and subsequently treated in the same way, as the first.

The normalisation of the conditions (time of exposure, neutron flux etc...) of activation analysis of Ag in emulsion is obtained by activating simultaneously a well known quantity of Ag.

We have found a density of Ag of $1.93 \times 10^{-3} g/cm^3$ in a K5 plate exposed to 0.5 rad of γ rays, while an unexposed plate gives an Ag density of $1.22 \times 10^{-4} g/cm^3$, from the background fog, when no special precautions were taken. Thus, the background can be thought to be equivalent to a 20 mrad predose.

Chemical washing technique is presently under study in order to eliminate the residual coloidal Ag to give a higher sensitivity in the low dose range. We are still working on the calibration.

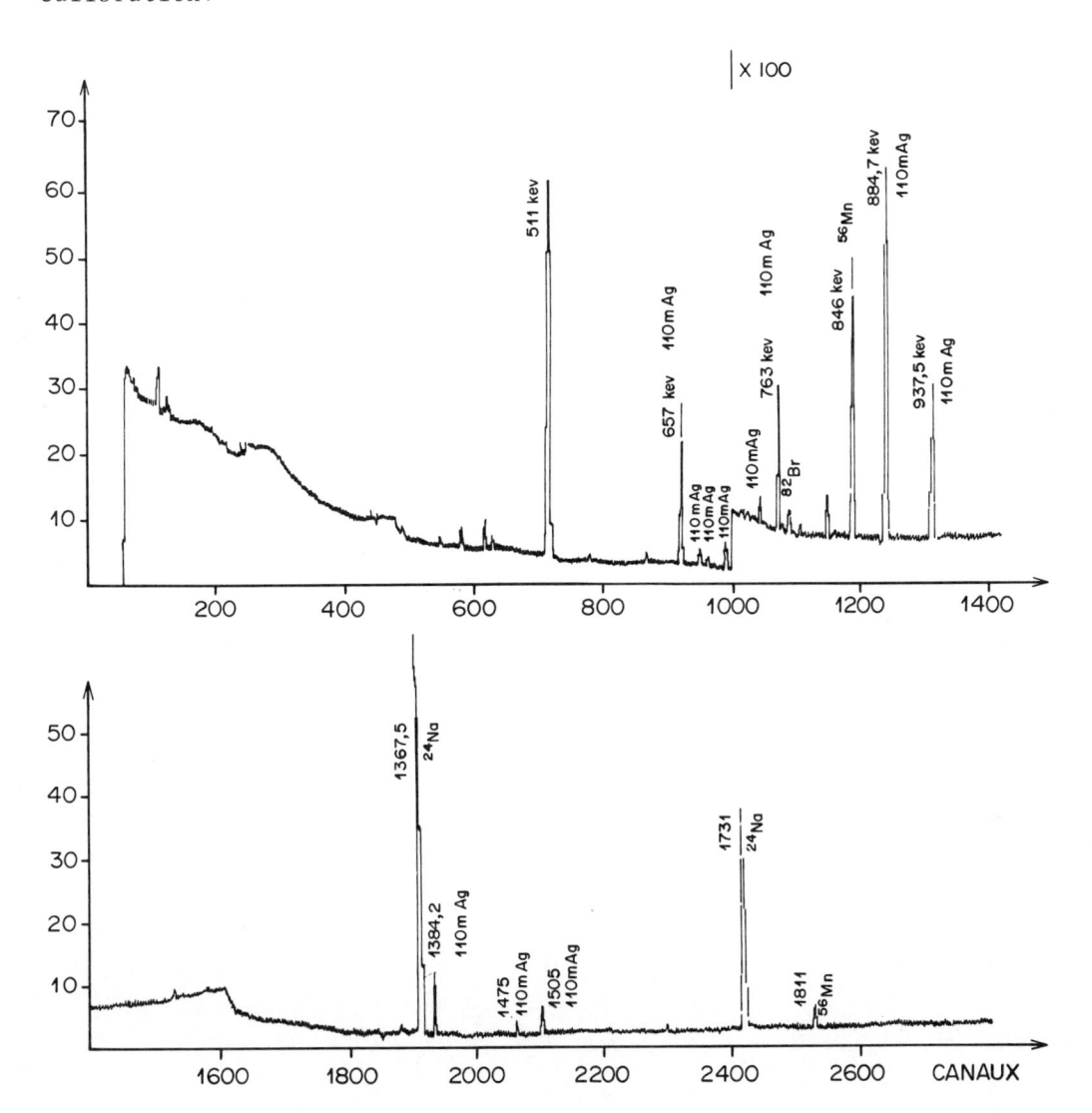

* This technic is registered Patent.
* The γ spectrometry has been made by A. STAMPFLER, C.R.N. STRASBOURG.

Session 9

Applications in Cosmic Ray Physics

Chairmen: A. J. Herz
P. H. Fowler

ULTRA HEAVY COSMIC RAY NUCLEI — ANALYSIS AND RESULTS

P.H. Fowler

H.H. Wills Physics Laboratory, Bristol, England

INTRODUCTION

For ten years cosmic ray scientists have been studying vigorously the heavy end of the cosmic ray charge spectrum, that is nuclei with $Z > 30$. Almost all results to date on present day cosmic rays have been obtained using sandwich stacks of emulsion and plastic track detectors, mostly Lexan. The detectors have often had areas ~ 10 m^2 and thicknesses varying between ~ 1.0 and 10.0 g cm^{-2}. Tracks of ultra heavy nuclei pass through both emulsion and plastic enabling comparisons to be made of the dramatic tracks that are seen.

First I will dwell on aspects of the tracks in the emulsion, and the steps necessary to get an estimate of charge and velocity, and then consider the same problems for the Lexan before finally giving a brief account of the present view of the primary ultra heavy cosmic ray fluxes.

Emulsion response

Fig. 1 is taken from Fowler et al. 1967 (ref. 1) and shows a historic and dramatic track, being the first present day ultra heavy track to be seen. The track is characterised by a high figure for the energy deposition at large distances from the track due to large numbers of long range δ-rays. The following steps are taken to arrive at an estimate of charge Z for such a track:-

a) Microphotometry at a radial distance r from the track for $10 < r < 100 \mu$ to yield the number of developed silver grains.

TEXAS 1966

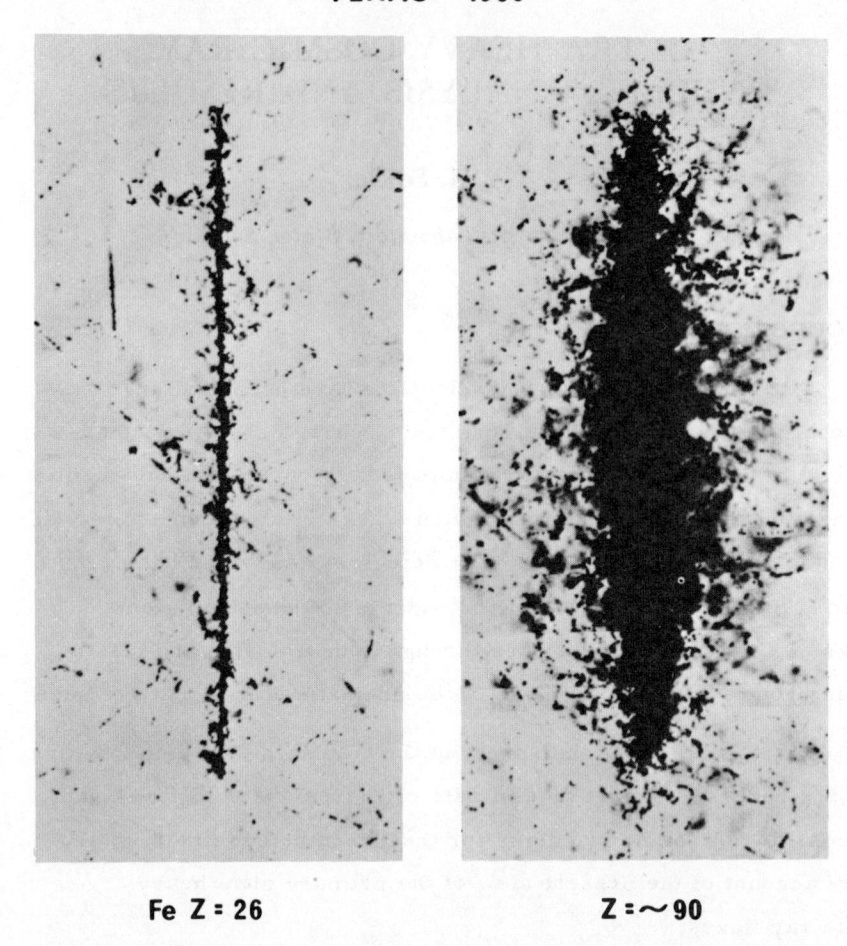

Fe Z = 26   **Z = ~ 90**

Comparison of tracks of Fe nucleus and that of a very heavy primary.

⊢ 50µ ⊣

Fig. 1. On the right a photomicrograph of the first track of a present day ultra heavy nucleus shown as it passed through an emulsion of thickness 200 µ. Note the large number of long range δ-rays and the transition effect at each end. On the left the track of an iron nucleus of the same geometry 1 cm away in the same emulsion is shown for comparison.

b) Calibration on Fe tracks of known range or velocity from the immediate environment of any important track.

c) Use of Mott scattering cross sections to obtain estimate of δ-ray yields versus Z and β.

d) An estimate of β for the track must be made either from direct measurement such as range or from geomagnetic considerations.

A comment on the use of the Mott cross section may be useful. The normal formulae for the knock-on spectrum is based either on the classical cross section, or with the use of the first Born approximation. They both yield the well known formula

$$dN_\delta = \frac{2\pi n e^4}{m_e c^2} \frac{Z^2}{\beta^2} \frac{dW}{W^2} \quad cm^2 g^{-1}, \quad \ldots (1a)$$

for $W \ll W_{max}$,

where n is the number of electrons per gm of material and W the kinetic energy of the knock-on electron. W_{max} is the kinematic limit and is given by the well known expression $W_{max} = 2 m_e c^2 \beta^2 / (1 - \beta^2)$. Eqn. 1 may be obtained by taking the appropriate scattering formula for electrons on positive point nuclei at rest and transforming the coordinate system so that the electrons of the emulsion are initially at rest and assumed unbound. The production of a δ-ray of kinetic energy W by the passage of a nucleus of charge Z moving with a velocity βc through the emulsion therefore corresponds to the single scattering, by a point nucleus, of an electron of the same velocity, βc, through an angle θ, where θ is given by the relations:

$$\sin^2(\tfrac{\theta}{2}) = \frac{1 - \beta^2}{2\beta^2} \frac{W}{m_e c^2} = \frac{W}{W_{max}} . \quad \ldots (1b)$$

For the first Born approximation to be valid we must have $Z/\beta \ll 137$. Likewise for the classical Rutherford cross section to be

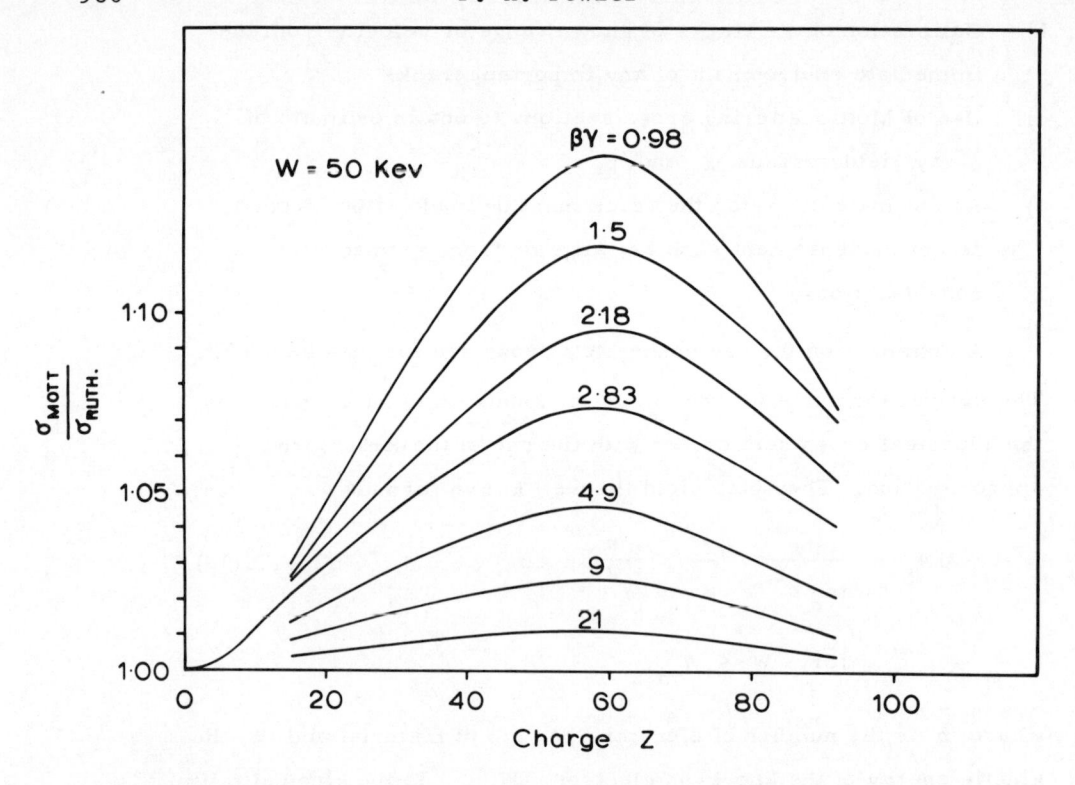

Fig. 2 The figure shows the ratio σ(Mott)/σ(Rutherford) versus charge for various values of momentum per nucleon, $\beta\gamma$, of the incoming nucleus for δ-ray production with W = 50 keV.

valid $Z/\beta \gg 137$. In many of our tracks Z/β is in fact close to 137, so that one must use the Mott cross sections which are valid at $Z/\beta \approx 137$. The ratios between the Mott and Rutherford cross sections are not large for high energies, and are displayed in Fig. 2. They assume that the signal at 10μ is determined by the knock-on cross sections at 50 keV. A new evaluation is being made and carried on to lower velocities where Mott cross-sections differ still more from those normally used. Professor Fleischer's review in this Conference refers to emulsion data for tracks of high energy and the agreement was fair.

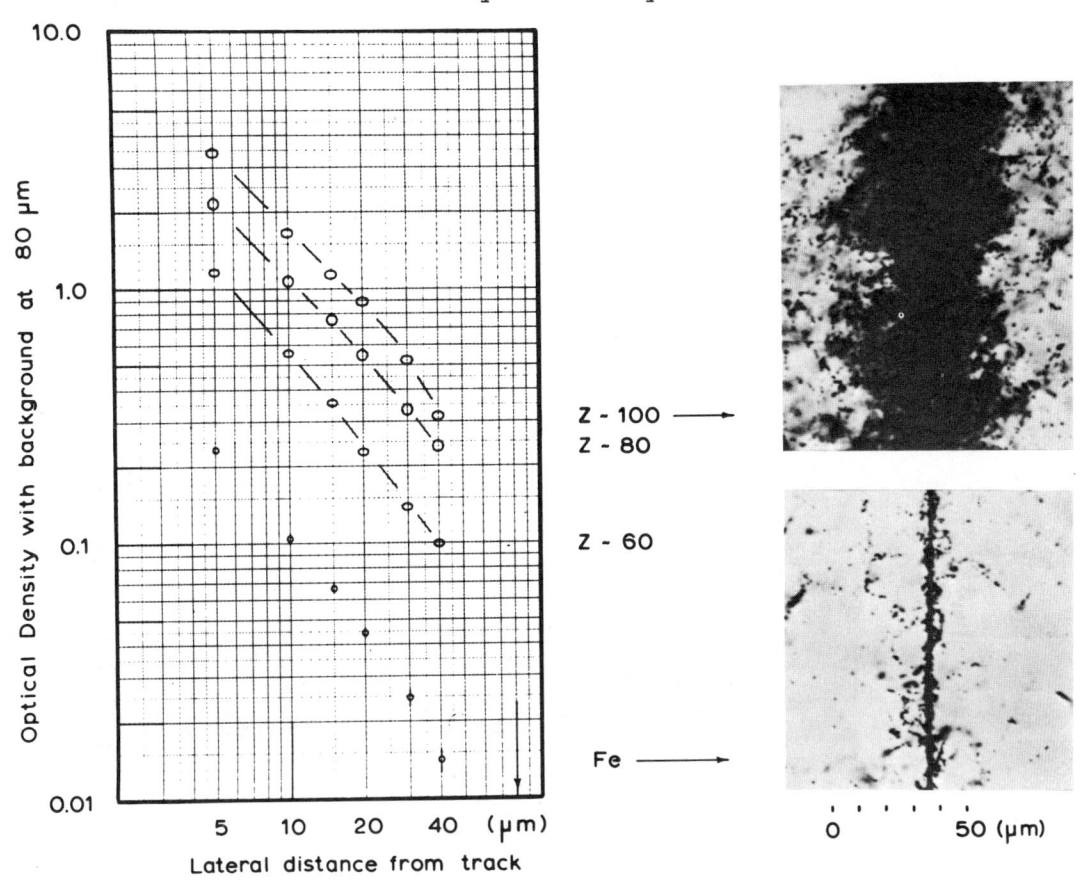

Fig. 3. The right hand side shows a short section of one of the early very heavy tracks and a segment of the track of a relativistic Fe nucleus for comparison. The left hand side displays the results of microphotometry on this heavy track and on two individual tracks of somewhat lower density as well as, on the bottom, average measurements on about 100 Fe nuclei. As indicated, background was taken at $\sim 80\,\mu$.

Distribution of Deposited Energy

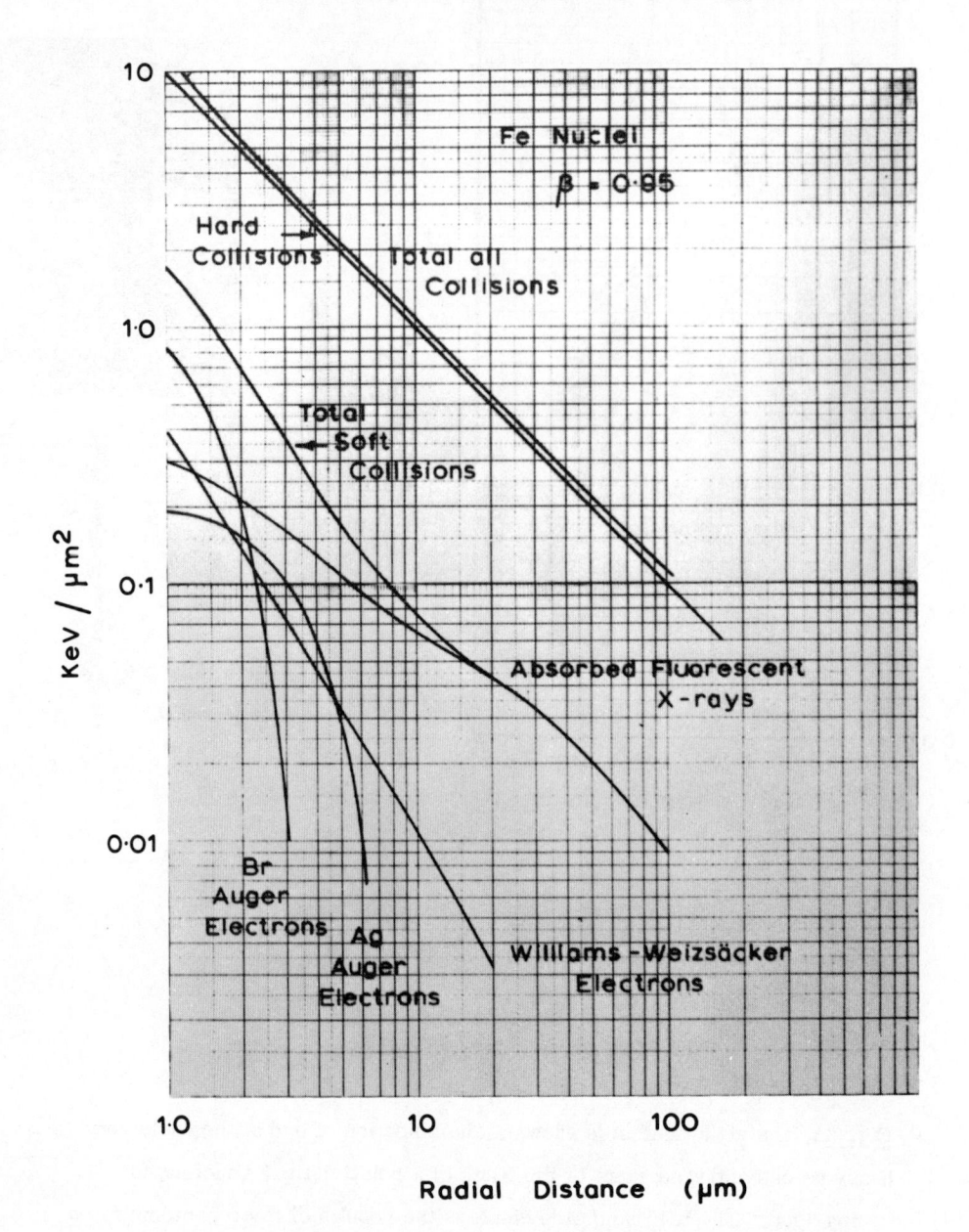

[Figure: Log-log plot of keV/μm² vs Radial Distance (μm), showing curves for Fe Nuclei β = 0.95: Hard Collisions, Total all Collisions, Total Soft Collisions, Absorbed Fluorescent X-rays, Br Auger Electrons, Ag Auger Electrons, Williams-Weizsäcker Electrons.]

Fig. 4 The figure shows the contributions of various atomic processes to the deposition of energy around the path of the particle by an Fe nucleus with β = 0.95. The ordinate is the total energy deposition per unit area around the track and is thus $2\pi r\rho$ where ρ is the energy density per unit volume. It is thus closely proportional to the projected energy density which is appropriate in microphotometry of moderately flat tracks.

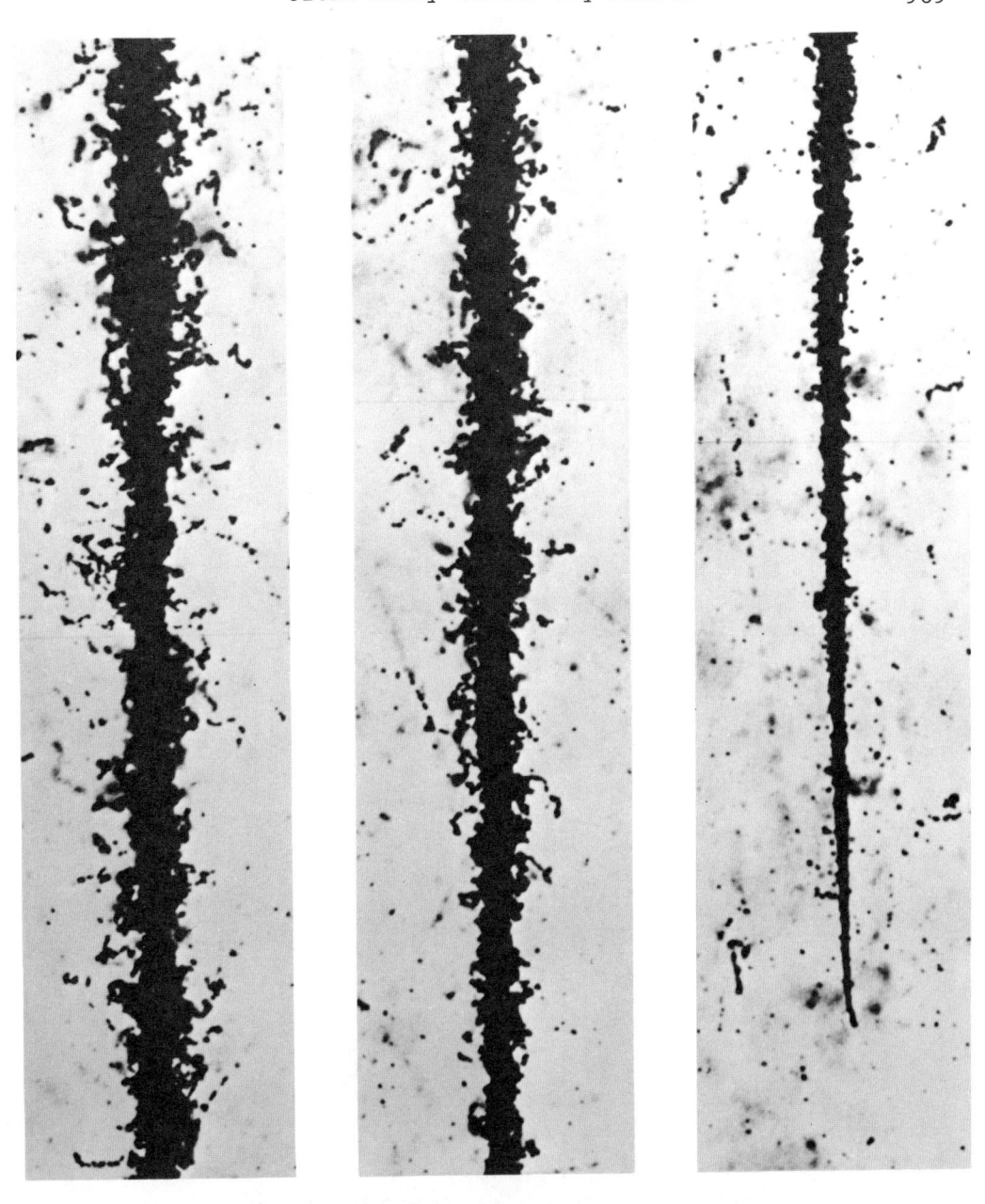

The last 630 μ

|← 50μ →|

Fig. 5. The figure shows photomicrographs of the last 630 μ of the track of an Fe nucleus as it came to rest in the emulsion.

SIOUX FALLS 1969

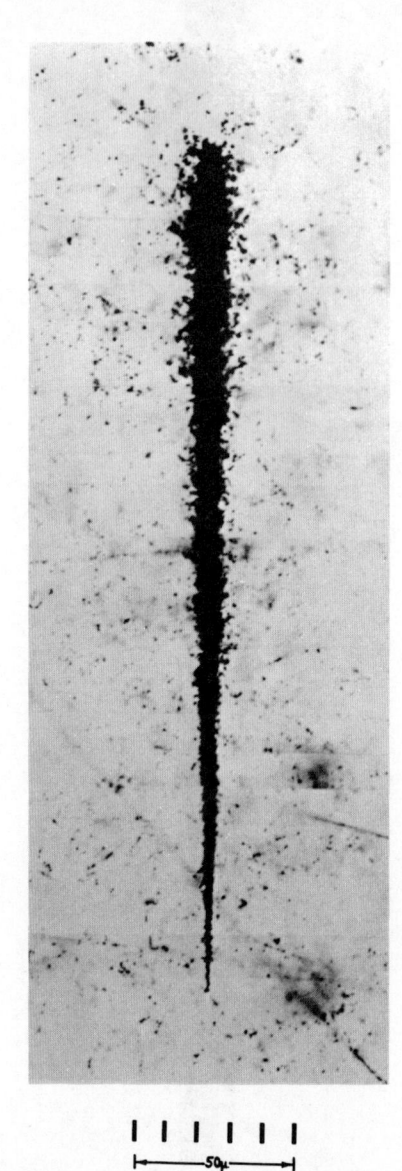

NUCLEUS OF Z~44 TAPERING IN AN EMULSION.

Fig. 6. The figure shows the only example in Bristol data of an ultra-heavy nucleus coming to rest in the emulsion. Note that the magnification in this figure is smaller than that for the Fe nucleus in Fig. 5.

The track structure that is apparent has the form

$$N \propto Z^2/\beta^2 r^2 \qquad \ldots (2)$$

for high β, where N is the number of developed grains, the radial dependence being understood in terms of the energy deposition versus distance. (See for example Katz (ref. 8) and this conference.)

Typical measurements are given in Fig. 3, showing r^{-1} dependence of projected density. An inverse square radial dependence of density will yield measured densities varying approximately as r^{-1} when the background is taken at around 80 µm in an emulsion of thickness 200 µm. In addition to the δ-rays there are some further processes, not always considered, that make a contribution to the structure of the emulsion track. They are the emission and readsorption of characteristic fluorescent K X-rays from silver and bromine, and the Auger electrons emitted following the ionisation of electrons in the K shell. The effects of these processes are given in Fig. 4; these processes have not been considered by Katz in his models and appear to contribute about 10% of the energy deposition at radial distance $r : 5 < r < 100$ µ from the track.

Emulsion response at low velocity

The discussion in the previous section has been made on the supposition that the range of the δ-rays at the kinematic limit was much greater than the value of r used.

The main and most striking feature of stopping heavy nuclei in emulsion has been well known for twenty-five years and is the characteristic tapering of the track. Figs. 5 and 6 show an example of a section of an Fe nucleus coming to rest and the only example of a stopping ultra heavy nucleus that we possess with the geometry enabling a satisfactory photograph to be made. We can understand the characteristic taper in terms of the effect of the kinematic limit of delta ray production on the radial distribution of energy deposition.

If there are no high energy δ-rays — as on a slow particle — there are no distant grains.

We find semi-empirically (Fowler et al. (ref. 2)) that a very good fit to track profiles at all β's is that N, the number of developed grains per unit volume, is given by:-

$$N \propto \frac{Z^2}{\beta^2 r^2} \exp\left(-\frac{r^2}{2\sigma_{max}^2}\right), \qquad \ldots (3)$$

where $\sigma_{max} = 0.22\, R_{max}^{\delta}$.

R_{max}^{δ} is the range of the δ-ray at the kinematic limit. The range of a δ-ray at the kinematic limit is approximately proportional to the residual range of the nucleus, thereby producing a cone-shaped envelope for the track, and hence the tapering of the track over the last few hundred microns of its range.

A year ago at the 14th International Cosmic Ray Conference at Munich, Buford Price told us of his monopole candidate (ref. 3), in which one of the planks used in the analysis was the estimate of velocity of the parent from the emulsion track profile:-

$$\beta = 0.50^{+0.10}_{-0.05}.$$

This point worried me and caused us to determine to what limits we could measure β on <u>individual</u> tracks or segments ~ 200 μ in length, of ultra heavy nuclei, at such high values of β. In contrast the recently published work of Jenson et al. and Behrnetz (refs. 4,5,6) has concerned itself with average behaviour with lower values of Z/β and ideal geometry. We have made use of all the very rare ultra heavy tracks of low velocity that we had available with such a high value for Z/β in the emulsion, and embarked on an extensive series of measurements that we have nearly completed. We measure the transmission profile out to 140 μ from the track and compare it with the transmission between 150 and 250 μ. The optical density at ~ 100 μ even on the heaviest tracks is only ~ 0.10 and background fluctuations on individual readings lie between 0.05 and 0.12 in differing developments and exposures.

Fig. 7 shows a summary of the results of our measurements on our tracks of ultra heavy nuclei which yield estimates of σ_{max} in eqn. 3. We may draw two conclusions from this figure:

1) As is apparent our points lie satisfactorily close to the expectation of eqn. 3.

2) We find it not possible by photometry of the emulsions to determine β for a particle from the track profile if $\beta \gtrsim 0.50$. Faster tracks have profiles indistinguishable from those of higher velocity if one only has available a short track length.

Thus indeed we still disagree strongly with the β assignment given by Price et al. on the track of the monopole candidate. A track with such a velocity should have a profile indistinguishable from that of a fast particle.

Lexan response

The tracks that have given these exciting emulsion pictures also penetrated many Lexan sheets giving dramatic etch cones in some cases. A typically beautiful case is in Fig. 8, it is "typical" in that it is the most photogenic event we have so far obtained! This illustrates dramatically the change in etch rate on an ultra heavy nucleus as it slows down, and the effects of a nuclear collision between L18 and L19 in a thin sheet of Fe. The secondary in its turn slows down and stops after L21. In such collisions it is reasonable to take the velocity of the secondary fragment immediately after the collision to be the same as that of the primary just before; this has been assumed in our analysis.

Track Structure measured on individual tracks

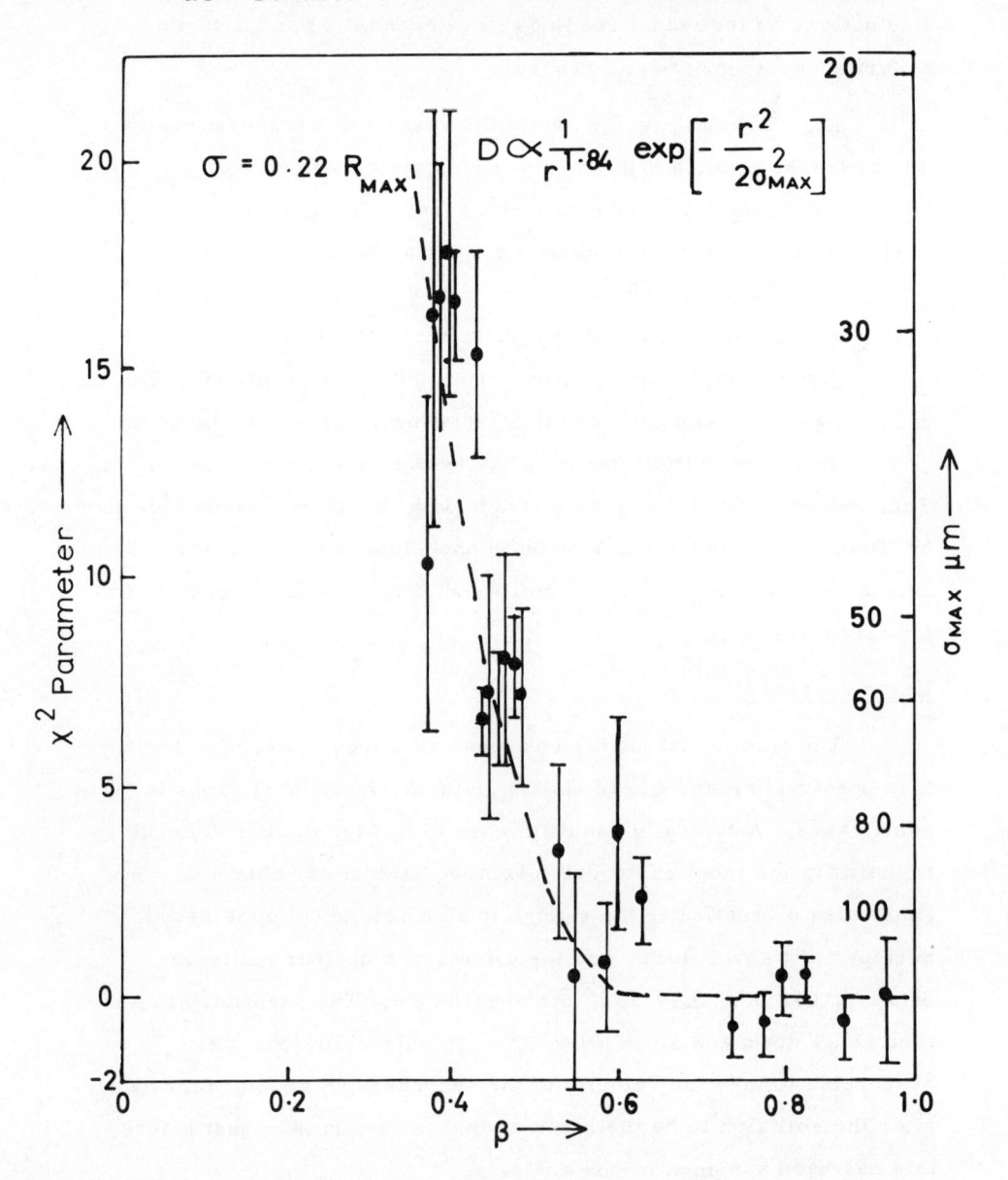

$\sigma = 0.22\, R_{MAX}$

$$D \propto \frac{1}{r^{1.84}} \exp\left[-\frac{r^2}{2\sigma_{MAX}^2}\right]$$

Fig. 7 The figure shows measured values of σ_{max} for individual nuclei. The track lengths available for measurement were typically only 200 μm. The dashed line shows the variation of σ_{max} with β from the expression $\sigma_{max} = 0.22\, R_{max}^{\delta}$. The index 1.84 in the formula was chosen to make the best fit to the profile of fast tracks.

Lexan 1   **Lexan 5**   **Lexan 11**

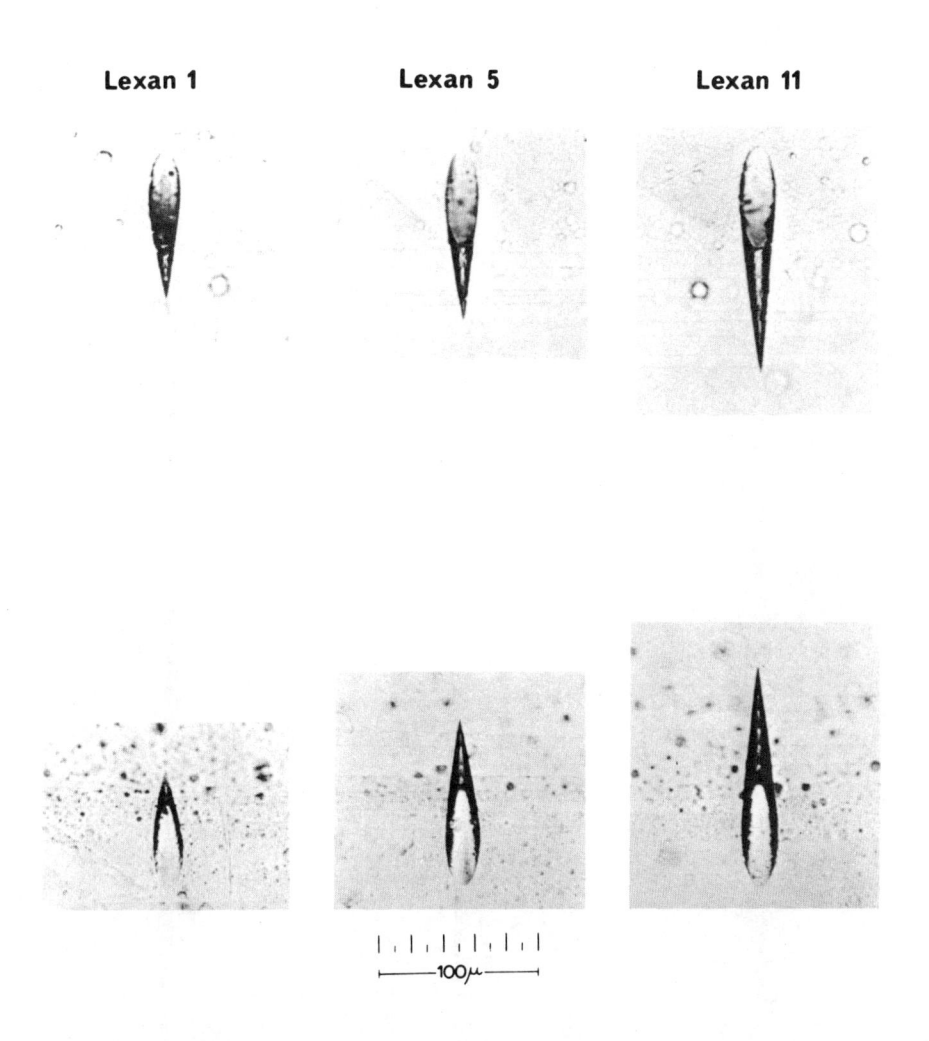

|—100μ—|

Fig. 8a, b and c.   These figures should be viewed together. They show prints of mosaics of photographs taken with a high power optical microscope. They illustrate the etch cones formed on the nucleus of a track of Z = 78 that entered the detector at a wide zenith angle $\theta = 69°$, so giving cones and emulsion tracks that were much more readily photographed than is usually the case.   .... Cont'd.

Emulsion 3 **Lexan 15** **Lexan 17**

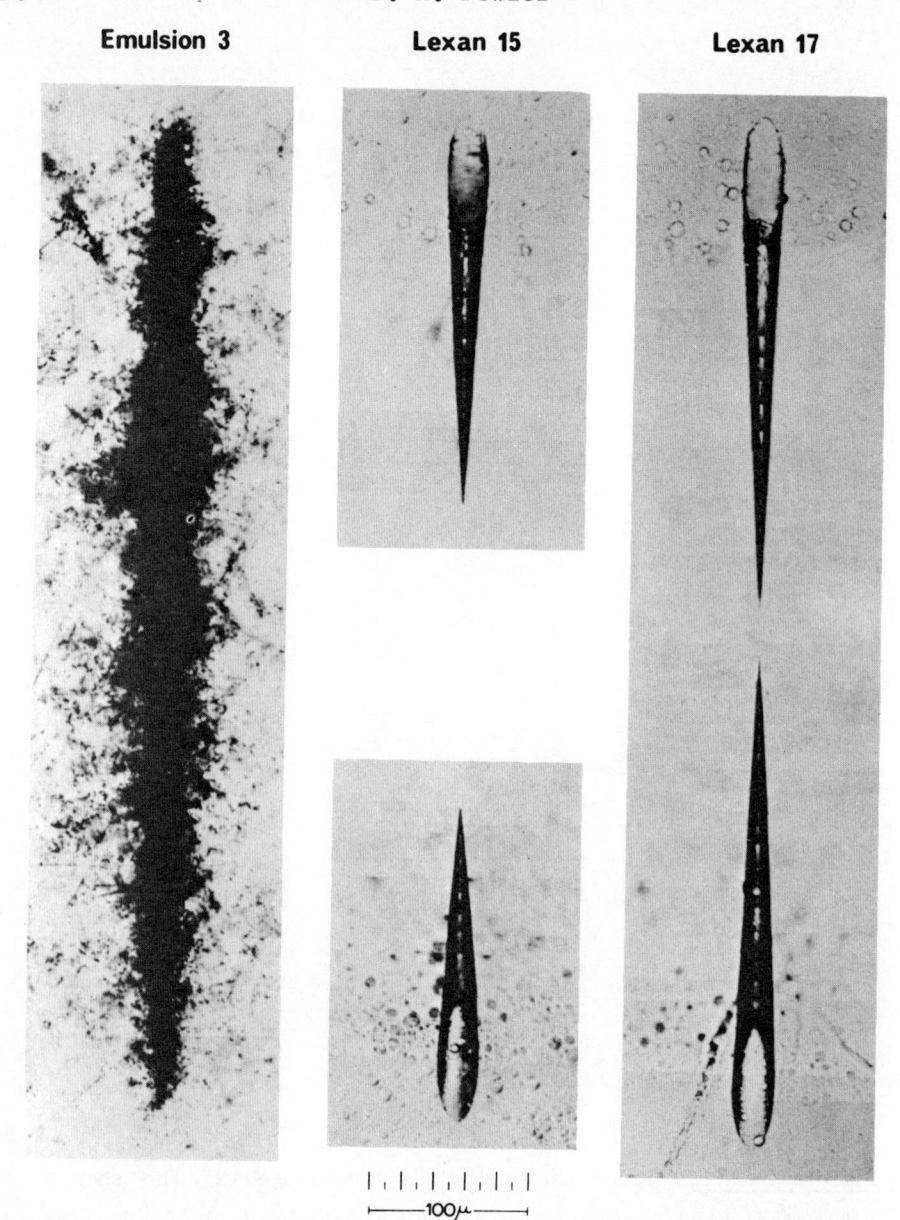

|—100μ—|

Fig. 8a, b and c. Cont'd. We show the track in selected Lexan sheets from a stack which contained Lexan sheets of thickness 250 μ alternating with thin sheets of iron (thickness $\sim$ 160 μ, 0.12 g cm^{-2} Fe). These thicknesses result in the iron sheets alone providing approximately 75% of the stopping power, but only 50% of the nuclear collision cross-section for such heavy nuclei. Some iron sheets were replaced by 200 μ thick emulsion on plastic backing. ....Cont'd.

Ultra heavy cosmic ray nuclei 997

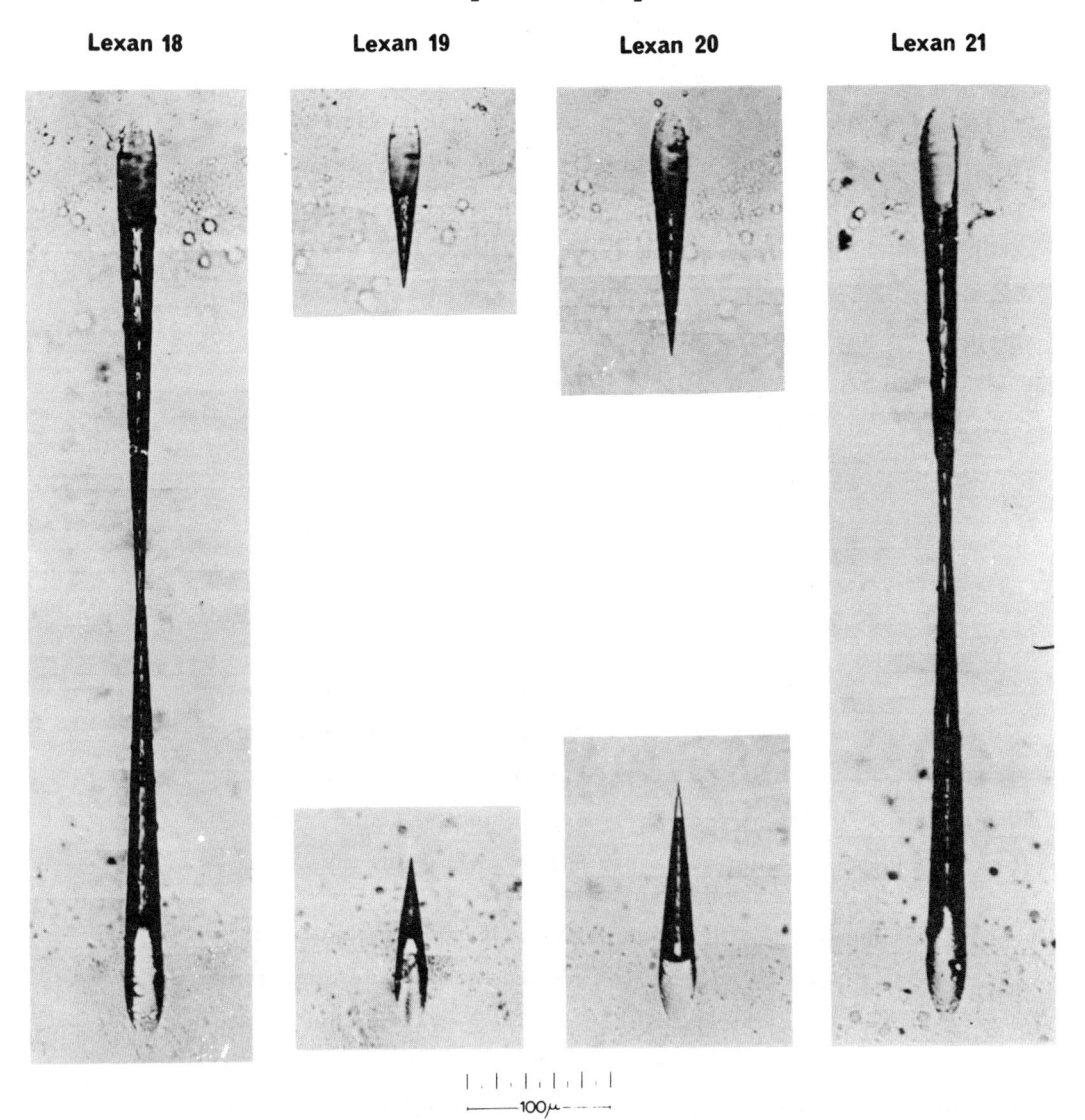

Lexan 18 **Lexan 19** **Lexan 20** **Lexan 21**

———100μ———

<u>Fig. 8a, b and c. Cont'd.</u> The particle entered with an energy $\sim$ 800 MeV/N and slowed down considerably in penetrating to Lexan no. 18. The characteristic increase in etch rate with penetration is well displayed. The 3-fold reduction in etch rate between Lexan 18 and 19 is caused by a nuclear collision in the intervening iron sheet. Thereafter the secondary nucleus slows down rapidly and comes to rest in the iron sheet immediately after Lexan 21. The measurements on this track are included in Fig. 9, with the points in the secondary circled for clarity.

A central point for the Lexan cone length measurements is their interpretation. They are measurements of great statistical weight, we may have 100 or more cone length measurements on a single cosmic ray track. But of what are they measurements? Fig. 9 shows measurements on representative tracks of stopping nuclei from some of our recent exposures. We have heard in the Conference, and a number of groups are agreed, that it is close to the truth to say that the track etch rate V_T is given by:-

$$V_T = f\left(\frac{Z_{eff}}{\beta}\right), \quad \ldots (4)$$

where $Z_{eff} = Z\left[1 - \exp-(130\beta/Z^{2/3})\right]$.

Empirically

$$V_T = a\left(\frac{Z_{eff}}{\beta}\right)^n. \quad \ldots (5)$$

The form of the function f of Eqn. 4 that is given in Eqn. 5 is substantiated by a lot of experimental data from many laboratories. The index, n, is found to lie between quite narrow limits, $4 \leq n \leq 5$, certainly for $0.2 < V_T < 2$ µ/hr. Expression (4) could result if Lexan response was due to energy deposition outside the core.

To obtain a charge estimate for a particle assuming (4) and (5) above - β must be known. In practice the particle often comes to rest so that the etch rate may be obtained as a function of range as shown in Fig. 9. Range energy relations are probably accurate enough at the present stage of development so that its errors make an unimportant contribution to the uncertainty in Z. Very often the function f is determined from measurements on Fe nuclei of the cosmic rays that come to rest, enabling the track of an unknown nucleus to be evaluated.

ETCH RATE vs RANGE   SIOUX FALLS 1972-74
SELECTED TRACKS

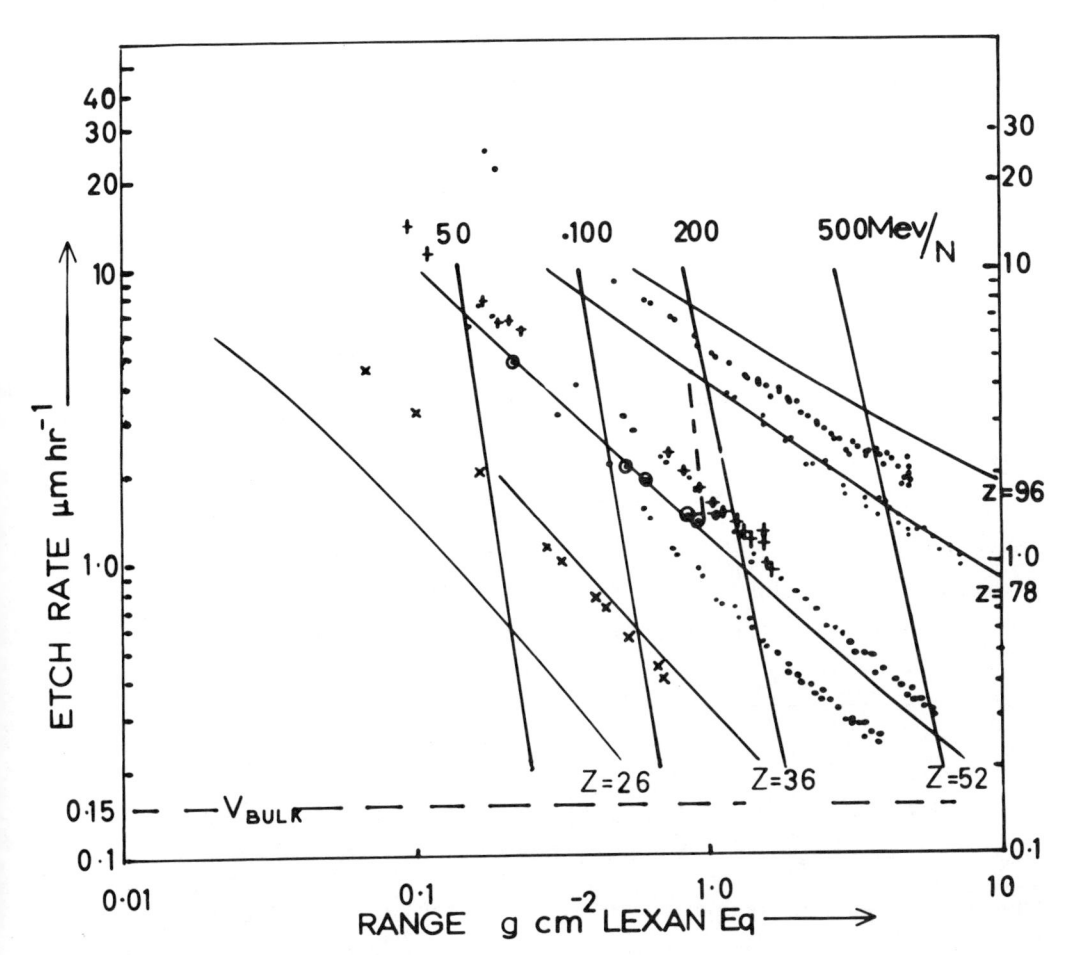

Fig. 9 The figure shows measured etch rates versus range for selected individual tracks stopping in the detector. The curve labelled Z = 26 applies to an average of our measurements on many iron nuclei used for calibration. The other curves have been calculated using Eqn. 5 with coefficients fitted to the iron curve.

Returning to fig. 9, different slopes are clearly apparent for tracks of differing charge. The curves drawn assume $V_T = f(Z_{eff}/\beta)$. The flatter slopes at higher Z are due to the onset of relativity at shorter range, not to any saturation in the function f. It is remarkable that for nuclei $Z \sim 52$ the etch rate is an almost perfect power law over a factor of 100:1 in etch rate - in both our own and the Irish and the data of Buford Price.

As far as we can see the simple functional form for the etch rate holds over the whole Z, β span ($Z > 26$, $\beta > 0.25$) mapped out in ultra heavy work in Lexan. I am not sure that this agreement carries over to the lower β values, $\beta < 0.25$ used for $6 < Z < 26$.

Determination of β

At high charge, $Z \geq 65$, where even relativistic particles are above the Lexan threshold, we have a serious problem in determining β because the particles either penetrate the detector or interact. We must measure the etch rate gradient, G, to determine β.

$$G = \frac{-1}{V_T} \frac{dV_T}{dX} \quad cm^2 g^{-1} , \quad \ldots (6a)$$

$$G \simeq \frac{n}{\beta^2 \gamma^3} \frac{(Z_{eff})^2}{A} \left(\frac{dE/dX}{m_p c^2}\right)_{Proton} cm^2 g^{-1} . \quad \ldots (6b)$$

The last expression assumes the relation between V_T and β given in Eqn. 5 to be correct. We see from this expression that at high energy G decreases as γ^{-3}. At low energies where $dE/dx \sim 1/\beta^2$, $G \sim 1/\beta^4$. The curves in Fig. 10 show the dependence of G with etch rate for various values of Z. We see that $G \sim 0.04$ $cm^2 g^{-1}$ at GeV/N falling rapidly at higher energy and thus presenting a problem for measurement. On such tracks with M approximately uniformally spaced cones over path length ΔX with a fractional standard deviation σ_c in the measured etch rate of an individual cone, we have a value for

ΔG due to statistical errors alone given by:-

$$\Delta G \sim \left(\frac{12}{M}\right)^{0.5} \frac{\sigma_c}{\Delta X} \text{ cm}^2 \text{g}^{-1}. \qquad \ldots (7)$$

In our experiments with $\Delta X \sim 4$ g cm^{-2} and $\sigma_c \sim 0.05$ we have $\Delta G \sim 4 \times 10^{-3}$. In the Skylab data I anticipate $\Delta G \sim 10^{-2}$ as ΔX is ~ 1 g cm^{-2} only. Such values of ΔG imply uncertainty in β estimate of $\Delta\beta \sim 0.03$, and ~ 0.05 respectively at high energy which embraces at least 50% of the tracks. These values of $\Delta\beta$ are the principle source of random error in charge estimation and so it is specially important to attend to their minimisation.

The points on Fig. 10 show the results of our latest series of measurements, details of which will be reported in the next paper. The measured G is plotted against the etch rate for penetrating and interacting tracks. Examples of heavy nuclei are found over a wide band of energies as one would expect. Representative errors computed using Eqn. 7 are shown for a few of the tracks.

Cosmic ray results

I will present a very brief summary of cosmic ray results that have been acquired in this exciting field involving plastics and emulsions. Charge spectra are obtained in the detector and then estimates are made of the degradation in penetrating through the overlying atmosphere. It is now well known that even the flux at the top of the atmosphere is degraded - as is witnessed by the huge abundance of Li, Be and B nuclei. It is natural to explain them as spallation fragments from C and O and heavier nuclei, and this requires paths with an average path length of ~ 5 g cm^{-2} interstellar hydrogen to fit the data. We accept this as quite likely to apply to the ultra heavy nuclei as well, and so attempt to obtain a source spectrum. If such a procedure was entirely illegitimate one could expect a stupid answer - such as negative abundances for some elements of the source - but no! One gets the following very interesting answers.

ETCH RATE vs GRADIENT IN ETCH RATE THRO' THE DETECTOR

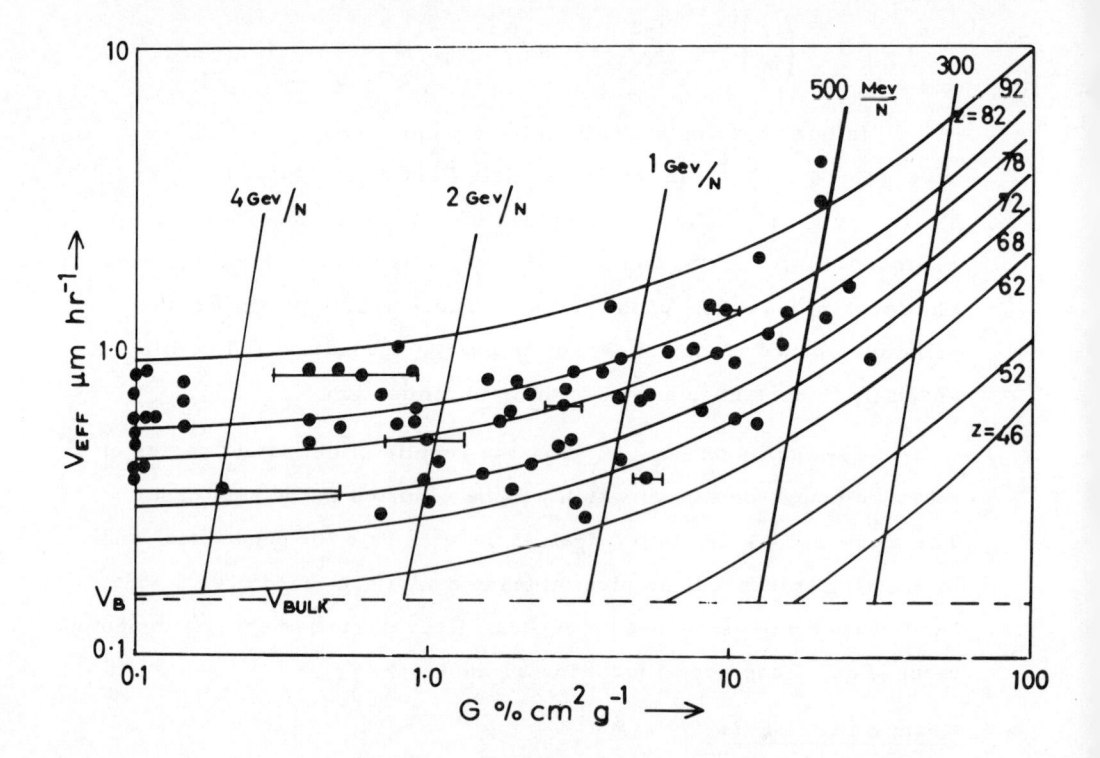

Fig. 10 The figure shows the expected variation of the etch rate gradient, G, with etch rate for various values of Z, together with experimental points for all our nuclei with Z ≥ 65 obtained in flights from Sioux Falls from 1971-1974 inclusive, for which measurements of G were necessary because the particle either penetrated or interacted within the detector.

Our knowledge of the cosmic ray spectrum at the top of the atmosphere up to the time of the Munich Cosmic Ray Conference (1975) may be summarised for the ultra heavy nuclei as follows:-

(1) The overall spectrum is very much like that of solar material when normalised to Fe. This is apparent in the table below when one looks at broad bands of charge estimated at the top of the atmosphere.

	Cosmic Rays (Top of Atmosphere) 10^6	Sun (Cameron 1973) ref. 7 10^6
$Z \geq 40$	100	88
$Z \geq 50$	30	42
$Z \geq 70$	11	10

(2) There are peaks present in the cosmic ray spectrum at $Z \sim 36$, 52 and 78. These peaks have been apparent in the data of many groups, but the peaks in solar material are at $Z \sim 40$, 56 and 82.

(3) There is a significant abundance of particles with $Z > 90$ in the cosmic radiation. Again the data of all groups, both that reported at the Denver Cosmic Ray Conference (1973) and that reported since, agree on this point.

(4) There is no evidence yet for superheavy nuclei $Z \sim 110$.

Using some of our own data reported at this conference and together for $Z > 65$ with data from Skylab (ref. 8) we have a new source spectrum containing only new data obtained since the Denver Conference. The source spectrum from this new data is displayed in Fig. 11. In this figure all the features outlines in (2) and (3) above present at detector level, are still more enhanced, and the whole is in good agreement with the Denver findings. Most of the nuclei with $Z = 60-75$ at the top of the atmosphere appear to be spallation fragments from collisions in interstellar space. While the source spectrum is still very similar to that of solar material it is a still better fit to nuclei made in the 'r' process alone. The 'r' process produces peaks at 52 and 78 and all the nuclei $Z \geq 90$ so it is natural to appeal to it.

Supernova outbursts, or the pulsars they often produce, seem to be the most plausible site not only for nucleosynthesis by the 'r' process but also for the generation of the cosmic radiation. The fit is good, it does not have to be perfect. In particular the abundance ratio between the long-lived radioactive nuclei and stable species depends upon the duration of element synthesis by the 'r' process prior to the formation of the Solar System, taken as being 4.6×10^9 years ago. From this time it is assumed that solar material has been an isolated system. In Fig. 11 the Cameron 1973 abundance estimates are given, just as corrected by the author for decay through 4.6×10^9 years, as well as an estimate of the yield of these elements by the 'r' process using a production time history suitable for the cosmic ray beam, that is production in a single event followed by 10^6 years transit time in which rapid radioactive decays take place in addition to nuclear collisions with interstellar material. This time history brings the observed cosmic ray abundance ratio $N(Z \geq 90): N(75 \leq Z \leq 79)$ close to that which might be expected if the cosmic radiation had been synthesised just prior to acceleration. It is indeed likely that we will learn much about conditions of nucleosynthesis in the source from detailed comparison of the abundances of nuclides in the solar system and in the cosmic radiation. The source conditions may well not be identical to those responsible for the production in the 'r' process of the heavy nuclides now present in the Solar System. The charge spectrum for $Z > 70$ will clearly give us by far the most direct information on the age of the cosmic radiation, a figure of considerable importance.

In the future I hope that scientists will be able to obtain still better exposures and certainly expect that plastic track detectors will play an important role. I believe that in time we will be able to calibrate the Lexan directly, and so no longer will we have a small but nagging uncertainty about the correctness of the charge scale at present employed.

Ultra heavy cosmic ray nuclei

The computed source spectrum for cosmic rays, of Z>35, normalised to unity for iron nuclei.

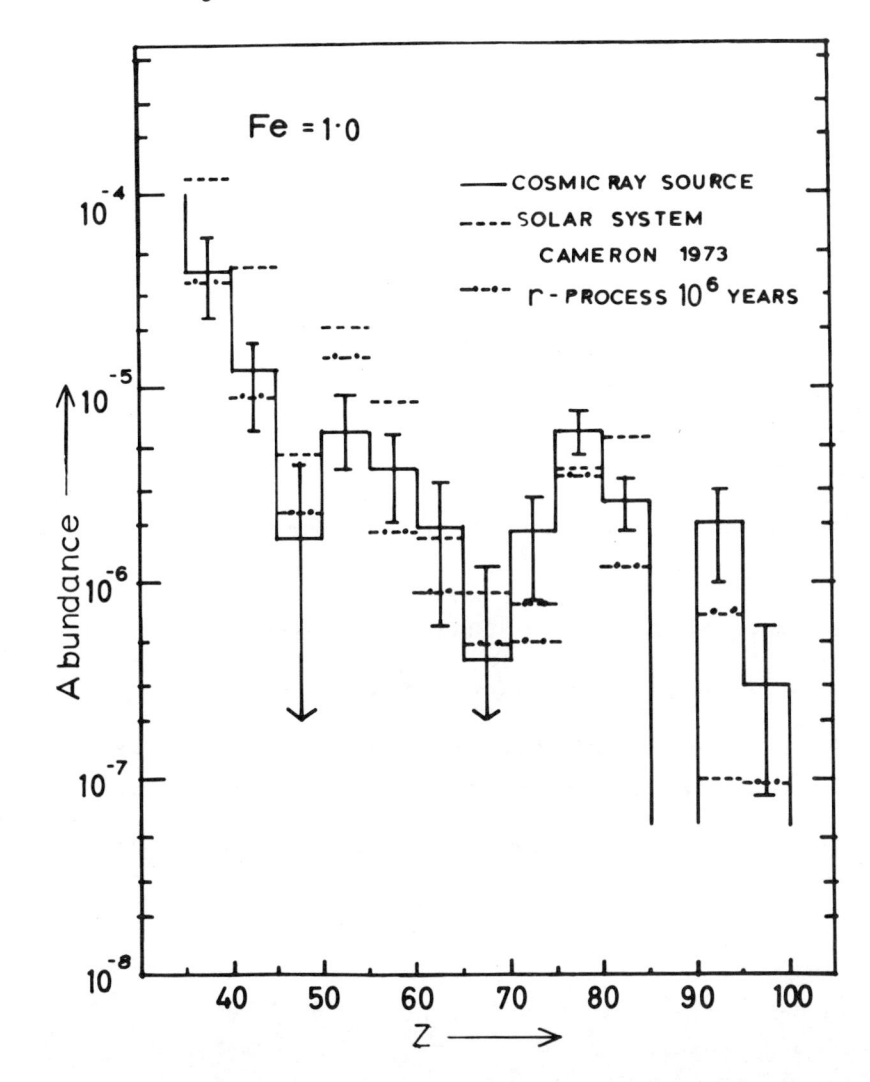

Fig. 11 The figure shows the computed source spectrum for cosmic rays for $Z > 35$. The ordinate is the sum of the abundance values for the five elements taken in each bin, with the normalisation set to unity for Fe.

The bin width taken is considered to be larger than the experimental error so that the features should be real. Also shown are estimates of total solar system abundance and abundance resulting from the 'r' process alone at an age of 10^6 years, based on the calculations of Schramm and Fowler (ref. 9).

REFERENCES

ref. 1. Fowler, P.H., Adams, R.A., Cowen, V.G. and Kidd, J.M. 1967 Proc. Roy. Soc. Lond. A, 301, 39. See also Fowler et al. Proc. Roy. Soc. Lond. A318 (1970) for discussion of emulsion measurements.

ref. 2. Evian Conference Report.

ref. 3. Price, P.B., Shirk, E.K., Osborne, W.Z., and Pinsky, L.S., Phys. Rev. Letters 35, 487 (1975) See also P.H. Fowler, Proc. 14th Int. Cosmic Ray Conf. p. 4049, Munich 1975 for discussion of this event.

ref. 4. Jensen, M., Larsson, L., Mathiesen, O. and Rosander, R., Physica Scripta Vol. 13, 65-74, 1976.

ref. 5. Jensen, M. and Mathiesen, O., Physica Scripta Vol. 13, 75-82, 1976.

ref. 6. Behrnetz, S., Nucl. Instr. Meth. 133, 113-119 (1976).

ref. 7. Cameron, A., Space Sci. Rev., 15, 121 (1973).

ref. 8. Katz, R. and Kobetich, E.J. Phys. Rev. 170, 391, (1968) See also Katz, this conference, for extensive bibliography.

ref. 9. Schramm, D.N. and Fowler, W.A. Nature 231, 103, (1971).

HIGH RESOLUTION STUDY OF NUCLEONIC COSMIC RAYS WITH Z ≥ 34

P. H. Fowler*, C. Alexander*, V. M. Clapham*,
D. L. Henshaw*, C.O'Ceallaigh**, D. O'Sullivan**
and A. Thompson**

*University of Bristol, England
**Dublin Institute for Advanced Studies, Ireland

Abstract

Preliminary results of the analysis of large area Lexan polycarbonate and nuclear emulsion sandwich stacks flown from Sioux Falls between 1971 and 1974 are given. The total exposure was ≃ 120 m² days at ≃ 3.8 g cm^{-2} atmospheric depth and 284 tracks of nuclei with Z ≥ 34 have been found to date, of which 97 have Z > 65. The charge distribution features a Platinum peak, a marked Actinide gap and a high Uranium group flux, but no example of a super heavy nucleus (Z > 110) was observed. The energy spectrum of nuclei with Z > 65 is "normal" confirming our earlier results[1].

1. Introduction   During the period 1971 to 1974 a series of exposures was made with thick sandwich stacks flown from Sioux Falls where the geomagnetic latitude is such that both high energy and low energy cosmic ray nuclei are present. The total exposure was ≃ 120 m² days at ≃ 3.8 g cm^{-2} atmospheric depth. The stacks were assembled from modules which varied in configuration from flight to flight and included both pure Lexan modules containing up to 200 sheets of 250 μm Lexan with one or two sheets of 200 μm G5 emulsion and composite modules in which Lexan sheets were interleaved with copper or iron degraders of thickness typically 0.12 g cm^{-2}. This latter type of module included up to four layers of emulsion. In all cases the total stopping power of the modules exceeded 4 g cm^{-2} Lexan equivalent. The stacks were thus thick in comparison with most others which have been exposed, enabling comparatively high energy nuclei (up to 800 MeV/N) to be brought to rest in the stack so that their tracks could be studied over a wide band of energy. A thick stack also, of course, enables one to determine the energy of penetrating particles with greater confidence.

2. __Experimental Method.__ For every particle the Lexan etch cones were formed by using standard etching procedures with 6.25 N NaOH saturated with etch products and with the addition of 0.05% Benax surfactant. Whenever possible, etching durations were chosen to optimise cone lengths while avoiding cylinder formation, so that in general three or four different etching times were used for a stopping nucleus of high charge. All cones were measured under $\simeq$ 3000X magnification and were used with associated ranges to compute sets of etch rate versus residual range for each event. In the case of penetrating nuclei in pure Lexan, up to 400 etch rate points were available for each event compared to typically 80 points for composite modules.

Two methods of scanning for tracks of ultra heavy nuclei were used, (1) a direct scan of the several emulsion layers with low power stereo microscopes and (2) a coincident cylinder scan in the Lexan sheets of the pure Lexan modules. The emulsion scan essentially located all the nuclei with $Z \geq 40$. It was a criterion of acceptance of such an event that it be associated with cones in the adjacent Lexan sheets having a vertical component of the track etch rate exceeding 0.20 μm/hr, which is about 1.3 times the bulk etch rate. The probability of acceptance is therefore strongly zenith angle dependent for tracks with low etch rates. At high charges the track etch rate, V, is sufficiently large for the acceptance probability to be near unity for all zenith angles, even when $\beta \rightarrow 1$. For $Z = 65$, V has fallen to $\simeq$ 0.22 μm/hr as $\beta \rightarrow 1$, thus such particles can only be accepted from a restricted solid angle near the zenith. At still lower charges, where V is still further reduced, particles with high β do not register, causing the detection efficiency for the element to fall dramatically. For example, when $Z = 78$ about 95% of all nuclei are accepted, for $Z = 65$ the relevant figure is still as high as $\simeq$ 70%, but it then drops rapidly to $\simeq$ 5% for $Z = 40$.

For $Z < 40$, emulsion scanning is so ineffective that only events obtained from a coincident cylinder scan in the pure Lexan modules will be accepted. The etch time and coincidence spacing were chosen, as usual, to reject the overwhelming majority of Fe group nuclei. Since this procedure can be carried out throughout the thickness of the stack, unlike the single layer emulsion scan, it detects at least 50% of the nuclei with $34 \leq Z < 40$ which stop in the stack. It does, of course, also detect stopping nuclei of higher charge, which acts as a check of the emulsion scanning. Similarly, all tracks of candidates located by emulsion scanning which stopped in the stacks were located independently by the Lexan coincident scan method within the necessary restrictions of the zenith angle.

3. **Track Measurements and Analysis**   The energy spectrum of nucleonic cosmic rays is such that in our thick detectors one expects three classes of events of comparable population:

(a)   Particles which stop in the detector.
(b)   Particles which penetrate the detector or interact with considerable change (>1.5:1) in track etch rate.
(c)   Fast particles which penetrate or interact with small or indiscernible change in etch rate.

Charge identification was based on the assumed relation:
$$V = A(Z_{eff}/\beta)^n \quad (1)$$
where the two constants were determined, as usual, from Fe peak calibration of the various batches of Lexan, assuming that the dominant isotope is ^{56}Fe. For all the batches of Lexan involved, n remained in the interval $4.0 < n < 4.6$. For stopping particles the residual range is known explicitly and the conditions for the charge assignment are most favourable. The widths of the Fe peaks suggest that $\Delta Z/Z \simeq 1.5\%$. The accuracy of charge determination for a stopping ultra heavy nucleus is unlikely to be better than this figure. Additional uncertainties of course arise for penetrating particles since the ranges can no longer be obtained from direct measurement. For class (b) events the effect of this uncertainty on ΔZ is not great, but for class (c) events the value of $\Delta Z/Z$ is largely determined by the accuracy with which β can be deduced.

A very convenient parameter for identifying class (b) and class (c) events, or indeed all three classes of events, is the fractional etch rate gradient
$$G = \frac{1}{V} \frac{dV}{dX} \quad (2)$$
where V is the etch rate and X the path length. In practice dV/dX

must be obtained from the chords $\Delta V_i/\Delta X_i$ and

$$G_{chord} = \frac{2(V_2 - V_1)}{(V_2 + V_1)\Delta X} \qquad (3)$$

where ΔX is the difference in range between the two points at which V_1 and V_2 are determined, which is of course known whether or not the residual range, R, is known. G_{chord} is associated with an effective track etch rate given by

$$V_{eff} = \frac{2V_1 V_2}{V_1 + V_2} \qquad (4)$$

For a track with 2n equally spaced cones the overall value of G is given by

$$\bar{G} = \frac{\sum_{i=1}^{n}\left[\frac{2(V_{2n+1-i} - V_i) W_i}{(V_{2n+1-i} + V_i)\Delta X_i}\right]}{\sum_{i=1}^{n} W_i} \qquad (5)$$

where the constituents of each cone pair are equally displaced on either side of the midpoint of ΔX, and where W_i is proportional to $(\Delta X_i)^2$ for class (c) events where $V_1 \simeq V_{2n}$. In practice this weighting is sufficiently accurate to be used for class (b) and class (a) events also. The overall value of V_{eff} is given by

$$\bar{V}_{eff} = \frac{n}{\sum_{i=1}^{n}\left[\frac{V_i + V_{2n+1-i}}{2(V_i)(V_{2n+1-i})}\right]} \qquad (6)$$

which is numerically equal to the harmonic mean of the 2n etch rate values.

The computed values of $\bar{G}$ and $\bar{V}_{eff}$ are, in practice, remarkably good approximations to G and V in Eq.(2). Using Eq.(1), G is a known function of β for a given charge and thus $\bar{G}$ and $\bar{V}_{eff}$ yield Z and β. It should be noted that all the information content of the set of etch rate points for each event is contained in one number pair $(\bar{G}, \bar{V}_{eff})$.

The errors in $\bar{G}$ due to statistical considerations alone are given approximately by $(6/n)^{\frac{1}{2}} \sigma/\Delta R_{max}$ where σ is about 0.05, the observed fractional standard deviation in etch rate. The fractional error in G therefore increases as $G \to 0$ and for these high energy particles the uncertainty in the estimate of β tends to a maximum and consequently the accuracy of the Z determination decreases. For example, a typical penetrating particle at 45° zenith and with 40 pairs of cones will have $\Delta R_{max} \simeq 6$ g cm^{-2}, yielding $\Delta G \simeq 3 \times 10^{-3}$, which is approximately the value of G at 3 GeV/N where $\beta \simeq 0.97$. Since $\Delta\beta/\beta \simeq \Delta Z/Z$, then the error in charge determination is about 3%. Whereas for 1 GeV/N where $\beta \simeq 0.87$, this value of ΔG yields $\Delta Z/Z \simeq 1.5\%$. Thus one must expect the principal source of error for particles with energies greater than 1.0 GeV/N to arise from uncertainties in G and hence in β. Such particles comprise over 60% of the observed nuclei with Z > 65.

4. Results   The scanning procedures give a sample of events which is strongly biased against lower charges as outlined in Section (2) above. Only for Z > 65 is the bias relatively unimportant. Fig.1 displays the charge spectrum of all selected events together with the probability of selection computed on the supposition that the energy spectrum at the top of the atmosphere is the same as that of low charge cosmic ray nuclei. The events with Z > 65 are replotted in Fig. 2 together with the detailed corrections for variation of efficiency with zenith angle and velocity of the detected particle. The most significant feature of this distribution is the prominent peak around Pt, as seen in earlier results. Nuclei with Z > 88 are clearly present and have a relative abundance of $\simeq 30\%$ of the Pt group (74 < Z < 82) at detector level. Correction for degradation in the overlying atmosphere does not alter this ratio significantly and it is higher than that observed in the Skylab exposure[2] where Price and Shirk obtained $\simeq 14\%$ in a sample of comparable size.

Although the results are preliminary, the distribution suggests that resolution of the Lexan in these experiments is indeed $\Delta Z/Z \simeq 3\%$. This resolution is concerned only with variation from track to track and does not include any systematic error in charge assignment which might follow, for example, from the assumption that the Lexan response is determined simply by $f(Z_{eff}/\beta)$ as in Eq.(1). Other systematic errors which may well be involved are incorrect range-energy relations for high charge nuclei and incorrect dependence of Z_{eff} on β where we have used

$$Z_{eff} = Z\{1 - \exp(-130\beta Z^{-2/3})\} \tag{7}$$

The position of the peaks and the presence of a gap in the region of the short-lived radioactive nuclides with $84 \leq Z \leq 89$ is perhaps the best indication that the overall systematic error is not large. Such a feature was observed in the Skylab data also.

We have emphasised above that our sample is fairly free from selection bias for $Z > 65$ (see Fig.2), so this is the best starting point for a _direct_ determination of the energy spectrum of this important cosmic ray component. The results presented are provisional since there remains an appreciable number of cone measurements to be made in some of the recent events. We have rejected the data from two flights reported at Denver[1] since the geomagnetic cut-off averaged throughout the flight was somewhat higher than that for the remaining flights. The remaining sample comprises about 80% of our total data. This sample has been refined further by rejecting events where the path length in the stack was insufficient for the adequate determination of G and hence of the energy. For example, we demanded a path length greater than 5.0 g cm^{-2} for events with kinetic energy above 2.0 GeV/N. The energy spectrum deduced for the top of the atmosphere is, of course, truncated at $\simeq 0.5$ GeV/N because the overlying atmosphere prevents the lower energy particles from reaching the stacks.

The results are given in Table 1, where they are compared with a normal spectrum and a steep spectrum such as that claimed by the Berkeley-MSC group[3,4]. It is apparent that our data are completely at variance with the steep spectrum. There is no evidence in our results that the energy spectrum for the $Z > 65$ component differs from that of the low charge component in this energy band. In our view it is clear that if one is unable to measure the energy spectrum effectively, then there is a great risk of obtaining a false charge distribution, unless the faster particles can be reliably rejected from the sample.

The analysis of the nuclei in the region $34 \leq Z \leq 65$ is still in progress and will be reported elsewhere.

5. Acknowledgements

All the stacks were launched and recovered by Raven Industries, Inc. to whom we are indebted. This series of flights included one which exceeded 120 hours at float altitude, the longest planned duration attained to date for a zero pressure polyethylene balloon flight.

We wish to express our thanks to Marian Cahill, Mary Herwig, Rose Maharaj, Catherine Murphy, Hilary O'Donnell, Emily Rankin, Rosaleen Toner and Jenny Whitley for their assistance.

6. References

(1) P. H. Fowler, R. T. Thorne, A. P. Muzumdar, C. O Ceallaigh, D. O'Sullivan, Y. V. Rao and A. Thompson: Proceedings of the 13th International Cosmic Ray Conference, Denver, Colorado, Vol 5, 3239 (1973)

(2) P. B. Price and E. K. Shirk: Proceedings of the 14th International Cosmic Ray Conference, Munich, Germany, Vol 1, 268 (1975)

(3) E. K. Shirk, P. B. Price, E. J. Kobetich, W. Z. Osborne, L. S. Pinsky R. D. Eandi and R. B. Rushing: Phys. Rev. D7, 3220 (1973)

(4) W. Z. Osborne, L. S. Pinsky, E. K. Shirk, P. B. Price, E. J. Kobetich and R. D. Eandi: Phys. Rev. Lett. 31, 127 (1973)

TABLE 1

The second column contains the numbers of detected nuclei from those flights which were accepted for the determination of the energy spectrum. In this column each particle is given an energy to minimise the error in charge assignment. For higher energies, where the particles either interact in, or penetrate the detector, an adequate path length for energy determination is demanded. The third column contains the numbers of events fulfilling this condition. In addition, allowance must be made for the lack of detection at large zenith angles for particles of low etch rate and for the failure of particles with low energy at the top of the atmosphere to penetrate to the detector, again at large zenith angles. The numbers of events fully corrected as above are given in the fourth column and correspond to an exposure of $\simeq 100$ m^2 days under 3.8 g cm^{-2} atmospheric depth. Comparison spectra normalised to the data are given in the fifth and sixth columns.

Energy Interval at Top of Atmosphere (GeV/N)	Number of nuclei detected	Selected Nuclei	Corrected Data	Normal Spectrum ($\gamma^{-1.5}$)	Steep Spectrum ($\gamma^{-3.5}$)
0.6 - 0.8	10	10	13	17	37
0.8 - 1.0	12	12	16	13	21
1.0 - 2.0	33	23	36	34	35
> 2.0	25	13	39	39	11

FIG. 1 Charge distribution, uncorrected, of 284 selected nuclei with $Z \geq 34$ from nine flights during the period 1971 - 1974. The total exposure was $\simeq 120$ m^2 days at $\simeq 3.8$ g cm^{-2}. The percentage figures are the probabilities of selection computed on the supposition that the energy spectrum at the top of the atmosphere is the same as that of low charge cosmic ray nuclei.

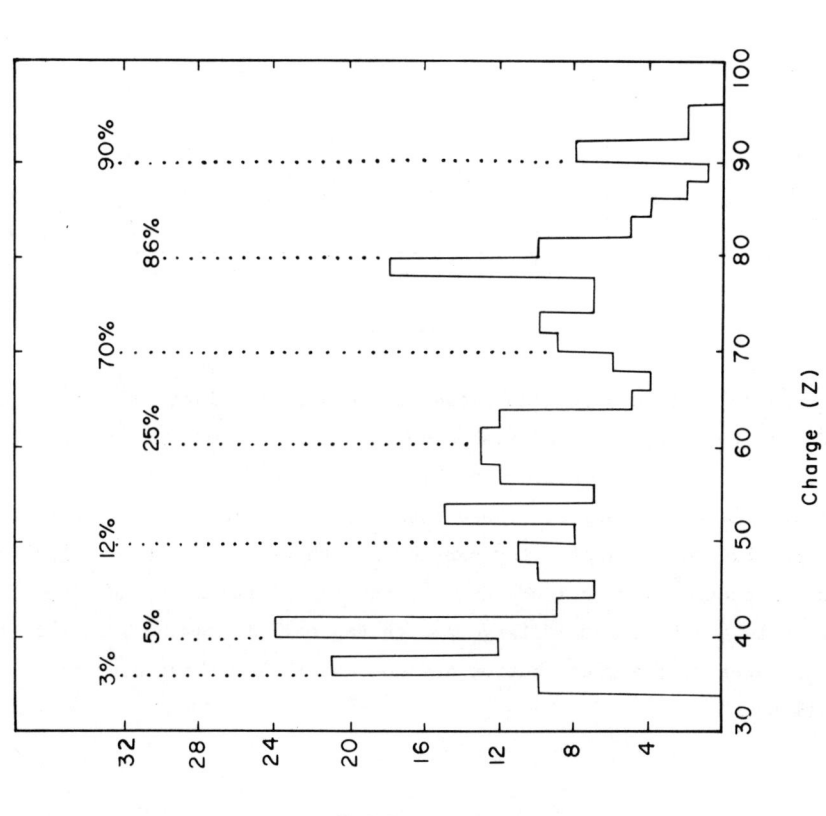

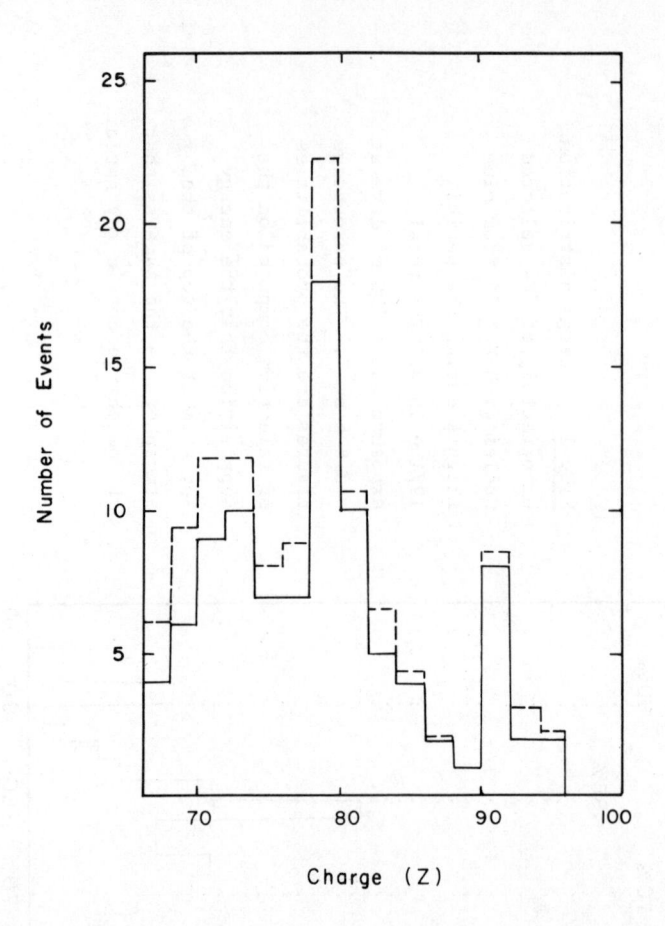

FIG.2 The solid line histogram shows the uncorrected data for all nuclei with Z > 65 for which the vertical etch rate exceeded 0.2 μm hr^{-1} at the first Lexan sheet. The dashed line histogram shows the data corrected for losses at wide zenith angle. These losses result from both the absence of tracks at wide zenith angles for particles with low etch rate and for the failure of particles with low energy at the top of the atmosphere to penetrate to the detector at large zenith angles. It may be seen that the major features of the distribution are not materially changed by the corrections.

MEASUREMENT OF THE COSMIC RAY ELEMENT ABUNDANCES BETWEEN ≃ 300 AND ≃ 750 MEV/N IN THE REGION FROM NICKEL TO KRYPTON USING LEXAN TRACK DETECTORS

P. H. Fowler*, D. L. Henshaw*, C. O'Ceallaigh**, D. O'Sullivan** and A. Thompson**

*University of Bristol, England
**School of Cosmic Physics, Dublin Institute for Advanced Studies, Ireland

ABSTRACT

This is an account of the preliminary results obtained in an experiment undertaken to determine the relative abundances of the elements with $28 \leq Z \leq 36$ in the Cosmic Radiation. Analysis of Lexan polycarbonate track detectors flown from Sioux Falls at an atmospheric depth of $\simeq 3.8$ gm/cm^2 has yielded 139 nuclei in this charge region so far. Three different scanning procedures were adopted, two of which were designed to eliminate most of the background of iron group particles.

1. Introduction

To date, information on the Cosmic Ray element abundances in the region $28 \leq Z \leq 36$ is limited and published data from nuclear track detectors are quite meagre. One of the main difficulties involves discrimination against the overwhelming preponderance of the iron group nuclei.

2. Experimental Procedure

The Lexan stacks used in this investigation were flown at various times between 1972 and 1974, at an average altitude of $\simeq 3.8$ gm/cm^2, from Sioux Falls, South Dakota. Details of these stacks are given in our first paper[1] presented at this Conference. The modules being studied have either 150 or 200 sheets of Lexan, each sheet 250 μm thick, along with two sheets of emulsion. The events examined so far can be conveniently divided into three categories depending on the type of scanning used to locate them.

(a) Batches of consecutive sheets of Lexan were etched for 96 hours at 40°C in 6.25 N NaOH solution saturated with etch products. The conditions of etching were such that each stopping particle with $Z \geq 26$ would produce a cylinder in at least one sheet. A sample of $\simeq 1500$ events, located by the ammonia scanning technique, were fully measured and a total of 43 events with $Z \geq 28$ were extracted. This procedure is very time consuming and was designed to select events in the Fe peak for isotopic analysis, but the sample with $Z \geq 28$ is completely free of any scanning biases.

(b) A second sample was located by requiring that a particle have at least two cylinders under the same etching conditions as in (a). Each 2-cylinder event was then subjected to further examination by measuring the dip angle and depth of penetration, z, of an etch cone 4 plates above the stopping point. By reference to a series of curves of z as a function of dip angle and position of the stopping point of the particle in the final plate, the bulk of iron peak events were eliminated by choosing only those that corresponded to $Z \geq 27$. This procedure increased the rate of data acquisition by a factor of $\simeq 6$.

(c) In order to locate the ultra-heavy events every third sheet of the Lexan stacks was etched and ammoniated to search for cylinder coincidences. We include here the candidates found in the region $30 \leq Z \leq 36$ with appropriate scanning efficiency corrections.

The total area of Lexan examined for each group, along with the mean cylinder density is given in Table 1.

TABLE 1

Sample	m^2	cylinders/m^2
(a)	3.7	$\simeq$ 680
(b)	8.7	$\simeq$ 1330
(c)	411.0	$\simeq$ 800

3. Analysis

By using the expressions for the track etch rate, v, namely

$$v = aJ^n \qquad (i)$$

where J is given by

$$J = \frac{Z^2_{eff}}{\beta^2} \{2\ln(\beta\gamma) - \beta^2 + K - \delta(\beta)\} \qquad (ii)$$

and assuming that the peak in the Iron group is ^{56}Fe, the constants in Eq.(i) were determined using a value of $K = 20$. Sample (a) events were analysed as described in reference (2). The resolution was found to be $\simeq 1$ AMU indicating good charge determination. The events selected by procedure (b) were measured in the same manner as sample (a) and the charge values for both samples are shown in Table 2.

TABLE 2

Sample \ Z	28	29	30	31	32	33	34	35	36
(a)	40	1	1		1				
(b)	64	6	4		1	1	2		
(c)			10	1		1	1	3	2
(a+b+c) normalised to Ni = 100	100	6.8±2.7	7.5±2.0	$0.19^{+0.26}_{-0.19}$	0.57±0.2				0.1
Binns et al [3]	100	9.9±3.2	8.7±3.1		1.1±0.28				
Solar	100	2	3.4		0.49				

The ratio of the number of events identified as Cobalt to those of $Z = 28$ is similar for both samples (a) and (b), namely 37/40 and 66/64, indicating that the selection procedure for sample (b) did not exclude any appreciable amount of Nickel while at the same time eliminating the majority of the "Fe peak" events. We estimate that the Gaussian tail of the Iron group events accounts for $\simeq 65\%$ of the nuclei identified as $Z = 27$.

The total number of events with $30 \leq Z \leq 36$ in sample (c) is also given in Table 2. The number of nuclei in the charge region $28 \leq Z \leq 36$

found to date for all three samples is 139. The nuclei with $Z < 30$ were chosen from (a) and (b) only.

4. Discussion

The fourth row of Table 2 displays the relative abundances of the various elements normalised to Ni=100 with scanning efficiency corrections applied. The results reported by Binns et al[3] who used a combination of Cerenkov counters, ionisation chambers and scintillators are also summarized, along with some solar values compiled by Cameron.

There is good agreement between our preliminary data and those of Binns et al[3]. The Zn/Ni ratios are similar, indicating a higher value for Cosmic Ray nuclei than for the solar system. For higher charges there is closer agreement with solar values but the sample is not yet large enough for this result to be very significant. It was possible to determine the Ni/(Z=25,26,27) ratio from sample (a) where $\simeq$ 1500 events were studied. The value obtained was 0.044 ± 0.009 which is consistent with that quoted by Benegas et al[4], namely 0.05 ± 0.004.

By continuing our scanning and analysis we hope to increase the statistics above $Z = 28$ and determine the abundances with greater precision and eventually provide information from the Iron to the Uranium group in conjunction with our ultra heavy studies.

5. Acknowledgements

We wish to thank M. Cahill, C. Murphy, H. O'Donnell, E. Rankin and R. Toner for their invaluable assistance throughout this project.

6. References

1. P. H. Fowler, C. Alexander, V. M. Clapham, D. L. Henshaw, C. O Ceallaigh, D. O'Sullivan, A. Thompson: The 9th International Conference on Solid State Nuclear Track Detectors, Neuherberg/Munchen, 1976

2. V. M. Clapham, P. H. Fowler, C. O Ceallaigh, D. O'Sullivan, A. Thompson: Proceedings of the 14th International Cosmic Ray Conference, Munich, 12, 4128, 1975

3. W. R. Binns, J. I. Fernandez, M. H. Israel, J. Klarmann, R. C. Maehl, R. A. Mewaldt: Proceedings of the 13th International Cosmic Ray Conference, Denver, 1, 260, 1973

4. J. C. Benegas, M. H. Israel, J. Klarmann, R. C. Maehl: Proceedings of the 14th International Cosmic Ray Conference, Munich, 1, 251, 1975

SEARCH FOR TRAPPED VAN ALLEN BELT PARTICLES HEAVIER THAN HYDROGEN IN SATELLITE-EXPOSED NUCLEAR EMULSIONS

R. C. Filz

Air Force Geophysics Laboratory, Hanscom AFB, Bedford, Mass. 01731, U.S.A.

Two recent "indirect" studies[1,2] suggest that energetic $Z>1$ nuclei populate the inner Van Allen belt at least in small numbers. The trapped heavy interpretation of Skylab Four light flash observations[1] can be criticized on grounds of inconclusiveness.[3] The plastic observations of the Chan et al[2] do not pinpoint where the heavy particles are detected in B,L space, and do not give any evidence that they are trapped. We examine here angular distributions of heavy nuclei ($Z \geq 6$) observed in oriented Air Force satellites and investigate whether at least some of these particles could have been detected within the trapped particle mirror plane within the South Atlantic anomaly. 18-70 MeV alpha particle observations with electronic counters are also examined to evaluate upper limits for the inner zone. Further studies of $Z \geq 6$ in the existing nuclear emulsions are suggested and possible means of obtaining further information by exposure of passive materials on the space shuttle are examined.

---

[1] L.S. Pinsky, W.Z. Osborne, R.H. Hoffman, J.V. Bailey, Science 188, 928 (1975).

[2] H. Chan and P.B. Price, Phys. Rev. Letters, V35, 539 (1975).

[3] P.L. Rothwell, R.C. Filz and P.J. McNulty, "Light Flashes Observed on Skylab 4 - The Role of Nuclear Stars" (Manuscript submitted to Science).

LUNAR AND METEORITIC MINERAL TRACK DETECTORS AND THE COMPOSITION OF THE GALACTIC COSMIC RADIATION

W. Krätschmer

Max-Planck-Institut für Kernphysik 6900 Heidelberg, Germany

Abstract

A method has been developed to identify heavy ions from their etchable tracks in feldspars, by measuring the track etching rates and the residual ranges of individual tracks. It was found that a resolution of one charge unit can be achieved if irregularly etched tracks are eliminated from evaluation. To study the long term average of the galactic cosmic ray composition, feldspar crystals of lunar and meteorite samples were etched according to this method. The results indicate that the average over the last 60 m.y. and the present day cosmic ray (Cr+Mn)/Fe abundance ratios are about equal.

Introduction

From studies of fossil tracks in minerals of lunar and meteoritic samples the elemental abundances of the track-forming cosmic ray ions (Price et al. 1971; Price et al. 1973) can be deduced. The history of cosmic ray composition can be investigated by comparing track data obtained from samples with different exposure histories.

In this report, methods of identifying the charges of heavy cosmic ray ions from their tracks in feldspars are described, and results of cosmic ray composition measurements on feldspars are presented.

Ion Identification

In contrast to pyroxenes, where charge assignments can be performed with some accuracy by length measurements on TINT tracks (Lal, 1969) for feldspars one is forced to determine the track etching rate as a function of residual range for ion identification (Krätschmer et al. 1975).
The increase if track length between two subsequent etching procedures has been used as a measure of the track etching rate. This increase has been determined by a marking method (Fig. 1). After a first regular etching one marks the high energy end of the track by means of a special etchant. For this purpose, an etchant was chosen which reacts much less vigorously with the radiation damage along the latent track. Hence, after the marking etching, the track is broader but not significantly longer. The increase of track length during a second regular etching id determined easily from the marking. The total length of the track has been

Fig. 1: The procedure of the marking method by which track etching rate and the corresponding residual range of the ion has been determined.

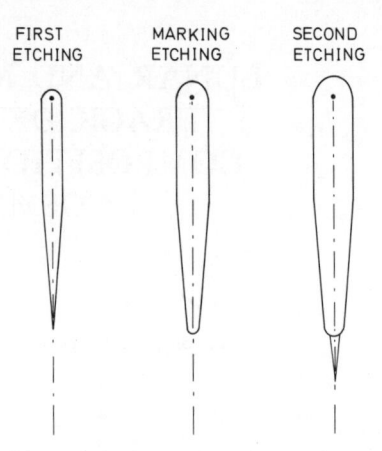

10 MeV/n Fe TRACKS IN BYTOWNITE

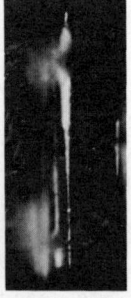

Fig. 2: Tracks of Fe ions in a terrestrial feldspar crystal, etched according to the marking method. Note the decrease of etching rate with increasing residual range.

10 MeV/n ION TRACKS IN BYTOWNITE

Fig. 3: The tracks of a Co(Z=27) and and Fe (Z=26) ion. Both kinds of ions can be distinguished by their different etching rate-residual range relationships.

used as a measure of the corresponding residual range. This technique has been checked and improved by studying tracks of artificially accelerated 10-MeV/n heavy ions in terrestrial feldspars.

Tracks of Fe ions in bytownite after etching with the marking technique are shown in Fig.2. The track etching rate gradually decreases with increasing residual range in a manner which is characteristic for Fe ions. Different ions can be distinguished by their different relationships between track etching rate and residual range. For example, Fig. 3 shows the track of an Fe and Co ion etched under similar conditions. The increases of their track length (etching rates) are about equal, while their total lengths are quite different.

During the studies on tracks formed by artificially accelerated heavy ions it was found that the track etching rate residual range distributions are generally much broader than expected. Since a contamination of the beam with other ions could be ruled out, the scatter is probably due to irregularities in the track etching process. In fact, it was observed that some of the tracks showed an unusually small increase of track length, which implies that interruptions in the track etching process took place. To eliminate such peculiarly etched tracks with sufficient efficiency from further evaluation, the marking and re-etching procedure on the tracks has been performed twice. By this double marking, the total length increase of a track has been divided into two, optically visible parts. Irregularities during etching therefore result in corresponding irregular sequences of the length increases. Fig. 4 demonstrates how this procedure helps to improve ion identification. The open circles represent results obtained in a scan from

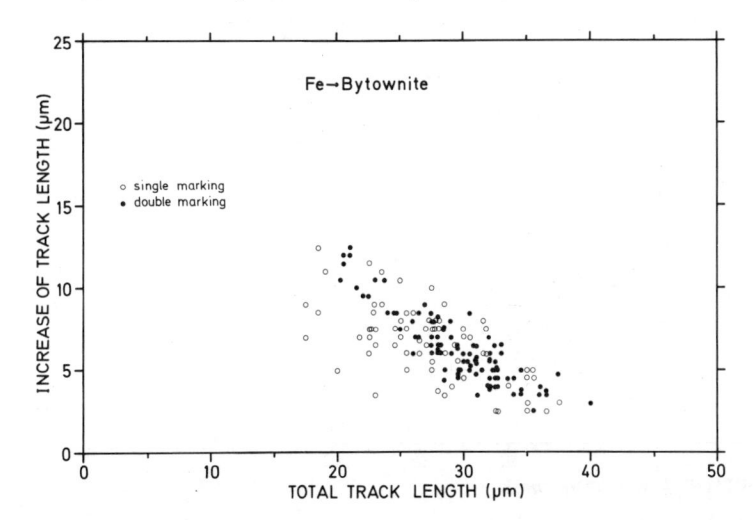

Fig. 4: The improvement of the charge resolution by the double-marking procedure.

singly marked Fe tracks and the closed circles refer to doubly marked tracks in the bytownite crystal. From the distribution of singly marked tracks alone, one might have concluded that the beam is contaminated with lighter ions, however, these tracks are in reality irregularly etched Fe tracks.

It was found that the double-marking method provides a charge resolution of about one charge unit.

Experimental Procedures

The terrestrial feldspar crystals (bytownite) were polished at two perpendicularly oriented surfaces and one surface was exposed to 10 MeV/n heavy ions. To generate TINTs on the heavy ion tracks, the other surface of the crystal was irradiated with 10 MeV/n Kr ions.

The polished meteorite and lunar rock sections were irradiated with 10 MeV/n Fe at 10-20° inclination angle to the surface. They served as a calibration for the feldspar crystals. Similar to the terrestrial samples, 10-MeV/n Kr tracks served as etching channels to develop simultaneously calibration and cosmic-ray tracks in the interior of the lunar crystals.

For regular etching a boiling solution of 7.5% by weight NaOH was used. The etching times were adjusted to the track density and ranged from 12 to 24 hr for the first and 4-15 hr for the second etching. For double marking, the re-etch time was divided into two equal intervals. A 0.5% HF solution served as a marking etchant. Etching times of 5-15 min at room temperature were found to be sufficient to achieve a satisfactory marking.

Results and Discussion

Figure 5 and Fig. 6 depict the distributions of track length increases (track etching rates) and total track lengths (residual ranges) measured on fossil cosmic ray and Fe calibration tracks on the meteorite Patwar and on lunar rock sample 75035, respectively. Calibration data on 10 MeV/n A, Fe, Co, and Cu ion tracks, obtained in terrestrial feldspar (bytownite) have been used to estimate the probable positions of Cr and Mn as they are indicated in Fig. 5. The most striking difference between the meteoritic and

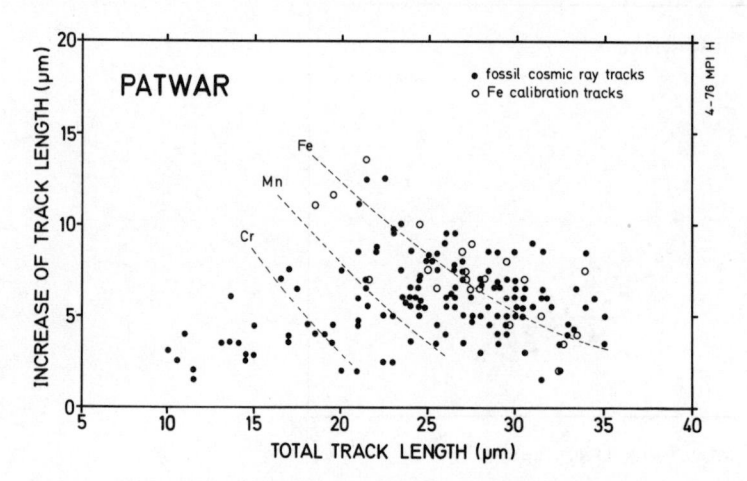

Fig. 5: The distribution of doubly marked galactic cosmic ray and Fe calibration tracks in feldspars of the meteorite Patwar.

the lunar distributions is the relative number of tracks in the Cr and Mn region compared to those of Fe. Provided that the fossil tracks in these samples have not been seriously affected by annealing effects, (Cr+Mn)/Fe abundance ratios of about 0.35 for the Patwar meteorite and 0.9 for the lunar sample can be deduced. The difference in the abundance ratios very probably is due to a difference in the exposure conditions of both samples rather than due to a long term change in the cosmic radiation. For the Patwar meteorite, the cosmic ray proton and track exposure age are in good agreement (60 m.y.), whereas for rock 75035 these two ages are rather different: 70 m.y. and 7 m.y. , respectively (Crozaz et al., 1974). The difference in the proton and track exposure ages is indicative for a complex burial history of the sample. Rock 75035 was probably exposed at such large shielding depths that the flux of Cr and Mn, produced by nuclear disintegration from Fe was significant.

The Patwar meteorite, however, seems to have had a rather primitive exposure history. The cosmic ray (Cr+Mn)/Fe ratio deduced from the data is in

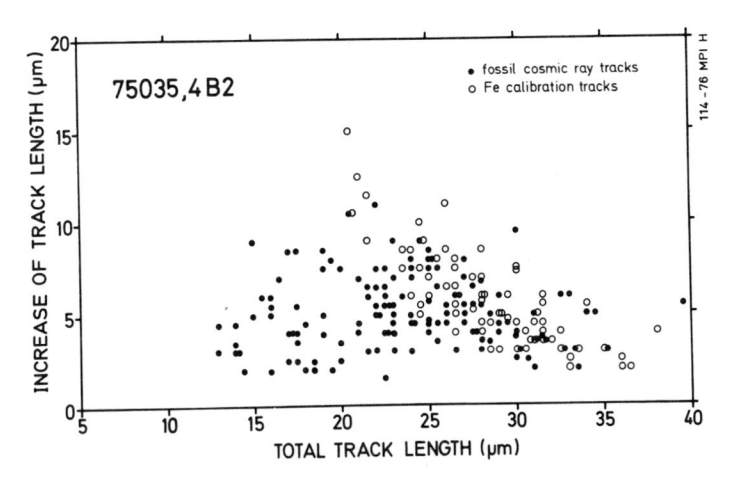

Fig. 6: The distribution of tracks measured on a sample of lunar rock 75035.

agreement with the ratio in the present day cosmic radiation. Annealing effects are assumed to be insignificant. Provided that most of the tracks ascribed to to Cr and Mn are in reality partially annealed Fe tracks, the abundance ratio measured in Patwar would be lower than the present day ratio. Although this possibility cannot be completely ruled out, at this state of investigation it seems to be more reasonable to argue that the tracks in the Patwar meteorite are not affected by annealing and that during the last 60 m.y. no significant compositional changes in the very heavy cosmic ray component have taken place.

Acknowledgements

The generous help of the LINAC staff of the University of Manchester is gratefully acknowledged.

References

Crozaz et al. (1974): Proc. Lunar Sci. Conf. 5th, p. 2475-2499

Krätschmer et al. (1975): Proc. Lunar Sci. Conf. 6th, p. 3577-3585

Lal (1969): Space Sci. Rev. 9, p. 623-650

Price et al. (1971): Ann. Rev. Nucl. Sci. 21, p. 295-333

Price et al. (1974): Earth Planet. Sci. Lett. 19, p. 377-395

STUDIES OF FRESH AND FOSSIL TRACKS IN METEORITIC HYPERSTHENE

R. K. Bull and S. A. Durrani

Department of Physics, University of Birmingham, Birmingham B15 2TT, England

Abstract

From measurements of the total recordable lengths of fossil tracks and Fe ion tracks in meteoritic hypersthene, the abundance of Fe relative to all VH ions ($20 \leqslant Z \leqslant 28$) in the time-averaged cosmic ray flux is found to be ~ 0.35. After allowance is made for the effects of fragmentation of nuclei at a depth of ~ 11 cm within the meteorite, this value is found to be in broad agreement with present day measurements.

These crystals have been irradiated with a number of heavy ions with a view to calibrating them for the identification of VVH cosmic rays by measurement of track etch velocity, V_T, as a function of residual range. A good fit to the calibration data is obtained by using a value for the constant K in the primary ionisation equation of 10.5 ($\pm$ 0.5). The relationship between V_T and primary ionisation J, appears to be approximately linear.

Introduction

Meteoritic and lunar rock crystals provide a means of studying the composition of heavy cosmic ray particles averaged over long periods of time. Both of these sources have their own different areas of usefulness. Lunar material contains an excellent record of low energy solar flare ions but interpretation of their track record is often complicated due to the multi-stage irradiation history of many lunar rocks. Many meteorites, however, appear to have had a single stage cosmic ray irradiation corresponding to the interval between the break-up of the parent body and capture by the earth. Meteorites therefore provide a useful means of detecting heavy cosmic ray particles with energies in the region of 1 GeV/nucleon.

We have studied the characteristics of tracks in meteoritic hypersthene, in an attempt to gain information on the time averaged composition of the heavy cosmic rays.

I VH COSMIC RAYS: TOTAL ETCHABLE TRACK LENGTHS.

The simplest method of charge assignment in crystals is to measure the total length over which a track is etchable. The use of this method rests on the assumption that each particle, on slowing down in a track storing medium, causes sufficient damage to form a track only when its rate of primary ionisation, J, exceeds a threshold level Jc. The length of trajectory over which this condition holds depends on the charge and mass of the incoming particle. Although recent work (1) has suggested that the concept of a sharp registration threshold is not strictly accurate, the total etchable track length method still works reasonably well in practice, as a track will generally etch out quite rapidly to a certain length and then grow much more slowly with further etching. Provided that the etching conditions for both fossil and calibration tracks are kept closely similar it should be possible to obtain an estimate of the charge of the track forming particles.

A convenient method of determining the total etchable length of a track is to etch out those entirely contained within the crystal by means of the track-in-track and track-in-cleavage (TINT and TINCLE) techniques (2).

For most of the work described in this paper hypersthene crystals from the diogenite Shalka were utilised. These crystals have dimensions of up to several millimetres and show fossil track densities up to 2×10^6 tracks.cm^{-2}. In a previous paper (3) we estimated the thickness ablated from this meteorite to be ~ 6 cm and the pre-atmospheric depths of seven sampling sites within the fragment BM33761 were calculated, ranging from 6 to 13 cm.

Also in this paper (3) we showed a reasonable correspondence between the main peak of the fossil track length distribution and the lengths of fresh Fe ion tracks, both being measured as tracks in cleavages. An estimate of the abundance of Fe relative to all VH ions ($20 < Z < 28$) of ~ 0.38 was obtained, in reasonable agreement with present day estimates (4) after allowance is made for nuclear fragmentation occurring within the stony meteorite material.

Relying on natural cleavages to etch out subsurface tracks is, however, a somewhat haphazard procedure and for further measurements poloshed crystals were irradiated with Fe and Kr ions at 9.6 MeV/nucleon from the Manchester University Linear Accelerator (LINAC) in directions at right angles to each other. In most cases the samples were covered with aluminium foils to reduce the residual ranges of the ions to ~ 30 µm.

The samples were then etched in 60% (by weight) NaOH solution, boiling under reflux, for 2 hours. The Kr etch channels (~ 5×10^5 cm^{-2}) acted as host tracks for the fossil and fresh Fe ion TINTS.

The track lengths were measured using a filar micrometer eyepiece attachment and a total magnification of 600x.

Results and Discussion

Figure 1 shows the length distribution of (a) fossil and (b) Fe ion tracks in Shalka hypersthene. The form of the fossil track distribution is similar to those obtained for TINCLES under similar etching conditions (3). A peak in the region 12-15 µm is apparent. The Fe ion track lengths show considerable spread but are peaked at 14-17 µm. The centre of this peak is displaced by about 2 µm from that of the most prominent fossil track peak. It seems reasonable to suppose that this peak is due to cosmic ray Fe and that a small degree of annealing has taken place.

If the extent of the proposed fossil Fe peak is taken to be from 12 to 15 µm, then the ratio $f(Fe)/f(VH)$ is found to be 0.35. This is close to the value of 0.38 obtained in our earlier work. As before this may be shown, after correction for fragmentation at a depth of 11 cm in stony meteorite material (using a fragmentation parameter of 0.25 (5)), to be compatible with the value of [Fe]/[VH] of 0.51 for the energy region > 850 MeV/nucleon calculated from the present day abundances found by Webber et al. (4).

However, the large spread in the length distribution of fresh Fe ion tracks suggests that this type of analysis may be subject to considerable errors. The fossil tracks of lengths up to 17 µm may be attributable to ions such as Ni and Co but it is not possible, from a study of the Fe ion tracks, to rule out cosmic ray Fe as the source of these longer fossil tracks.

Total track length measurements, whilst giving some crude estimate of the relative abundance of VH nuclei, are less useful in studying VVH tracks. The total etchable lengths of tracks due to ions of Z >> 30 are likely to be large, typically several hundred microns, and the requirement that they be fully etched out is often difficult to fulfil as crystals tend to disintegrate after long etch times.

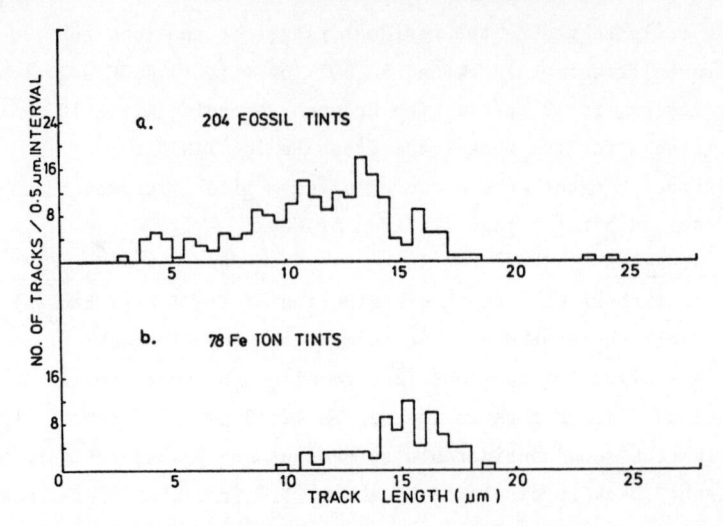

Fig. 1. TINT length distributions for (a) fossil and (b) fresh Fe ion tracks in Shalka hypersthene.

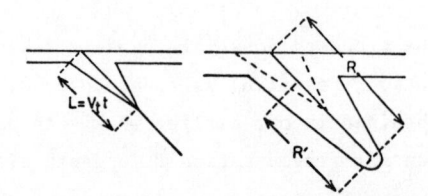

Fig. 2. The initial etch stage gives the track etch rate. To a first approximation this applies to a residual range $R'(=R-L/2)$ as shown.

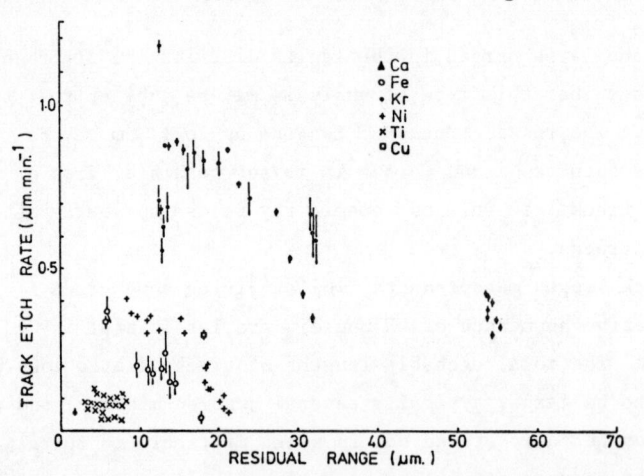

Fig. 3. Track etch velocities against residual range for 6 ions in Shalka hypersthene. Points without error bars are data from L-R plots and each such point represents a single track.

II ETCH RATE VERSUS RESIDUAL RANGE.

It is well known that the rate of chemical attack along a track varies with the residual range and is a function of the primary ionisation of the track forming particle (6).

In plastics the most usual method of measuring the etch rate as a function of residual range is to measure the lengths of etch cones at a number of different surfaces after the particle has passed through several sheets of plastic. An analogous procedure is not generally possible in crystals but by etching a crystal surface and measuring track etch cones, values of the etch velocity V_T may be found. The residual range is then found by etching the crystal for a much longer time, until the tracks are fully etched, and then re-locating the track measured, after the initial etching stage. These measurements should, in principle, be sufficient to identify the track forming particle. The advantage of this method is that only a fairly small portion of the total etchable range need be present in the crystal for particle identification to be made.

Figure 2 illustrates the method. The total cone length L is found simply from measurement of the projected track length s, and the depth of the cone tip z from $L = (z^2 + s^2)^{\frac{1}{2}}$. The use of this simplified formula introduces only small errors provided $V_T \gg V_G$, typically < 2%.

Calibration of the meteorite crystals

The Shalka hypersthene crystals were irradiated with various heavy ions, Fe, Kr, Ni and Cu from the Manchester LINAC and Ca and Ti from the heavy ion accelerator at Dubna. In order to obtain etch rate data at a number of different residual ranges two methods were used. Samples were covered by aluminium foils of various thicknesses from 10 to 50 μm to obtain reduced energies. The beam energies usually obtained for the Manchester irradiations were 9.6 and 4.14 MeV/nucleon and from Dubna, 5 MeV/nucleon. Secondly, samples were polished at a shallow angle to the surface, after irradiation. This is the L-R plot method described by Price et al. (1). In this way a continuous distribution of residual ranges may be obtained by scanning across the crystal.

Samples were etched as before with boiling 60% NaOH under reflux. Freshly prepared solutions were used in all cases and each solution was replaced after each 30 minutes of etching.

LR plots were usually made after 30 minutes of etching. For

the other samples initial cone lengths were measured at an etch time of 30 minutes and residual ranges determined after prolonged etching, usually 1-3 hours. Measurements were made, as described in section I, to a precision of ± 0.3 μm.

In the case of the lighter calibration ions (Ca, Ti) and for Fe ions at long residual ranges the condition that $V_T \gg V_G$ is not so clearly satisfied as for VVH tracks and in order to calculate the etched length L the equations

$$L = \frac{s - B \tan \tfrac{1}{2} (\delta - \theta)}{\cos \delta}$$

and

$$\sin \theta = \frac{B \cos \delta}{s - B \tan \tfrac{1}{2} (\delta - \theta)}$$

derived by Henke and Benton (7) were solved simultaneously. s is the measured projected length, δ the dip angle is known (for calibration tracks) and the layer etched from the undamaged surface B (= V_G x etch time) was computed using a value of $V_G = 0.02$ μm.min^{-1}. This value of V_G was determined by measuring the diameters of VVH track openings, these being very nearly equal to $2V_G t$. Even for Ca ion tracks, however, this determination of L differs by only ~ 5% from an estimate from the formula $L = (s^2 + z^2)^{\tfrac{1}{2}}$.

Results and Discussion

Track etch rates are shown as a function of residual range in Figure 3. There appears to be a real separation of Fe and Ni data indicating a possible charge resolution of about ± 2 charge units. Two different LR plots for Ti gave significantly different results, possibly due to crystallographic effects.

From this data it is possible to construct a master curve of V_T vs primary ionisation J. It is first necessary to convert the measured ranges to real residual ranges, i.e. to take account of the range deficits due to non-etchability of the low energy end of the track. Experiments indicate a range deficit of 2.8 μm for Kr. Although measurements on range deficits for other ions have not been made, they are estimated at 3.3, 3.2 and 3.1 μm for Fe, Ni and Cu. Fortunately errors in these quantities do not greatly affect the total residual ranges.

Using the range energy data of Northcliffe and Schilling (8), track etch velocity V_T has been plotted against primary ionisation using different values for the constant, K. A value of this constant is chosen such as to give a smooth fit between data for different ions. A value of K = 9 is clearly too low and the best estimate is K = 10.5 (± 0.5) (see Figure 4).

It would seem that the data may adequately be represented by a linear relationship between V_T and J, over the range of values studied so far.

It will be noted that the etch rate rises sharply and defines a threshold value of 24.4 (± 0.2), using this threshold predicts range deficits quite close to those estimated above.

Studies of fossil VVH tracks

VVH cosmic ray tracks in hypersthene from the Shalka and Patwar meteorites have been studied. This work is still in a very early stage and as yet no charge spectrum has been determined.

The Shalka crystals show track density ratios ρ_{VVH}/ρ_{VH} of $\sim 10^{-4}$ and for Patwar samples $\sim 6 \times 10^{-4}$. Guided by the results shown in Figure 3, VVH tracks were taken as those with lengths ~ 15 μm after an initial etch time of 30 minutes.

The difference in ratios arises because the Shalka samples were shielded by between $\sim 6-13$ cm (3) and the Patwar samples by only ~ 2 cm of material (9). Heavy ions are more readily stopped than lighter ions and are enriched in the upper layers of the stopping medium.

The Patwar crystals will therefore be more useful in studies of VVH tracks, although since the calibration work was carried out on crystals from Shalka it will be necessary to check for any systematic differences in response between hypersthene from Shalka and Patwar.

Conclusions

Studies of fossil TINTs and TINCLEs in Shalka hypersthene have revealed no evidence for gross changes in the chemical composition of the VH cosmic rays over the last $\sim 2 \times 10^7$ yrs (the cosmic ray age of Shalka). The total track length method does not seem very promising for studies of fossil VVH tracks. Measurements of track etch rate as a function of residual range may provide a better means of identifying these particles.

A start has been made in calibrating Shalka hypersthene crystals and an approximately linear relationship between V_T and primary ionisation has been indicated. It will now be possible to compute

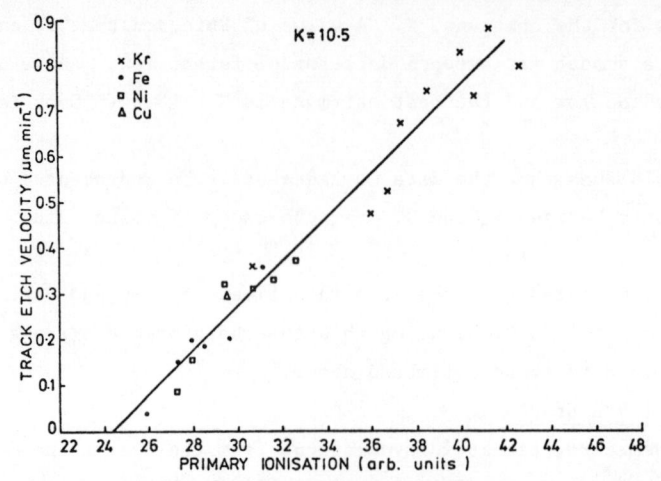

Fig. 4. V_T plotted against primary ionisation J, assuming a value for K in the primary ionisation equation of 10.5.

curves of cone length against residual range (for a given etch time) for all ions. Before fossil track data can be assessed however, the possible effects of long-term low temperature annealing on the fossil tracks must be determined.

Acknowledgements

We wish to thank Dr. R. Hutchison of the British Museum (Natural History) for providing samples of Shalka and Professor D. Lal for providing samples from Patwar. Also we would like to acknowledge the financial support of the S.R.C.

References

(1) P.B. Price, D. Lal, A.S. Tamhane and V.P. Perelygin, Earth and Planet. Sci. Lett. 19 (1973) 377.

(2) D. Lal, R.S. Rajan and A.S. Tamhane, Nature 221 (1969) 33.

(3) R.K. Bull and S.A. Durrani, Earth and Planet. Sci. Lett. To be published.

(4) W.R. Webber, S.V. Damle and J. Kish, Astrophys. and Space Sci. 15 (1972) 245.

(5) R.L. Fleischer, P.B. Price, R.M. Walker and M. Maurette, J. Geophys. Res. 72 (1967) 331.

(6) P.B. Price and R.L. Fleischer, Ann. Rev. Nucl. Sci. 21 (1971) 295.

(7) R.P. Henke and E.V. Benton, Nucl. Instrum. and Meth. 97 (1971) 483.

(8) L.C. Northcliffe and R.F. Schilling, Nucl. Data Tables. A7 (1970) 233.

ISOTOPIC COMPOSITION OF COSMIC RAY NUCLEI

W. Enge

*Institut für Reine und Angewandte Kernphysik, University of Kiel,
23 Kiel, West Germany*

Abstract

A review will be given on the role of cosmic ray isotopes as tracers of the astrophysical nucleo-synthesis. The products of every nuclear burning chain are first of all isotopes and not elements. Thus, it is the study of the isotopes rather than that of the elements that responds to the questions on these nucleo-synthetic reactions. The problems concerning the solar system isotopic abundances and the cosmic ray isotopic abundances as well as a comparison between both will be presented. Furthermore the present stage of the experimental techniques and the latest results will be discussed.

I Evolution of the galaxy and stars

In recent years a great deal of work has been put into a new field of cosmic ray research. Since it has become possible to measure besides the elemental also the isotopic composition of these cosmic radiation, a new important branch could be added to cosmic ray research to study astro- and nuclear physical phenomena. Here a brief review will be given on the role of cosmic ray isotopes as tracers of the stellar nucleo-synthesis. The products of every nuclear burning chain are first of all isotopes not elements. Thus it is the study of the isotopes rather than that of the elements, which responds to the questions on these nucleo-synthetic reactions.

This review starts with a description of the development of elements and isotopes in the genesis of stars and galaxies. After the beginning of the universe, as it is calculated in a big-bang-model by Wagoner (1) only elementary particles up to charges of $Z = 2$ (He) or maybe $Z = 3$ (^{7}Li) are available. Out of interstellar clouds of this origional matter stars and galaxies are formed in gravitational collapses. After stars have entered the main sequence of the stellar evolution nuclear burning starts which results in creating by and by all heavier elements and isotopes.

This happens continuously over several star generations in a recycling mode of interstellar matter (fig. 1). Under very much simplified assumptions as indicated in fig. 1 the elemental development of the galaxy has been calculated in a model where the galaxy is considered a closed, homogeneous, well mixed gas volume (2,3).

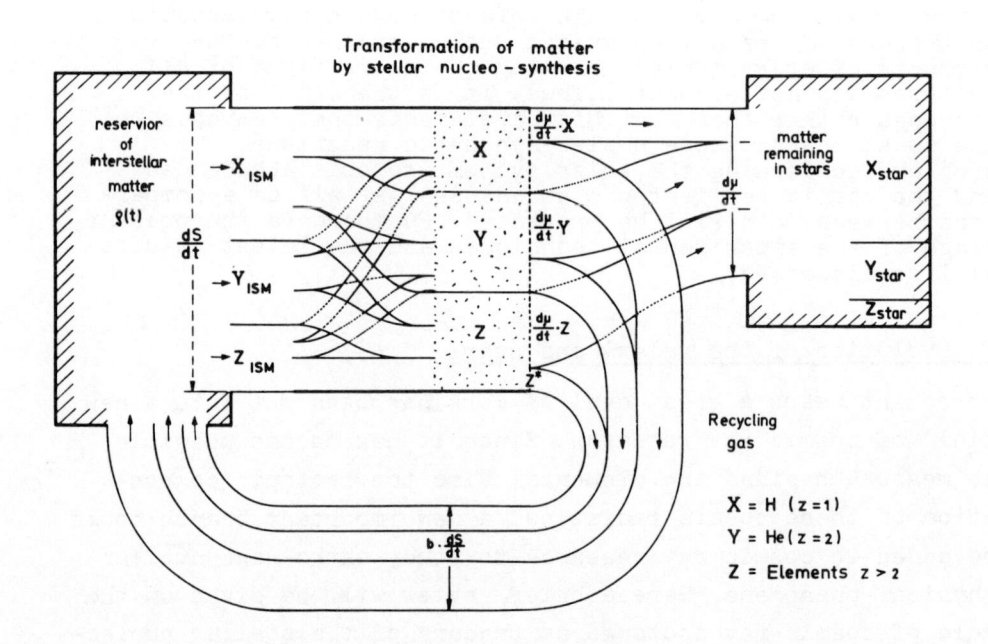

Fig. 1. Recycling of interstellar matter in the stellar nucleo-synthesis

Audouze (4) have obtained the temporal change of $^{12}C/^{13}C$ and $^{14}N/^{16}O$ as well in the galactic center (GC) as in the solar neighbourhood (⊙ and SS), see fig. 2.

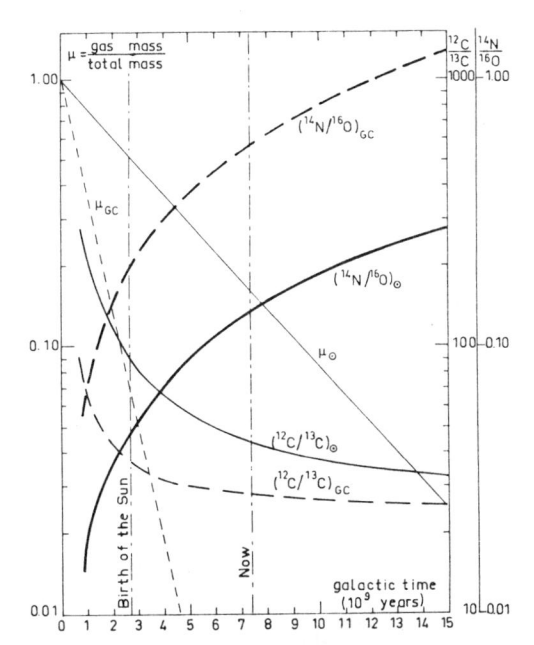

Fig. 2. Time evolution of $^{12}C/^{13}C$ and $^{14}N/^{16}O$ in the interstellar gas close to the sun (solid line) and at the galactic center (dotted line) (ref. 4).

This very simplified calculation nevertheless shows several effects in agreement with the observations: (a) the time variation of $^{12}C/^{13}C$ by about a factor 2 in the solar vicinity between the birth of the sun $4.6 \cdot 10^9$ years ago and now, (b) the strong spatial gradient of $^{14}N/^{16}O$ between solar vicinity and galactic center and (c) the lack of such a strong gradient for $^{12}C/^{13}C$. This different behaviour of the isotopic ratios results from different creation processes of the isotopes which is included in the above calculations. Some of the review experimental data of Audouze (4) may be summarized: $^{12}C/^{13}C(ISM) = 36$ and $^{12}C/^{13}C(SS) = 89-100$; $^{16}O/^{18}O(ISM) = 400$ and $^{16}O/^{18}O(SS) = 500$; $^{17}O/^{18}O(ISM) = 0.25-0.28$ and $^{17}O/^{18}O(SS) = 0.11-0.33$. For a comparison with fig. 2 the values ISM refer to the "birth of the sun" and the values SS (solar system) refer to "now", both for the solar vicinity curve. The authors (4) conclude from experimental data alltogether that an increase in the ISM of ^{17}O and ^{15}N in analogy to ^{13}C seems possible since the birth of the sun, such an increase maybe probably not significant for ^{18}O. Furthermore the $^{12}C/^{13}C$ ratios exhibit no significant variation with galactocentric distance.

II Nucleo-synthesis in stars

After this more general considerations I will briefly recall the important aspects of stellar nucleo-synthesis. This theory was the most important step forward in understanding the physical processes in stars and the development of chemical elements. In fig. 3 the situation is summarized (5).

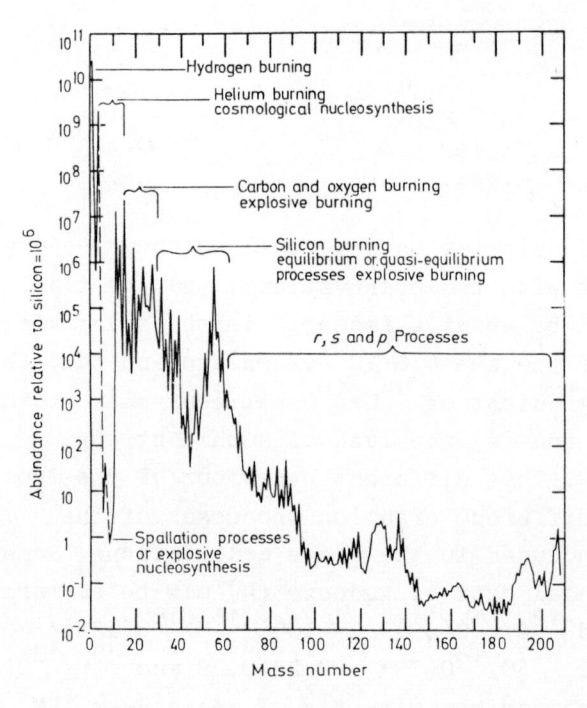

Fig. 3. Abundances of the nuclides as a function of their mass number according to Cameron (ref. 5, 6).

For cosmic ray aspects - as will be seen later - the most interesting nucleo-synthetic processes are those for the elements C, N, O, Ne etc. and for the Fe-region elements.

After the regular He-burning has created enough heavier elements like C, O, Ne etc. up Ar, the well-known CNO-cyclus transforms more H into He and also into ^{14}N whereby C and O react as catalysators.

Isotopic composition of cosmic ray nuclei

All nuclear burning continues in centric shells of the stars while the most advanced process burns in the stellar center. The heavier the mass of a star the more advanced to heavier elements the nuclear burning may proceed. Fig. 4 (7) is the attempt to demonstrate this situation. Elements of the

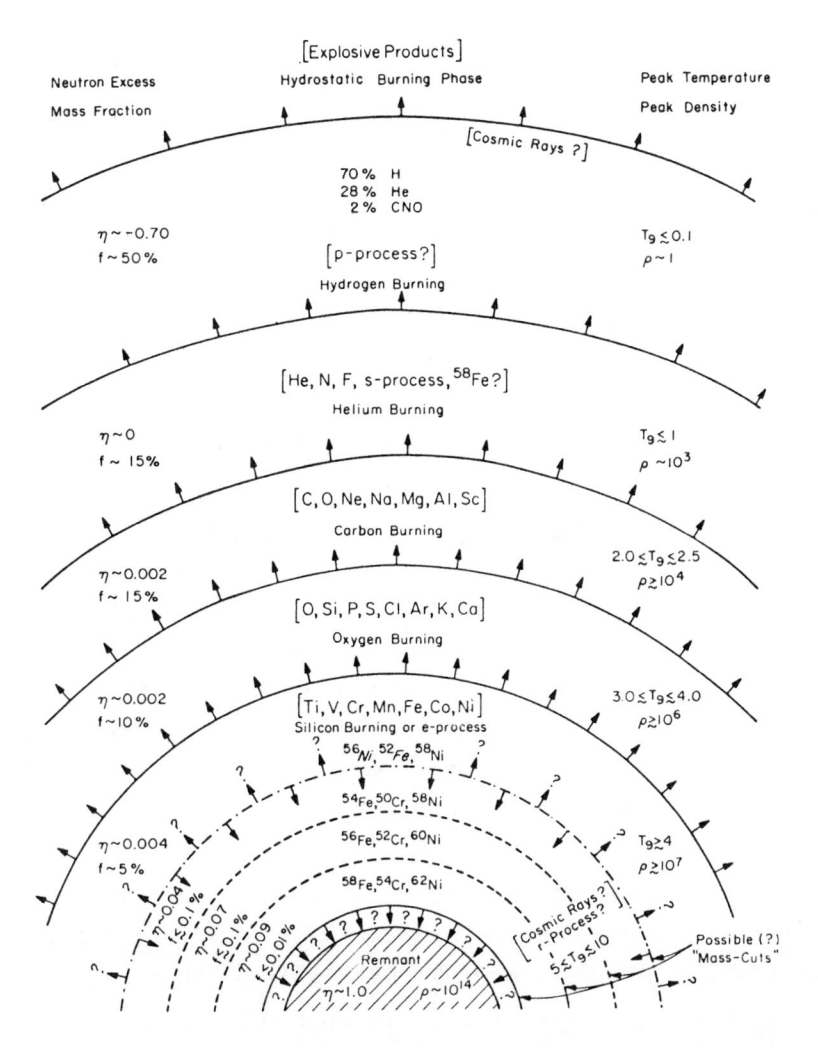

Fig. 4. "Onion shell model" of a star just before its super nova explosion. It is assumed that after the explosion a neutron star remanent exists. For "mass cut" which separates the remanent from the expanding gas cloud several possibilities are shown. The abreviations are: $T_9 = 10^{9}°K$; $\rho = [g/cm^3]$; $\eta = (N-Z)/(N+Z)$ and f = mass fraction in % (ref. 7).

Fe-region are the most stable nuclei concerning the nuclear binding energy. Here all regular burning chains - they may be hydrostatic or explosive - stop. Heavier elements are created by the r-, s- or p-process.

The reaction chains in the C, N, O-region are highly influenced by the amount of available fuel like H from outer shells and He from inner shells. Thus the theories of the development of the more unusual isotopes like ^{13}C, ^{15}N, ^{17}O, ^{18}O and maybe ^{22}Ne are based on the idea of disturbed and mixed shells (8). I like to emphasize that every nucleosynthetic chain is governed first of all by reactions between isotopes where nuclear stability reasons take care of an about equal number of protons and neutrons in the newly created nuclei.

In the Fe-region the neutron excess $\eta = (Z-N)/(Z+N)$ - where Z = number of protons and N = number of neutrons - is the dominant factor as will be seen later in fig. 15.

III Origin and propagation of cosmic ray nuclei

Everything said up to here refers to elemental or isotopic abundances in stellar objects or interstellar matter. These abundances, however, cannot be related without possible drastic changes to the nucleonic component of the cosmic rays. Cosmic ray nuclei propagate free in the interstellar space with kinetic energies of more than 50-100 MeV/Nuc up to 10^{20}-10^{21} eV, while the above mentioned nuclei are either cought in stars or are interstellar matter at rest.

It is here impossible to cover the whole field of theories of the origin and the propagation of cosmic ray nuclei, thus only briefly the main aspects can be touched.

In fig. 5 a schematic view of several recycling modes and of two possible theories of cosmic ray sources are given. Interstellar matter passes through different phases of stellar evolution over the stadium of red giants to supernova explosions. The matter blown off is recycled into the reservior

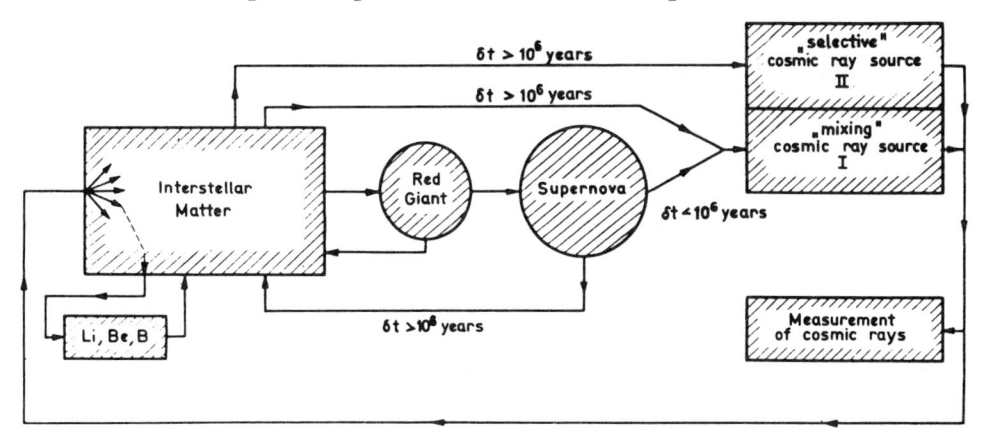

Fig. 5. Schematic view of recycling processes of interstellar matter showing stellar evolution, cosmic ray sources and the production of Li, Be, B by cosmic rays.

of interstellar matter or is accelerated to become cosmic rays. In theory I supernova explosions accelerate newly created nuclei after they have been mixed in an appropriate way with interstellar matter (9, 10) while in theory II a selective mechanism of ionisation and acceleration of primarily neutral interstellar matter takes place (11, 12).

While cosmic ray nuclei propagate and diffuse through interstellar magnetic fields to the earth, their abundance is influenced by various processes like e.g. fragmentation and energy loss in the interstellar matter and in the earth atmosphere and deceleration and disturbance by solar modulation and geomagnetic fields.

I cannot go here into any detail of the great effort that has been made to deduce from experimental data the cosmic ray source composition. One attractive method, however, how cosmic ray isotope measurements are used in this field, I like to mention. Radioactive isotopes of the Fe-region with special K-decay modes (ref. 13) may help to determine the time of acceleration in theory I, while radioactive isotopes like the well-known ^{10}Be, ^{26}Al or ^{36}Cl which are secondarily produced in fragmentations give us information on the propagation time and the spatial structure of the interstellar space.

IV Technique of measurement of cosmic ray nuclei and isotopes

A high energy, charged particle is characterized by its mass M, its charge Z and its velocity β. These desired parameters, however, can only be obtained indirectly measuring the kinetic energy E, the energy loss dE/dx, the range R, the yield of Cerenkov-light C and the magnetic rigidity P.

The wide field of nuclear detectors, which serve in cosmic ray measurements may be subdivided and classified in various ways. Nuclear detectors may be active or passive, they may record permanent the spatial track of the particle or they may give a temporally short charge or light signal, they may be of large area with large collecting factors or small with higher isotopic resolution, they may be analysed visually or electronically etc.

Not all nuclear detectors, however, are suitable for cosmic ray measurements. To measure cosmic ray nuclei not fragmentated the experiments have to be carried by balloons, rockets or satellites to the top of the earth atmosphere or outside of it. Thus strong restriction are put to experimental equipment according to its weight, size, energy supply, reliability etc.

Furthermore the wanted nuclei which arrive at random time, direction, energy and composition at the earth, have to be collected in a sufficient number in the limited time of a balloon flight (10-30 hours), in a rocket (minutes) or in a satellite (years). Fig. 6 (14) shows which exposure time is necessary to collect 1000 particles in large area passive balloon experiments (m^2) of plastic-detectors or emulsions or in small area electronic satellite experiments (cm^2).

Isotopic composition of cosmic ray nuclei   1047

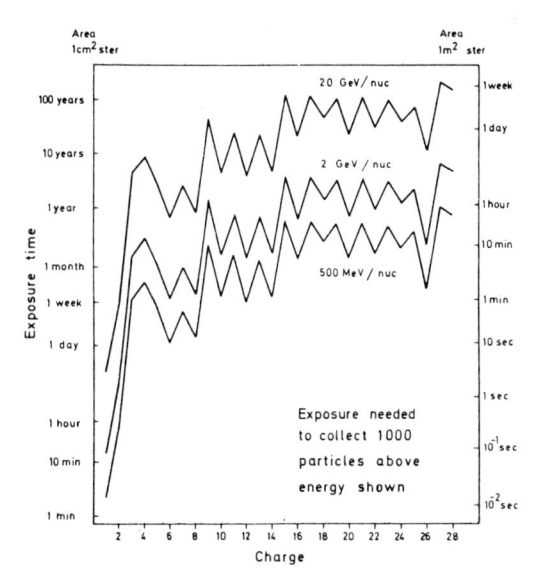

Fig. 6. Needed exposure times to collect 1000 particles in experiments with collecting areas of 1 $cm^2 \cdot$ ster (left) and 1 $m^2 \cdot$ ster (right) (ref. 14).

In fig. 7 is shown how the evaluation of a typical dE/dx - R detector like plastic-detectors work according to particle identification and fig. 8 shows a set of measurements in the CNO-region (15).

The three unknown values M, Z and β are determined by two equations and one restriction which is valid in most of the practical cases.

$$\frac{dE}{dx} = f(Z, \beta) \quad \text{(cone length in fig. 7)}$$

$$R = g(Z, M, \beta) \quad \text{(range in fig. 7)}$$

$$M = 2 \cdot Z \text{ or } 2Z - 3 \leq M \leq 2Z + 3$$

A detector combination of e.g. E - $\frac{dE}{dx}$ - C allows a unique solution without any restriction.

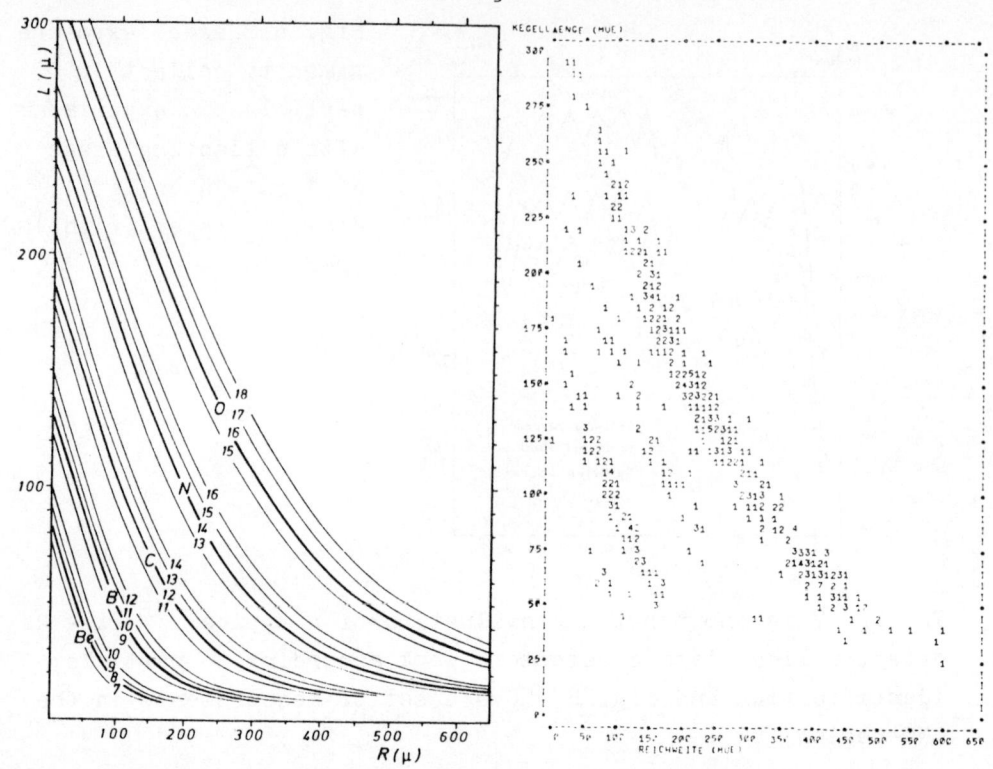

Fig.7. Conelength-range-curves set(dE/dx-R) for a Daicel-cellulose-nitrate plastic-detector (ref. 15)

Fig.8. Measurement of cosmic ray B,C,N,O-particles belonging to the curves set of fig. 7 (ref. 15)

In fig. 9 (16) a summary of the isotopic resolution which can be presently obtained, is given. As adjacent isotopes of equal abundance, which is the most optimistic assumption, can be resolved as two peaks only if the standard deviation is smaller than 0.44 amu, presently only the isotopes of lower charge ($Z \leq 14$) can be realy separated. It is, however, also without the above resolution possible to determine isotopic abundances from the distribution itself if a sufficient number of particles has been measured or if the most abundant isotopes are two mass units apart like it seems to be the case for ^{54}Fe and ^{56}Fe (fig. 14). In fig. 9 the presently best, but still preliminary results of fig. 13 are not included.

Isotopic composition of cosmic ray nuclei

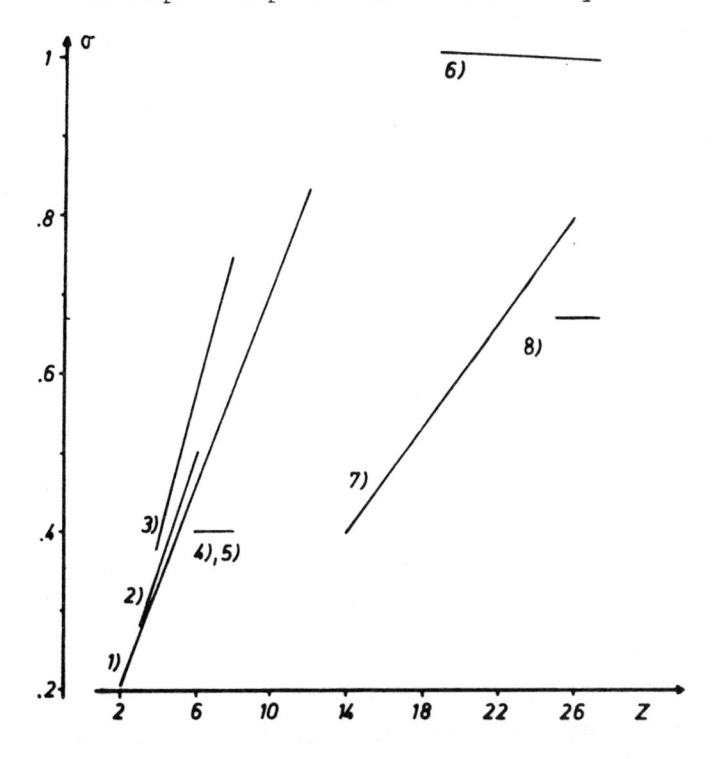

Symbol	Authors	Group	Method	Energy MeV/Nuc.	Ref.
1	Preszler et al. 1975, S.243	New Hampshire	$\frac{dE}{dx} - E$	100 - 300	31
2	Garcia-Munos et al., 1975, S.325	Chicago	$\frac{dE}{dx} - E$	50 - 200	22
3	Hagen et al. 1975, S.361	GSFC	$\frac{dE}{dx} - R - (E)$	150 - 450	32
4	Bjarle et al. 1975, S.337	Lund	Emulsion	200 - 400	33
5	Mewaldt et al. 1975, S.349	Caltec	$\frac{dE}{dx} - E$	6 - 12	34
6	Claphan et al. 1975, S.400	Dublin	Plastic	250 - 500	28
7	Webber et al. 1973	New Hampshire	C - E	300 - 700	27
8	Siegmon et al. 1976	Kiel	Plastic	150 - 400	30

Fig. 9 Present status of isotopic resolution. σ in atomic mass units (amu). (ref.16)

Let me finally mention some future experiments. Two Cerenkov-detectors with a thick absorber inbetween are planed for an experiment by the two groups in Saclay (France) and Lynby (Danemark) (17, 18).

A plastic-emulsion experiment of the groups in Kiel (Germany) and Lund (Sweden) has been flown and is in progress (19).

Three similar experiments combining electronic dE/dx or C-detectors with range measurements in plastic or emulsions are in progress in Berkelay (USA) (20), in Minnesota (USA) (21) and in a cooperation between Durham (USA) and Kiel (Germany).

V Recent results of cosmic ray isotopes

In this section I will describe the recent results of elemental and isotopic abundances of cosmic rays and compare them with those of the solar system.

First of all I will mention some problems related to the solar system abundances. They have been assembled by Cameron (6) and discussed by Trimble (3). These elemental abundances are primarily taken from meteorite measurements, whereas all isotopic abundances of Cameron (6) are of terrestrial origin except those of H, He, Ne, Xe and ^{40}Ar (3). There are, however, no indications that these values (except boron) are not valid all over the solar system. Extraterrestrial isotopic abundances can be obtained by using spectra of molecules only in the HCNO-region and by some further singular spectroscopical values.

Thus only measurements on cosmic ray isotopes can presently supply extensive informations on isotopic abundances outside the solar system for elements heavier than oxygen.

Fig. 10 (16) shows a comparison between solar system and cosmic ray elemental abundances. Both distributions are somewhat arbitrarily normalized to $Si = 10^6$, but a normalisation to Mg or Fe or even C and O would not give much different results. It is clearly to be seen that the values of the odd elements, the Li, Be, B group and the elements between Ca and Fe are filled in by secondarily produced nuclei. Thus information on the stellar nucleo-synthesis can only be obtained from isotopes of nuclei which are not very much effected in this way like C, O, Ne, Mg, Si, S (maybe Ar and Ca), Fe and Ni.

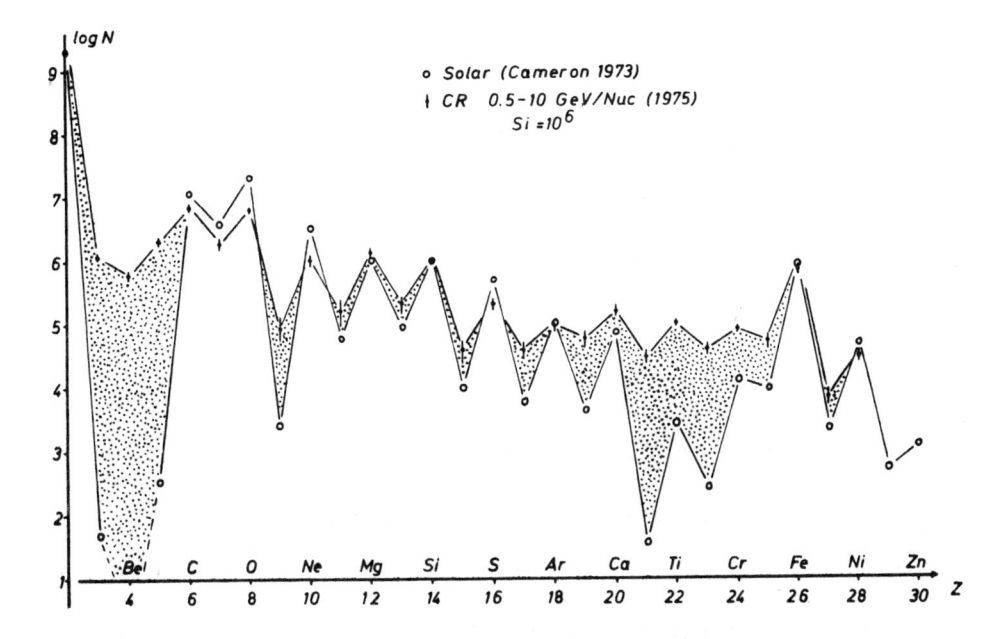

Fig. 10. Comparison of the elemental abundances of the solar system (6) and of cosmic rays (ref. 16).

Fig. 11 shows the present status of measurements in an energy-charge and isotope-diagram. The three sections are given by experimental reasons as indicated.

Now I will summarize the results of isotopic cosmic ray measurements. The isotopes of the Li, Be, B group are well understood from fragmentation processes in the interstellar medium, see e.g. fig. 12 (22). Applying homogeneous propagation models J.P. Meyer (23) has summarized in his rapporteur talk that present measurements of the radioactive isotope ^{10}Be, which are not in good agreement with eachother, and to some extend ^{36}Cl are consistent with an interstellar density

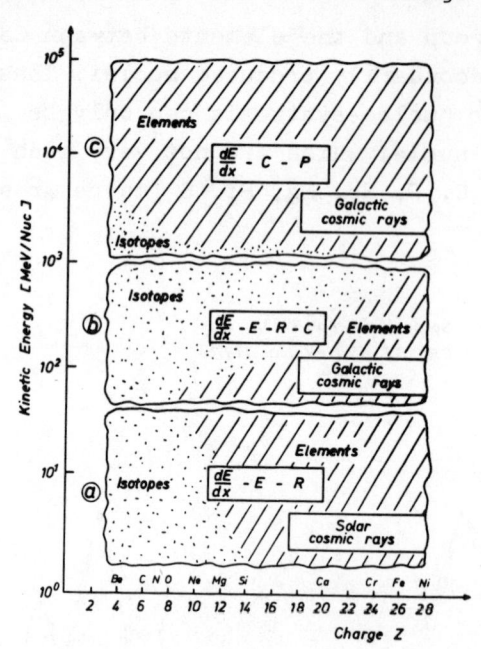

Fig.11. Present status of measurements in an energy-charge and isotope-diagram.

and propagation time T of

$$0.1 < \xi < 1.3 \text{ H-atoms/cm}^3$$
$$33 > T > 2.5 \cdot 10^6 \text{ years}$$

Using a more realistic two-zone-model (24) even wider limits have to be allowed.

The best presently measured ratios of the isotopes $^{13}C/C$ and and $^{15}N/N$ (see ref. 23) are also well understood from fragmentation processes. The ratios ($^{17}O + {}^{18}O$)/O, however, where up to now only some few measurements exist (23), indicate some excess of $^{17}O + {}^{18}O$ to the expected amount of about 4 % (25). If this excess which is still to be regarded as preliminary, stands up in later measurements, special chains of nuclear burning have to be allowed for ^{17}O and ^{18}O which puts quite interesting new aspects to the scene of the nucleosynthesis.

In the region of Ne, Mg and Si the presently measured mean masses show still such a wide spread that no nucleo-synthetic

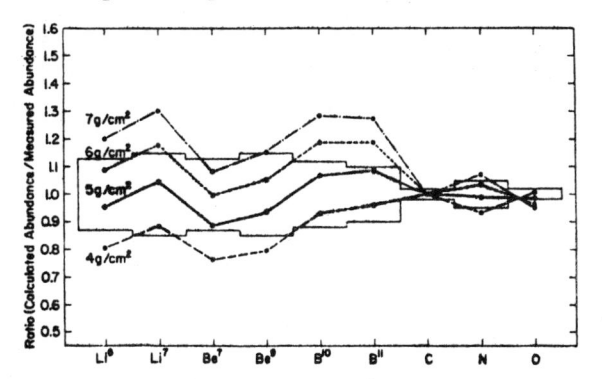

Fig. 12. Comparison of calculated and measured isotopic abundances for 4-7 g/cm^2 interstellar matter (ref. 22).

conclusions may be drawn. The most recent results of the Durham-group (26) using the C-E-method look very promising (fig. 13), but they are still preliminary, so that I do not go into further details.

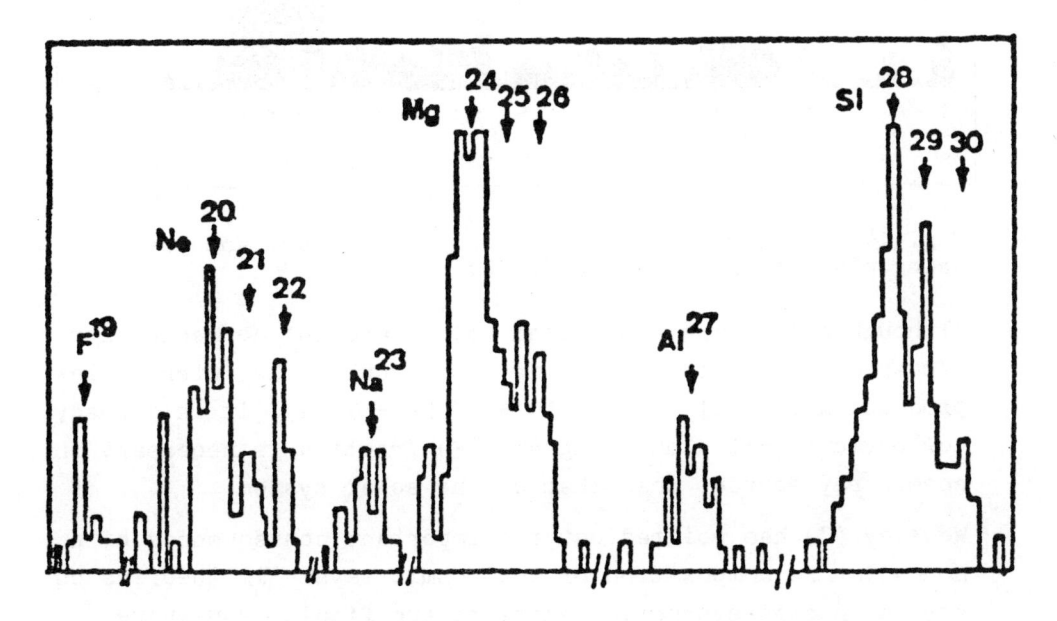

Fig. 13. Measurements of F to Si cosmic ray isotopes using the C-E-method (ref. 26).

In the Fe-region, however, some newer very interesting results have been obtained. Several experiments (27,28,29,30) achieve with only very small deviations the same results for the following fractions: $(^{52}Fe)+(^{53}Fe)+^{54}Fe = 31 \pm 5$ % $(^{55}Fe)+^{56}Fe+^{57}Fe = 50 \pm 7$ % and $^{58}Fe+(^{59}Fe)+(^{60}Fe) = 19 \pm 9$ %. The brakets indicate radioactive isotopes. Neglecting these isotopes and ^{57}Fe results in the ratio $^{54}Fe/^{56}Fe = 0.6 \pm 0.2$. A solar system source composition expects at the earth $^{54}Fe/^{56}Fe = 0.09$ (25). The high ^{54}Fe cosmic ray abundance could be now measured directly in a plastic-detector-experiment with a good peak-to-valley resolution, see fig. 14 (30).

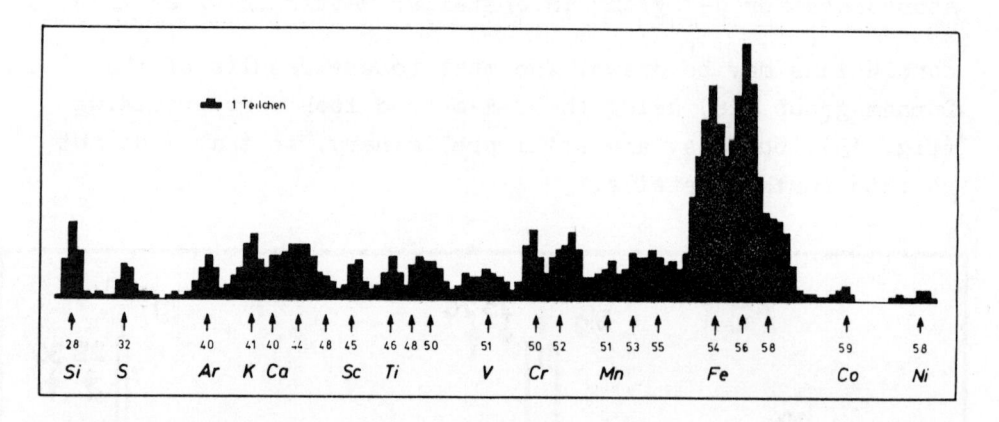

Fig. 14. Measurements of Si to Ni cosmic ray isotopes using plastic-detectors (ref. 30).

Although all of the above used values are not corrected for atmospheric effects, which may about double the interstellar produced amount of ^{54}Fe of 10 % (^{56}Fe = 90 %), it is already now obvious that a much higher ^{54}Fe fraction is necessary in cosmic ray sources than that of the solar system.

Woosley (7) has pointed out the important consequences of a non-solar Fe composition of the cosmic rays. All theories on stellar nucleo-synthetic reactions are finally somewhere adjusted to observations. Thus the cosmic ray data - if different from solar data - can be used as a second set of parameters at high energy which can put new restraints to the processes of the nucleo-synthesis.

The above ratio $^{54}F/^{56}Fe = 0.6 \pm 0.2$ has been put into fig. 15 (7), where also solar system values are shown. In this way two new neutron excess regions are determined $\eta = 0.013-0.018$ and $\eta = 0.055-0.062$ which indicates (see fig. 4), that those cosmic ray nuclei may originate from regions of higher neutron excess in super nova explosions and that the "mass-cut" (fig. 4) may be put to deeper regions.

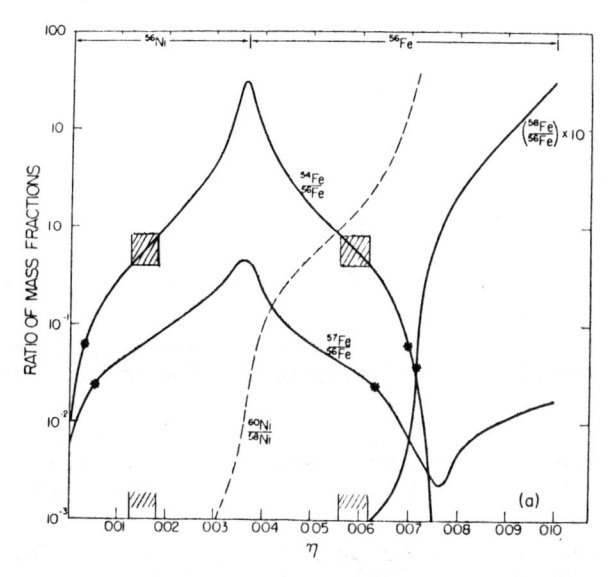

Fig. 15. Ratio by mass of the isotopic abundances of Fe as a function of neutron excess η. Asterisks denote points where the given ratio corresponds to its solar value. The dashed areas correspond to the cosmic ray mass ratio $^{54}Fe/^{56}Fe = (54 \cdot 0.31)/(56 \cdot 0.5) = 0.6$, see text (ref. 7).

In this brief review I am not able to discuss all possible consequences of the above measurements and maybe it is a little too early to squeeze these still preliminary data too much. One thing, however, has been shown already here: Future cosmic ray isotope measurements will contribute an important part to the knowledge of stellar nucleo-synthesis.

References

1) Wagoner, R.V., Ap.J. <u>197</u>, 343 (1973)
2) Searle L., Sargent W.L.W., Ap.J. <u>173</u>, 25 (1972)
3) Trimble V., Rev. of Mod. Physics <u>47</u> 877 (1975)
4) Audouze J., Lequeux J., Vigroux L., Astron. a. Astrophys. <u>43</u>, 71 (1975)
5) Lang K.R., Astrophysical Formulae, Springer-Verlag Berlin-Heidelberg-New York (1974)
6) Cameron A.G.W., Space Sci. Review <u>15</u>, 121 (1973)
7) Woosley S.E., Ap. a. Space Science <u>39</u>, 103 (1976)
8) Arnett W.D., Ap.J. <u>195</u>, 727 (1975)
9) Scott J.S., Chevalier R.A., Ap.J. Letters <u>197</u>, L5 (1975)
10) Hainebach K.L., Norman E.B., Schramm D.N., Ap.J. <u>203</u>, 245 (1976)
11) Kristiansson K., Astrophys. Space Sci. <u>30</u>, 417 (1974)
12) Cassé M., Goret P., Cesarsky C.J., 14th CRC München, S.646 (1975)
13) Cassé M., Soutoul A., 2nd Cosmic Ray Isotope Symposium, Durham, N. Hampsh. (1974)
14) Rasmussen I.L., Origin of Cosmic Rays, S.97, ed.: J.L. Osborne, A.W. Wolfendale, D. Reidel Publish. Company (1975)
15) Beaujean R., Sagebiel H., Enge W., Isotopic Composition of Low Energy Cosmic Ray Particles with Charges Z=5-8, 9th Int. Conf. on Solid State Nucl. Track Detectors, München (1976)
16) Beaujean R., Summary-report of the 14th CRC München, Inst. f. Kernphysik Kiel, (Jan, 1976)
17) Cantin M., Goret P., Jorrand J., Juliusson, E., Koch L., Maubras Y., Metsreau P., Petrou N., Soutoul A., 14th CRC München, S.3205 (1975)
18) Meyer P., Gaulier F., 14th CRC München, S. 3199 (1975)
19) Scherzer R., Enge W., Beaujean R., Hertzman S., Kristiansson K., Soederstroem K., Study on cosmic ray iron isotopes in an emulsion-plastic detector 9th Int. Conf. on Solid State Nucl. Track Detectors, München (1976)
20) Ahlen S.P., Cartwright B.G., Crawford H., Price P.B., Tarlé G., 14th CRC München, S.3256 (1975)
21) Gilman C.M., Waddington C.J., 14th CRC München, S.3166 (1975)

22) Garcia-Munoz M., Mason G.M., Simpson J.A., 14th CRC München, S.25 (1975)
23) Meyer J.P., Rapp.Paper, 14th CRC München, S.3698 (1975)
24) Simon M., Scherzer R., Enge W., (Ex 68), DPG-Frühjahrstagung, Freiburg (1976) (paper in preparation)
25) Shapiro M.M., Silberberg R., Tsao C.H., 14th CRC München, S.532 (1975)
26) Simpson G., Kish J.C., Preszler A.M., Webber W.R., 14th CRC München, S.389 (1975)
27) Webber W.R., Lezniak J.A., Kish J., Ap.J. $\underline{183}$, L81 (1973)
28) Clapham V.M., Fowler P.H., O'Ceallaigh C., O'Sullivan D., Thompson A., 14th CRC München, S.400 (1975)
29) Henke R.P., Benton E.V., 14th CRC München, S.395 (1975)
30) Siegmon G., Bartholomä K.-P., Enge W., Composition of Fe-isotopes in cosmic rays, 9th Int.Conf.on Solid State Nucl.Track Detectors, München (1976)
31) Preszler A.M., Kish J.C., Lezniak J.A., Simpson G., Webber W.R., 14th CRC München, S.243 (1975)
32) Hagen F., Fisher A.J., Ormes J.F., Arens J.F., 14th CRC München S.361 (1975)
33) Bjarle C., Herrström N.-Y., Jacobsson L., Jönsson G., Kristiansson K., 14 CRC München, S.337 (1975)
34) Mewaldt R.A., Stone E.C., Vidor S.B., Vogt R.E., 14th CRC München, S.349 (1975)

CRC = Cosmic Ray Conference

COMPOSITION OF Fe-ISOTOPES IN COSMIC RAYS

G. Siegmon, K.-P. Bartholomä and W. Enge

*Institut für Reine und Angewandte Kernphysik, University of Kiel,
23 Kiel, West Germany*

Using the ultraviolet irradiation of Lexan polycarbonate plastic for the enhancement of mass resolution the isotopic composition of cosmic ray nuclei with $Z \succeq 14$ was measured. An UV-exposure equipment was built, which enables us to irradiate 12 600 cm^2 of plastic in one run. A new method of particle identification was evaluated. The accuracy of mass determination in plastic detectors will be discussed. In this experiment we estimate a resolution of ± 0.7 AMU for the iron region. Based upon the measurement of about 530 particles with energy 150-500 MeV/nuc. we present the mass distribution of Fe-nuclei.

1. Introduction

In recent years much work has been devoted to the study of the charge composition of cosmic radiation. Apart from a lot of physical processes that may transform cosmic ray composition during their propagation in space the elementally determined abundance pattern of the elements has constituted the primary input and chief constraints upon theories of nucleosynthesis. In this regard isotopic ratios are of particular importance because they are insensitive to chemical separation effects and thus place severe restrictions upon nucleosynthesis models. Complications such as contamination by secondary spallation, ionisation energy losses and rigidity dependent leakage from the galaxy are negligible in the case of source iron; an element whose modification by secondary processes of heavier elements has been small and which is heavy enough to have very similar values of A/Z.

In order to achieve the required mass resolution in cosmic ray measurements new detector systems have been developed or are still under construction.

In this paper we present the spectrum of cosmic ray isotopes measured by means of a balloon born plastic detector experiment being flown at 1.8 g/cm^2 for 10 h. The investigation was made on nuclei which stopped in the stack consisting of 18 Lexan sheets of 250 μm thickness, in total an area of 2 800 cm^2. We used the multible-dE/dx-R method by measuring the etched cone length L as a function of the residual range R of UV-enhanced tracks. The problem of reducing the two-dimensional track data of the LR-plot to a one-dimensional charge and mass histogram was solved by simple integration of a L(R)-fit over a fixed cone length interval.

2. Experimental procedure

It is known that UV-irradiation of exposed Lexan sheets leads to an increase of the track etching rate [1]. We used this enhancing effect to improve the resolution capability of the detector [2]. For that purpose we built a UV-equipment which allows the sensitation of an area of 12 600 cm^2 plastic in one run with a homogeneity better than 0.5 %. Four individual runs of UV-irradiation and etching procedure were performed. The etching was done in an ultrasonically agitated, mechanically stirred solution of 6.0±0.05 n NaOH with 0.05 % Benax surfactant at 70±0.03°C.

3. Track evaluation

The information of mass and charge of the stopped particles is stored in the developed configuration. After the scanning procedure which was performed by means of a stereo-microscope, the cone lenths were measured with a digitized microscope using linear displacement transducers for numerical readout. Depending on the angle, charge and

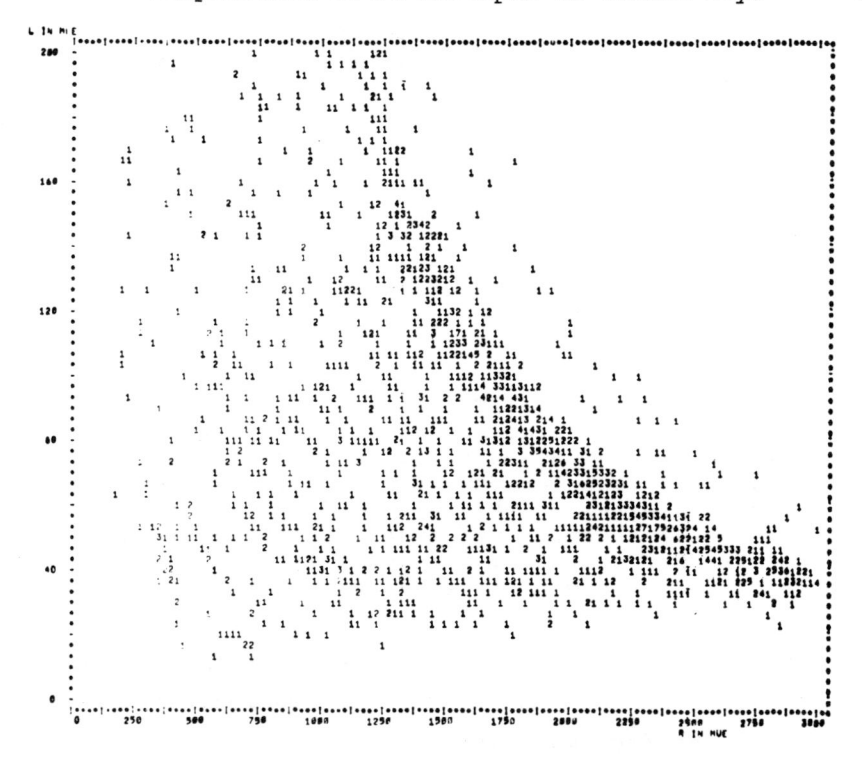

Fig.1
Cone length vs residual range of UV-enhanced cosmic ray tracks in Lexan

stopping position of the incident particle up to 14 cones could be measured. Thus each event was represented by several data points (L,R) in an LR-plot. The points spread around curves characterizing the particles charge and mass. Fig.1 shows a sample of measured tracks ranging from nuclei with charge Z=28 down to Z=14. In order to reduce the track data to a single parameter for each event the following procedure was performed:

1. The sets of LR-data of each event were fitted by means of analytical functions of the form

$$L(R) = A_0 + \sum_{k=1}^{g-1} A_k \frac{(-1)^{k+1}}{2k-1} (R-c)^{-\alpha \cdot k + 1} \qquad \begin{array}{l} \alpha = 1.8 \\ c = 0.45 \\ R \text{ in mm} \end{array}$$

The number g of derived coefficients was adjusted to the number n of measured cones according to

$$g = \text{entier}(n/2) \quad ; \quad g \in \{2,3,4\}$$

In order to discriminate against nuclear interactions two selection criteria were used:
(1) values of L which deviate by more than 5 μm (= 2 standard deviations) from the fit were eliminated and (2) at least 4 cones and 75% of the original cones must be good to accept the track.

2. The resulting coefficients A_k which contain the track information were then aggregated to a single parameter by simple integration. As only cone lengths were actually measured the integration was performed over a fixed L-interval being unchanged for the whole analysis. For practical reasons this value was replaced by a mean range $\bar{R}$ as indicated in fig.2 .

Thus, by using the LR-measurements as input data the application of the described procedure leads to a single parameter for each event.

Apart from the adopted selection criteria only those events could be accepted whose data sets contain cones with length $L > L_1$ and $L < L_2$. In total 335 particles were left; 1/3 was excluded from analysis, especially the short tracks of nuclei with $Z < 16$. Thus it is not reasonable to compare elemental abundances in the lower charge region and we therefore confined ourselves to the study of groups of neighbouring isotopes.

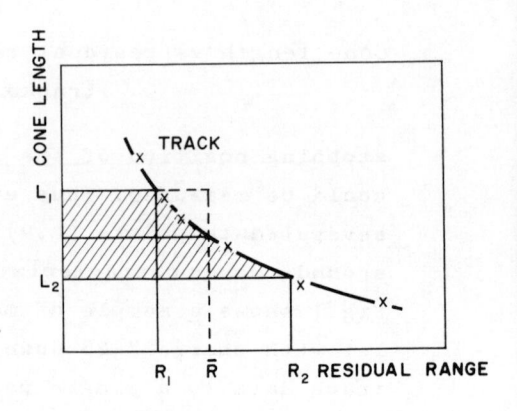

4.   Calibration

To calibrate the $\bar{R}$-histogram we made the following assumptions:

I. The highest charge nuclei of greatest abundance are of iron and furthermore the most abundant iron isotope is the nucleous ^{56}Fe,

II. The mean mass of the Si-mass-distribution is $\bar{A} = 28$.

III. The model of the Restricted Energy Loss (REL) as proposed by Benton [3] is found to be an adequate theory which describes the modified energy loss as being relevant for the track forming processes.

We chose the adjustable parameter w_o of the REL model to 1 000 eV, a value which is consistent with the assumptions I. and II. A short outline of the procedure how to fix the discret set of charges Z and masses A at the correct $\bar{R}$-positions we give in another paper at this conference [4]. Fig.3 presents the histogram of 335 particles of galactic cosmic rays measured in this experiment. The numbers of the scale give $\bar{R}$ in μm and the charges are assigned according to the calibration.

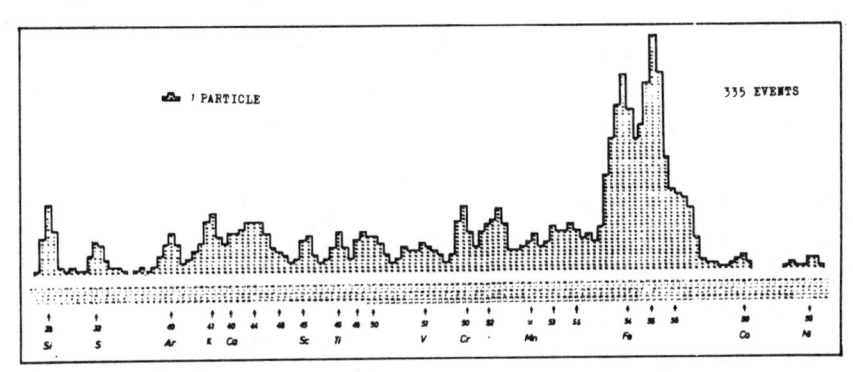

Fig.3 : Spectrum of Si (Z=14) up to Ni (Z=28) for E=150-500 MeV/nuc.

5. Accuracy of the measurements

There are several sources of error in the mass identification procedure. The most obvious error source is due to fluctuations in the cone lengths. The x- and z-projections which were determined with an accuracy of

~1 µm yield an error in the Fe-region of 0.6 AMU.

The error of the track-fit, in the least square sense, is shown in fig.4 . The expectation value of the distribution leads to an error of mass determination which is close to that derived from measurement errors (0.56 AMU for Fe).

Another source of error originates in the location of the mass lines themselves. The uncertainty in the calibration at the ^{56}Fe peak is believed to give an absolute error of ~0.1 AMU in the location of the mass scale. This uncertainty will lead to a systematic displacement of the observed mass distribution relative to the calculated scale. The ramaining systematic errors depend on the adopted track theory REL including its adjustable parameter w_o.

To estimate the established resolution the histogram of the iron isotopes was fitted by a least-square technique to 3 Gaussian distributions centered at masses A=54,56,58. The best fit indicates a mass resolution of 0.7 AMU. According to the decreasing number of cones for lower charges the resolution increases up to 2 AMU for Si.

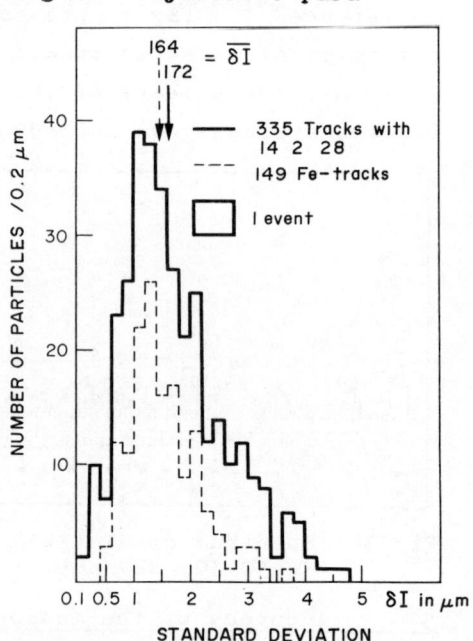

6. Results and discussion

In Tab.I we summarize the cosmic ray mass distribution observed under 1.8 g/cm^2 of atmosphere. The results are compared with other experimentors and with theoretical values. Although there are some discrepancies

Composition of Fe-isotopes in cosmic rays

which may be of statistical origin the results of the different experiments agree within 30 % on an average.

Fe-isotope	Kiel 1.8 g/cm^2		NHam [5] 3 g/cm^2		Dublin [6] 3.5 g/cm^2		San Francisco [7] 2 g/cm^2		Mean	Theoretical 2.3 GeV solar [8]
	events	%	events	%	events	%	events	%	%	%
(52+53)+54	56	38	15	28	166	29	16	30	31±5	8
(55)+56+57	71	48	23	43	276	48	32	60	50±7	91.7
58+(59+60)	22	15	16	30	133	23	6	10	19±9	0.3

Table 1

Usually cosmic ray abundances are compared with solar abundances as elaborated by Cameron [9]. In order to study differences between solar system abundances and measurements, a first approximation starts with a solar composition at the source of galactic cosmic rays which is then transformed during its propagation in space before being detected. In the case of iron which shows comparable abundances of ^{54}Fe and ^{56}Fe there is obviously a significant difference to a solar source composition. Although the data have not been extrapolated to the top of the atmosphere it is impossible to explain the discrepancy with atmospheric fragmentations. Thus the Fe- -measurement accounts for the existence of large fractions of the isotope ^{54}Fe and, less significant of ^{58}Fe, in the cosmic ray sources themselves. Following the assumption that super nova events are the sources of galactic cosmic rays, the source composition may be regarded as a signature of nucleosynthetic processes in such events. Although there is a variety of possible iron isotopic configurations in cosmic rays which have been extensively discussed by Woosley [10] an excess of ^{54}Fe would imply that cosmic ray iron originates in an environment that was more neutron-rich than the source of solar iron. The interpretation of ^{58}Fe requiring the ejection of an inner ^{58}Fe-rich zone with a very high neutron excess is ambiguous as well as difficult to measure.

Acknowledgements

The authors are grateful to Prof. E. Bagge for his support of the work. This work was supported by the Deutsche Forschungsgemeinschaft.

References

1) Crawford W.T., DeSorbo W., Humphrey J.S., Nature 220, 1313-1314, (1968)
2) Siegmon G.-F., Bartholomä K.-P., Enge W., Nucl. Instr. Meth. 128, 461, (1975)
3) Benton E.V., A Study of Charged Particle Tracks in Cellulose Nitrate, U.S. Naval Radiological Defense Laboratory, S.Francisco, (1968)
4) Siegmon G.-F., Koehnen H.J., Bartholomä K.-P., Enge W., The dependence of the mass-identification-scale on different track formation models, 9th Int.Conf. on Solid State Nucl.Track Detectors, München, in prep.(1976)
5) Webber W.R., Lezniak J.A., Kish J., Nucl. Instr. Meth. 111, 301, (1973)
6) Clapham V.M., Fowler P.H., O'Ceallaigh C., O'Sullivan D., Thomson A., 14th CRC München, 400, (1975)
7) Henke R.P., Benton E.V., 14th CRC München, 395, (1975)
8) Tsao C.H., Shapiro M.M., Silberberg R., 13th Int.Cosmic Ray Conf., Vol.1, 107, (1973)
9) Cameron A.G.W., Space Sci. Rev. 15, 121, (1973)
10) Woosley S.E., Astrophys. Space Sci. 39, 103, (1976)

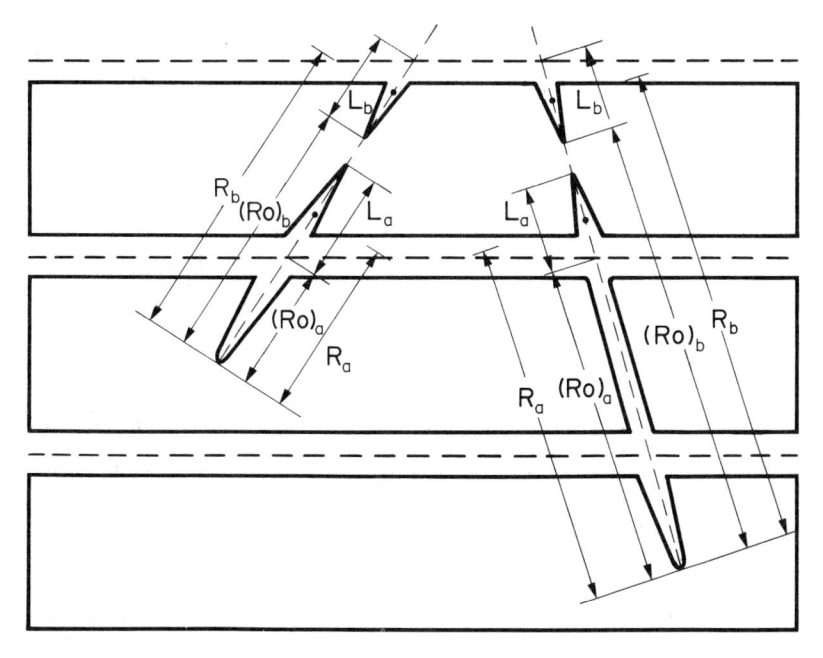

$R_a = (Ro)_a + L_a/2$
$R_b = (Ro)_b + L_b/2$

Fig. 5.

STACK OF PLASTIC SHEETS FOR ETCHING PROCESS

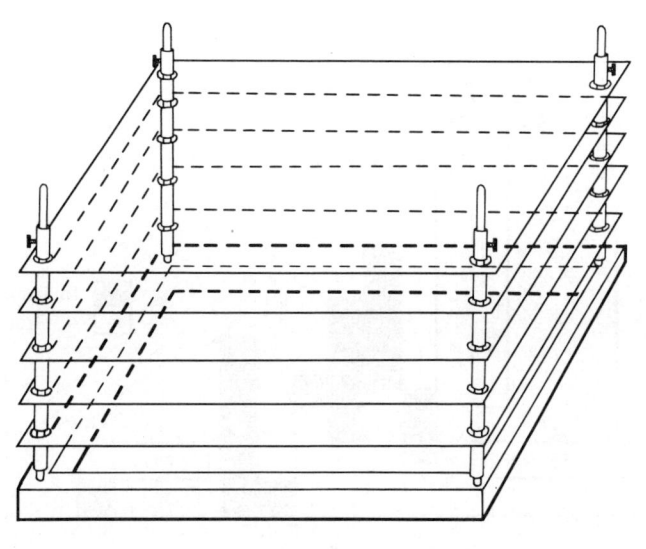

Fig. 6.

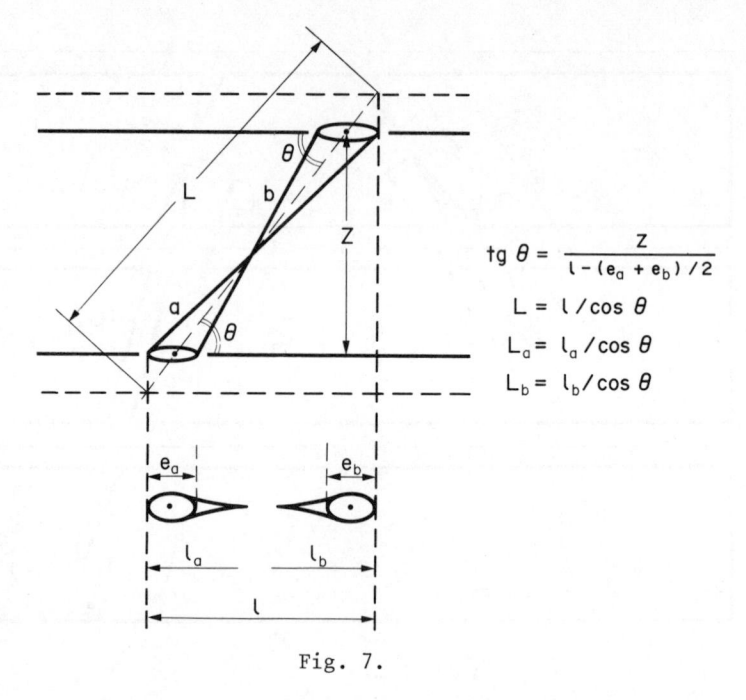

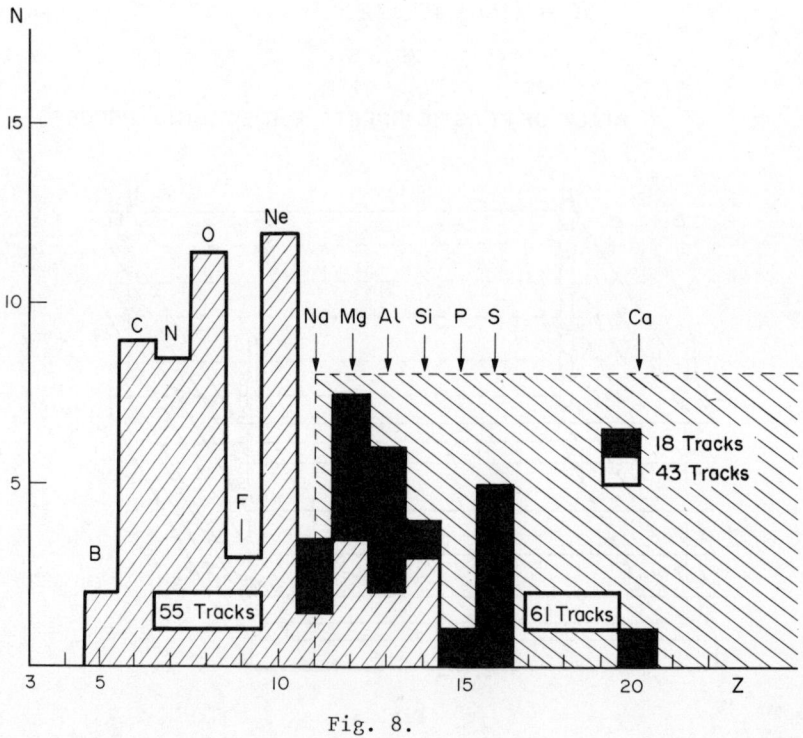

$$\mathrm{tg}\,\theta = \frac{Z}{l - (e_a + e_b)/2}$$
$$L = l/\cos\theta$$
$$L_a = l_a/\cos\theta$$
$$L_b = l_b/\cos\theta$$

Fig. 7.

Fig. 8.

ISOTOPIC COMPOSITION OF LOW ENERGY COSMIC RAY PARTICLES WITH CHARGES Z = 5 - 8

R. Beaujean, H. Sagebiel and W. Enge

*Institut für Reine und Angewandte Kernphysik, University of Kiel,
23 Kiel, West Germany*

A stack of 250 μm Daicel cellulose nitrate sheets was flown from Ft. Churchill under 2.7 g/cm^2 atmospheric depth for 13.25 hours. The charge and mass determination on 847 stopping particles was done by means of the cone length versus residual range method using the REL criterion. The isotopic ratios are measured in the top part of the stack and under additional 2 g/cm^2 in the bottom part. Averaged numbers for C^{13}/C and O^{17+18}/O are 7.4±1.9 and 15±1.8 respectively [%].

1. Introduction

We have shown that cellulose nitrate nuclear track detector sheets can be used to measure the isotopic composition of low energy cosmic ray particles even in the absence of a clear separation of adjacent isotopes (1,2). Hence we carried on our early measurements with an improved evaluation technique on an enlarged number of particles.

The analysis of more than 1000 stopping particles in the energy region 100-200 MeV/nuc was performed in a stack of Daicel cellulose nitrate which was flown from Ft. Churchill under 2.7 g/cm^2 for 13.25 hours. The stack consists of 100 sheets of a size 250 μm x 20x40 cm^2. The upper 56 sheets (equivalent to 2 g/cm^2) were etched at 40 °C in a 6n NaOH solution with 0.05 % Benax added for 240 min. The residual range and cone length measurements along the tracks were done under a Leitz optical microscope with a 100x objective.

2. Data evaluation

The scanned stack volume was divided into two parts. One part includes particles stopping in sheets 1 to 15 (further refered to as the top part), the second part includes those particles stopping between sheets 41 to 56 (refered to as the bottom part). These two sets are separated by a mean matter of 2 g/cm^2 for particles with $45°$ zenith angle.

It was detected during the analysis that the response of the detector sheets varies from the edge towards the centre of the stack. The reason for this behaviour is not cleared but may be due to a temperature effect since the edge was not completely protected against heating or cooling at ceiling level during the balloon flight. The main change in response occurs within 1 cm from the edge whereas the inner part shows a reasonable constant response. As a compromise between a large number of particles and a somewhat constant response on a single detector sheet all particles were rejected from the following analysis with tracks in the area 2 cm or closer to the edge. Thus 75 % of the particles remained within an area of 16x36 cm^2.

Furthermore a different response was detected for the top and bottom parts which can be easily seen in the cone length versus residual range diagrams of the accepted particles (Fig. 1+2). Further studies are in progress to clear whether a temperature gradient or a diffusion process causes such spatial changes in the response.

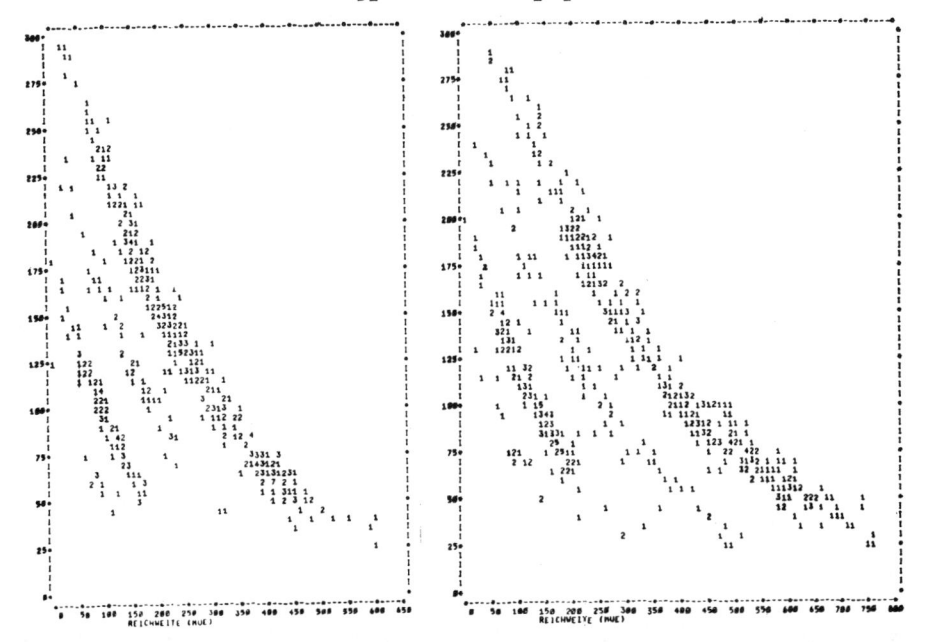

Fig. 1 + 2. Cone length versus residual range measurements in the top (left) and bottom part of the stack.

3. Results

The charge and mass identification of the accepted particles with nuclear charge Z=5-8 was done following the procedure described in ref. (3). The track etching rates for the top and bottom parts were calibrated using the predominant peaks of C^{12} and O^{16}. The mass histograms in Fig. 3 and 4 are based mainly on single cone events. From some available multi-cone events a fluctuation of 0.6 amu for the single cones was deduced. To extract the isotopic abundances from Fig. 3 and 4 the following methods were applied. The amount of secondary C^{11} and O^{15} produced in the matter above and in the stack was calculated (4,5) and substracted on the left side of the corresponding distributions. Assuming symmetric Gaussian distributions for C^{12} and O^{16} the amount of C^{13} and O^{17+18} is thus obtained as the excess on the right side. For Boron and

Nitrogen the mean mass of the distribution yields the portion of B^{10}, B^{11}, N^{14} and N^{15} (the calculated number of N^{13} was taken into consideration).

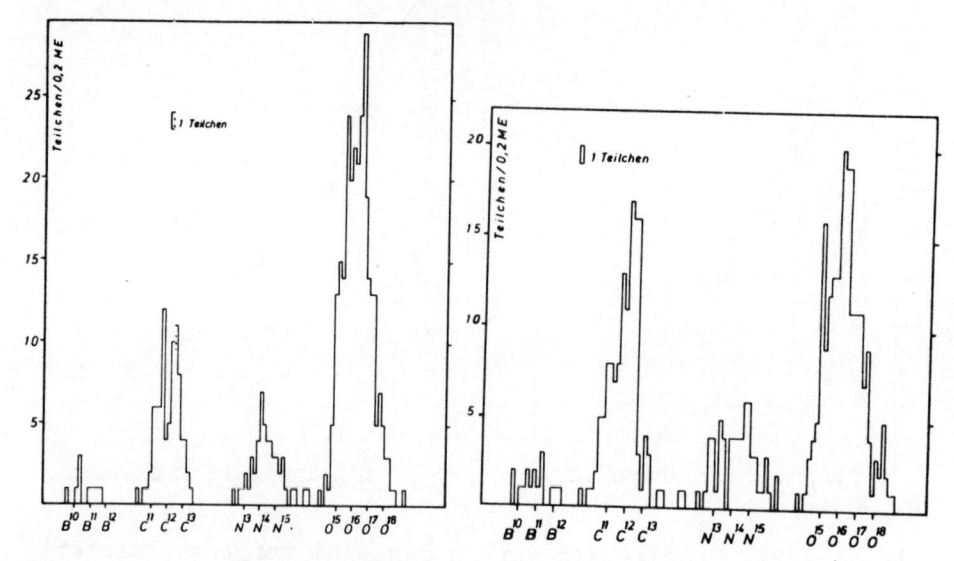

Fig. 3 + 4. Mass histogram for the top (left) and bottom part. The number of particles per 0.2 amu is plotted.

Table I
a) calculated numbers (%)

	C^{11}/C	N^{13}/N	O^{15}/O
top	3 ± 1.8	0.7 ± 0.7	2.2 ± 0.9
bottom	4 ± 1.7	1.3 ± 1.3	3.7 ± 1.3

b) measured numbers (%)

	B^{11}/B	C^{13}/C	N^{15}/N	O^{17+18}/O
top	70 ± 26	7.7 ± 3	40.2 ± 8.9	16.9 ± 2.5
bottom	78.3 ± 22	7.2 ± 2.3	43 ± 8.2	13.1 ± 2.5

4. Discussion

Table I summarizes the calculated numbers of the secondaries and the resulting numbers of the cosmic ray particles. All numbers are valid at detector level. The cosmic ray abundances are in good agreement with results reviewed in ref. (6). The amount of C^{13} can be explained by interstellar and atmospheric fragmentations while the measured O^{17+18}/O ratio is larger than expected from pure fragmentation processes.

Acknowledgement

We like to thank Prof. Dr. E. Bagge for his support and the staff of Raven Ind. for performing the balloon flight. This work was supported by the Deutsche Forschungsgemeinschaft.

References

1) R. Beaujean and W. Enge, Zeitschr. Physik 256 (1972), 416
2) K. Fukui, W. Enge and R. Beaujean, Z. Physik A227 (1976), 99
3) R. Beaujean and W. Enge, Fragmentation and isotope measurements on accelerator Neon and Argon particles, to be published
4) R. Silberberg and C.H. Tsao, Astroph. J. Suppl. No.220(I), 25 (1973) 315
5) P.J. Lindstrom, E. Greiner, H.H. Heckman, B. Cork and F.S. Bieser, Lawrence Berkeley Lab. report LBL-3650 (1975)
6) J.P. Meyer, Rapp. paper 14th CRC München 1975, p. 3698

STUDY ON COSMIC RAY IRON ISOTOPES IN AN EMULSION-PLASTIC DETECTOR

R. Scherzer*, W. Enge*, R. Beaujean*, S. Hertzman**,
K. Kristiansson** and K. Söderström**

*Institut für Reine und Angewandte Kernphysik, University of Kiel,
23 Kiel, West Germany
**Department of Physics, University of Lund, Lund, Sweden

Abstract

Combining cellulose nitrate plastics and nuclear emulsions a detector system for mass measurement in the iron group has been designed. The detector operates in an energy interval of 500-700 MeV/Nuc and has been flown in two balloon flights. In the emulsion range and track width are measured. Cone lengths and range are measured in the plastics. The charge of a particle can be determined both in emulsion and in plastics. The mass is determined by relating cone lengths to residual range. One advantage of this detector stystem is the high rejection of interacting particles. The present status is reported.

1. Introduction

It has become more and more obvious that cosmic rays and especially its elemental and isotopic component can play an important role in understanding phenomena of the origin and the history of this radiation and in understanding the process of stellar nuclear-synthesis. In recent years great effort has been put into improvement of measurements; it is, however, still a field where experimental progress is necessary. Most isotopic studies of cosmic rays have so far been made on elements lighter than neon, but recently great efforts have been started in reaching satisfactory resolution in the Fe-region where presently only cosmic rays can supply information on isotopic abundances outside the solar system.

2. Detector system

In this report we describe a detector system which is intended for isotopic abundance studies of the iron group elements in primary cosmic radiation in balloon experiments. The detector (fig.1) consists of a central stack of

Ilford G5 nuclear emulsion with cellulose nitrate plastic stacks on both sides. Outside these plastic stacks there are two further thin emulsion stacks.

In the central emulsion stack we select stopping iron group particles which have passed through one of the plastic stacks in such a way that there are etched cones along the paths. In the emulsion we measure the residual ranges and the track widths. In the etched plastic sheets we measure cone lengths and ranges. The charge of a particle can be determined independently from the track width-range relation and the cone-length-range relation. The mass can be determined by relating the measured cone lengths to the residual range. Information about the mass can also be obtained from the tapering part of the track in the emulsion.

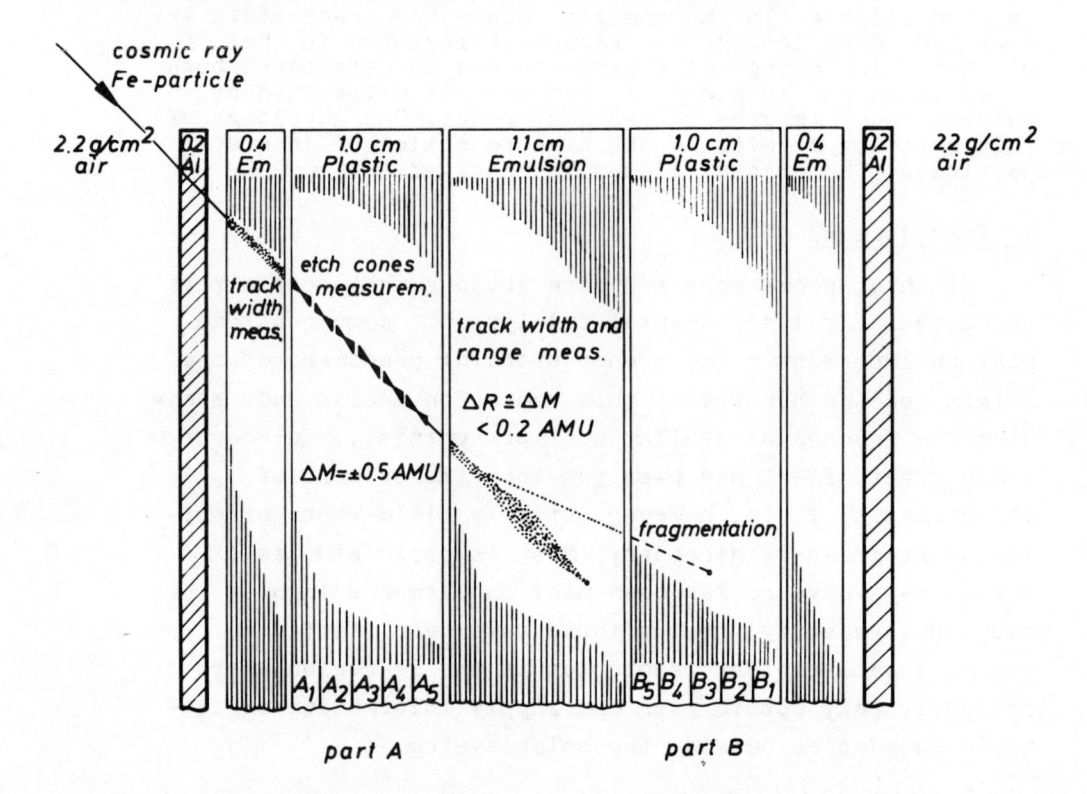

Fig.1. Schematic diagram of the emulsion-plastic experiment

An important advantage of the detector is the possibility to detect and analyse or exclude particles which have produced nuclear interaction within the detector. This is possible to achieve not only in the emulsion stacks but also in the plastic sheets. A discontinous change in the cone lengths along the track indicates an interaction in which the charge of the particle is changed. (1)

Another advantage of this detector is that the number of possible measurements in a track is more than sufficient for the charge and mass determinations and allows testing the consistency in the identification of the particle.

As shown in fig.1 65% of the particles' path (g/cm^{-2}) under $45°$ through the matter of overlaying air, the container of the experiment during the balloon flight and through the detector material are controlled for nuclear interactions. This refers in an average to 0.25 uncontrolled interaction mean free path lengths of Fe-ions.

3. Experimental procedure

Some short information on the stack arrangements may be given. The size was 10 cm width and 20 cm height. A central stack of 18 nuclear emulsions (CE) is covered on both sides with stacks of cellulose nitrate CN plastics (A and B). Outside the latter there are two additional thin stacks, each of 3 emulsions (AE and BE). Ilford G5 nuclear emulsions of 600 um thickness are used in all the emulsion stacks. The two plastic stacks consist of 5 equal blocks of CN plastics, each block having 10 Kodak-CN sheets of 100 µm thickness and 5 Daicel CN sheets of 250 µm thickness. Direct contact between emulsion and CN is avoided by 10 µm thick polyester films. The whole stack is pressed together with plates of aluminium of 2 mm mean thickness.

The stack was exposed twice in balloon flights from Fort Churchill in Canada on July 3-4 and July 14-15 with a total time of 26 h at a ceiling altitude corresponding to 2.2 g/cm^2 residual atmosphere. During the second flight the stack was exposed "upside down". This allowes us to determine whether a particle was registered during the first or the second flight.

The two components of the stack, nuclear emulsion and plastic, were processed and analysed as described in ref.(1). The error of the range measurements in the emulsions results in an error of mass determination of < 0.2 amu, while the error resulting from typical spreads of cone lengths arises to about 0.5 amu. These values do not include the systematic errors described later.

4. Present status of the experiment

The analysis of the tracks in the emulsions and in the plastic sheets have followed the line of evaluation described in ref.(1). It is, however, too early to present final results. Among other things an effect which has been just recently discovered (2) and which may effect all high accuracy plastic detector experiments if no special care has been taken, has been found here too and will be described.

Fig.2 shows a cone length-range diagram of tracks measured in the plastics. Each track has been measured either in part A or part B (fig.1). All measurements of part B could be summarised by curves shown. The tracks of part A, however, are individually plotted. Part A (and of course also part B) consists of 5 subdivisions, called block A_1 to A_5. In each block 10 cone lengths in 5 adjacent plastic sheets have been measured. The mean of these 10 cone length values are plotted in fig.2. The errors of the measurements are smaller than the size of symbols. Some tracks could be measured in all five blocks, others in less than five, depending on the position of the track in the stack and on the angle of the track. All in fig.2 shown tracks are supposed to be Fe-tracks.

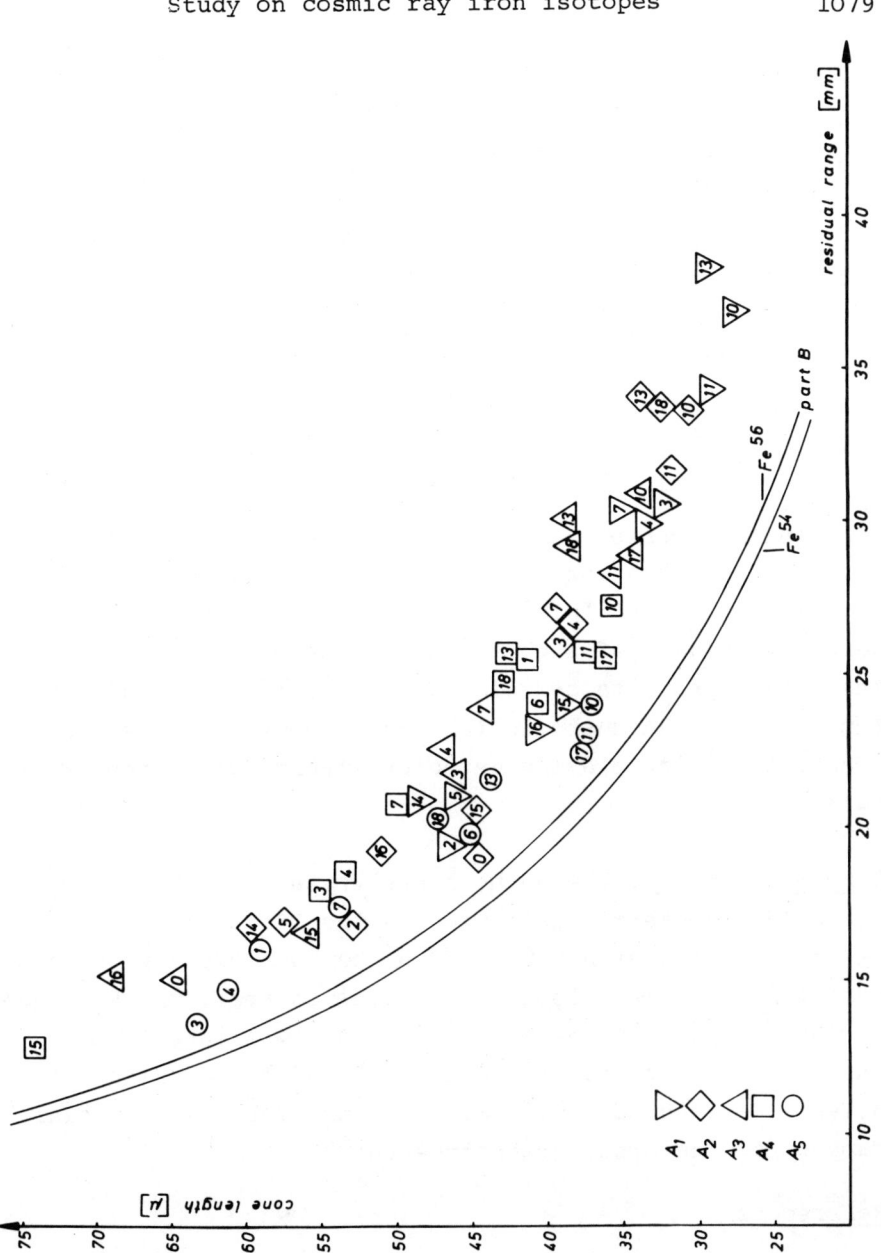

Fig. 2 Cone-length-range diagram of cosmic ray Fe-tracks measured in the blocks A1 to A5 of part A. The curves denote all tracks of part B.

Now, in fig.2 an obvious tendency can be seen. Tracks of part A and part B behave different. Furthermore all measurements of block A_5 and A_4 tend to the left side of the field. This tendency is not so pronounced for the blocks A_3, A_2 and A_1 although it is still there. The A_5 measurements seem to approach the curves obtained from part B where such an effect is not to be seen.

We believe that the tendency in the blocks A_1 to A_5 is due to a possible temperature gradient in the plastic sheets during the exposure. Such a possible temperture effect has been found also – as already mentioned above – in an other plastic detector experiment (2).

In the present experiment we think to overcome this effect by calibrating each block individually using cosmic ray oxygen tracks. This work is in progress together with detailed studies of such a possible temperature effect. If the effect proves to be related to temperature gradients future experiments should take care to avoid temperature gradients inside plastic detector stacks during the exposure.

Acknowledgements. The authors are grateful to Prof. Meyer at the Enrico Fermi Institute and the Office of Naval Research, Washington D.C. for the balloon exposure of the detector. The Kiel group is grateful to Prof. Dr. E. Bagge for his support of the work. This work was supported by the Deutsche Forschungsgemeinschaft and the Swedish Natural Science Research Council. R. S. thanks for the financial support of the Graduiertenförderung.

References

1) Enge W., Beaujean R., Scherzer R., Otte H., Hertzman S., Kristiansson K., Mathiesen O., Söderström K., 14 th Int. Cosmic Ray Conference München, S.3251, (1975)

2) Beaujean R., Sagebiel H., Enge W., Isotopic Composition of low Energy Cosmic Ray Particles with Charges Z=5-8 – this conference – München (1976)

USING NUCLEAR EMULSIONS AND PLASTICS IN A LONG EXPOSURE SATELLITE EXPERIMENT

N. S. Ivanova, D. G. Baranov, V. V. Varyukhin,
Yu. F. Gagarin, V. N. Kulinkov, V.E. Myshkin,
K. M. Romanovskaya, I. G. Khilyuto
and E. A. Yakubovsky

*A. F. Ioffe Physicotechnical Institute, Academy of Sciences
of the U.S.S.R, Leningrad, U.S.S.R.*

The multicharged component of the primary cosmic rays (nuclei with $Z \geqslant 3$) constitutes ∿ 1% from the total cosmic ray flux. A study of this component at high Z - values requires long exposure times and large detector volume and area.

The exposure time in nuclear emulsion experiments is limited, the maximum possible time for relativistic emulsions being 4 days. Using low sensitivity emulsions permits to increase substantially the exposure time.

Installed on board the "Kosmos - 613" satellite was a 14-1. nuclear emulsion chamber combined with plastic detectors. The exposure time was 1441 hrs. (60 days) at an altitude of 200 - 300 km.

Previously, some properties of various emulsions and plastics had been investigated to choose the type of detectors suitable for long-time exposition.

The necessary conditions for long exposures are the absence of track fading and constant detector sensibility throughout the irradiation.

Track fading was studied during several months for various types of low sensitivity emulsions[1]. The best results were obtained for NIKFI A-2 emulsion which did not reveal track fading 140 days after irradiation with alpha-particles and carbon ions. Hence emulsion of this type may be used in long experiments. As claimed by the Manufacturer, its sensibility does not change during 6 months. The only limitation may come from the background which is primarily due to low-energy charged secondaries forming in the interactions of cosmic rays with the nuclei of the emulsion.

Track fading was studied also for various plastics: cellulose triacetate[1], lavsan (polyethyleneterephthalate)[1,2] and vyniproz[2] based on polyvynilchloride. For this purpose, the plastics were irradiated with ions of various energies in a cyclotron. The irradiated plastics were stored at room temperature and etched in lots 1, 3, 5, 7, 8 and 16 months after irradiation. Track length and diameter at the beginning of the track were measured. In the case of cellulose triacetate we observed considerable track fading during the first eight months after irradiation, no fading being revealed after that (from 8 to 16 months). This plastic cannot be used for quantitative estimates.

In lavsan and vyniproz, no fading was observed in 2 to 16 month interval. Thus these plastics can be employed in experiments of long exposure (up to

∼ 1,5 years) in the detection of superheavy nuclei (Z ⩾ 70), there being no background limitations for these plastics.

A technique suitable for etching these plastics was first developed[2,3]. Adding wetting agents to the etching solution resulted in a considerable improvement of etching quality for lavsan and vyniproz, the sheets becoming more transparent after etching. Ultraviolet irradiation of lavsan and vyniproz reduces the threshold level of energy losses by a factor of ∼ 1,5 for nuclei entering the plastic at normal incidence[2,3]. The rate of etching along the track V_t in these plastics increases under UV irradiation 3 to 10 times (at a fivefold increase of energy loss relative to the threshold at the entrance of the nucleus into the plastic) and is only weakly dependent on the UV irradiation dose. UV irradiation increases also the surface etching rate V_G. The results obtained permitted to choose optimum duration of UV irradiation for the plastics of not more than 15 - 30 hrs.

The emulsion chambers in the "Kosmos - 613" satellite experiment represented stacks of low-sensitivity NIKFI A-2 emulsion sheets (with the proton sensitivity threshold of about 30 Mev) 250 μm thick measuring 20×10 cm^2 and 15×10 cm^2. Four emulsion chambers were exposed, three of them being $20 \times 20 \times 10$ cm^3 in size, and one, $15 \times 15 \times 10$ cm^3. The absorbing layer above the chamber was ∼ 1,5 g/cm^2. Stacks of plastic detectors covered 3 faces of each chamber. Besides the plastics, each stack contained 3 emulsion sheets.

Fig. 1 shows the arrangement of the emulsion sheets and plastic stacks in the chamber. The emulsion sheets in the chamber were arranged in such a way that cosmic rays impinging on the working face (with the thinnest absorbing layer above the chamber) pass through the sheets at a small dip angle.

X-ray marks were traced to assist in matching the tracks passing from the stack into the emulsion chamber. In addition, mechanical marking by holes was also used.

The emulsion sheets were developed glued on glass in a normal regime. The development of the emulsion was carried out in the High-Energy Laboratory of the Joint Institute for Nuclear Research.

A study of the developed emulsion showed that despite a long exposure (60 days) and considerable background, high-Z particles (relativistic nuclei with Z ⩾ 26) can be reliably discriminated. The microphotograph presents tracks of a relativistic iron nucleus (a) and of a stopping iron nucleus (b).

The plastic detectors were subjected to the following treatments.

The cellulose triacetate (CT) was etched four months after the end of exposure during 10 hrs. at $40 \pm 0,1°C$ in a pure 6N solution of KOH. The energy loss detection threshold of CT was determined from the maximum range of low-energy iron nuclei found in CT and analyzed to their stopping in the emulsion chamber; it corresponds to the energy deposition of relativistic nuclei with Z ∼ 45. At high energy depositions in CT, one observes after etching on both sheet surfaces, besides the track, also concentric haloes 0,5 to 1,5 mm in diameter. In such cases the point of particle passage can be located with unaided eye.

Vyniproz and lavsan were irradiated prior to etching for 12 hrs. by UV (3100 - 3200 Å) simultaneously in both sides.[2] Etching was carried out 7 months after the end of exposure on the satellite in a 6 N solution of KOH at $40 \pm 0,1°C$

with a wetting agent added and with saturation by the etching products. Vyniproz was etched in a single regime (for 36 hrs), and lavsan, in two regimes (L-1 for 25 hrs, and L-2 for 15 hrs). The resulting surface etching rate V_G for vyniproz at 36 hrs of etching was $\sim 0,28$ μm/hr, for lavsan-1 $\sim 0,35$ μm/hr, and for lavsan-2 $\sim 0,38$ μm/hr.

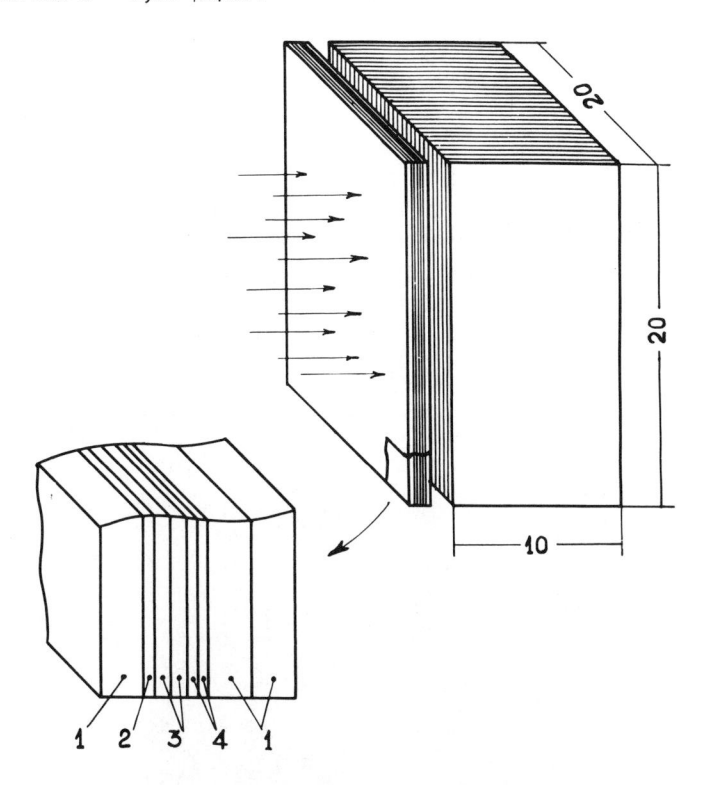

Fig. 1. The arrangements of emulsion sheets and stack of plastic detectors in the chamber.
The stack of plastic detectors consists of emulsion sheets, 250 μm thick (1); cellulose triacetate, 160 μm thick (2); vyniproz V-1 and V-2, 100 μm thick (3); lavsan L-1 and L-2, 60 μm thick (4).

The iron group nuclei identified in the emulsion were used to calibrate the plastics and estimate their threshold sensibilities for the etching regimes chosen. To do this, we measured the length of the cones in the plastic produced by the passing iron nucleus, its energy deposition at the given point being determined by the residual range of nucleus in the emulsion chamber. The cone length dependence on the charge of the relativistic nucleus releasing an energy equivalent that of a low-energy iron nucleus at the point of particle passage through plastic was determined. The threshold of heavy nuclei detection for the etching regimes chosen corresponds to relativistic nuclei of $Z \sim 60$ for vyniproz, $Z \sim 65$ for lavsan-1 and $Z \sim 70$ for lavsan-2.

1084   N. S. Ivanova *et al.*

The calibration graphs can be used to evaluate the charge of very heavy relativistic nuclei passing through the emulsion and the plastic sheets from cone length measurements in the latter independently of the emulsion measurements.

The study of the exposed emulsions is carried out at present along the following lines: a) a detailed investigation of the iron group nuclei at the low energies E = 125 - 280 Mev/nucl; b) an investigation of the charge and energy distributions of heavy nuclei with $Z \lesssim 33 \div 35$.

Conclusions

1. Low-sensitivity emulsions may be used in long exposure satellite experiments (60 days) in studies of the multicharged cosmic ray component.

2. Plastic detectors (lavsan and vyniproz) may be employed in much longer satellite experiments (up to $\sim 1,5$ years) in charge spectrum studies of very heavy nuclei ($Z \lesssim 70$).

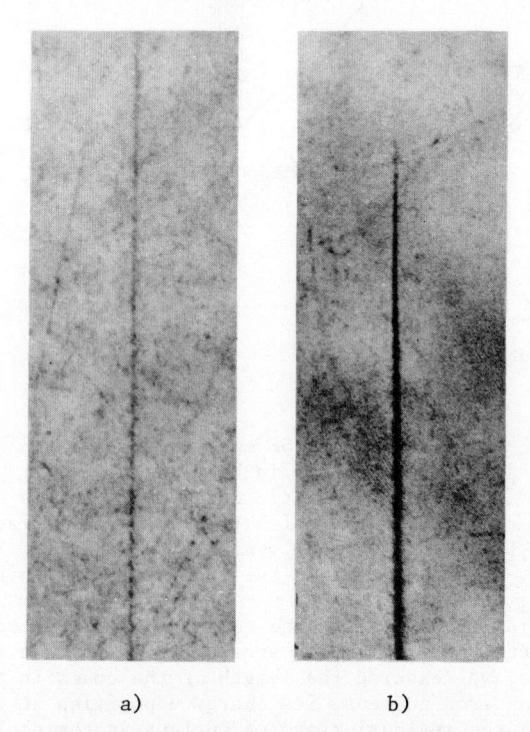

Fig. 2. The microphotograph of tracks of an iron nucleus in emulsion exposed during 60 days: a) relativistic nucleus, b) stopping nucleus.

The authors are grateful to V. A. Budina, N. N. Kotlova and L. A. Isakova for the measurements, and to researchers in the JINR for the emulsion development.

References

1. Gagarin Yu. F., Ivanova N. S., Khilyuto I. G., Izv. Acad. Nauk SSSR, ser. fiz., 37 (1973), 1506.
2. Gagarin Yu. F., Ivanova N. S., Prib. i Tekhn. Exsper., 3 (1976), 53.
3. Gagarin Yu. F., Gusinskii G. M., Ivanova N. S., Lemberg I. Kh., Preprint FTI 500, Leningrad, 1975; Prib. i Tekhn. Exsper. 1976 (in press).

LOW ENERGY HEAVY COSMIC IONS CHARGE DISCRIMINATION WITH PLASTIC DETECTORS[⊕]

J. Sequeiros*, J. Medina*, A. Durá*, M. Ortega*,
A. Vidal-Quadras*, F. Fernandez* and R. T. Thorne**[+]

*Laboratorio de Física Corpuscular, Departamento de Física
Fundamental, Universidad Autónoma de Barcelona, Bellaterra, Spain
**H. H. Wills Physics Laboratory, University of Bristol, England

A method for high resolution (average $\Delta Z \simeq 0.3$) charge discrimination of very heavy ($20 \leq Z \leq 29$) low energy ($E \sim 100$ MeV/N) ions in plastic detectors is presented. The discrimination is based on an empirical relation between etch rate (V_t) and ionization ($J(Z, \beta, k)$) of the form $V_t = A((B/J)+1)^{-n}$ valid for high ionization at low residual ranges, where the etch rate shows saturation.

The results presented are from the partial annalysis of a stack of 1,240 cm^2 of area and a thickness of 9.28 g/cm^2 copper equivalent, composed of 225 layers of Lexan of about 250 μm thickness and three G-5 Ilford nuclear emulsions of 200 μm thicness. The exposure took place in Sioux Falls (South Dakota, U.S.A) reaching an altitud of 4 g/cm^2 of residual atmosphere where it remained for about 82 hours.

Each batch has been calibrate independently and the values of the parameters A,B,k and n have been obtained by a least square fit of the above expression to an average of 10 tracks (about 100 pairs V_t, R) for each batch. The tracks selected for calibration are considered to be produced by iron ions (Z=26) because their values of V_t v.s. R lie in the highest density zone of the plot.

Each of the track used have been identified and the corresponding ion charges obtained with probability and error.

The charge spectrum shows that the abundances of Cr (Z=24), Mn (Z=25) and Co (Z=27) are more than three times higher at low energy than at high energy.

[⊕] Work partially supported by the Comision Nacional de Investigación del Espacio (CONIE).

[+] Present address: Reactor Physics Department. Bradwell Nuclear Power Station. Essex. England.

SOME RESULTS OF CHARGE SPECTRUM MEASUREMENTS WITH A ROCKET EQUIPMENT USING PLASTIC DETECTORS

B. Sojka and H. Röhrs

Institut für Reine und Angewandte Kernphysik, University of Kiel, 2300 Kiel, West Germany

On 22 Jan. 1972 in Kiruna two rockets were launched six hours apart into a strong PCA. The apogee was 130 km approx. A tray arrangement was used to expose CN-plastic foils to the radiation. The duration of exposal was about 170 sec. Charge identification of low energy heavy nuclei is obtained by the following method: After having etched the CN-sheets the cone length is measured; then the material is etched for another time till the stopping end is visible. The relation between cone length and residual range allows to estimate the charge. The method will be discussed. First results will be presented.

Concerning to the determination of alpha particle flux it was found that the tracks of low energy alpha particles hardly were to be recognized. It was tried to modify the etching procedure to decrease the noise to signal ratio. Method and first results will be reported.

CALIBRATION OF TWO PLASTIC DETECTORS AND APPLICATION ON STUDY OF HEAVY COSMIC RAYS*

J. Tripier and M. Debeauvais

SADVI, Centre de Recherches Nucleaires, 67037 Strasbourg, Cedex

ABSTRACT

In order to study the charge distribution of heavy cosmic ray ions, we have achieved the calibration of Makrofol and Kodak cellulose nitrate by the method of measurements of etched cone lengths as function of the residual range. We give some results obtained by this method in a french - sovietic experiment aboard the Cosmos-782 satellite.

---

* This work was sponsored by the French Centre National d'Etudes Spatiales and the authors wish to thank Doctor BOST for his support.

I - CALIBRATION OF MAKROFOL AND CELLULOSE NITRATE

A - Principle of the method

If an ion enters in a plastic detector sheet, it produces, along its path, some damages. The importance of which depends on the LET (or energy-loss dE/dx) of this ion. These damages, then etched by a chemical reagent, give rise to a track. The track etching rate depends on the LET of the ion, which varies along the path. Now, if this ion passes through several detector sheets, the track fragments in the sheets have different values of dE/dx as the variation of the residual range. In other words, the etched track fragments (or cone lengths L_i) for a given etching time, depend on dE/dx, therefore on the ion residual range R_i. This function is a characteristic of the ion, particularly of the ion's charge.

Consequently, the experimental determination of a set of points (L_i, R_i), permits us to identify an ion by means of the knowledge of the $L_i = f(R_i)$ function for this ion [1, 2, 3, 4].

B - Establishment of the $L_i = f(R_i)$ curves

At first, we have to know a relation between the track etching rate V_T and the energy-loss dE/dx, for given etching conditions.

1°) Relation between V_T and dE/dx

We have studied two plastic detectors : The Makrofol polycarbonate and the Kodak cellulose nitrate. The etching conditions for these two materials are the following ones :

- Makrofol : NaOH-5N , T = 60° c
- Cellulose nitrate : NaOH-2.5 N , T = 40° c

Calibration was achieved by ions of accelerators. Taking account into the relations between energy, range and energy-loss we determined previously [5], we have obtained for makrofol the curve of the mean track etching rate $\bar{V}_T$ in function of $(\overline{dE/dx})$ shown in Fig. 1. The track etching rate in makrofol can be expressed by a power function :

$$\overline{V}_T = a \, [\overline{dE/dx}]^b$$

$$a = 1.95 \times 10^{-4} \; ; \; b = 2.55$$

For cellulose nitrate, the curve $\overline{V}_T = f(\overline{dE/dx})$ shown in Fig. 2 can be expressed by the following equations :

- for $1.3 \leq \overline{dE/dx} \leq 2.5$ MeV.mg^{-1}.cm^2 :

$$\overline{V}_T = a \cdot \exp[b \cdot (\overline{dE/dx})] \qquad \begin{array}{l} a = 1.86 \cdot 10^{-3} \\ b = 1.97 \end{array}$$

- for $2.5 < \overline{dE/dx} \leq 4.1$ MeV mg^{-1} cm^2 :

$$\overline{V}_T = [a \, (\overline{dE/dx}) - b]^{1/2} \qquad \begin{array}{l} a = 0.26 \\ b = 0.585 \end{array}$$

- for $\overline{dE/dx} > 4.1$ MeV^{-1}.cm^2 :

$$\overline{V}_T = a \, [1 - \exp(-b(\overline{dE/dx}))] \qquad \begin{array}{l} a = 1.24 \\ b = 0.2 \end{array}$$

2°) <u>Curves $L_i = f(R_i)$</u>

If we consider an ion with given mass, charge and velocity, we know its range R_i and energy-loss $(dE/dx)_i$, from which we determine $\overline{V}_T (R_i)$ by the above relations. For each value of R_i, we can thus calculate an etchable cone length L_i such as :

$$L_i = \int_0^t \overline{V}_T (R_i) \, dt$$

During a given time Δt, the etched cone length is :

$$L_i = \overline{V}_T (R_i) \cdot \Delta t$$

So we have calculated for all charges the whole set of curves of cone length as function of residual range shown in fig. 3 for makrofol, in fig. 4 for cellulose nitrate.

According to the way of establishing these diagrams (L_i, R_i), we have to measure the residual range R_i at the center of the etched cone of length L_i such as :

$$R_i = (R_o)_i + L_i/2$$

in which $(R_o)_i$ is the range from the stopping end of the ion trajectory to the point where the cone appears (see fig. 5).

II - APPLICATION OF THE METHOD TO THE BIOBLOC EXPERIMENT

We relate a part of a French-sovietic biological experiment. A stack of nuclear emulsions, biological subjects and plastic detectors was flown aboard the Cosmos-782 satellite. Within a collaboration between the authors and the russian physicists V.I. POPOV, A.M. MARENNYI and Y.A. VINOGRADOV, we had to investigate the charge spectrum of cosmic ray ions by means of a stack of seventy-five-100 µm thick-sheets of Kodak cellulose nitrate.

1°) Etching process

Sheets have been etched by stacks of five ones held in a frame like it is shown in Fig. 6. The separation of each sheet is about 2 cm and this distance is kept by plastic tubes held on stainless steel bars. The etching solution was NaOH - 2.5 N at 40° c and the etching time was 5 hours.

2°) Observation

For the track observation under an optical microscope, we have realized the following stack : from the bottom to the top, a holder in plexiglass, then a millimetric grid and five etched detector sheets each above the other and the whole stack was sandwiched between two plastics layers, the edges of which were soldered under vacuum. We have obtained so very good correlations between tracks fragments.

Tracks have been measured as shown in fig. 7 where Z is the deepness, L the true track length, ℓ the projected length and e_a, e_b the ellipse lengths.

3°) Results

a) Density of events

- for the first sheet, we have found 511 tracks in 59.21 cm^2 (8.6 tracks/cm^2).
- in the stack of sheets 1 → 5 : 569 tracks in 59.21 cm^2 (9.6 tracks/cm^2), 76 of which are tracks coming to rest in this stack (25.6/cm^3).
- in the stack of sheets 6 → 10, we have found 505 tracks in 61.41 cm^2 (8.2 tracks/cm^2), with 86 tracks coming to rest (28.0/cm^3).

b) charge spectrum

In order to obtain preliminary results on the charge spectrum, we have selected a surface of 14.40 cm^2 placed at the center of the stack of sheets 26 → 30. We have measured all the tracks which crossed through this surface by travelling them in the whole stack from the first sheet to the 75th one (about 1/4 of the stack). We have observed 116 tracks, among of them :

- 55 tracks coming to rest within the stack have been identified.
- 18 tracks coming to rest without but near the stack have been estimated by the change in dE/dx along their trajectory[6].
- 43 tracks crossing through the whole stack without any possible identification (37 % of the tracks).

The charge spectrum obtained is shown in fig. 8. This distribution is in general agreement with the measurement of cosmic ray abondances by others, but our stack was not thick enough to identify high charge ions. In order to get over this disadvantage, we are studying longer tracks with russian physicists whose stack was adjacent with the our one.

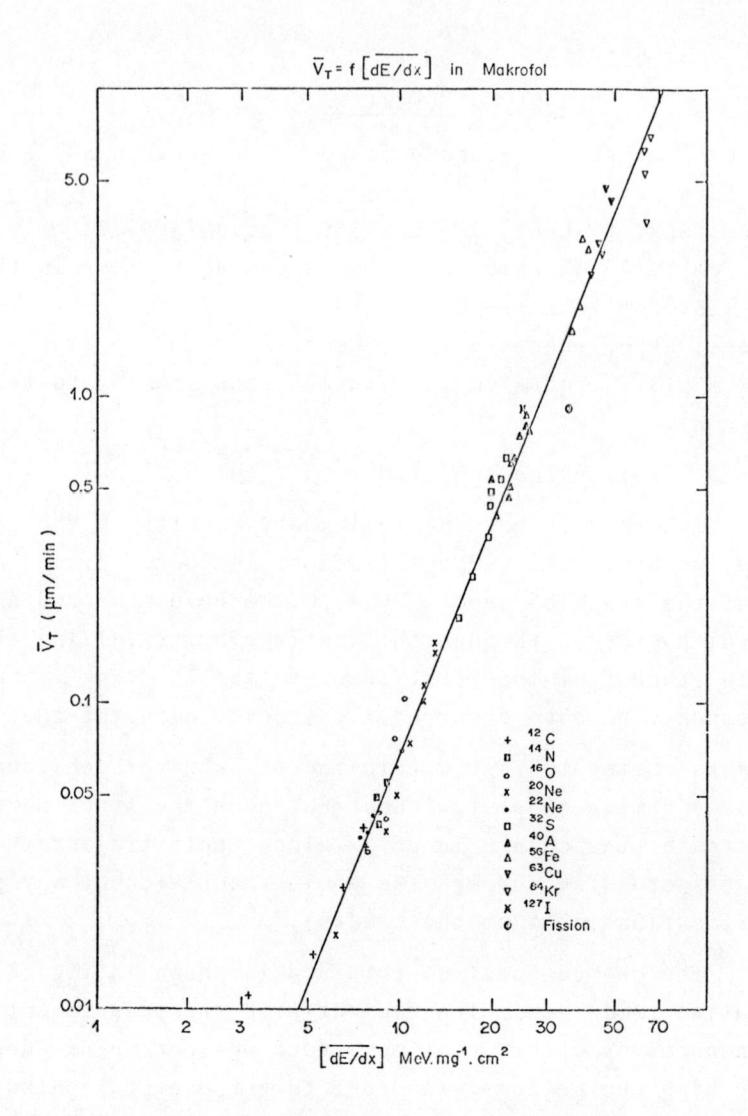

Fig. 1 : The curve of $\bar{V}_T$ plotted versus $(\overline{dE/dx})$ in makrofol

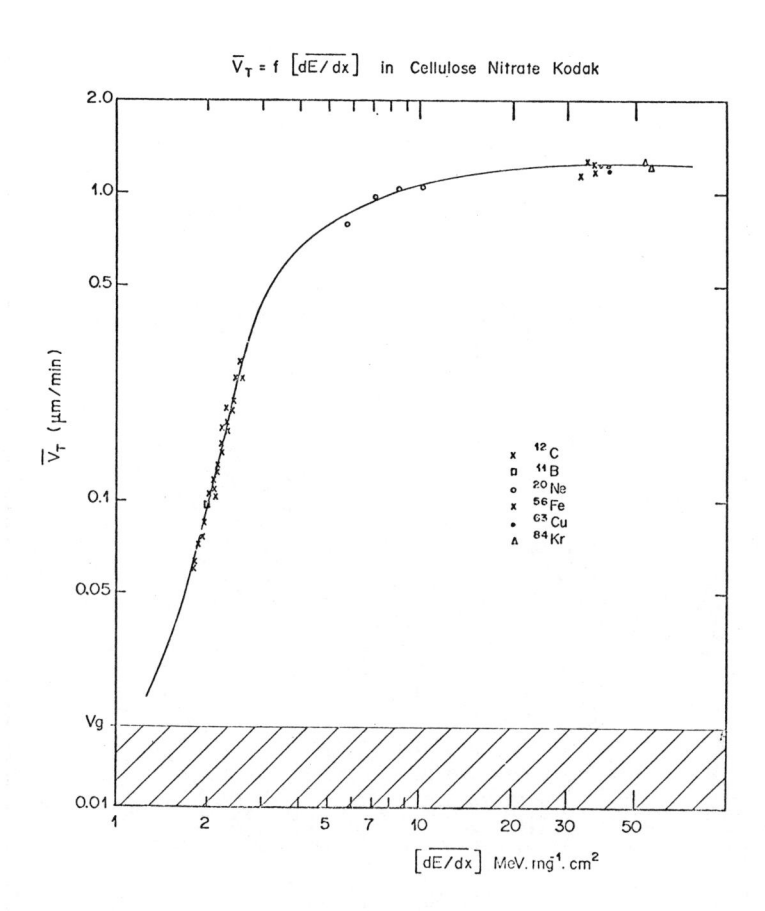

Fig. 2 : The track etching rate $\overline{V}_T$ in cellulose nitrate

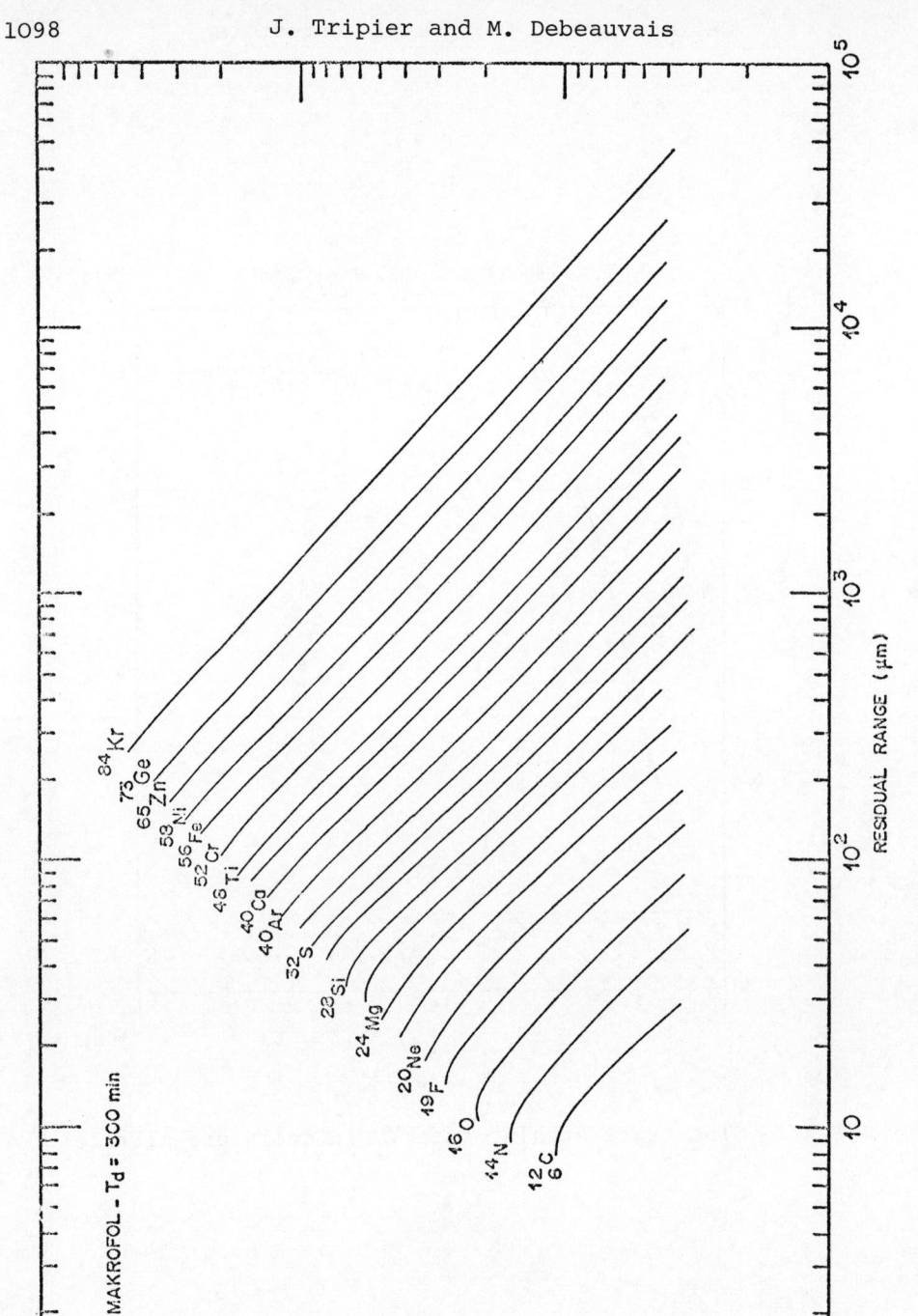

Fig. 3 : Cone length versus residual range for makrofol

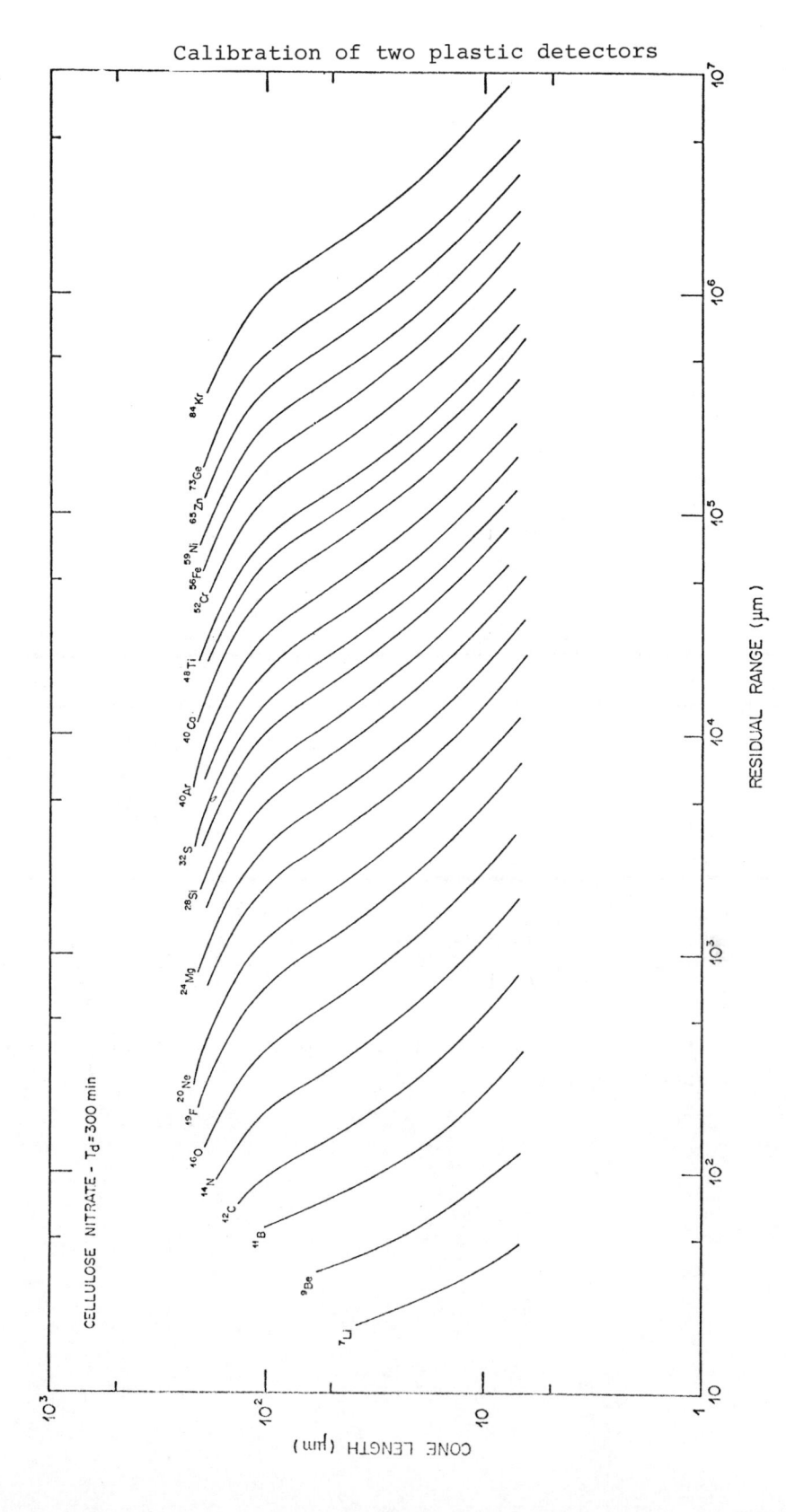

Fig. 4 : Cone length versus residual range for cellulose nitrate

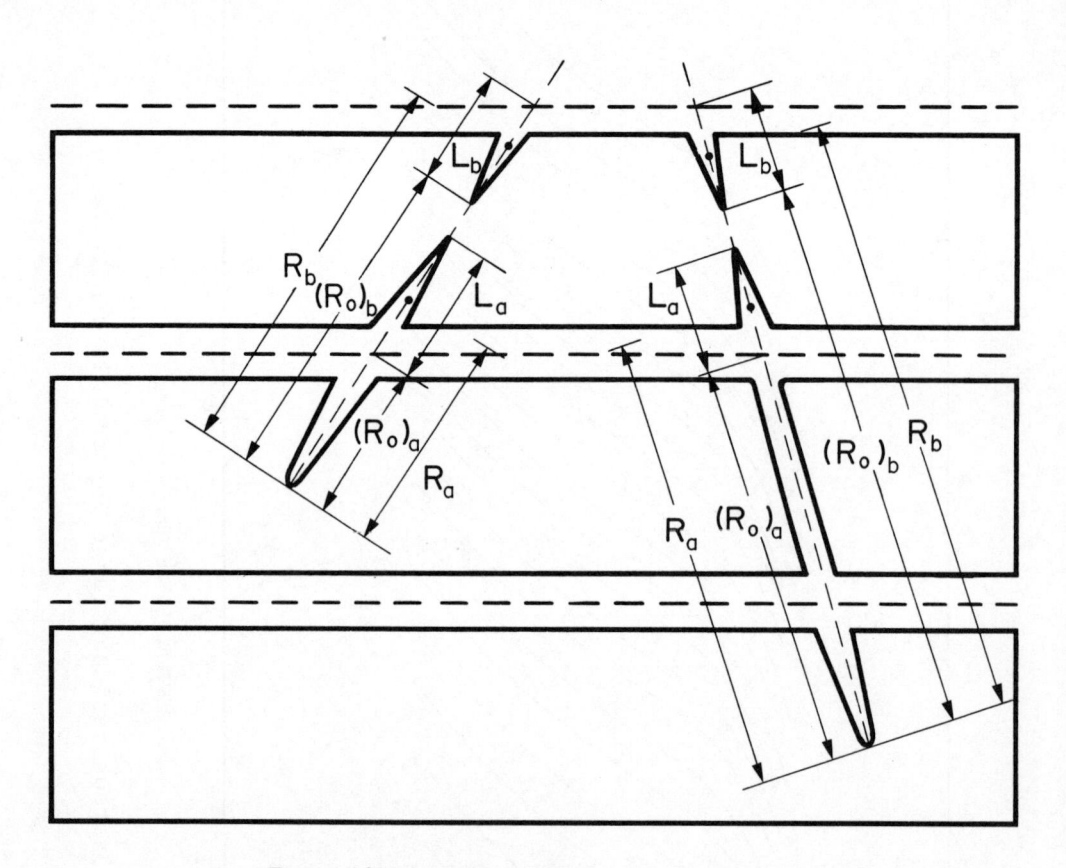

$$R_a = (R_o)_a + L_a/2$$
$$R_b = (R_o)_b + L_b/2$$

Fig. 5 : Measurement of the residual range R_i in connection with a cone length L_i.

STACK OF PLASTIC SHEETS FOR ETCHING PROCESS

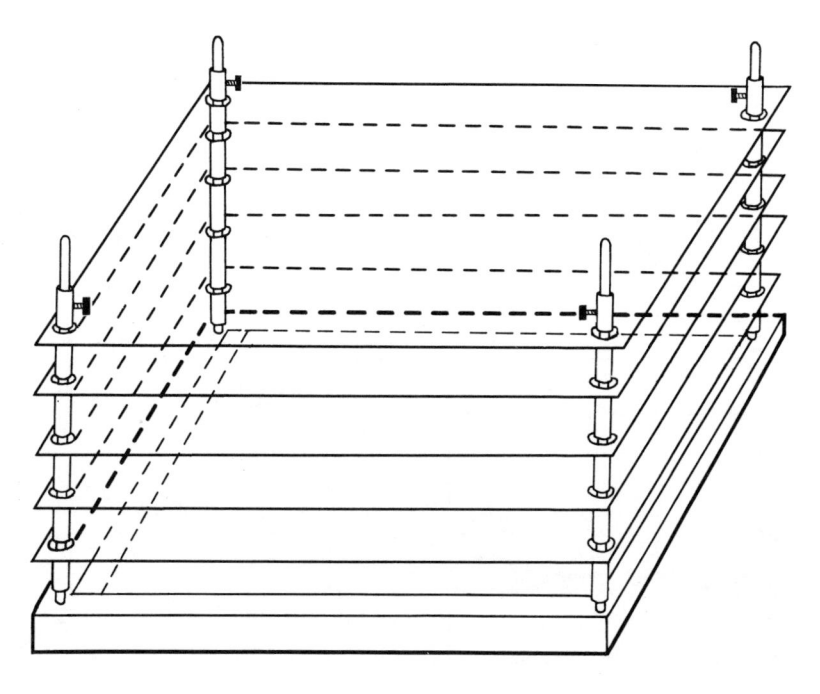

Fig. 6 : Frame to hold plastic sheets for etching process.

$$\operatorname{tg} \theta = \frac{Z}{l - (e_a + e_b)/2}$$

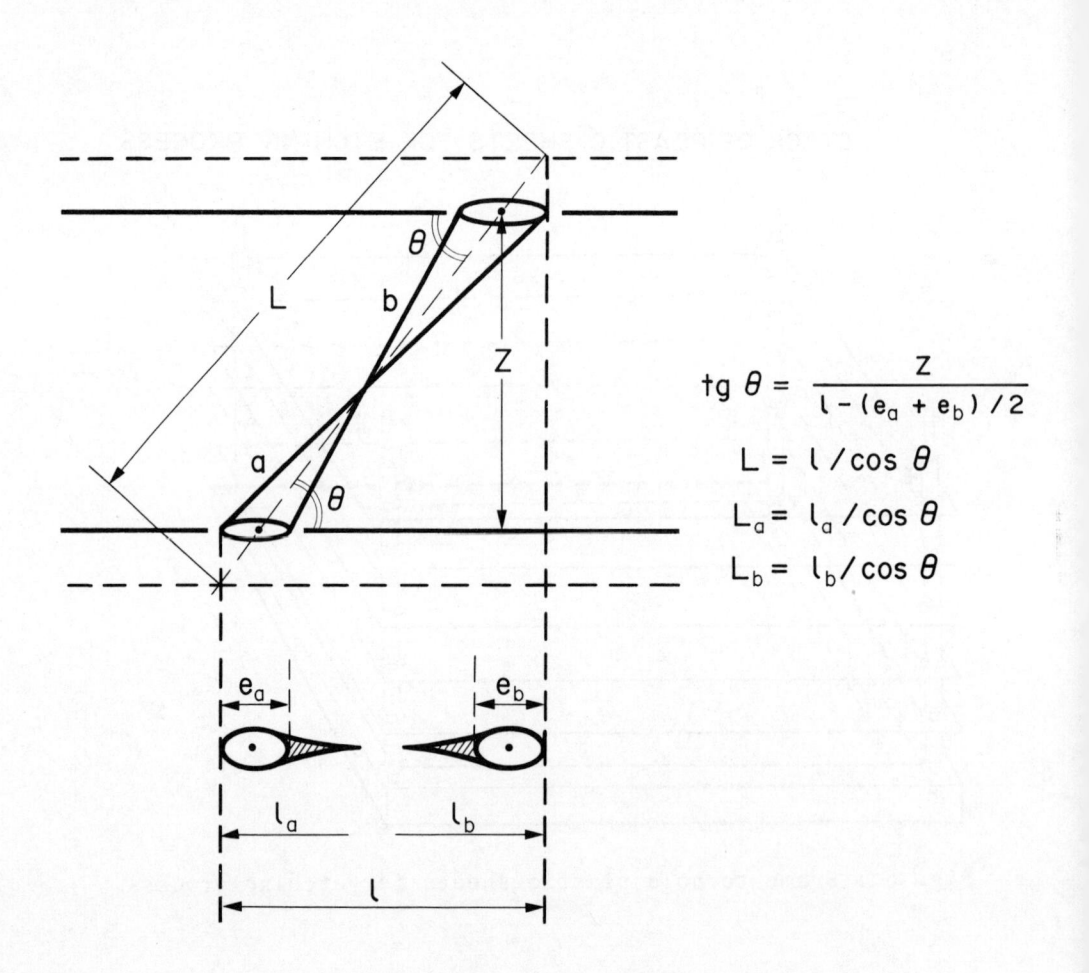

$$L = l/\cos\theta$$

$$L_a = l_a/\cos\theta$$

$$L_b = l_b/\cos\theta$$

Fig. 7 : Track measurements.

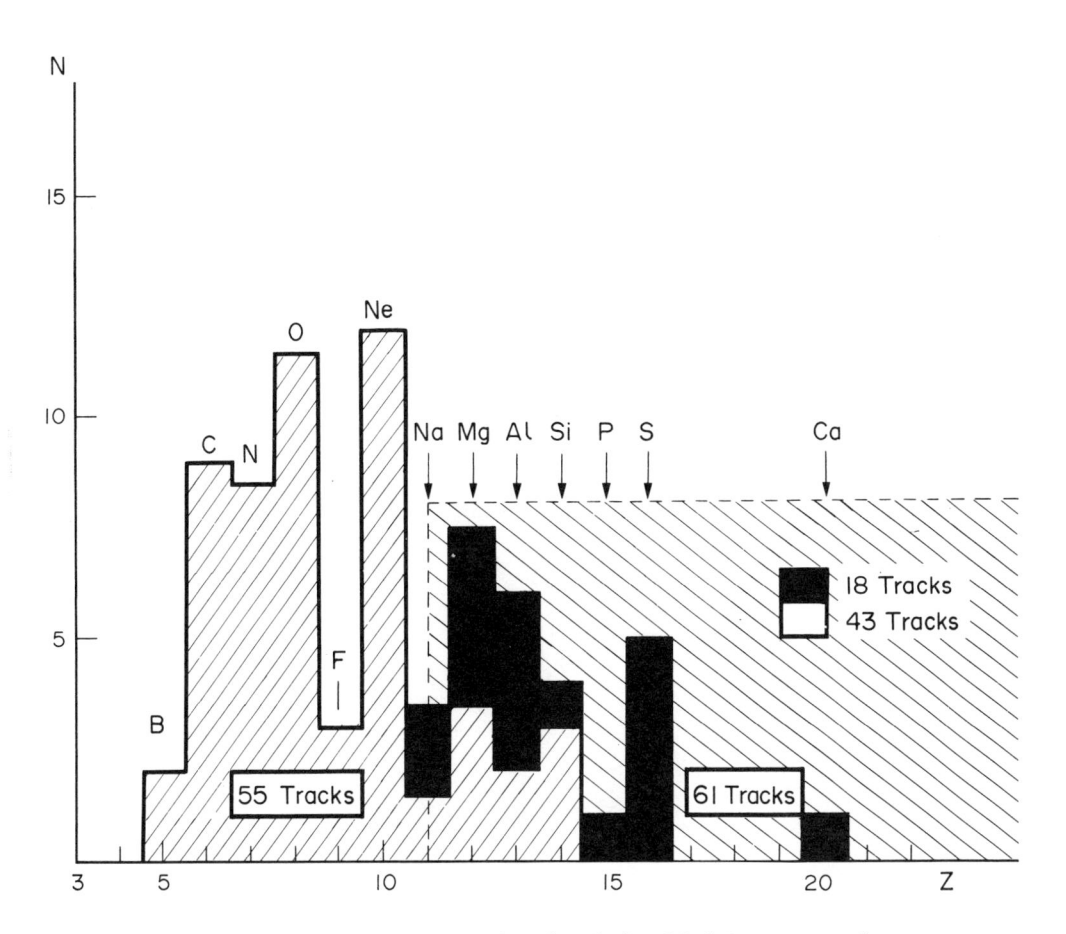

Fig. 8 : Charge spectrum obtained in Biobloc experiment.

ACKNOWLEDGEMENTS

The authors are grateful to Professor PLANEL for his support in the french-sovietic collaboration.

REFERENCES

1) P.B. PRICE, R.L. FLEISCHER, D.D. PETERSON, C. O'CEALLAIGH, D. O'SULLIVAN and A. THOMPSON, Phys. Rev. Let. 21 (1968) 630.
2) W. ENGE, K.P. BARTHOLOMÄ, K. FUKUI, Proc. 12th Int. Conf. on Cosmic Rays (1971).
3) K. FUKUI, W. ENGE, R. BEAUJEAN, K.P. BARTHOLOMÄ, Use and processing of Plastics as particle detectors, Report AFCRL-TR-75-0223 (1975).
4) D.D. PETERSON, Thesis, New-York (1969).
5) J. TRIPIER, G. REMY, J. RALAROSY, M. DEBEAUVAIS, R. STEIN and D. HUSS, Nucl. Inst. and Meth. 115 (1974) 29.
6) E.V. BENTON and R.P. HENKE, A method for charge determination of heavy cosmic ray particles, Report USNRDL-TR-67-112 (1967).

Session 10

Applications in Nuclear Physics

Chairmen: E. Schopper
I. Otterlund
R. Schmitt

APPLICATIONS OF SSNTDs IN HIGH ENERGY PHYSICS

I. Otterlund

*Department of Cosmic High Energy Physics, University of Lund,
Sölvegatan 14, S-223 62 Lund, Sweden*

Abstract. Different applications of the emulsion technique in high energy physics are given. Investigations of heavy ion and proton-nucleus reactions with the conventional emulsion technique will be presented together with a short interpretation of recent results. Methods of using nuclear emulsion with embedded targets will be discussed. Emulsion stacks in hybrid systems with electronic tagging suggest a new and interesting application of the emulsion technique. This method and experiments now under way using hybridized emulsions will shortly be reviewed.

1. Introduction.

In this talk, applications of solid state nuclear track detectors (SSNTD) will be limited to applications of nuclear emulsions which have so far been the SSNTDs most used in high energy physics. Applications of silver chloride track detectors, developed by Schopper et al. (1), and plastic track detectors, especially experiments by Price et al. (2), have produced interesting results from high energy heavy ion reactions. However, in order to limit the scope of my talk, applications of these techniques will not be reviewed.

In most emulsion experiments, high energy physics have used a conventional emulsion technique. However, recent improvements in the emulsion technique, due to identifiable nuclear targets embedded in the emulsion, may increase the accuracy of the experimental data and stimulate emulsion experiments.

In high energy heavy ion physics, the emulsion technique seems to be useful, especially in studies of the emission of fast target fragments (search for shock wave phenomena), of beam nucleus fragmentation, and of correlations between the projectile and target fragmentation (3-10).

The emulsion technique is advantageous in that the interactions can be observed event by event. Thus, it is possible to select reactions with only target fragmentation or projectile fragmentation from more central collisions, to investigate reactions without break up or excitation of the target nucleus or reactions with high pion multiplicities, etc. These properties of the detector are of special importance in studies of heavy ion collisions owing to the complex nature of the reaction.

In investigations of high energy hadron-nucleus reactions, emulsion detectors have given significant contributions to the investigations in this field. When high energy protons and pions became available at Serphukov and at Fermilab, the interest in the emulsion technique increased. Since then the interest in emulsion experiments has continued to grow. The first results from emulsion experiments showed that the intra-nuclear cascade in the hit nucleus failed to appear at high energies (11). It was then believed that the nucleus can act as a suitable detector for the space time evolution of multiparticle production, which can hardly be observed from the asymptotic states of produced particles in hadron-nucleon reactions. Now the questions have also been focused on the problems of what really happens when a hadron hits a nucleus. Are there independent reactions between the impinging hadron and target nucleons, or does the incident hadron react coherently with several target nucleons?

During the last two years we have been witnessing a remarkable number of important discoveries: the ψ and ψ' states; the observation of a sudden rise in the ratio R of hadrons to muon pairs produced in e^+e^- annihilations at $\sqrt{s} \simeq 3.1$ GeV (12) respectively $\sqrt{s} \simeq 3.7$ GeV (13), and further structure in R for $\sqrt{s} \sim 4$ GeV (neutral state with mass 1865 ± 15 MeV/c² and charged state with mass 1876 ± 15 MeV/c² (14)), $\mu^+\mu^-$ events in the Fermilab neutrino experiments (15), etc. The leading interpretation of this phenomenon is that production of charmed particles is responsible for most of the observed structure. Clearly, it is now of the utmost importance to see whether one can directly observe the production and weak decay of charmed particles. Experiments now under way, using emulsion stacks in hybrid systems with electronic tagging, is a relatively simple and efficient method of searching for charmed particles.

2. Notations in emulsion experiments

In emulsion experiments, the emulsion is in general used as both target and detector. The targets consist of hydrogen, a light (L) group of CNO-targets and a heavy (H) group of AgBr--targets. The mean mass of emulsion nuclei is 60 (16). The light nuclei have a mean mass equal to 14; and the heavy nuclei have a mean mass equal to 94.

The average number of encounters between an incident hadron and the nucleons in the target nucleus, $\bar{\nu}$, is defined from the equation

$$\bar{\nu} = \frac{A\,\sigma_{hp}}{\sigma_{hA}} \qquad (1),$$

where σ_{hp} and σ_{hA} are the inelastic cross-sections for hadron-proton and hadron-nucleus interactions (16,17). Experimental values of the inelastic cross-sections give (16)

$$\pi\text{-A interactions } \bar{\nu} = 0.74\,A^{0.25}$$
$$p\text{-A interactions } \bar{\nu} = 0.70\,A^{0.31} \qquad (2)$$

For proton-emulsion nucleus reactions $\bar{\nu}_L = 1.6$ and $\bar{\nu}_H = 2.9$.

The particles emitted in emulsion interactions are classified according to the ionization produced along the tracks. Normally, we do not identify the particles. Consequently, we simply call them black, grey, heavy track-producing particles, and shower-particles. In high energy p-A reactions we define:

n_b = the number of black track-producing particles. These particles have an ionization of $I > 6.8\,I_o$, where I_o is the ionization of the primary proton. This ionization range corresponds to protons with energies $\lesssim$ 30 MeV. Black track-producing particles are mainly fragments emitted from the excited target. They not only have energy but also angular distributions typical of thermal processes.

n_g = the number of <u>grey track-producing particles</u>. The ionization of these tracks are $6.8\, I_o > I > 1.4\, I_o$, corresponding to protons in the energy range 30-400 MeV. Grey track-particles are believed to be associated with the recoiling particles.

$N_h = n_g + n_b$ = the number of <u>heavy prong particles</u> (having $\beta < 0.7$). N_h is interpreted as the number of charged fragments emitted from the target.

n_s = the number of shower-particles. These particles have $\beta \geq 0.7$ ($I \leq 1.4\, I_o$), and are mainly pions produced in the reaction.

$\langle n_{ch} \rangle$ = the average number of charged particles observed in a collision of a hadron with a proton.

When investigating hadron-nucleus reactions, the quantity
$R = \dfrac{\langle n_s \rangle}{\langle n_{ch} \rangle}$, which is the ratio between the shower-particle multiplicity in hadron-nucleus reactions and the charged particle multiplicity in pp reactions, is frequently used to interpret the data.

In emulsion experiments on multiparticle production, emission angles θ are measured. From these measurements it is possible to obtain rapidity distributions by using the pseudorapidity variable:

$$Y_{lab} \approx \eta = -\ln \text{tg}\, \theta/2$$

A comparison between Y_{lab} and η is given, for example, in ref. 16.

3. Applications in heavy ion physics

During the first experiments in high energy heavy ion reactions, emulsion detectors and nuclei from cosmic radiation were used. Many of the observations made in these early experiments have now been confirmed in experiments with beams of artificial accelerated heavy ions. Especially, there are the following observations from the "cosmic ray period", which may be of special importance for the interpretation of the reaction mechanism:

i   Fragments with energies much above the fermi energy are often emitted from the target (18).

ii Events occur with high multiplicity of fast He-nuclei from the target nucleus (Table 1).

Table 1. Results from cosmic ray heavy ion reactions ($12 \leq Z_{inc} \leq 26$, $E > 2$ GeV/nucleon) (19).

Number of He-fragments $E_{He} \geq 36$ MeV	Number of interactions	$<E_{He}>$ MeV	$<\theta_{He}>$ degrees
0	45		
1	19	93±17	79±8
2-3	12	109±17	62±8
4-7	6	165±31	64±8

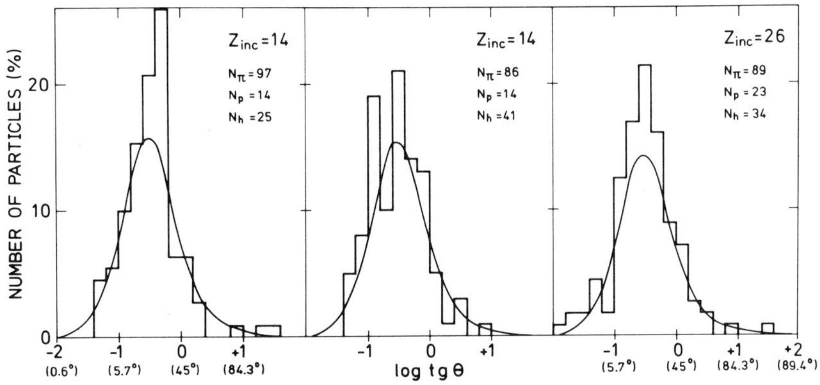

Fig. 1. Experimental logtg θ distributions of shower particles in three large cosmic ray stars. The curves are pion distributions from 10 GeV p-p interactions (20).

iii There exist reactions with small impact parameters where no protons emitted from the incident nucleus can be observed within a narrow forward cone in the lab. system thus indicating large transverse momentum transfer (20). This is clearly seen in Fig. 1 where the excess (= the difference between the histogram and the curve) is in the range of $5°$ - $35°$ (20).

iv The pseudo-rapidity distribution (log tg θ) of relativistic singly charged particles has a surprisingly small standard deviation in interactions where both of the interacting nuclei are almost totally disintegrated (20).
v   There may exist reactions with pion multiplicities much higher than expected from a simple superposition of hadron-nucleus reactions (20).

Table 2. The production cross-sections (barn) for emission of 1-4 He nuclei in ^{16}O-emulsion nucleus interactions.

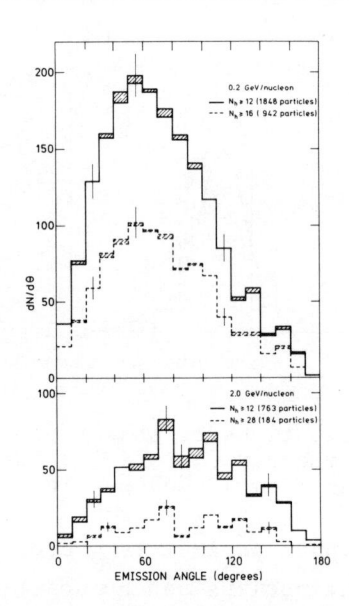

Number of He nuclei from the projectile nucleus	E = 0.2 GeV/n Kullberg et al. (8)	E = 2 GeV/n Jakobsson et al. (7)	E = 2 GeV/n Judek (6)
1	0.25 ± 0.03	0.19 ± 0.03	0.20 ± 0.02
2	0.32 ± 0.04	0.19 ± 0.04	0.22 ± 0.03
3	0.22 ± 0.03	0.27 ± 0.04	0.20 ± 0.02
4	0.04 ± 0.01	≈ 0.01	0.03 ± 0.02

Fig. 2. Angular distributions of all particles with REL > 44 MeV/cm in ^{16}O-AgBr reactions at 0.2 and 2 GeV/nucleon (3).

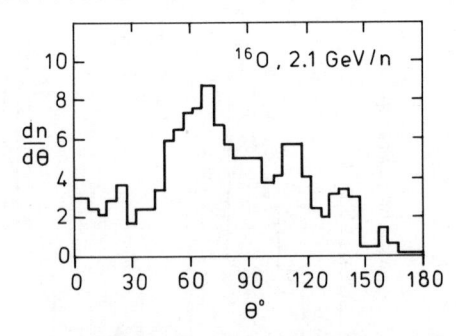

Fig. 3. Angular distributions of reaction products in ^{16}O+Ag collisions recorded as stars in AgCl-detectors; > 15 prongs (1)

In emulsion experiments, using heavy ion beams from the Bevatron-Bevalac at Berkeley, fragmentation of the incident nuclei and emission of fast target fragments has been the main subject of study (Figs. 2 and 3, Table 2). Of special interest are the results on angular distributions of high energy He nuclei, emitted from the target in central ^{16}O-AgBr interactions. It has recently been found that the angular distribution is highly forward peaked at 0.2 GeV/nucleon but almost isotropic at 2 GeV/nucleon (Figs. 2 and 3).

Figs. 4 and 5 illustrate two different models designed to explain the emission of fast fragments from the target. Hydrodynamic calculations predict the formation and propagation of shock waves when the nuclear sound velocity (v ~ 0.2 c) is exceeded. Some of these models predict comparatively narrow peaks at a straight angle to a conical shock front (Fig. 4)(1). In the nuclear fireball model, the nucleons which are swept out from the target and projectile form a hot, quasi-equilibrated fireball, which decays as an ideal gas, Fig. 5 (21).

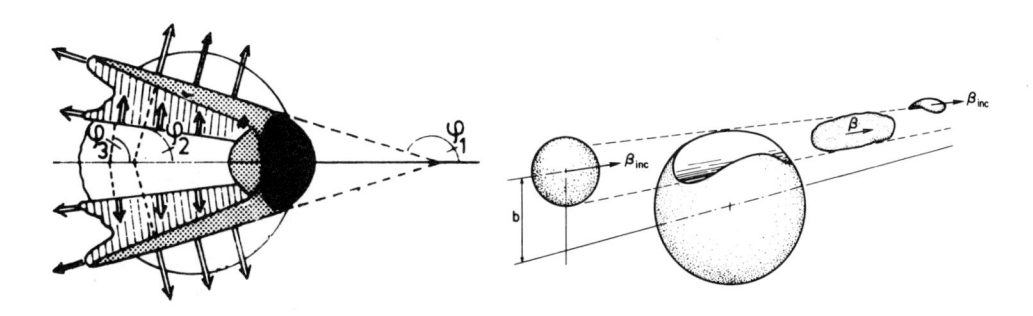

Fig. 4. The final stage of a central penetration of a light nucleus into a heavier one. The main Mach shock front moves in the direction given by φ_2 (1).

Fig. 5. The nuclear fireball model (21).

The technique with embedded wires in emulsions (see chapter 4) should be suitable for heavy ion physics and may give significant information about, for instance, pion production, correlations between projectile and target fragmentation, fragmentation cross sections in the hundreds of MeV/nucleon regions, etc. All of which properties are not very well known.

4. **Applications in experiments on multiparticle production (hadron-nucleus reactions).**

4.1. Experiments using the conventional emulsion technique.

There have been comprehensive emulsion experiments at Serphukov and Fermilab, all of them in the field of multi-hadron reaction studies. The emulsion technique allows an event-to-event study of the reactions, which is a great advantage. The high spatial resolution makes it possible to measure emission angles with high accuracy (angles ~ 10^{-3} are measured with an error of 10-20%). The registration of very slow charged fragments and electrons from nuclear decay permits the experimentalists to select with confidence events without break up or excitation of the nucleus.

Coherent p-A reactions have therefore been extensively studied with the emulsion technique. Multiplicities in reactions without any target fragments are shown in Fig. 6. The peaks contain coherently produced pions from the following reactions:

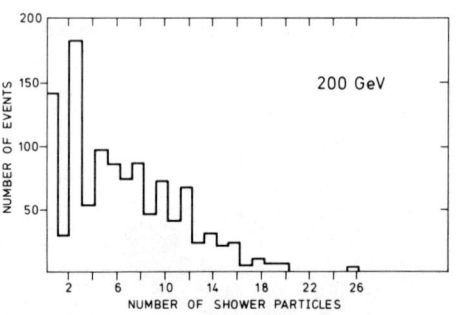

Fig. 6. Multiplicity distribution in reactions without break up or excitation of the target nucleus ("clean events") (22).

$$n_s = 1 \begin{cases} p + A \to p\pi^o + k\,\pi^o + A \\ p + A \to n\pi^+ + k\,\pi^o + A \end{cases}$$

$$n_s = 3 \begin{cases} p + A \to p\,\pi^+\pi^- + k\pi^o + A \\ p + A \to n\,\pi^+\pi^+\pi^- + k\pi^o + A \end{cases}$$

etc.

$k = 0,1,2, \ldots$ \hfill (4)

In order to understand the space-time development of hadronic matter, multiplicity distributions and rapidity distributions (ln tg θ/2) have been measured. Figs. 7 and 8 show multiplicity and rapidity distributions at different incident energies.

A comparison of the shower-particle multiplicities in p-A and pp reactions gives $R_{em} \approx 1.8$ (Table 3) and this value stays fairly constant with energy. The low multiplicity observed cannot be explained by any one-step-process where the secondaries are produced instantaneously (Fig. 9).

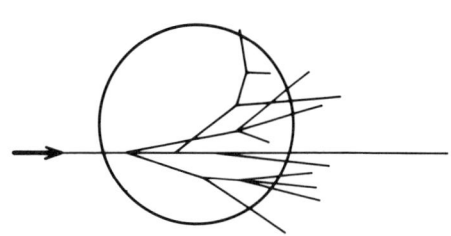

Fig. 7. Multiplicity distributions at 24,50,67,200 and 300 GeV (23-26).

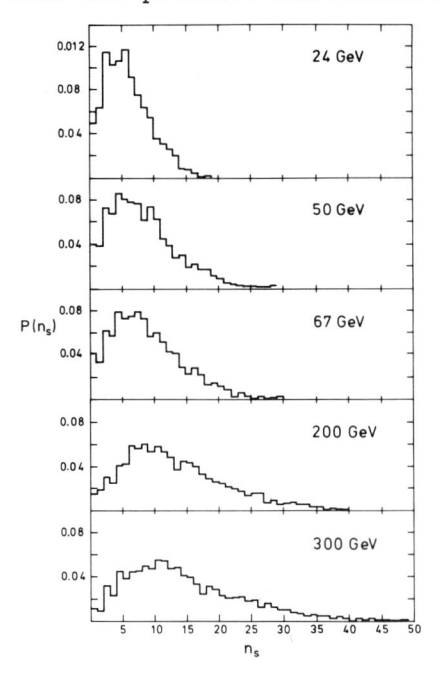

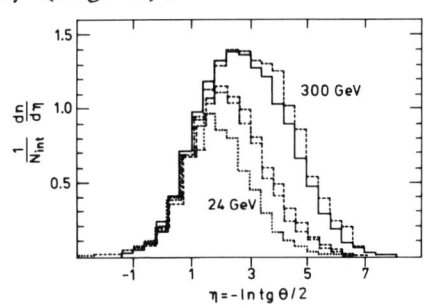

Fig. 8. Rapidity distributions at 24,50,67, 200 and 300 GeV (23,25).

Fig. 9. Intra-nuclear cascade process.

Table 3 (27,28)

Energy (GeV)	$<n_s>$	$R = \dfrac{<n_s>}{<n_{ch}>}$
200	13.2±0.1	1.72±0.03
300	15.1±0.2	1.78±0.03
400	16.5±0.5	1.83±0.07

If we assume that the hadron interacts independently with the target nucleons the first stage of a hadron-nucleus collision can be considered as a simple collision of a hadron with a nucleon. The behaviour of the hadron matter just after the first collision must be determined and different possibilities for this behaviour have been suggested (29). If intermediate states are generated by diffractive excitation of the incident proton and the target nucleons (Fig. 10), the average multiplicities are predicted by the formula

$$R = \frac{\langle n_s \rangle}{\langle n_{ch} \rangle} = 1 + 0.5\,(\bar{\nu}-1) \qquad (5)$$

In Fig. 11, multiplicities predicted from this formula are compared with experimental results.

It has also been suggested that the incident hadron may interact collectively with nucleons in the path of the incident particle (31). The target nucleons seen by the incident hadron are thus considered as a single object. The hadron-nucleus reaction should therefore have the same properties as a hadron-nucleon reaction at the same centre of mass energy. For example, the scaled multiplicity distribution should be the same in pp and p-A reactions and this prediction agrees with experimental observations (Figs. 13 and 16).

Fig. 10. Independent particle model.

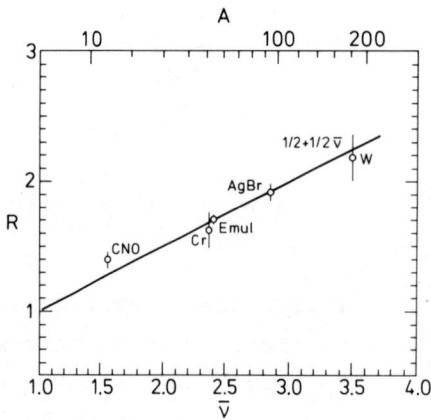

Fig. 11. R as a function of $\bar{\nu}$ (30).

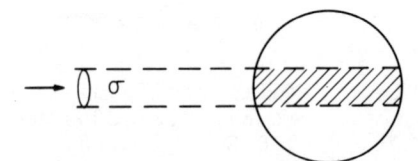

Fig. 12. Coherent Tube Model (31).

In hadron-nucleon reactions the rapidity distributions are often divided into three parts; target fragmentation, projectile fragmentation and pionization regions (Fig. 14). The rapidity distributions in Fig. 8 show that the rapidity

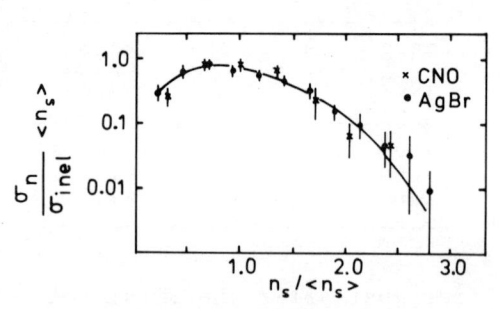

Fig. 13. Slattery pp-curve (solid curve) compared to scaled multiplicity distributions in p-CNO and p-AgBr interactions (25).

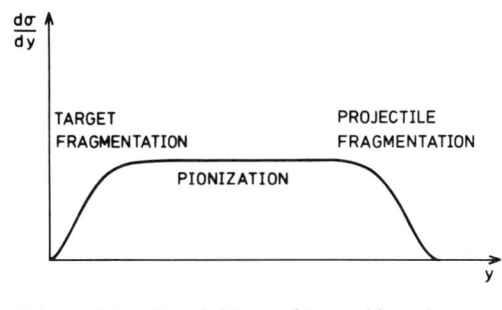

Fig. 14. Rapidity distribution in pp reactions.

distribution in the target fragmentation region is energy-independent in pA-reactions. On the other hand, the rapidity distribution in the projectile fragmentation region is independent of the target mass and equal to the pp distribution (Fig. 17). This observation can be understood in the picture of intermediate states production.

It is now very important to determine the number of recoiling protons among the shower-particles. This information is necessary not only to obtain a correct interpretation of the emulsion results but also to draw conclusions as to what extent the suggested theoretical models can reproduce the experimental observations.

4.2. Studies of hadron-nucleus interactions using nuclear emulsion with embedded targets

Despite the many favourable features nuclear emulsions offer to physicists, there are also a few disadvantages. The most conspicuous one of these is intrinsic to their nature as an inhomogeneous medium, denoted by the great variety of nuclei in the emulsion. Consequently, when the emulsions are used to study the interactions of incoming particles with the nuclei of the emulsions, it is not possible to determine, with any degree of certainty, the identity of the nucleus involved.

One method used for the study in emulsions of the events produced in an element has been to load emulsions with wires. The first incorporation of wires into nuclear emulsions was described by Meulemans et al. (32) and Danysz and Yekutieli (33). The most important problem they encountered in processing such plates was the elimination of the distortion introduced by the presence of the wires. The distortion arises from volume changes of the emulsion and movements of the wires during the processing:

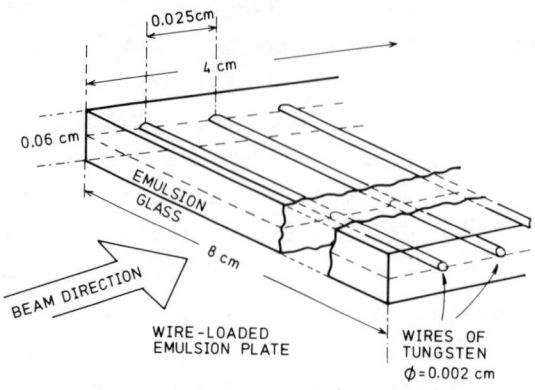

Fig. 15. Nuclear emulsions with embedded wires.
B. Lindkvist (34).

A new method of introducing known targets into nuclear emulsions has recently been developed by B. Lindkvist (34). In this method, fine wires are embedded in the median plane of nuclear emulsions. Figure 15 shows a nuclear emulsion with embedded wires. Two emulsion pellicles - one stripped emulsion and one emulsion on glass - are laminated with fine wires in a grid between the emulsions. The method makes it possible to develop the emulsions without any distortion.

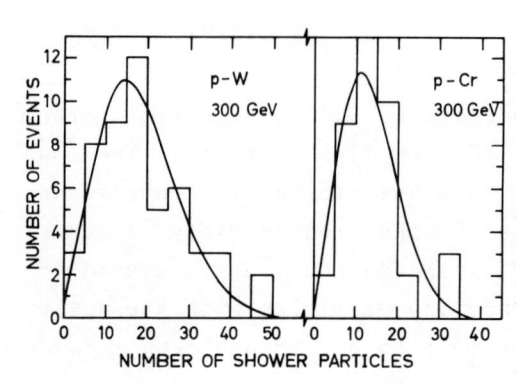

Fig. 16. Shower-particle multiplicity distributions for tungsten and chromium. The solid curves are obtained by a KNO-like scaling of the charged particle multiplicity distribution in proton-proton interactions (30).

A similar method has been used by Lord et al. (30). They use a mixture of one gram of powder and about 100 cc of water, which is quickly poured over a 200 μm emulsion on glass. A second 200 micron emulsion layer is finally added and the resultant sandwich is dried. The average granule diameter in their loaded emulsions is about 15 microns.

Figures 16, 17 and 18 show results obtained with the technique of loaded emulsions. Figs. 16 and 17 show shower-particle multiplicity distributions and rapidity distributions in p-W and p-Cr reactions obtained by Florian et al. (30).

Fig. 18 shows $R_{eq} = \frac{<n_s>}{<n_{ch}> - 0.5}$ as a function of N_h. The black points show the emulsion data of 300 GeV (35). The triangles are from p-Cr reactions (30). The consequence of the limited size of the target is obvious. When the number of heavy prongs approaches the number of protons in the hit nucleus, the linear correlation between R and N_h disappears (30).

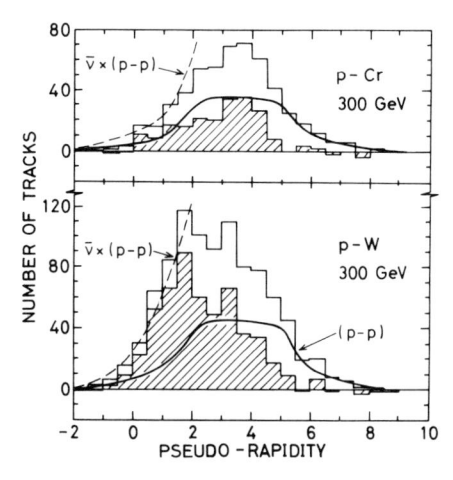

Fig. 17. Pseudo-rapidity distributions for chromium and tungsten events. The solid curve represents hydrogen data (30).

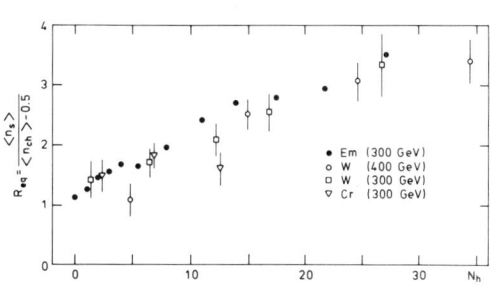

Fig. 18. R_{eq} as a function of N_h.

Data from p-W reactions obtained at 300 GeV (30) and 400 GeV (36) are also shown in Fig. 18. The p-W results are similar to the emulsion results and they also show a slightly linear behaviour.

We conclude that the correlation between R and N_h is not only energy-independent but also seems to be comparatively independent of the target mass, at least when $N_h \ll Z$.

5. Applications in particle physics.

5.1. The location and analysis of neutrino interactions in nuclear emulsions

The first neutrino experiment where emulsion stacks are used in hybrid systems with spark chambers was reported by Burhop et al. (37) in 1965. In this experiment a stack of nuclear emulsion pellicles of 10-litre total volume was placed together with two spark chambers in the neutrino beam of the CERN proton synchrotron (Fig. 19). The tracks of secondary particles from neutrino interactions in the emulsion were observed in the spark chamber. The spark chamber tracks were used to locate those produced by the same particles in the emulsion. These tracks were then followed back in the emulsion to the neutrino interaction in which they originated.

It was pointed out by Burhop et al. (37) that the high resolution of emulsions might be advantageous for studying neutrino interactions, and the technique might enable one to detect short-lived intermediate particles whose lifetime is too short to permit their detection by other techniques. Table 4 shows the range of particles as a function of decay rate. In nuclear emulsions it should be possible to distinguish decays occurring within a few μm from the primary interaction.

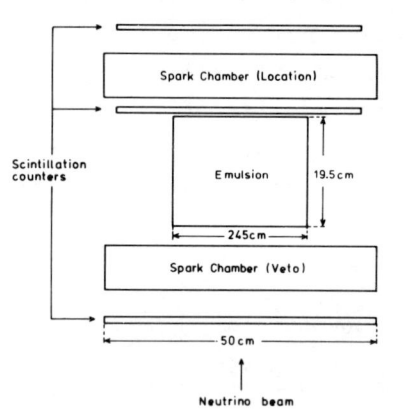

Fig. 19. The arrangement of the emulsion stack in hybrid system with spark chambers (37).

Table 4 (38)

Mass (GeV)	τ (sec)	$c\tau$ (μm)
2	3×10^{-13}	90
5	3×10^{-15}	1

A remarkable number of important discoveries (ψ and ψ' states, sudden rise in $R = \sigma(e^+e^- \to had))/\sigma(e^+e^- \to \mu^+\mu^-)$ at $\sqrt{s} \approx 3.1$ GeV and $\sqrt{s} \simeq 3.7$ GeV, structure in R for $\sqrt{s} > 4$ GeV, $\mu^+\mu^-$ production in neutrino experiments, appearance of muons at high $P_\perp$ etc.) have intensified the search for charmed particles. From the SPEAR data one infers that the lightest charmed bosons should have a mass ~ 2 GeV and the lifetime of such an object is expected to be 10^{-13} - 10^{-14} sec. (39). Nuclear emulsion will thus be a simple and efficient detector in the search for charmed particles.

5.2. Search for short-lived particles produced in neutrino interactions using emulsion-spark chamber hybrid systems (38).

In this experiment the following apparatus for the location of neutrino interactions in emulsion is used:
- wide gap chambers in which the charged secondaries of the neutrino interactions are seen. The thin tracks will allow one to localize the positions at which secondaries leave the emulsion stack.
- "veto" counters and a narrow gap spark chamber are placed before the emulsion stack in order to ensure that the observed events are actually produced by neutral particles.

To analyse secondary products of the interactions, the following apparatus is used:
- A muon spectrometer which determines the sign and momenta of muons emerging from neutrino interactions in the emulsion.
- A hadron calorimeter determines the energy in the hadronic component of the secondaries from the neutrino interaction.
- A shower detector detects the electrons emitted in the decay of short-lived particles and determines their energies.

To obtain 500 neutrino interactions for 2×10^{18} protons incident on target, requires a volume of about 20 liters of emulsion. Of the 500 reactions expected about 450 are charged current reactions, and about 45, neutral current reactions. This neutrino experiment is in progress and no results have been published yet.

5.3. Search for charmed hadrons produced by muon deep inelastic scattering in tagged nuclear emulsions (40).

The main idea in this experiment is that deep inelastic muon scattering provides a promising source of charmed particles. It has been estimated that 10-20 % of the events with $q^2 \gtrsim 1$ GeV^2/c^2 would contain charmed final states. The muons which undergo deep inelastic interactions in the emulsion are identified by a muon spectrometer and tagged by a system of proportional and drift chambers.

Since the incident muon energy is precisely known, and the scattered muon's momentum can be precisely measured, it is possible, in this experiment, to select events in those kinematic regions where an enhanced yield of interesting events is expected. By precise triangulation with drift and spark chambers the volume of emulsion within which a deep inelastic event occurs can be located within < 10 mm^3.

A second aim of this experiment is to determine the variation of mean multiplicity with q^2, thereby providing further insight into the space-time structure of multiparticle production.

5.4. Emulsion stacks combined with bubble chamber and counter techniques to study "new short-lived particles" produced by neutrinos (41,42).

The feasibility of combining emulsion, bubble chamber, and counter techniques to investigate the nature of dimuon events and to search for new short-lived particles produced by neutrinos has been discussed by M. Conversi (41) and M. Conversi et al. (42). Fig. 20 shows the layout of the proposed experiment. A signal from the VCS (= veto coincidence counter system) and the indication that two muons have reached the EMI (= external muon identifier), at the same time as that signal, indicates a possible dimuon event. By inspection of the BEBC photographs the reactions will be localized to the emulsion stack and all possible information will be extracted from the photograph and the emulsion. A dimuon event would appear as illustrated in Fig. 21. This dimuon event is assumed to be caused by the production and subsequent decay of a $\bar{D}^0$ charmed meson (41).

$\nu_\mu + N \rightarrow \mu^- + \bar{D}^0 +$ ordinary hadrons
$\hookrightarrow \mu^+ + K^- + \nu_\mu$

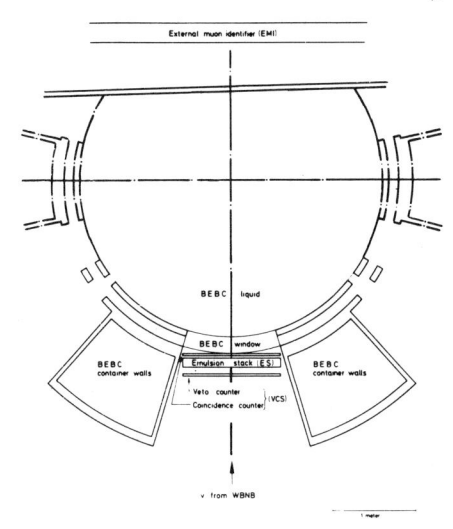

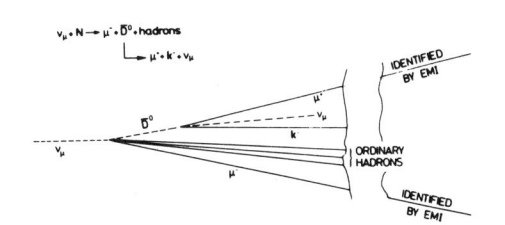

Fig. 21. Sketch showing the production and decay of a hypothetical charmed meson appearing as a dimuson event (41).

Fig. 20. Schematic view of experimental set up proposed to investigate dimuon events and to search for short-lived particles.
WBNB = wide band neutrino beam,
BEBC = Big European Bubble Chamber (41).

5.5. Nuclear emulsion as neutrino track-sensitive target (TST) in bubble chamber.

L. Voyvodic (43) has considered the possibility of mounting emulsion stacks inside bubble chambers. Fig. 22 shows how an emulsion stack might be mounted in the nose cone of the 15-foot bubble chamber at Fermilab. Neutrino interactions in the emulsion will be selected from the bubble chamber photo-graphs. Tracks of charged particles from the interaction are observed on photographs, and as many

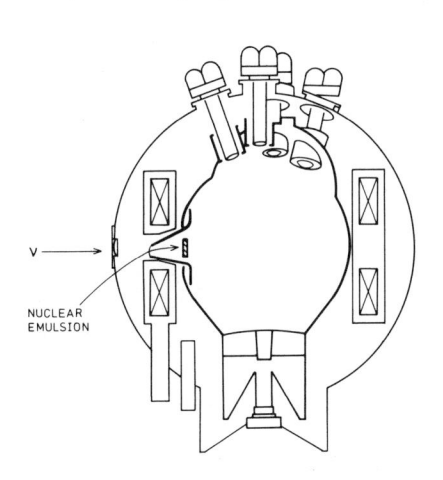

Fig. 22. An emulsion stack mounted in the nose cone of the 15-foot bubble chamber at Fermilab (43).

as possible of the reaction products are identified. The extrapolation of the tracks on the photo locates the reaction vertex in the emulsion stack.

It seems to be possible to use not only chambers operating near room-temperature but also cryogenic bubble chambers. In the latter case, one needs to choose large-grain emulsion such as NIKFI Type R or Ilford Type G5, which have been demonstrated to be usable at liquid hydrogen temperature (44).

A method for trigged record of particles in AgCl-crystals (45). Finally I would like to mention an interesting method to detect heavy particles in AgCl-crystals which has been suggested by E. Schopper (45). A sketch of the apparatus is shown in Figure 23. The detector consists of multiwire chambers (MWC), Cerenkov detectors, AgCl-crystals and light emitting diods. Signals from the multi-wire chambers (X_1Y_1, X_2Y_2) determine the path of the particle in the detector and the position of the latent track of the particle in the AgCl-crystal. The aim of the Cerenkov detectors is to select particles in certain velocity and charge intervals.

In order to get the latent track in the AgCl-crystal stable against fast fading, its formation has to be supported by additional electrons by simultaneous irradiation with yellow light. This important feature allows to select the record of tracks by will. In this detector the irradiation comes from light emitting diods. The diod close to the latent track in the crystal is turned on by a signal from the MWC:s when the conditions for particle registration are satisfied.

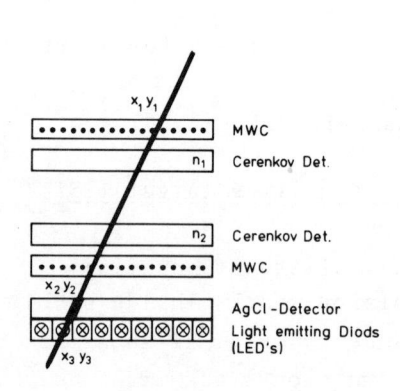

Fig. 23. Trigged record of a particle in AgCl (45).

This detector is suitable when long exposures are required, for example in experimental equipments placed in space aircrafts. The background in conventional track detectors will in most cases be too heavy. The suggested detector records only particles to be analysed, and the background will be substantially suppressed.

A similar technique applied to heavy ion reactions has also been suggested (45). Here the intention is to detect alpha-particles in the reaction $^{16}O(\alpha,\alpha')4\alpha$.

After the exposure the AgCl-crystals are revealed simply by irradiation with a flux ($\sim 10^{18}$ quanta/cm^2) of ultraviolet light. This dry procedure does not imply any shrinkage or deformation.

This summary of recent application of SSNTDs in high energy physics has necessarily been confined almost exclusively to the emulsion techniques. However, the examples given should indicate that SSNTDs are still very important in high energy physics.

Acknowledgments.
In preparing this talk I had a good help from many colleagues. In particular I wish to thank G. Baroni and J.J. Lord for providing me with valuable information about the hybridized emulsion technique, L. Voyvodic for clarifying discussions about emulsion stacks in bubble chambers and E. Schopper for fruitful conversations about applications of AgCl-crystals. I am grateful to all who sent me results relevant to this talk.

References.
1. H.G. Baumgart, J.U. Schott, Y. Sakamoto, E. Schopper, H. Stöcker, J. Hofmann, W. Scheid, and W. Greiner, Z. Physik A237, 359 (1975).

2. J.D. Sullivan, P.B. Price, H.J. Crawford, and M. Whitehead, Phys. Rev. Letters, 30, 136 (1973).
3. B. Jakobsson, R. Kullberg, and I. Otterlund, Preprint LUIP-CR-76-04, Lund 1976.
4. K.K. Gudima, and V.D. Toneev, Preprint JINR, E4-9765 (1976).
5. P.J. Lindstrom, D.E. Greiner, H.H. Heckman, Bruce Cork, and F.S. Bieser,
H.H. Heckman, D.E. Greiner, P.J. Lindstrom, and H. Shwe, Proc. of the 14th Int. Cosmic Ray Conf. Munich, August (1975).
6. B. Judek, Proc. of the 14th Int. Cosmic Ray Conf. Munich, August (1975).
7. B. Jakobsson, R. Kullberg, and I. Otterlund, Lettere al Nuovo Cimento 15, 444 (1976).
8. R. Kullberg, K. Kristiansson, B. Lindkvist, and I. Otterlund, LUIP-CR-76-03, Lund 1976.
9. G.M. Chernov, K.G. Gulamov, U.G. Gulyamov, Sh. Z. Nasyrov, and L.N. Svechnikova, Paper submitted to the Topical Meeting on Nuclear Production at Very High Energies, Trieste (1976).
10. B.P. Bannic, A. El-Naghy, R. Ibatov, J.A. Salomov, G.S. Shabratova, M. Sherif, K.D. Tolstov, Paper sumbitted to the XVIII Int. Conf. on High Energy Physics, Tbilisi (1976).
11. I. Otterlund et al., Barcelona-Batavia-Belgrade-Bucharest--Lund-Lyons-Montreal-Nancy-Ottawa-Paris-Rome-Strasbourg--Valencia collaboration, Proc. of the 5th Int. Conf. on High-Energy Physics and Nuclear Structure, Uppsala (1973).
12. J.J. Aubert et al., Phys. Rev. Letters 33, 1404 (1974).
J.E. Augustin et al., Phys. Rev. Letters 33, 1406 (1974).
13. G.S. Abrams et al., Phys. Rev. Letters 33, 1453 (1974).
14. G. Goldhaber et al., Phys. Rev. Letters 37, 255 (1976).
I. Peruzzi et al., Phys. Rev Letters 37, 569 (1976).
15. A. Benvenuti et al., Phys. Rev. Letters 34, 419 (1975).
16. W. Busza, Proc. of the VIth Int. Conf. on High Energy Physics and Nuclear Structure, Santa Fe and Los Alamos, June 1975 and references therein.
17. A. Białas and W. Czyż, Phys. Lett. B51, 179 (1974).
18. R. Kullberg and I. Otterlund, Z. Physik 259, 245 (1973).
19. B. Jakobsson, R. Kullberg, and I. Otterlund, Z Physik 268, 1 (1974).
20. B. Jakobsson, R. Kullberg, and I. Otterlund, Z. Physik A272, 159 (1975).
21. G.D. Westfall, J. Gosset, P.J. Johansen, A.M. Poskanzer, W.G. Meyer, H.H. Gotbrod, A. Sandoval, and R. Stock, Preprint LBL-5072, Berkeley (1976).
22. Z.V. Anzon et al., Alma-Ata-Leningrad-Moscow-Tashkent collaboration. Preprint 1973.
23. S.A. Azimov, K.G. Gulamov, U.G. Gulyamov, L.P. Chernova, G.M. Chernov, V.S. Navothny, V.I. Petrov, N.S. Skripnik, and T.P. Trofimova, Paper submitted to the Topical Meeting on Nuclear Production at Very High Energies, Trieste, June 1976.
24. Alma-Ata-Leningrad-Moscow-Tashkent collaboration, Preprint No 9, Moscow 1974.

25. Hebert et al., Barcelona-Batavia-Belgrade-Bucharest-Lund--Montreal-Nancy-Ottawa-Paris-Rome-Strasbourg-Valencia collaboration. Proc. of the 14th Int. Cosmic Ray Conf. Munich, August (1975). Paper submitted to the VI Int. Conf. on High-Energy Physics and Nuclear Structure, Santa Fe and Los Alamos, New Mexico, U.S.A. (1975).
26. J. Babecki, private communication.
27. I. Otterlund, invited talk at Tropical Meeting on Multiparticle Production on Nuclei at Very High Energy, Trieste, June 1976.
28. Belgrade-Lund-Nancy-Ottawa-Paris-Rome-Valencia collaboration, to be published.
29. K. Gottfried, Phys. Rev. Lett. $\underline{32}$, 957 (1974).
E.M. Friedlander, Lett. al Nuovo Cimento 9, 349 (1974).
P.M. Fishbane and J.S. Trefil, Phys. Rev. D9, 168 (1974).
B. Andersson and I. Otterlund, Nucl. Phys. $\underline{B95}$, 237 (1975).
M. Miesowicz, Proc. of IX Int. Conf. on Cosmic Rays, Budapest 1967, Progress in Elementary Particle and Cosmic Ray Physics $\underline{10}$, 128 (1971),
J. Babecki, Kraków report INP No 911/PH (1976).
30. J.R. Florian, M.Y. Lee, J.J. Lord, J.W. Martin, R.J. Wilkes, R.E. Gibbs, and L.D. Kirkpatric, Preprint, September (1975).
31. Y. Afek, G. Berlad, G. Eilam and G. Dar, Preprint (1976). A. Dar, invited talk at Topical Meeting on Multiparticle Production on Nuclei at Very High Energy, Trieste, June (1976).
32. G. Meulemans, G.P.S. Occhialini, and A.M. Vincent, Nuovo Cimento $\underline{8}$, 341 (1951).
33. M. Danysz and G. Yekutieli, Phil. Mag. $\underline{42}$, 1183 (1951).
34. B. Lindkvist to be published. The use of wire loaded emulsions in the Bucharest-Cern-Cornell-Lund proton-emulsion nucleus experiment was suggested by A.J. Herz, CERN. The grids were manufactured by R. Lorenzi, CERN, using special equipment developed in the CERN NP Div.
35. I. Otterlund et al., Belgrade-Lund-Nancy-Ottawa-Paris--Rome-Valencia collaboration.
Univ. of Lund Preprint LUIP-CR-74-12 (1974).
36. Bucharest-Cern-Cornell-Lund collaboration, private communication.
37. E.H.S. Burhop, W. Busza, D.H. Davis, B.G. Duff, D.A. Garbutt, F.F. Heymann, K.M. Potter, and J.H. Wickens, Nuovo Cimento $\underline{39}$, 1037 (1965).
38. Bruxelles-Dublin-Fermilab-London-Rome-Strasbourg collaboration, Fermilab Proposal for Exp. 247 (February 1975).
39. M.K. Gaillard, B.W. Lee, and J.L. Rosner, Rev. Mod. Phys. $\underline{47}$, 277 (1975).
40. K. Gottfried et al., Fermilab Proposal for Exp. 382 (1975).
41. M. Conversi, N.P. Internal Report 75-17 (1975).
42. Ankara-Brussels-CERN-Dublin-London-Strasbourg-Turin collaboration, CERN/SPSC/76-41 (1976).
43. L. Voyvodic, Fermilab Report FN-289 (1976).
44. M. Debauvais-Wack, Nuovo Cimento 10, 1590 (1953), 1er Colloque International de Photographie Corpusculaire, Strasbourg, 125 (1958).
R.W. Waniek, Bull, Amer. Phys. Soc. $\underline{1}$, 219 (1956).
A.W. Johnson, C.J.D. Hébert, J. Hébert, and J.L.G. Lamarche, Can. Journ. of Phys. $\underline{49}$, 2524 (1971).
45. E. Schopper, private communication.

FAST STABLE PARTICLES FROM LIGHT NUCLEI OF EMULSION FOLLOWING THE INTERACTION OF 1.5 GeV/c K⁻MESON

A. Waheed* and M. Jurić**

*Physics Department, Gomal University, D. I. Khan, Pakistan
**Institute of Physics, P. O. Box No. 57, Belgrade, Yugoslavia

Abstract

From measurements on grey tracks of 1214 light nuclear interaction stars of K⁻ meson, the emission probability for proton, deutron and triton is found respectively to be:

$$0.31^{+0.02}_{-0.03}, \quad 0.25^{+0.04}_{-0.03}, \quad 0.075^{+0.013}_{-0.009}$$

It has been estimated that inelastic scattering of positive pion and Σ^+ hyperon, and absorption of pion, Σ^+ hyperon contribute to proton emission in 5.9 %, 2.3 % and 15.6 %, 5.8 % of the cases respectively. About 60 % of protons are emitted due to the elastic scatterings taking place in the light(CNO) nuclei. It has been concluded that deutron emission is due mainly to pick-up process and pion absorption, and that absorption of pion give rise also to the emission of triton of energy upto 200 MeV. About 10-15 % of triton with energy exceeding 200 MeV may be expected from the absorption of incident K⁻meson on alpha cluster in light nuclei.

I. Introduction

In experiments with fast(300-800 MeV) K⁻meson[1-5] it was established that fast nucleon(kinetic energy exceeding 27 MeV) are emitted due to non-mesonic(K⁻2N) processes, stimulated decay of Λ hyperon in hypernuclei, or the inelastic scattering and absorption of pion and Σ hyperon. Similarly production of fast deutron(kinetic energy more than 54 MeV) was explained in terms of Pick-Up mechanism of the nucleon cascade[6]. On the basis of different experimental results[7-14], obtained by using beams of pion, proton and cosmic rays, Yasin[15] concluded that triton(with kinetic energy exceeding 80 MeV) emission may take place due to the nucleon cascade and pion absorption on alpha cluster. In a similar manner Todorović and Jurić[16] predicted that pion and K⁻meson absorption on alpha cluster of heavy nuclei(Ag,Br) give rise to fast triton.

Evidently no significant data is available for the emission of proton, deutron or triton from complex nuclei when the interactions of K⁻meson are in question in GeV region. Also it is known that disintegration process of complex nuclei is the same at high energies no matter whether the incident particle is K⁻meson or any other particle[17]. We have, therefore, made an attempt to analyse the interactions of 1.5 GeV/c K⁻meson, which took place in light nuclei(C,N,O), to study the emission mechanism of fast stable particles and to gauge the extent to which different mechanisms, discussed in earlier works[1-16], are effective in the emission of fast proton, deutron and triton.

II. Measurements and Results

II(a) Experimental Procedure: A stack of Ilford G5 emulsions was used and 7750 interaction stars were located by systematic area scanning. From these stars 1214 were separated as the light nuclear events. Criterion for prong analysis and for the classification of stars, in respect of K⁻meson interactions in light and heavy nuclei, is given in paper I[18]. Fast proton, deutron and triton produce grey tracks($0.234 \leq \beta \leq 0.7$) in emulsion. Relativistic tracks($\beta > 0.7$) were identified for pion production and grey tracks for emission frequency of fast stable particles by standard techniques also discussed in paper I. About $59.4^{+3.4}_{-4.1}$ % of grey tracks were due to stable particles with relative proportion as given in Table I.

II(b) Proton Emission: Energy distribution of fast protons is displayed in Fig.1. Only grey tracks were taken into account and geometrical correction was applied to the steeping tracks (dip angle $\theta > 45°$). Assuming that emission frequency of proton among identified($\theta < 45°$) and unidentified tracks is nearly the same, the observed proton number was corrected. Emission probability for proton was thus found as $0.31^{+0.02}_{-0.03}$.

Table I

Stable Particle	Relative Proportion	
	This Experiment	From pion absorption in light nuclei(7,8)
Proton	$48.47^{+1.03}_{-1.62}$ %	49 ± 2 %
Deutron	$38.86^{+0.90}_{-0.24}$ %	31 ± 2 %
Triton	$11.35^{+0.46}_{-0.96}$ %	20 ± 4 %
$_2He^3$	$1.32^{+0.27}_{-0.32}$ %	-

The following mechanisms were considered and estimates were made of their contribution to fast proton emission:
1) Non-Mesonic Interactions: Energetic protons may be emitted in the non-mesonic processes like $K^-+N+N=Y+N$. From energy spectra of proton(Fig.1) it becomes evident that 23 % of protons have energy exceeding 150 MeV. Such protons were put to kinematical test and it was found that in 2 % of the cases proton kinematics were consistent with production in the non mesonic interactions. Having applied some geometrical and statistical corrections it was found that about 2.2 ± 0.2 % of K^- meson interactions were of non-mesonic nature. This percentage is too low to account for proton emission and hence the contribution from non mesonic processes seems to be negligible.

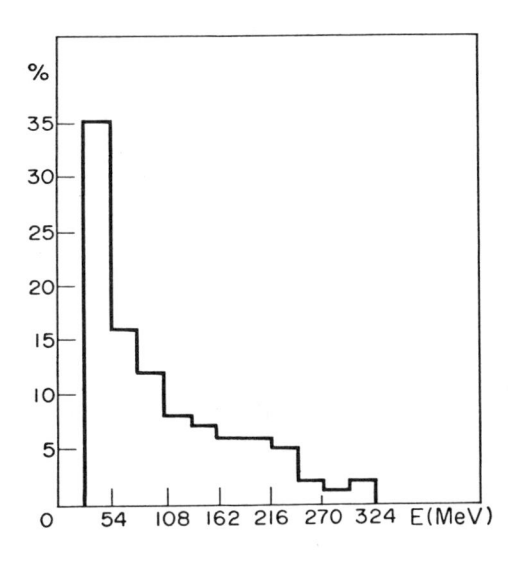

Fig.1 Fast Proton Energy Distribution

2) **Inelastic Scattering of π^+ Meson and Σ^+ Hyperon**: Fast protons may also be emitted as a result of the collisions of one of the secondary products(pion and Σ hyperon) on its way out of the nucleus. It is believed that π^+ mesons, produced at low energy in the K^- meson interactions, require at least 80 MeV energy to eject protons with energy values exceeding 30 MeV[19]. From the energy distribution of pions, which was observed in this experiment, it was estimated that 35 % of π^+ mesons had energy values ranging between 80 MeV and 340 MeV. Having applied corrections to observed number of π^+ mesons, production probability for 80 MeV π^+ meson was estimated at 0.057/star. Assuming that all such π^+ mesons may give rise to fast protons, and taking into account the total emission rate of observed protons, an upper limit of 5.9 % was calculated for probable emission of fast protons as a result of the inelastic scattering of π^+ mesons.

It has been found that 8–13 % of Σ^+ hyperons, produced on account of the interaction of 1.5 GeV/c K^- meson, experience inelastic scattering inside light nuclei[20]. Considering this percentage and the total proton emission rate, it was found that 2.3 % of fast protons might have been emitted on account of the inelastic scattering of Σ^+ hyperons in light nuclei.

3) **Absorption of $\pi^{\pm}$ Meson and Σ^+ Hyperon**: Absorption of pions and Σ^+ hyperons in nuclei may also be responsible for the emission of fast protons in the following type of $\pi^{\pm}$ meson, Σ^+ hyperon interactions:-

$$\pi^+ + n = p + \pi^0$$
$$\pi^+ + n + n = p + n$$
$$\pi^- + p + p = n + p$$
$$\pi^+ + n + p = p + p$$

$$\Sigma^+ + n = \Lambda + p$$

The probability of pion($\pi^{\pm}$) and hyperon($\Sigma^{\pm}$) absorption in light nuclei, following the interaction of 1.5 GeV/c K^- meson, is found to be, $W_{\pi^{\pm}} = 0.276^{+0.023}_{-0.016}$, $W_{\Sigma^{\pm}} = 0.58^{+0.10}_{-0.07}$, respectively[18].

From pion absorption probability($W_\pi\pm$), and the total rate of fast proton emission, it has been estimated that 15.6 % of fast protons may be emitted due to pion($\pi^\pm$) absorption.

The charge ratio for Σ hyperons was found in the present experiment to be, $\Sigma^-/\Sigma^+ = 2.8^{+0.21}_{-0.15}$. On the assumption that relative absorption rate of Σ^-, Σ^+ hyperon is proportional to their production rate, the absorption probability of Σ^+ hyperon was taken, in terms of $W_{\Sigma\pm}$, to be, $W_{\Sigma^+} = 0.19^{+0.03}_{-0.18}$. On the basis of '$W_{\Sigma^+}$', and the total emission rate of fast protons, it was concluded that emission of 5.8 % of fast protons might have taken place because of Σ^+ hyperon absorption in light nuclei.

No contribution to fast proton emission is expected from stimulated decay of Λ hyperon, because hypernuclei from light nuclei are emitted hardly in 1.81±0.41 % of the interactions of 1.5 GeV/c K^- meson[21]. One may therefore conclude that at the most 30 % of fast protons are emitted due mainly to the inelastic scattering and absorption of pions and Σ^+ hyperon. In more than 60 % of the cases elastic scattering of nucleon in light nuclei may have resulted in fast proton emission.

II(c) Deutron Emission: Energy distribution of fast deutrons, given in Fig.2, extends upto 432 MeV. Making corrections in the observed deutron number, emission probability of such deutrons turned out to be $0.25^{+0.04}_{-0.03}$.

Emission of fast deutrons may be attributed to pick-up processes of the following type:-

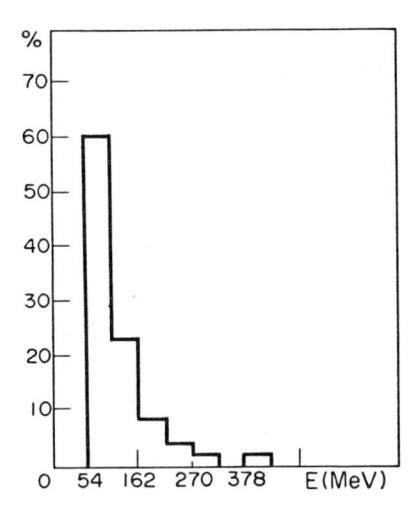

$$p + p = d + \pi^+ \quad \ldots\ldots (1)$$
$$p + n = d + \pi^o \quad \ldots\ldots (2)$$

Charge independence indicates that: $\sigma_1 = \sigma_3 = 2\sigma_2$. From observed energy distribution of protons(Fig.1) it was assumed that the average proton energy would be, $\langle T_p \rangle = 150$ MeV. Following Kamal et al[6] the interaction probability of protons was approximated at 0.6/star. The average deutron production cross-section, $\bar{\sigma}_d = 6.9$ mb, was calculated by averaging over the yield excitation curve given, for σ_1 in pp collision, by Cocconi et al[22]. Allowing for cascade nucleon and averaging over the above reactions, the expected number of deutrons becomes 0.15/star. This number cannot fully explain the experimental yield of $0.25^{+0.04}_{-0.03}$.

Table II

Nature of Stars	$\langle N_b \rangle$	$\langle N_g \rangle$	$\langle n_s \rangle$
Stars with deutron	2.90±0.11	2.18±0.07	0.41±0.03
Stars without deutron	3.01±0.09	1.42±0.13	0.65±0.01

From the charge multiplicities, given in Table II, it becomes evident that for the interaction stars, which have the deutron track, there is an apparent increase in mean grey track number $\langle N_g \rangle$, and corresponding decrease in mean number of relativistic track $\langle n_s \rangle$. This gives the idea that pion absorption may somehow contribute to deutron emission.

II(d)Triton Emission: Energy distribution of observed tritons is shown in Fig.3. Corrections were made in the observed number of tritons for steeping and unidentified tracks to obtain the emission rate of fast tritons as $0.075^{+0.013}_{-0.009}$

As regards the mechanism of fast triton emission, the evaporation process of the

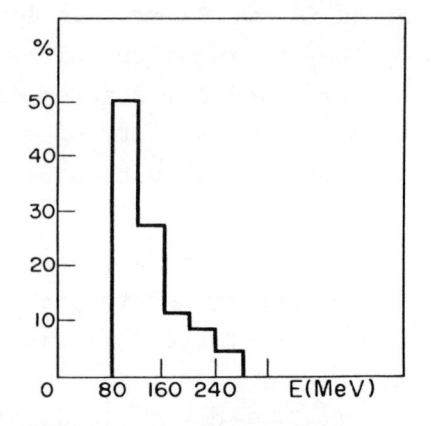

Fig.3 Energy Distribution of Fast Triton

residual excited nuclei do not make any contribution[23,24].
Fast triton may, however, be emitted in the following modes:-

1) **Interaction of Cascade Nucleon**: From kinematical point of view nucleon need 250 MeV energy to emit fast(70 MeV) triton in the following manner,

$$p + d = t + \pi^+ \quad , \quad n + d = t + \pi^0$$
$$p + {}_2He^4 = t + p + p \quad , \quad n + {}_2He^4 = t + d$$

Observed number of protons, having 250 MeV energy or more, was 0.012/star(Fig.1). Taking neutron distribution to be identical to proton energy distribution[25], and keeping in view the probability of above reactions, it was found that cascade nucleon play negligible role in triton emission.

2) **Pion Absorption**: Fast triton may be emitted in pion(π^-,π^0) absorption on alpha cluster as, $\pi^- + {}_2He^4 = t + n$ or p. Almost 90 % of negative pions were observed in this analysis to have energy values upto 100 MeV. With a knowledge of the π^-/π^0 ratio obtained from K^-N channels[18], and in view of the probability of pion absorption in alpha cluster(Table.I), it was estimated on kinematical basis that most of the fast tritons with 70-200 MeV energy might have been emitted due to the absorption of pions(π^-,π^0) having 20-100 MeV energy.

Triton energy spectrum(Fig.3) extends beyond 200 MeV. Energetic tritons(T > 200 MeV) may be assumed to have been emitted in K^- meson interactions on alpha cluster in light nuclei. From energy distribution of tritons it was concluded that 10-15 % of tritons might have been given through the processes of type: $K^- + {}_2He^4 = Y^* + t, \; Y^* = \Lambda + \pi^0$.

Contribution due to cascade nucleon being small, it may be expected that cluster absorption of mesons is the most important phenomena so far as the emission of fast tritons is concerned.

III. Conclusions

Discussion of various mechanisms leads to the conclusion that pion absorption is responsible in most of the cases for the emission of fast stable particles. Deutron emission may also take place in the nucleon interactions on alpha cluster, like $p + {}_2He^4 = {}_2He^3 + d$, $n + {}_2He^4 = t + d$, though probability is small, It is also reasonable to believe that tritons may be emitted in the interactions of K^- meson directly with alpha cluster. The life time of alpha cluster in light nuclei(C,N,O) varies from 5×10^{-22} to 4×10^{-23} seconds[26]. The flight time of 1.5 GeV/c K^- meson in light nuclei being 2.2×10^{-23} seconds, it is possible that K^- meson may find alpha cluster on its way and induce an interaction to give fast triton. Alpha structure of light nuclei therefore play the most effective role in the emission of fast stable particles.

IV. References

1) Lagnaux and Lemonne .....Bull.Inst.Phys.,Brussels, No.18,1964
2) M.Bleu, M.Canlton .....Phy.Rev. 96 (1954) 150
3) A.Filipowski et al .....N.Cimento 25 (1962) 1
4) A.Arni et al .....Phys.Lett. 11 (1964) 174
5) P.Renard et al .....Nucl.Phys. 70 (1965) 609
6) A.A.Kamal et al .....N.Cimento 43A (1966) 91
7) A.O.Viesenberg et al .....JETP 47 (1964) 1263
8) P.J.Castleberry et al.....Phys.Lett. 34B (1971) 57
9) V.I.Ostroumov .....JETP 37 (1959) 643
10) V.I.Ostroumov .....JETP 36 (1959) 367
11) V.I.Ostroumov .....JETP 39 (1960) 105
12) H.Dubost et al .....Phy.Rev. 136 (1964) 1618
13) S.O.C.Sørensen .....Phil.Mag. 42 (1951) 188
14) M.Yasin .....N.Cimento 28 (1963) 673
15) M.Yasin .....N.Cimento 34 (1964) 1145
16) Z.Todorović, M.Jurić .....FIZIKA 4 (1972) 77
17) J.Hudis, J.M.Miller .....Ann.Rev.Nucl.Sci. 9 (1959) 59
18) A.Waheed et al .....Proc.9th Intern.Conf.SSNTD, Munchen,1976(to be published)
19) K^-Collaboration .....N.Cimento 13 (1959) 690
20) A.Waheed .....Ph.D Thesis,1973,Belgrade Univ
21) A.Waheed, M.Jurić .....FIZIKA 4 (1972) 87
22) G.Cocconi et al .....Proc.Sienna Conf.1963, V.I,p.608
23) V.I.Ostroumov .....JETP 10 (1960) 459
24) K.J.Le Couteur .....Proc.Phy.Soc. 63A (1950) 259
25) P.Renard et al .....Bull.No.19,University of Brussels,1963
26) P.E.Hodgson .....Nucl.Phys. 8 (1958) 1

PROBABILITY DISTRIBUTION FOR K⁻p AND K⁻n CHANNELS IN THE INTERACTION OF 1.5 GeV/c K MESON IN LIGHT EMULSION NUCLEI

A. Waheed*, M. Jurić** and V. Zlatarov***

*Physics Department, Gomal University, D. I. Khan, Pakistan
**Institute of Physics, P.O. Box No. 57, Belgrade, Yugoslavia
***Institute of Sciences, Vinča, Box 550, Belgrade, Yugoslavia

Abstract

By analysing 1214 light nuclear interactions of 1.5 GeV/c K⁻ meson, 149 K⁻p and 114 K⁻n probable interaction channels have been identified. Estimates of the absorption rates of Σ hyperon and pion have been made as, $W_{\Sigma^\pm} = 0.58$ and $W_{\pi^\pm} = 0.276$, respectively. Applying corrections for the secondary processes, which may give rise to the absorption of particles in the nucleus, and for pion contamination in K⁻ meson beam, 119 K⁻p and 101 K⁻n channels have been finally selected. Relative probabilities of various groups of K⁻p and K⁻n channels seem to be in agreement with the expected probabilities. The experimental yield ratio,

$$K^-p/K^-n = 1.21^{+0.05}_{-0.03},$$

is close to the expected yield value of $K^-p/K^-n = 1.32 \pm 0.02$.

I. Introduction

Several groups[1-11], working with bubble chambers, studied K⁻p interactions in zero to 3 GeV/c region, while for the K⁻n interactions the investigations were made indirectly in 700 to 1200 MeV/c interval[12]. Studies were also carried out at higher energies, but certain specific channels were taken into consideration[13,14]. Attempt, however, have not been made to use emulsion technique for such studies in GeV region, in spite of the fact that in nuclear emulsion both K⁻p and K⁻n interactions are recorded, and that at high energies K⁻ meson interactions occur (in more than 90 % of the cases) on single nucleon[15]. The most probable reason being in that more than 75 % of K⁻ meson interactions take place in heavy (Ag,Br) nuclei[16], in which size of the secondary interactions is quite large.

In this work an attempt has been made to analyse K⁻meson interactions taking place in light nuclei(C,N,O) of nuclear emulsion. These nuclei have clear structural individualities and the secondary processes involved in them are much small. Results so obtained for the K⁻p channels are comparable with the expected values based on K⁻p cross-section. Probabilities for certain K⁻n channels have been given in GeV region.

II. Experimental Procedure

A stack of Ilford G5 emulsions was used in the present investigation. The stack was exposed(at CERN PS) to the K⁻meson beam having composition, $K^- : \pi^- : \mu^- = 73\% : 11\% : 16\%$. A total of 7750 interaction stars were located by area scanning. Tracks of the outgoing particles were classified as black($\beta < 0.234$), grey($0.234 \leq \beta \leq 0.7$), or relativistic($\beta > 0.7$), according as their normalized grain density, g^*, was exceeding 10 g^o, it was between 1.5 g^o and 10 g^o, or less than 1.5 g^o, respectively[17]. The value of 1.5 g^o is nearly equal to about 33 grains per 100 micron[18]. The interaction stars were then categorized in respect of K⁻meson interaction on light and heavy nuclei if one of the following conditions were satisfied:

Heavy Nuclei=(a) $N_h > 7$, where N_h is the number of
(Ag,Br)   heavy(black,grey)tracks,

(b) $N_h \leq 7$, and a recoil of very short range(R < 5 micron),

(c) $N_h \leq 7$, and total charge on heavy tracks, $\sum_{i=1}^{n} Z_i > 7$,

Light Nuclei= $N_h \leq 7$, no short recoil, and total
(C,N,O)   charge, $\sum_{i=1}^{n} Z_i \leq 7$.

Charge values of the black track producing particles were found by track width measurement[19]. In all 1214 interaction events were found to have taken place in the light nuclei. It corresponds to 15.6 % of the total K⁻meson interactions, which is in agreement with the percentage given at the same energy by Cuevas et al[16].

All kinds of tracks were put to test for identification of the particles producing them. The relativistic tracks, dipping at less than 30° in emulsion, were identified by grain counting[20], combined with the range of particle. The scattering method[21] was also applied to such relativistic tracks with statistical error upto 10 %. The complementary method of g*-scattering was utilized to discover the identity of some of the relativistic tracks($\theta < 30°$), which left the stack or resulted in an interaction without reaching the end of their range. Method of mean gap[22], $\bar{G}$, was tried for grey tracks having dip angle upto 45°. The statistical error in such cases was less than 6 %. The identification of those grey tracks which did not come to rest in the stack, was subject to the information obtained from comparison of $\Delta\bar{G}$ with the range ΔR of the observed segment of the track.

Black track producing particles are fragments of the disintegration of residual light nucleus[23]. To identify an interaction channel, therefore, nature of every grey and relativistic track of the interaction star was to be known. Total 263 events were analysed completely such that the identity of each grey and relativistic track could be found. Not more than three pions can be expected in the final channel at the energy considered. Keeping in view the products of interaction, and having made decision about the nature of target nucleon(as a proton or the neutron) in terms of the charge conservation, 149 K^-p and 114 K^-n probable channels were finally selected. Certain corrections were needed, however, to obtain the correct proportion of these channels.

III. A Simple Absorption Model

The compton wave length($\lambda = \hbar/p$) of 1.5 GeV/c K^- meson is of the order of 1.3×10^{-14} cm. Taking average nucleon separation to be nearly 10^{-13} cm[24], it may easily be assumed that K^- meson initially collides with a single nucleon in the nucleus.

Let the impact parameter for collision of K^- meson with the nucleus be 'p'. The probability for interaction in the strip dx of nuclear matter may have the distribution[25],

$$P_x = Q\, n\, \tilde{\sigma}_a \exp(-n\, \tilde{\sigma}_a\, x)\, dx \quad \ldots\ldots\ldots\ldots (III.a)$$

where $Q = \tilde{\sigma}_g / \tilde{\sigma}_T$, $\tilde{\sigma}_g$ is the geometrical cross-section for interaction, $\tilde{\sigma}_T$ is the absorption cross-section of K^- meson, $n = 3A/4\pi R^3$ the nucleon density and $\tilde{\sigma}_a$ the average cross-section per nucleon equal to,

$$\tilde{\sigma}_a = \frac{Z\, \tilde{\sigma}_{K^-p} + (A-Z)\, \tilde{\sigma}_{K^-n}}{A} \quad \ldots\ldots\ldots\ldots (III.b)$$

The probability distribution for the interaction of the secondary particle (π, Σ) in another strip dy of the nucleus may be given in a similar manner by:

$$P_y = n\, \tilde{\sigma}_s \exp(-n\, \tilde{\sigma}_s\, y)\, dy \quad \ldots\ldots\ldots\ldots (III.c)$$

where $\tilde{\sigma}_s$ is the cross-section for the interaction of secondary particles inside the nucleus.

Total probability for the primary and the secondary type of interactions will then be:

$$P_x P_y = Q\, n^2\, \tilde{\sigma}_a\, \tilde{\sigma}_s \exp\left[-n(x\, \tilde{\sigma}_a + y\, \tilde{\sigma}_s)\right] dx\, dy$$

Multiplying this expression by $2\pi p\, dp$ ($0 \leq p \leq R$) and integrating over x, y, p, the cross-section for the successive collisions in the nucleus will come out to be:

$$\tilde{\sigma}_m = Q\, \tilde{\sigma}_g \left[1 + \frac{\tilde{\sigma}_s}{\tilde{\sigma}_a - \tilde{\sigma}_s} f(KD) - \frac{\tilde{\sigma}_a}{\tilde{\sigma}_a - \tilde{\sigma}_s} f(n\, \tilde{\sigma}_s\, D)\right]$$

Absorption probability of the secondary particles may be obtained from the relation, $W = \tilde{\sigma}_m / \tilde{\sigma}_g$, which after substitution for $\tilde{\sigma}_m$, takes the following form:

$$W = \frac{\tilde{\sigma}_g}{\tilde{\sigma}_T}\left[1 + \frac{\tilde{\sigma}_s}{\tilde{\sigma}_a - \tilde{\sigma}_s} f(KD) - \frac{\tilde{\sigma}_a}{\tilde{\sigma}_a - \tilde{\sigma}_s} f(n\, \tilde{\sigma}_s\, D)\right] \ldots (III.d)$$

where $D = 2R$, $R = r_o A^{1/3}$, $\tilde{\sigma}_g = \pi R^2$, and

$$f(n\,\sigma_s\,D) = 2/(n\,\sigma_s\,D)^2 \left[1-(1+n\,\sigma_s\,D)\exp(-n\,\sigma_s\,D)\right],$$
$$f(K\,D) = 2/(KD)^2 \left[1-(1+KD)\exp(-KD)\right].$$

The absorption cross-section is given by[26]:

$$\sigma_T = \pi R^2 \left[1-\left\{1-(1+2KR)\exp(-2KR)\right\}/2\,K^2R^2\right]$$

Taking $K=2.26^{+0.34}_{-0.20} \times 10^{12}$ /cm [27], $r_0=1.3$ fm, the probabilities for absorption of secondary particles were calculated from Eq.(III.d). The distributions are displayed in Fig.1.

IV. Results and discussion

Proportion of π^-p, π^-n interactions among respective K^-p, K^-n channels was found using relation: $N = \sigma.j.f.V$. Fraction 'f' of nuclei/cm^3, and volume 'V' of emulsion, were the same for both π^-, K^- meson interactions. Calculations were therefore done considering the flux 'j' of π^-, K^- meson in terms of the beam composition (Sect.II), and cross-sections for 1.5 GeV/c K^- meson [1-12], and π^- meson [28].

The following corrections were then made on the basis of proposed absorption model (Sect.III):-

1) <u>Σ Hyperon Absorption</u>: The Σ hyperon, produced in 1.5 GeV/c K^- meson interactions, take away on the average 576^{+40}_{-38} MeV (1253^{+53}_{-21} MeV/c). Cross-section at this energy for $\Sigma + N = Y^0 + N$ being [29]

$\sigma_{\Sigma^{\pm}}(\Lambda,\Sigma^c) = 53^{+32}_{-27}$ mb,

corresponding absorption probability was found from Fig.1 to be:

$W_{\Sigma^{\pm}} = 0.58^{+0.10}_{-0.07}$.

Frequencies of various K^-N channels were then corrected with the knowledge of '$W_{\Sigma^{\pm}}$' and the relative yield of respective channel.

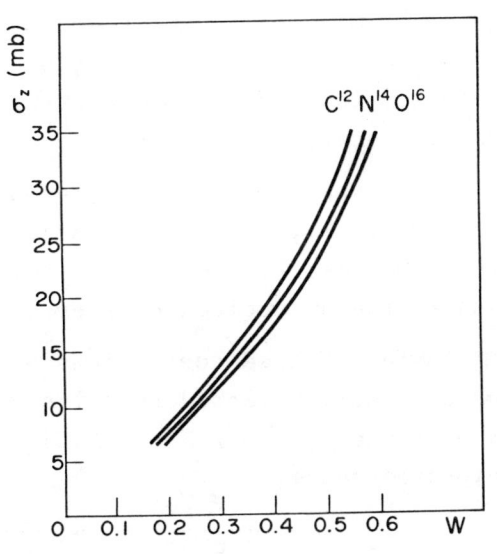

Fig.1 Absorption probability of secondary particles

2) **Pion Absorption**: When produced in the interaction of 1.5 GeV/c K^- meson, the pion on the average have 452^{+31}_{-17} MeV energy. Taking $\pi^- N$ interaction cross-section, $\overline{\sigma_{\pi N}}$, at this energy[28], the probability for pion($\pi^{\pm}$) absorption was obtained from Fig.1 as, $W_{\pi^{\pm}} = 0.276^{+0.023}_{-0.016}$. Observed channel frequency was corrected for pion($\pi^{\pm}$) absorption on the basis of $W_{\pi^{\pm}}$, and the relative yield of respective channel. Assuming that absorption probability of π^o meson is nearly proportional to its production rate in the ratio, $\frac{\pi^o}{\pi^{\pm}} = 0.5$[18], in terms of $W_{\pi^{\pm}}$, correction was also applied for π^o meson absorption in reactions like, $\pi^o + p = \pi^+ + n$.

Pion absorption may result in multiple pion production, in processes like $\pi + N = N + i\pi$ (i=1,2,3), at a threshold of 160 MeV. Rate of pion production 'N_s' was calculated from:

$$N_s = W(\overline{\sigma_{\pi s}} / \overline{\sigma_{\pi N}})$$

where $\overline{\sigma_{\pi s}}$ is the cross-section for multiple pion production, η the average pion number produced in K^- meson interaction, and ε the fraction of secondary pions with kinetic energy exceeding 160 MeV. Calculation of ε was based on kinematics of 1.5 GeV/c K^- meson interaction. The absorption probability $W_{\pi^{\pm}}$, and $\overline{\sigma_{\pi N}}$ being known, and having calculated $\overline{\sigma_{\pi s}}$, η, on cross-sectional basis[28], the multiple pion production rate was found to be 0.016/star. Correction for this process was therefore negligible.

The probability for Σ hyperon production, in $\pi + N = \Sigma + K$, is very small, because the secondary pion need to have at least 900 MeV to induce such reaction. Rate of pion absorption on nucleon cluster is also small[18].

Having applied corrections 119 K^-p and 101 K^-n channels were finally selected. Relative yield for total K^-p and K^-n channels, $K^-p / K^-n = 1.21^{+0.05}_{-0.03}$, was close to the expected value,

$$\frac{\overline{\sigma_{K^-p}}}{\overline{\sigma_{K^-n}}} = \frac{20.1 \mp 1.4}{15.13 \pm 1.08} = 1.32 \pm 0.002$$

Corrected relative probabilities were then found and, as in Table 1, compared with expected values based on cross-section

Distribution of K⁻p and K⁻n channels

Table 1

K⁻N Channel	Experimental Probability	Expected Probability
$K^-+p=Y^0$+neutrals	$0.231^{+0.013}_{-0.016}$	0.226 ± 0.01
$K^-+p=\Sigma^\pm+\pi^\mp$+neutral	$0.21^{+0.006}_{-0.007}$	$0.185^{+0.009}_{-0.005}$
$K^-+p=Y^0+\pi^\pm+\pi^\mp$+neutral	$0.079^{+0.006}_{-0.004}$	$0.212^{+0.008}_{-0.009}$
$K^-+p=\Sigma^\pm+\pi^\mp+\pi^\pm+\pi^\mp$	0.033 ± 0.004	0.0186 ± 0.001
$K^-+p=K^-+n+\pi^+$+neutral	$0.062^{+0.010}_{-0.009}$	0.069 ± 0.017
$K^-+p=K^-+\pi^++\pi^\pm+\pi^\mp$	$0.0084^{+0.0003}_{-0.0016}$	0.006 ± 0.0001
$K^-+p=K^-+p$+neutral	$0.051^{+0.011}_{-0.007}$	$0.045^{+0.006}_{-0.004}$
$K^-+p=K^0+p+\pi^-$	$0.040^{+0.003}_{-0.003}$	0.091 ± 0.005
$K^-+p=K^0+n$+neutral	$0.246^{+0.022}_{-0.021}$	0.224 ± 0.02
$K^-+n=Y^0+\pi^-$+neutrals	$0.34^{+0.018}_{-0.025}$	0.394 ± 0.009
$K^-+n=\Sigma^-$+neutrals	$0.181^{+0.006}_{-0.008}$	0.11 ± 0.002
$K^-+n=\Sigma^-+\pi^++\pi^-$+neutral	$0.075^{+0.008}_{-0.009}$	0.044 ± 0.007
$K^-+n=Y^0+\pi^++\pi^-+\pi^-$	$0.038^{+0.005}_{-0.012}$	0.036 ± 0.001
$K^-+n=K^-+n+\pi^++\pi^-$	$0.038^{+0.002}_{-0.003}$	0.0008 ± 0.00002
$K^-+n=K^-+n$+neutrals	$0.041^{+0.003}_{-0.002}$	0.096 ± 0.005
$K^-+n=K^0+n+\pi^-$+neutral	$0.218^{+0.010}_{-0.005}$	0.247 ± 0.012
$K^-+n=K^-+p+\pi^-$+neutral	0.011 ± 0.002	0.012 ± 0.0017

for particular group of channels. The upper and lower limits come from uncertain events. Results of Armenteros et al[12] were extrapolated for K⁻n interaction cross-sections.

It should also be mentioned that the probability for Σ hyperon production in, $\Lambda+p=\Sigma^++n$, is negligible[30]. Moreover the decay of secondary Σ hyperon and pion is not possible inside the nucleus, because the half lives of the Σ hyperon ($\sim 10^{-10}$ sec) and pion ($\sim 10^{-8}$ sec) are longer than their respective time of flight, $T_\Sigma \sim 10^{-13}$ sec, and $T_\pi \sim 10^{-21}$ second, even for as low energy as 40 MeV.

The experimental and expected probabilities, given in Table 1, are pretty close in some cases. However, in some cases there is an apparent disagreement, which may be attributed to the difficulties in identifying the pion tracks, and Σ_ρ events. For accurate results analysis on much larger statistics are needed. Probabilities for the K^-n channels may give some idea about K^-n interaction cross-section.

V. References

1) M.B.Watson et al..............Phy. Rev. 131 (63) 2248
2) P.L.Bastien & J.P.Berge....Phy. Rev. Lett 10 (63) 188
3) W.Graziane & S.G.Wojcicki..Phy. Rev. 128 (62) 1868
4) W.A.Cooper et al ....Proc.Sienna Conf.Ele.Particles 1963,Volume I,page 154,160
5) G.M.Pjerrou ....Phy. Rev. Lett. 9 (62) 114
6) Humphery & Ross ....Phy.Rev. 127 (62) 1305
7) Ferro-Luzzi et al ....Proc.CERN Conf.1962,page 376
8) Button-Shaffer et al ......Proc.CERN Conf.1962,page 303
9) M.H.Alston et al ....Proc.CERN Conf.1962,page 311
10) Bertanza et al ....Prof.CERN Conf.1962,page 279
11) Gelsema et al ....Proc.Sienna Conf.Ele.Particles 1963,Volume I,page 134,143
12) R.Armenteros et al ....Nucl.Phys. B18 (70) 425
13) Barloutaud et al ....Nucl.Phys. B 9 (69) 493
14) Yen et al ....Phy.Rev.Lett. 22 (69) 963
15) L.Culhane et al ....Proc.Irish Acad. 63A (64) 61
16) J.Cuevas et al ....Nucl.Phys.B1 (67)411
17) C.F.Powell et al ....Study of elementary particles by phtographic method,1959, page 427,Pergamon Press
18) A.Waheed ....Ph.D Thesis,Belgrade Univ. 1973
19) M.Jurić & S.Popov ....Nucl.Phys. B8 (68) 529
20) P.H.Fowler & D.H.Perkins...Phil. Mag. 46 (55) 587
21) P.H.Fowler ....Phil. Mag. 41 (50) 169
22) W.H.Barkas ....UCRL 9181
23) A.Waheed ....Proc. 8th Intern.Conf.Nucl. Photo.Bucharest,1972,page 125
24) Frisch & Thorndike ....Elementary Particles,1964, page 87
25) Z.Todorović ....Ph.D Thesis,Belgrade Univ. 1972
26) S.Fernbach et al ....Phy. Rev. 75 (49) 1352
27) A.Waheed et al............FIZIKA 4 (72) 87
28) E.Flaminio et al ....CERN/HERA No.70-(5,6,7),1970
29) V.Bisi et al ....Phys.Lett. 10 (64) 252
30) G.Alexander et al ....Phy.Rev.Lett. 7 (61) 348

HYPERNUCLEI IN SOME HAMMER-LIKE EVENTS RECORDED IN A PHOTONUCLEAR EMULSION IRRADIATED WITH STOPPING K⁻MESONS

M. K. Jurić and S. B. Drndarević

Institute of Physics, 11001 Belgrade, SFR, Yugoslavia

Abstract

In a nuclear emulsion stack exposed to K^--mesons a total of 675 events were found in which incident stopping particles interacted with emulsion nuclei to produce the ^8Li or ^8B nuclei (hammer-like events). It was established that 70% of the events pertain to the K^--meson interaction, while 18.25% belong to the stopping π^--meson interaction. Comparing these productions it has been found that in about 5% of events π^0-meson hypernuclei occur in hammer-like events produced by stopping K^--mesons.

Data are reported on one case of a π^--mesonic hypernucleus decaying in such a way that, in addition to the emission of a π^--meson, a hammer-like event is produced, i.e. two short-range par-ticles and an electron are emitted. This event may be acribed to the decay of a $^5_\Lambda Li$ hypernucleus.

In studying the process of K^--meson absorption by nuclei it is necessary to know the products of interaction and de-excitation of the nuclei. Identification of low-energy nuclear fragments, however, is difficult because the fragment ranges are short and the variation of specific ionization in relation to the mass and charge of the particle is negligible. Unstable fragments characterized by their specific decay are an exception in this respect. Thus the ^8He, ^8Li, ^9Li and ^8B nuclei decay characteristically into two alpha-particles. Fragments discriminated in such a manner facilitate the study of the processes of decay and fragmentation of the nucleus. Thus e.g. a first example of the $^{10}_\Lambda Be$ hypernucleus (1) has been identified by the $(\pi^-, ^1H, ^1H, ^8Li)$ decay mode and the binding energy of the lambda hyperon in this nucleus has been found to be $B_\lambda = 9.30$ MeV. In addition to the mesonic decay a channel of non-mesonic decay $(p, ^8Li, n)$ has subsequently been established (2). In both cases the decay could be identified uniquely only due to the discriminated ^8Li fragment.

In this work it has been examined whether all the hammer-like events produced by the interaction of stopping K^--mesons with nuclear emulsion nuclei pertain to the characteristic ^{8}Li or ^{8}B nuclei. This is in so far more interesting as a decay of a hypernucleus e.g. by emission of a π^o-meson may be similar in its configuration in nuclear emulsion to a hammer event.

Experimental Procedure and Results

Experimental data on the interaction of stopping K^--mesons with nuclear emulsion nuclei by the emission of a fragment having a hammer-like configuration have been obtained in a portion of an Ilford K5 emulsion stack (exposed at the Brookhaven AGS to a beam of K^-mesons) analyzed within the framework of a study on the binding energy of the λ-hyperon in hypernuclei of A>5 made by the European K Collaboration. The plates were area-scanned to locate the K^--meson interaction. For each established K^- interaction all black prongs were systematically followed over a range projected on the emulsion plane of at least 100 µm in order to detect the presence of a secondary star (1). Secondary stars with a hammer-like configuration were separately noted. By a hammer-like event is meant a secondary star with only two short, approximately colinear prongs.

In order to establish whether a hammer-like prong belongs to a star originating from a K^--meson interaction, in particular the direction and characteristics of each incident particle track were analyzed. Among about 2000 established mesonic hypernuclei the following number of hammer-like events were found (Table I):

Table I

Primary star	RK^-	FK^-	$R\pi^-$	$R\Sigma^-$	RO	Total
Number of events	498	19	123	13	19	673

where RK^- and FK^- denote K^- meson interactions at rest and in flight respectively, while $R\pi^-$, $R\Sigma^-$ and RO designate interactions caused by the π^--meson, Σ^--hyperon and a neutral particle respectively. A star originating from interaction of a stopped strange particle may have a secondary star belonging to a hypernucleus, whereas a hammer-like star associated with a $R\pi^-$ interaction pertains to only one of the following events:

I) $^8_\Lambda\text{Li}$ z$^-$

$^8_\Lambda\text{B}$ z$^+$ $^8\text{Be}^* \rightarrow 2\text{He}^4$

$^8_\Lambda\text{He}$ z$^-$z$^-$

IIa) $^9_\Lambda\text{Li}$ z$^-$ $^9\text{Be}^* \rightarrow$ $^8\text{Be}^* + n \rightarrow {}^4\text{He} + {}^4\text{He}$

$\searrow {}^5\text{He} + {}^4\text{He}$

IIb) $\rightarrow {}^4\text{He} + n$

In these decay two alpha particles and an electron are emitted, the tracks of the alpha particles from events I) and IIa) being colinear and of the same range. The range distribution, the difference between the ranges of a pair of alpha particles ($\Delta R = R_1 - R_2$) and the deviation from colinearity ($\Delta = 180 - \gamma$) in hammer-like $\overline{RK}$ and $\overline{R\pi}$ events from Table I are in agreement with those in hammer-like events obtained from interactions induced by other projectiles (5,6), where the measurements were made with standard biological Zeiss microscopes ($\Delta R < 3\mu m$, $\Delta < 30°$).

According to I) and II), a hammer-like event may also be characterized by the emission of a β-particle. Therefore careful search has been made for the presence of electrons in every hammer-like event. The possible existence of an electron in every event has carefully been checked in each questionable case by a physicist and two experienced microscopists. Except for cases recorded at the bottom or top of emulsion, it has been found that in $\overline{RK}$ and $\overline{R\pi}$ events respectively $(8.5^{+1.4}_{-1.2})\%$ and $(3.5\pm1.8)\%$ of hammer-like events are not accompanied by the emission of an electron. The result for $\overline{R\pi}$ events is also in agreement with previously reported data on hammer-like events (5). From comparison of these two results it may be stated that about 5% of hammer-like events in $\overline{RK}$ interactions are not associated with decay modes I) and II) and may be ascribed to hypernuclear decays.

In order that we might gain more knowledge on these hammer-like events without the emission of an electron due to interactions induced by K^--mesons we have measured the ranges of prongs and the angles between them in primary stars as well as in secondary ones and made a kinematic analysis of the processes on the basis of the data obtained. This analysis involved also $\overline{RK}$ interactions with dubious emission of an electron in hammer-like events, i.e. cases where the distance from the hammer star to the first grain of electron track was larger than 7 μm (the mean number of grains on an electron track is (19.4±2.2) per 100 μm). The total number of such events for the analysis was found to be 70.

Table II summarizes the numbers of cases (N) according to the number of prongs of the primary star (N_p) and inducates number of events with a prong belonging to a π meson (N_π) or to a fast baryon (N_g).

Table II

N_p	2	3	4	5	≥6	Total
N	5	35	18	10	2	70
N_{π^+}	-	3	-	-	-	
N_{π^-}	-	3	-	-	-	
$N_{\pi^\pm}$	-	8	6	3	-	
N_g	-	10	8	4	1	

We have tried to identify the hypernucleus by usual methods for kinematic analysis of products arising from the processes of production or decay of the hypernucleus (8,1,7). The criterion used for the selection of hypernuclei was the agreement between the binding energy values (B_λ) obtained for a hypernucleus in question and standard tabular values (1). It is assumed that only the interactions of K⁻ mesons with light emulsion nuclei occur and that the emitted nuclei are stable isotopes.

All secondary and primary processes involving π^- and π^+ mesons were kinematically analyzed. The results of this analysis are presented in Table III. (Column 2 of Table III

Hypernuclei in a photonuclear emulsion 1149

gives data on track ranges of particles emitted in the primary process, i.e. those of $\pi^{\pm}$ mesons, hammer prongs and third particle. Column 5 contains data on particle tracks and angles between them in the secondary process, while columns 4 and 7 give the corresponding calculated binding energies of the λ-hyperon in the hyperfragment).

Table III

No	Primary process			Secondary process	
	prongs (μm)	decay	B_λ (MeV)	prongs (μm)	decay and B_λ(MeV)
31/10	25408 5,6 33,4	$^{16}O(\pi,^{+12}_\lambda Be,^3He,n)$ $(\pi,^{+9}_\lambda B,^6Li,n)$	-17 ± 10 $+37\pm9$	4,9μm 39μm 188°	$^{9,10,11}_\lambda B(Li,He,\pi^\circ)$ $B_\lambda = \sim14$
24/10	19650 13,1 1254,0	$^{12}C(\pi,^{+8}_\lambda Li,^3H,n)$ $^{14}N(\pi,^{+10}_\lambda B,^3H,n)$	$1,16\pm5,8$ $16,6\pm8,5$	4,3μm 5,1μm 175,5°	$^{9,10,11}_\lambda B(Li,He,\pi^\circ)$ $B_\lambda(^{10}_\lambda B)=10\pm12$
44/24	16220 21,5 46,0	$^{12}C(\pi,^{+8}_\lambda He,^3He,n)$ $^{12}C(\pi,^{+7}_\lambda He,^4He,n)$	$21\pm4,9$ $5,8\pm5,1$	18,4μm 19,6μm 175°	$^7_\lambda Li(^3H,^4He,\pi^\circ)$ $B_\lambda=11,2\pm2,5$
36/07	12000 2 250,7	$^{12}C(\pi,^{-7}_\lambda Be,^4He,n)$ $^{14}N(\pi,^{-9}_\lambda B,^4He,n)$	$12,9\pm18,1$ $12,3\pm24$	4,5μm 5,2μm 182°	$^{9,10,11}_\lambda B(Li,He,\pi^\circ)$ $B_\lambda(^{10}_\lambda B)=7,9\pm12,7$
33/27	16680 4 8507	$^{12}C(\pi,^{-9}_\lambda B,^2H,n)$ $^{14}N(\pi,^{-11}_\lambda C,^2H,n)$	$6,6\pm6$ $5,1\pm8,5$	4,5μm 4,5μm 188°	$^{9,10,11}_\lambda B(Li,He,\pi^\circ)$ $B_\lambda(^{10}_\lambda B)=5,6\pm13$
26/9	13220 6 3	(π^-)	-	10,1μm 7,5μm 180°	$^4_\lambda He(^2H,^2H,\pi^\circ)$ $B_\lambda = 4,4\pm2,9$

The kinetic analysis shows in general that the hypernucleus identification is insufficiently unambiguous. Errors in determination of B_λ are large not only in the analysis of secondary but also of primary processes in which also a π^- or π^+-meson is emitted (see Table III). In concrete cases in general one can not completely rule out the possibility that a ^8Li or ^8B nucleus is emitted together with two or more neutrons instead of a hypernucleus. Table III also gives a uniquely identified case (No 26/09). In this case three charged particles, one of which being a π^--meson, are emitted in the primary process. The primary process could not be identified, but the analysis of the secondary process leads to only one interpretation, viz. that the hammer-like event is the hypernucleus decay $^4_\lambda He(^2H,^2H,\pi^0)$ with $B_\lambda = (4.2\pm2.9)$ MeV. The analysis of other secondary processes does not provide so unique solutions, but in all the cases the third particle appears to be a π^0-meson.

Here it is interesting to note that in seaching for hammer-like configuration among secondary stars a mesonic hypernucleus has been found which decays not only by the emission of a π^--meson but also by the emission of two particles having short ranges in emulsion and, in addition, an electron is emitted from the hypernucleus. The characteristics of this hypernucleus are given in Table IV. The hypernucleus has been identified uniquely in the usual way (1) as the $^5_\lambda He$ hypernucleus with the λ-hyperon binding energy of $B_\lambda=(3.5\pm2.3)$ MeV. Since also an electron is emitted from the hypernucleus, the process may be described by the following scheme :

$$^5_\lambda Li \rightarrow ^5_\lambda He \rightarrow \pi^- + p + ^4He$$

Thus this event may be assumed to be a β-decay of the hypernucleus.

Table IV

No.	range of prong	$<(R_1, R_2)$
36/100	R_π = 16641 μm	145,8°
	R_1 = 8,2 μm	
	R_2 = 6,7 μm	
	β	

Conclusion

In this experiment it has been found that in the process of absorption of stopping K^--mesons in nuclear emulsion among interactions in which a hammer-like fragment is emitted about 5% of the fragments may represent a π^0-mesonic hypernucleus. One case of a π^--mesonic hypernucleus has been detected which emits an electron and is ascribable to a β-decay of the $^5_\lambda Li$ hypernucleus.

It is a pleasure to thank to Mrs N. Antanasijević for the very careful and tenacious scanning.

References

1) M. Jurić et.al.: Nucl. Phys. B52 (1973) 1.
2) M. Jurić, Proc. 8th Inter. Conf. on Solid State Nuclear Track Detec., Bucurest 1971.
3) D. Abeledo et.al., Nuovo Cimento, 22 (1961) 1171.
4) European K^- Collaboration, unpublished data
5) W. Gajewsky et.al., Nucl. Phys. 37 (1962) 226
6) G.C. Deka et al., Nucl. Phys. 23 (1961) 658.
7) W. Gajewski, Universite Libre de Bruxelles, Bull. no 29, novembre 1966.
8) G. Bohm et al., Nucl. Phys. B12 (1969) 1.

METHOD FOR A DETERMINATION OF LOW MOMENTUM ANTINEUTRON FLUX

M. Brun and H. Annoni

*Laboratoire de Physique Nucléaire, Université de Clermont-Ferrand,
B.P. 45, 63170 Aubiere, France*

A new method was elaborated in order to determine the flux of low momentum (less than 150 Mev/c) antineutrons. The nuclear emulsions were exposed near an intersection at the I.S.R. of CERN.

All stars with at least three white tracks, observed in a surface scanning, were measured. But only the angles of the tracks in the plane of the emulsion were measured. So we chose three different groups.

(I) Stars with only one most possible incident charged particle. These stars have a specific "shower tracks" aspect.

(II) Stars without a possible incident charged particle track. Those stars also have a "shower tracks" aspect.

(III) All remaining stars.

By means of a simulation performed by a computer program it was shown that the probability of finding antineutron annihilation stars in the third group is roughly double of that of the other groups.

This methodology may be interesting for a precise determination of antineutron flux or in space research.

SOME ASPECTS OF 400 GeV PROTON INTERACTIONS IN NUCLEAR EMULSIONS

P. S. Young*, K. Fukui**† and Y. V. Rao***

*Mississippi State University, Mississippi 39762, U.S.A.
**Air Force Geophysics Laboratory, Bedford, Mass. 01731, U.S.A.
***Dublin Institute for Advanced Studies, Dublin, Ireland

ABSTRACT

Ilford G-5 nuclear emulsions were exposed to 300 GeV and 400 GeV proton beams at the National Accelerator Laboratory. Total 350 interactions were found and analyzed. The study was concentrated on non-white stars because of insufficient scanning efficiency for white stars. The mean free path for 300 GeV proton is 36.9±2.6 cm which is in good agreement with other authors. A linear relation is observed between the charge shower particle multiplicity and the star size indicating that the number of shower particles is independent of the target nuclei. Black prongs for smaller stars seems to show higher value of forward-backward ratio suggesting larger motion of nucleus after collisions.

INTRODUCTION

In order to study interactions between high energy protons and complex target nuclei in emulsion, stacks of nuclear emulsion were exposed to proton beams of highest energy available today at the National Accelerator Laboratory, Batavia. Studies of interest cover star size distribution, charged shower multiplicity, the ratio of black and gray tracks as well as the ratio of forward and backward ejected

---
†NRC-AFSC Resident Research Associate

tracks from the stars as a function of the proton incident energy.

EXPOSURE, PROCESSING AND SCANNING

Several stacks of Ilford G-5 nuclear emulsions, 600 microns thick, were exposed to 300 GeV proton beams at the Fermi National Accelerator Laboratory, Batavia, Illinois, in October 1973. The flux of the proton beams was in the order of 10^5 per cm^2. Another stacks were exposed to 400 GeV proton beams at the same laboratory in November 1975, and the flux was in the order of 10^4 per cm^2.

The emulsions exposed to the 300 beams were unmounted pellicles. They were processed in full strength. Those exposed to the 400 GeV beams were glass mounted plates, and they were processed in slightly weaker strength in order to avoid heavy background. The development of the unmounted plates turned out uniform and has made it easy to observe particle tracks of low grain density such as left by the shower particles.

In this work only area scanning with a 150x magnification was used. To avoid stars created by secondary interactions, only those stars produced by the protons within less than 1 degree from the original beam direction were taken. Also the scanning was limited to the area close to the entrance edge of the plates. Namely for the 300 GeV plates the area between 4 mm and 22 mm, and for the 400 GeV plates the area between 10 and 29 mm from the entrance edge were used. This method limits the contamination by secondaries to less than 5%.

The study on white stars without prongs, and semi-white stars with one prong are of importance for the analysis of p-p collisions. However, this scanning was not aimed to find substantial amount of white stars. Higher magnifications and the following track method should be applied for systematic study of white stars. In this preliminary report only the study on stars with more than one prongs was carried out.

The selected stars as identified as interactions due to primary protons and target nuclei in our emulsion stacks amounted to the 154 and 196 from the 400 GeV and 300 GeV plates respectively.

ANALYSIS

1. The Star Size Distribution and Inelastic Mean Free Path

The star size distribution obtained from the 300 GeV and 400 GeV plates is shown in Fig. 1. N_h indicates the number of total prongs including both black and gray tracks. In this distribution, the percentage of events for $N_h \leq 8$ is about 50% of the total number. It is reasonable to assume that a half of these events and the events with larger stars are due to Ag or Br nuclei. Most of the events with $N_h \leq 8$ are assumed to have originated from the N, C, and O nuclei. Only a few percentage of these events are attributed to the p-p collisions.

From theoretical consideration, it is believed that only 4% of the interactions could take place in hydrogen, thus the amount of small stars not included in this distribution is in the order of no more than 8% of entire events.

The inelastic mean free path was calculated. The value obtained for the 300 GeV plates is 36.9±2.6 cm, and this is slightly higher than the value given by Herbert et al.[1] If we consider the amount of loss mentioned above, this value may be lowered to 34.0 cm. It becomes in good agreement with the other authors. Therefore we see no significant difference in the mean free path for the 300 GeV and 200 GeV protons.

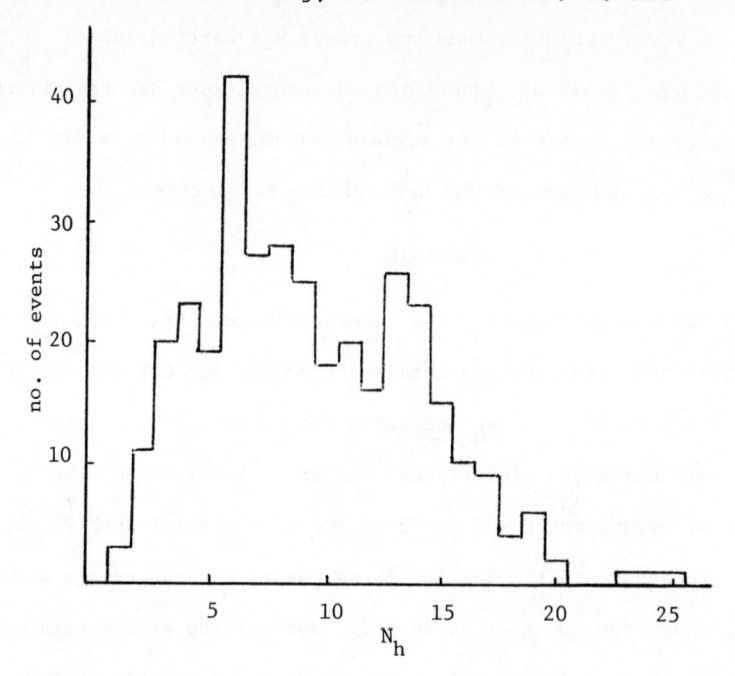

Fig. 1. Histogram of the combined N_h distribution from 300 GeV and 400 GeV plates.

2. Charge Shower Particle Multiplicity as a Function of Star Size

The charge shower particle multiplicity $\langle n_s \rangle$ as a function of star size N_h is plotted in Fig. 2 together with the values by other works. Agreement with other works are within the statistical error and no drastic change due to the energy of incident proton beams is observed. We observe a similar tendency for this work and for the works by others. Namely, there is clearly a leanear dependence between $\langle n_s \rangle$ and N_h. This fact indicates that the number of shower particles is not depending on the target nuclei, but rather it depends on the process of breaking up targets. The amount of shower particles is apparently decided by the complex process of creating black and gray prongs inside the target nuclei after collision.

400 GeV proton interactions

Fig. 2. Charged shower particle multiplicity $\langle n_s \rangle$ v.s. the star size N_h. ▲ 400 GeV plates and △ 300 GeV plates of this work. ○ Paris, ✕ Lund, ● Ottawa, ☐ Bucharest, all from 200 GeV proton (ref. 1)

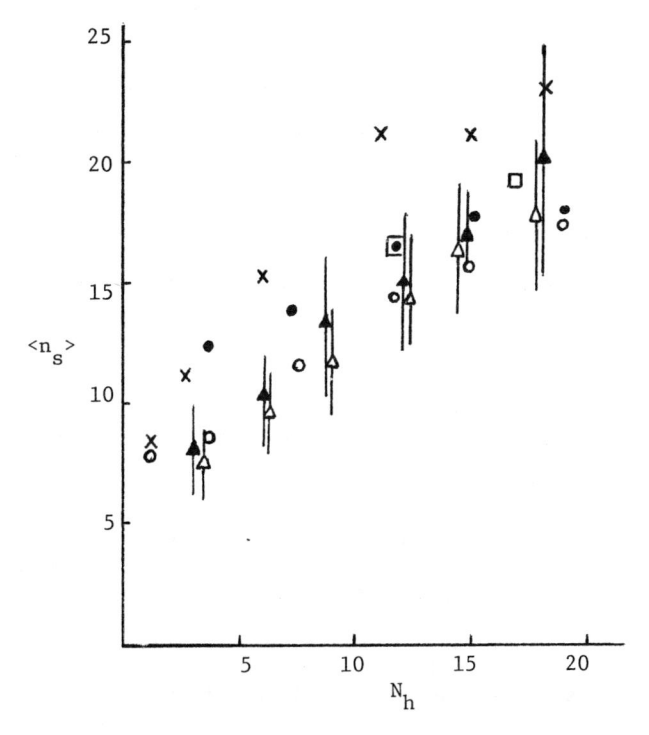

Table 1 shows a comparison of the results of the average number of black, gray, and shower-particle tracks oberserved from the stars produced by the protons of 300 and 400 GeV and lower energies.

Table 1. Composition of Events

E(GeV)	$\langle N_b \rangle$	$\langle N_g \rangle$	$\langle n_s \rangle$
6.2*	5.68±0.21	3.58±0.11	2.21±0.04
22.5*	5.22±0.29	3.38±0.14	5.08±0.12
200+	7.0 ±0.1	2.79±0.15	12.9 ±0.15
300	5.42±0.19	3.92±0.16	12.99±0.29
400	5.32±0.17	3.93±0.14	12.03±0.25

*: Winzeler, +: Hebert et al. after ref 1.

The numbers $\langle N_b \rangle$ and $\langle N_g \rangle$ of this work are in good agreement with the works of lower energies, but somewhat different from the values at 200 GeV. On the other hand $\langle n_s \rangle$ for this work seems to agree better with the 200 GeV result. Even considering the loss of white stars in this work the value of $\langle n_s \rangle$ seems to increase as the energy of incident protons increases. However there is no apparent difference between the 300 and 400 GeV results.

3. Ratio of Forward and Backward Tracks

Table 2 shows the ratios of forward and backward ejected prongs for black, gray, and heavily ionizing tracks.

The black tracks (with grain density 10 g_p) are attritubed to the evaporation process of highly excited nuclei, and the gray tracks are mostly due to recoil nucleons. Evaporation particles are believed to be ejected after the target nuclei are set in motion as a result of interaction. The motion for heavier nuclei such as Ag or Br should be thus smaller and the ratio (F/B) should be also smaller than the ratio for the smaller nuclei such as C, N, O. Since the stars with $N_h \geq 9$ are mostly due to Ag or Br and those with $N_h \leq 8$ are due to lighter nuclei, we should observe a larger ratio for the stars with $N_h \leq 8$. This is clearly demonstrated by the first column of Table 2. So far as the gray tracks are concerned, no similar features were observed (see Table 2).

Table 2. The Ratio of Forward and Backward Tracks

	$(F/B)_b$	$(F/B)_g$	$(F/B)_h$
300 GeV	1.22±0.08	1.19±0.09	1.21±0.06
400 GeV	1.18±0.08	1.22±0.11	1.19±0.06
$N_h \leq 8$	1.32±0.11	1.18±0.12	
$N_h \geq 9$	1.15±0.06	1.21±0.08	

Subscripts b, g, and h stand for black, gray, and heavily ionizing tracks. Note that $N_h = N_b + N_g$.

CONCLUSION

No significant differences have been observed between the interactions of this work and other works at lower energy. However it appears that the value of $<n_s>$ increases with the proton energy. In order to be sure, we should re-examine the number of gray and black tracks.

The ratio of forward and backward tracks should be compared with the result of plastic detectors whose components consist of only C, N, O, and H. Such data may be reported in this conference.

Further study and scanning of the events including white stars should be continued.

The authors thank Dr. Voyvodic of the National Accelerator Laboratory for the exposure of the emulsions.

REFERENCE

1. J. Hebert et al., AIP Conf. Proceed. no. 12, 131, 1973.

THE SCATTERING CONSTANT FOR MULTIPLY-CHARGED PARTICLES IN EMULSIONS

Y. V. Rao*, K. Fukui**† and P. S. Young***

*Dublin Institute for Advanced Studies, Dublin, Ireland
**Air Force Cambridge Research Laboratories, Bedford, MA. 01731, U.S.A.
***Mississippi State University, Mississippi 39762, U.S.A.

The scattering constant K has been determined in G5 and K5 emulsions from multiple scattering measurements on beam tracks of 2.1 GeV/amu Nitrogen and Oxygen ions and, 1.05 GeV/amu Helium ions. By confining measurements in selected regions virtually free from distortion, it has been possible to obtain reliable values of K for cell lengths in the range 1 to 3 mm. The estimation of K at these cell lengths, suggests that it is fairly constant in this region and lies below the theoretical curve. The correlation coefficients for the higher order differences in multiple scattering are shown to be higher than those given in the frame work of Moliere's theory. The trend of the increased values of the correlation coefficient is found to be consistent with the calculations of Dado and Rosendorff's theory.

1. Introduction:

Momenta of high energy charged particles in nuclear emulsions can be determined by measuring the multiple Coulomb scattering suffered by these particles. If $\bar{D}_c$ is the average second difference due to Coulomb scattering, as measured by the procedure suggested by Fowler [1] then the product of momentum p and velocity β is given by

$$p\beta = KZt^{3/2}/18.1\,\bar{D}_c \qquad \ldots \quad \ldots \quad (1)$$

In the above relation the cell length t is in units of mm, $\bar{D}_c$ in μm, pβ in GeV/c and Z the charge of beam particle. In order to use this relation effectively, one needs to know the value of the scattering constant K as a function of cell length t. The value of K can be calculated theoretically. This has been done, for example, by Fichtel and Friedlander [2] using Moliere's theory of multiple Coulomb scattering [3]. Experimentally, the scattering constant has been determined for small cell lengths by Backus et al [4]. In their experiment Backus et al exposed emulsions to Beryllium and Carbon ions which were accelerated to 1000 MeV at the University of Chicago Synchrotron [5]. However with the availability of high energy heavy ion beams at Bevlac [6], it has become

† NRC-AFSC-Resident Research Associate.

possible to make reliable multiple scattering measurements in emulsions and to determine scattering constant at long cell lengths. It has been customary to use a value $K = 32$ for all measurements on particles having $Z > 1$, but there appears to be no justification for this in the literature. The purpose of the present investigation is to determine K as a function of cell length in emulsions exposed to heavy ion beams.

2. Theory:

In Moliere's theory the mean of the absolute value of the projected scattered angle Φ, measured by differences of order K, is given by

$$<|\Phi|>_k = \chi_c/\sqrt{\pi} \sqrt{C^k_{i,i} B_k} \{1 + 0.982/B_k - 0.117/B_k^2 + \dots \} \quad ..(2)$$

where B_k is a generalization of Moliere's parameter B, which is defined by the transcendental equation

$$B - \ln B = e\chi_c^2 / \gamma^2 \chi_a^2 \quad \dots \quad \dots \quad (3)$$

where

$$\ln \gamma^2/e = 0.154 \quad \dots \quad \dots \quad (4)$$

$$\chi_c^2 = 4\pi N t e^4 Z^2 z/(pv)^2 \quad \dots \quad \dots \quad (5)$$

The screening angle χ_a had been calculated by Moliere using Fermi-Thomas polential

$$\chi_a^2 = (\hbar/ap)^2 \{1.13 + (3.76\, Zze/137v)\} \quad \dots \quad \dots \quad (6)$$

Here Ze is the charge of the scattered particle, p its momentium, and v its velocity. Z is the atomic number of the scattered medium, N is the number of atoms per unit volume, and t is the path length in the scattering medium, a is Fermi-Thomas radius. Equation (2) can be written as

$$<|\Phi|>_k = K_k (Z\sqrt{t}/pv) \quad \dots \quad \dots \quad (7)$$

where K_k is called the scattering constant of the Kth difference and is given by

$$K_k = 2 z^2 e^2 \sqrt{N} \sqrt{C^{(k)}_{i,i} B_k} \{1 + 0.982/B_k - 0.117/B_k^2 + ..\} \dots \quad (8)$$

The scattering constant K for the tangent case is obtained by putting $B_k = B$ and $C^{(k)}_{i,i} = 1$. Thus K_2, the scattering constant of the second difference, is given by

$$K_2 = K \sqrt{2/3}\, F_{21} \quad \dots \quad \dots \quad (9)$$

where
$$F_{21} = \sqrt{B_2/B_1} \left[\frac{1 + 0.982/B_2 - \ldots}{1 + 0.982/B_1 - \ldots} \right] \qquad \ldots \quad (10)$$

Finally, the correlation coefficient ρ_{mm}, the ratio of the mth difference, is given by (2)

$$\rho_{mm} = <|\Phi|>_m / <|\Phi|>_n = K_m/K_n = \sqrt{C_{i,i}^{(m)}/C_{i,i}^{(n)}} \; F_{mm} \qquad \ldots \quad (11)$$

where the corection factor

$$F_{mm} = \sqrt{B_m/B_n} \left[\frac{1 + 0.982/B_m - 0.117/B_m^2 + \ldots}{1 + 0.982/B_n - 0.117/B_n^2 + \ldots} \right]$$

$$C_{i,i}^{(m)} = \left[\frac{m(m-1)}{3} \right] \left[\frac{(2m-4)!}{\{(m-1)!\}^2} \right] , \quad \text{for } m > 1$$

$$C_{i,i}^{(m)} = 1 \qquad \text{for } m = 1$$

Φ_m projected scattering angle for the mth difference
B_m Moliere parameter B for the mth difference
K_m Scattering constant for the mth difference.

The cut-off correction factor F_{mn}^{co} is defined by

$$\rho_{mn}^{co} = K_m^{co}/K_n^{co} = \sqrt{C_{i,i}^{(m)}/C_{i,i}^{(n)}} \; F_{mn}^{co} \qquad \ldots \quad (12)$$

3. Experimental Procedure:

The emulsions used in this experiment is a part of a stack of Ilford G5 emulsions with pellicles dimensions (75x75x0.6) mm exposed to a beam of 2.1 GeV/amu Nitrogen ions and 1.05 GeV/amu Helium ions and, Ilford K5 emulsions of size (105x105x0.4) mm exposed to a beam of 2.1 GeV/amu Oxygen ions. According to Bevlac group these ions at such energies are fully ionized. All the tracks that were flat and oriented to the emulsion edge were considered as beam tracks. The tracks were picked up mostly within 2 cm from the leading edge of the emulsion. Using Fowler's co-ordinate method [1] multiple scattering measurements were made on the tracks of 2.1 GeV/amu Nigrogen ions with a Reichert microscope; while on 2.1 GeV/amu Oxygen ions measurements were made with a Koristka R5 microscope. Similarly measurements were made on Helium ion tracks using Reichert microscope. In order to investigate the variation of scattering constant, a total track length of 8 meters was scattered using X100 oil objective with basic cell length of 1 mm, and independent cells for 2 and 3 mm were constructed. Further, relative scattering measurements were made on closely spaced pairs of beam tracks on a total track length of 3.5 meters. Measurements of relative scattering have the important advantage that they are unaffected by 'stage noise'. Further, so long as the two tracks of a pair remain separated by only a short distance, the measurements are little affected by distortion. The total noise (stage, grain and reading noise) found by the method of Biswas et al [7] was found to be 0.14 µm. The total noise was subtracted quadratically from the observed signal. The resultant signal for Coulomb scattering from relative scattering measurements was found to be in good agreement with that for measurements on single tracks.

4. Results:

The scattering constant K has been determined using eq. (1). In order to eliminate abnormally large deflections which are mostly due to single scatters a $4\bar{D}$ cut-off procedure has been used - a procedure in which all the deflections larger than $4\bar{D}$ are truncated and the mean is computed from the resultant distribution. The observed distribution of D's (second differences) for cell 2 mm for Nitrogen data is shown in Fig. 1. The theoretical curve (Gaussian) is also shown for the tail up to $4\bar{D}$ cut-off. The fit with Gaussian curve is excellent. A convenient test for Gaussian distribution is provided by using Pearson's criterion. This consists of evaluating the parameters β_1 and β_2 and comparing them with the theoretical values which are 0 and 3 respectively. Here $\beta_1 = \mu_3/\mu_2^{3/2}$ and $\beta_2 = \mu_4/\mu_2^2$, where μ_n is the nth moment about zero. From Table I it is seen that the observed values are in remarkable agreements. In Table II are shown the values of scattering constant with $4\bar{D}$ cut-off (K_{co}) for various cell lengths. Also, reproduced in the table are the values due to Backus et al [4] for 250 µm and 500 µm respectively. The theoretical values for scattering constant for multiply-charged particles in emulsions as worked out by Fichtel and Friedlander [2] are given in the last column of Table II. Fichtel and Friedlander pointed out that the value of the scattering constant with cut-off is rather lower than the assumed value of 32 in many earlier experiments. Based upon the theory of Moliere, they computed numerical values for the scattering constant. However, our results suggest that the scattering constant is lower than the computed values of Fichtel and Friedlander for multiply-charged particles.

In Table III are shown the values of correlation ratios for differences in various orders. Dado and Rosendorff [8] calculated correlation ratios using eq. (12). It is clear from eg. (12) that the correction factor is by several percent greater than one. In Moliere's original treatment the correction factor F_{mn} was taken as unity. Rosendorff and Eisenberg [9], and Dado and Rosendorff [8] have included the B terms in their calculations and re-evaluated the correction factor F_{mn}. This has the consequence of raising F_{mn} by a few percent. The expected values of F_{mn} are inferred from the work of Dado and Rosendorff. The mth difference for Coulomb signal can be found by a quadratic subtraction of noise from the observed value in the corresponding difference. The expected value of noise in the mth difference can be calculated [10] by multiplying $\bar{D}_N$, the arithmetic mean of noise in the second difference, by the quantity $1/m!\sqrt{(2m)!/6}$. The correlation ratios for differences in various orders are displayed in Table III. The expected values on the basis of Moliere's theory are also presented in Table III.

The results suggest that ρ_{32} is higher by 10% compared to theoretical values 1.22 given by Moliere's theory in its approximate form. Similarly the ratios ρ_{42} and ρ_{52} are also higher than the theoretical values 2.00 and 3.53 respectively. According to the calculations of Dado and Rosendorff, correction factor for the neighbouring differences is small, for example, the correction factor for ρ_{43} and ρ_{54} (for $\Omega = 10^4$, where Ω is the effective number of collisions in a given segment of the scattering medium) is 1.0038 and 1.0028 respectively. On the other hand, the differences which are not consecutive, the correction factor is relatively larger for a given Ω, for example, the correction factor for ρ_{52} is 1.023. The trend of the increased ratios is compatible with our observations.

One of the authors, Y. V. Rao, wishes to express his gratitude to Professor C. O'Ceallaigh for much encouragement and support. He also wishes to thank Mr. J. Daly for setting up the equipment. It is a pleasure to thank Dr. H. H. Heckman for the loan of emulsions exposed to Helium and Oxygen ions and to Dr. I. Otterlund for the loan of 'Liner' to estimate the stage noise.

References:
1. P. H. Fowler, Phil. Mag. 41(1950)169.
2. C. Fichtel and M. W. Friedlander, Nuovo Cimento, 10(1958)1032.
3. G. Moliere, Zeits. f. Naturf. 2a(1947)133, 3a(1948)78, 10a(1955)177.
4. M. Backus, J. J. Lord and M. Schein, Phys. Rev. 88(1952) 1431.
5. C. N. Chou, W. F. Fry and J. J. Lord, Phys. Rev. 87(1952)671.
6. H. A. Grunder, W. D. Hartsough and E. J. Lofgren, Science, 174(1971) 1128.
7. S. Biswas, B. Peters and Rama, Proc. Ind. Acad. Sci. 41(1955)154.
8. S. Dado and S. Rosendorff, Nuovo Cimento, 50(1967)238.
9. S. Rosendorff and Y. Eisenberg, Nuovo Cimento, 7(1958)23.
10. P. J. Lavakare and E. C. G. Sudarshan, Suppl. Nuovo Cimento, 26(1962)251.

TABLE I

Break down of Data on Multiple Scattering

Cell length in mm	Nitrogen Data				Oxygen Data			
	$<D>_c$ (μm)	ρ_{32}	β_1	β_2	$<D>_c$ (μm)	ρ_{32}	β_1	β_2
2	1.030 ± 0.023	1.347 ± 0.032	0.087 ± 0.003	2.87 ± 0.01	1.108 ± 0.023	1.433 ± 0.034	0.150 ± 0.003	3.50 ± 0.02
3	1.784 ± 0.049	1.332 ± 0.042	0.167 ± 0.005	2.91 ± 0.02	1.843 ± 0.047	1.399 ± 0.042	0.091 ± 0.005	3.31 ± 0.02

TABLE II
Scattering constant for various cell lengths

Authors	Cell length in mm	Ions	Energy	K (Experimental)	K (Theoretical)
Backus et al[6]	0.25	Carbon	10-1000 MeV	24.8 ± 1.4	27.6
Backus et al[6]	0.5	Carbon	10-1000 MeV	26.2 ± 2.2	28.5
Present experiment	1	Helium	1.05 GeV/amu	26.50 ± 0.94	29.40
Present experiment	2	Nitrogen	2.1 GeV/amu	27.68 ± 0.61	30.20
Present experiment	2	Oxygen	2.1 GeV/amu	29.78 ± 0.61	30.20
Present experiment	3	Nitrogen	2.1 GeV/amu	26.17 ± 0.72	30.63
Present experiment	3	Oxygen	2.1 GeV/amu	26.96 ± 0.69	30.63

TABLE III
Correlation ratios for Differences in various orders

Correlation ratio with cut-off	Nitrogen data	Oxygen data	Moliere's Theory	Dado & Rosendorff
ρ_{32}	1.347 ± 0.031	1.433 ± 0.035	1.22	1.25
ρ_{43}	1.684 ± 0.035	1.676 ± 0.036	1.63	1.66
ρ_{54}	1.820 ± 0.036	1.822 ± 0.037	1.77	1.77
ρ_{42}	2.268 ± 0.035	2.402 ± 0.036	2.00	2.04
ρ_{52}	4.126 ± 0.036	4.376 ± 0.037	3.53	3.64

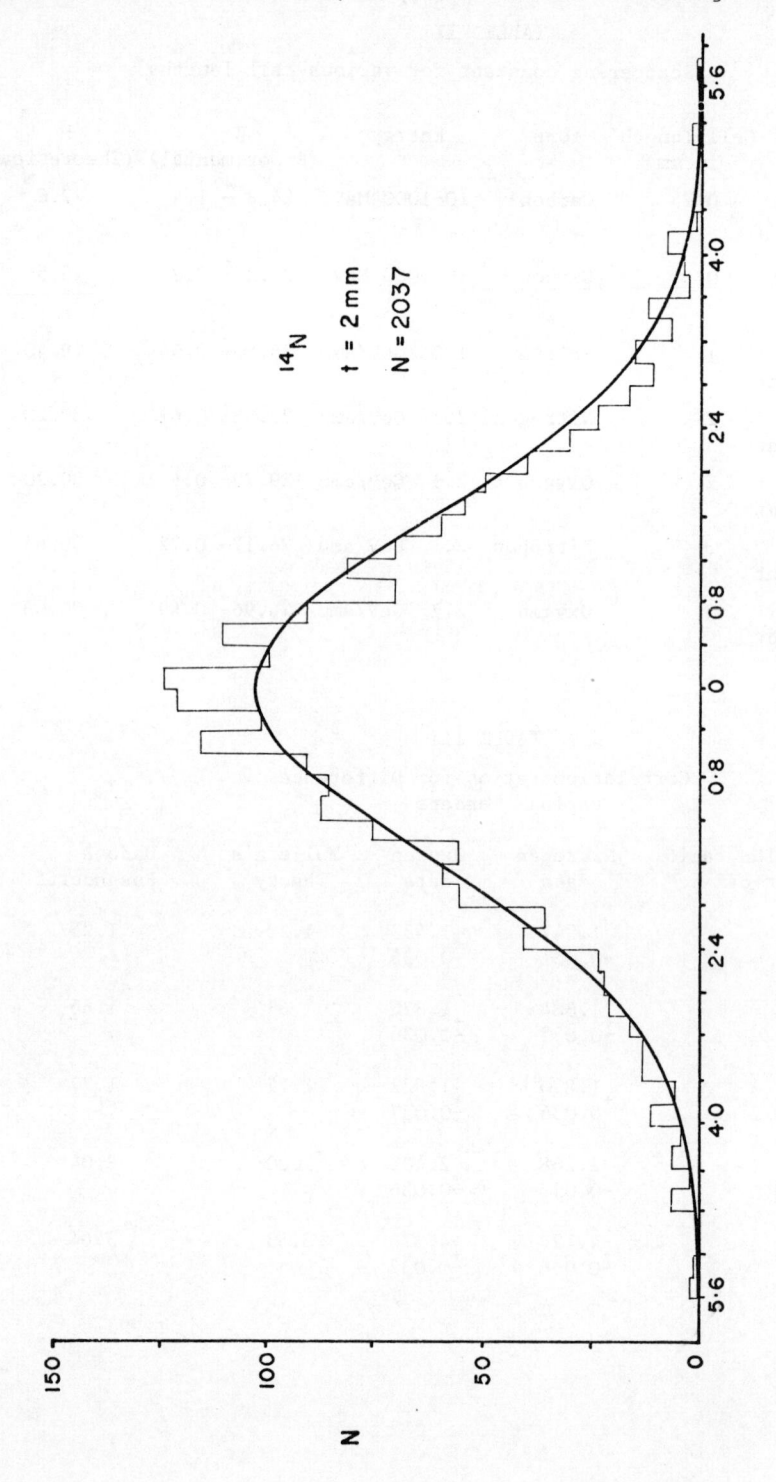

FIGURE 1. DISTRIBUTION OF SECOND DIFFERENCES (t = 2 mm) FOR NITROGEN DATA

SPALLATION PRODUCTS INDUCED BY ENERGETIC NEUTRONS IN PLASTIC DETECTOR MATERIAL

K. Grabisch, R. Beaujean, R. Scherzer and W. Enge

*Institut für Reine und Angewandte Kernphysik, University of Kiel,
23 Kiel, West Germany*

Abstract.

Cellulose nitrate plastic detector sheets were irradiated with secondary neutrons of the 22 GeV/c proton beam at the CERN accelerator. He, Li and Be particles which are produced in nuclear interactions of the neutrons with the target elements C, N and O of the plastic detector material are measured. Preliminary angle and range distributions and isotopic abundances of the secondary particles are discussed.

1. Introduction

In recent years a lot of work has been done to determine the spallation process of various targets induced by energetic nuclei (1, 2). Most of these studies were performed using nuclear emulsions. Because of the limited resolution for the low energy secondary He, Li and Be particles these measurements, however, were restricted to an analysis in charge groups. Only the so called hammer tracks due to ^{8}Li particles allowed an isotopic analysis.

Plastic track detectors of cellulose nitrate provides a reasonable isotopic resolution for low energy light nuclei (3). Thus we exposed stacks of cellulose nitrate sheets to high energy neutrons and studied the induced secondary light particles. The advantages of this method are:

a) The detector allows isotopic resolution, b) the detector serves as target material, c) the target nuclei C, N and O are very similar. The mass, angular and energy distribution of the secondary particles are measured.

2. Selection criterion

By the nuclear interactions of the high energy neutrons with the CNO target nuclei low energy fragments are produced. In this work only fragments heavier than protons are studied.

After etching the image of a spallation depends on the distance of the interaction from the surface and on the angle and energy of the emitted light nucleus. The following images occur (Fig. 1).

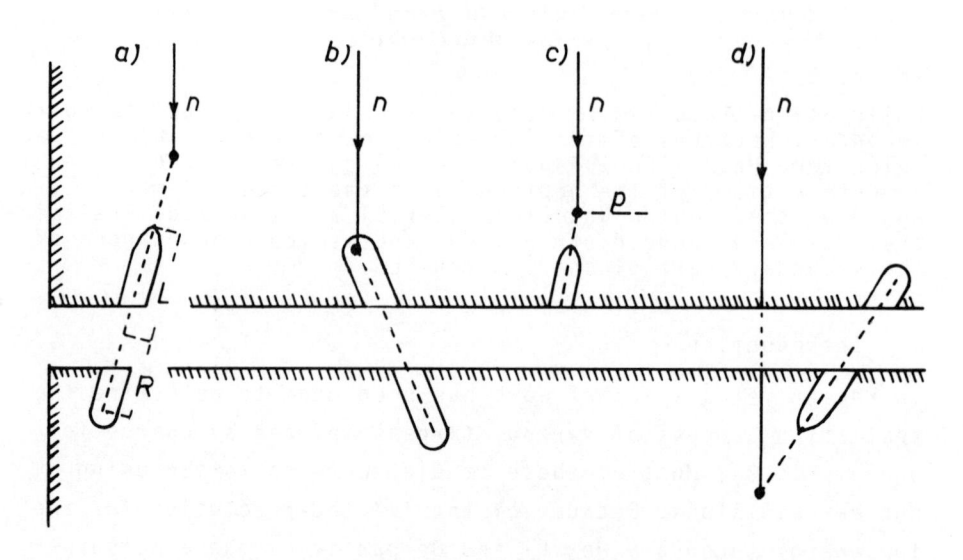

Fig. 1: Images of spallations in plastic detectors

For type a) particles the registration range intersects the detector surface and the particle is emitted in forward direction. The cone length L and the residual range R which are measured to identify the charge and mass are indicated. Type b) is similar, however, the cone has reached the point of interaction and is therefore etched round. Type c) particles stop in that volume near the surface which is etched away and thus no identification can be made. Type d) particles are equivalent to type a) but emitted in backward direction. If the emitted particles have enough energy to penetrate more than one sheet and intersect two surfaces two or three cones are available and the mass identification is improved. Such large energies were detected only in forward direction.

3. Experimental data

The stack consists of 50 sheets of 100 μm thick Kodak Pathé cellulose nitrate (CA 80-15). It was irradiated at the CERN accelerator with secondary neutrons from the 22 GeV/c proton beam for 40 min resulting in a neutron fluence of about $2 \cdot 10^9$ cm^{-2}. The neutron beam penetrated the stack perpendicular with respect to the detector surfaces.

Four sheets were etched at 50°C for 120 min in a 6n NaOH solution with 0,05% Benax surfactant added. The exposed area of about 3.6 cm^2 was scanned under 250x magnification for forward and backward emitted particles of type a) and d). Only those particles were accepted for cone length and residual range measurements which have clearly identified rounded stopping ends and a corresponding pointed cone. The result of these measurements for single cone events in the forward direction is plotted in Fig. 2. A clear charge separation for He, Li and Be is to be seen.

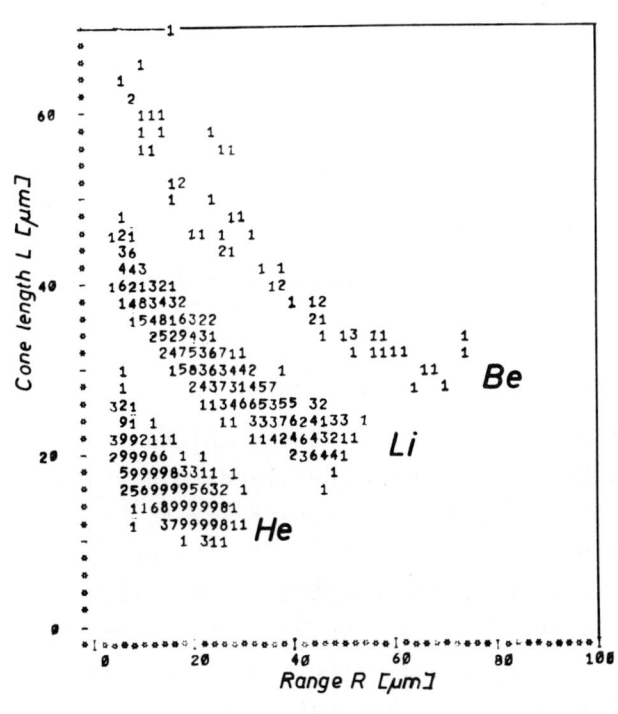

Fig. 2:

Cone length versus residual range measurements. The figure '9' indicates nine or more cones per interval.

4. Charge and mass identification

The track etching rate of the Kodak material was calibrated using the clearly separated peak of ^{7}Be. The calibration procedure and the mass identification method by means of calculated L-R curves is described in ref. (4). Fig. 3 shows the mass histogram for forward emitted Li and Be deduced from Fig. 2.

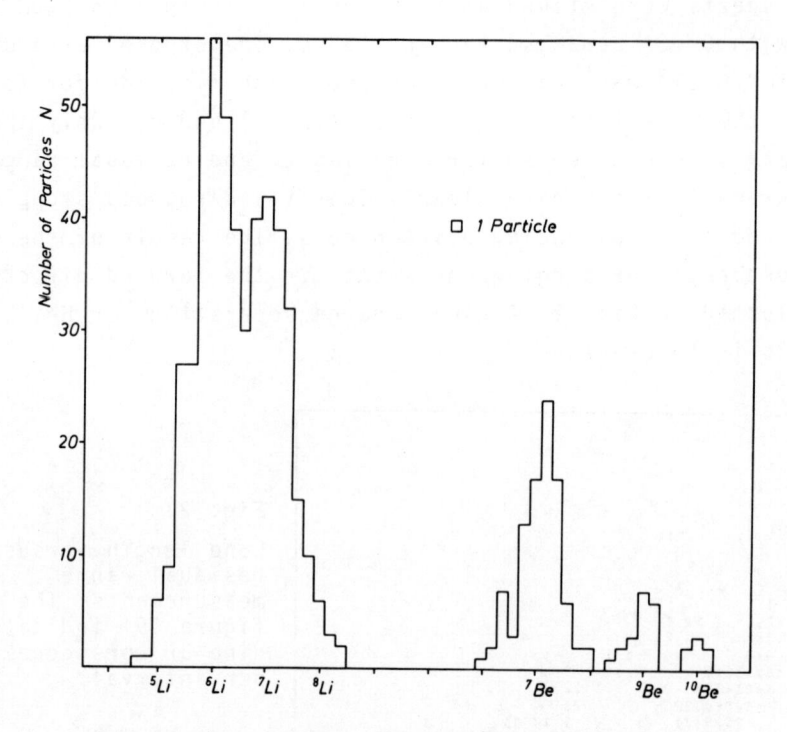

Fig. 3: Mass histogram for secondary Li and Be from nuclear interactions

The detected numbers are not corrected to losses neither from scanning nor from the etching procedure. The standard deviation for the ^{7}Be peak is 0.39 amu. A least square fit to Li yields a standard deviation of 0.42 amu and a ^{6}Li/^{7}Li ratio of 1.28 ± 0.06. This ratio is in good agreement with the value 1.27 calculated from cross sections (5) and weighted according to the CNO contributions in the target material. Corresponding numbers for the measured Be particles are not significant due to the applied method. The emission of a secondary particle with a high mass and a sufficient high energy to be accepted in the analysis is (seldom) unlikely and thus high masses are underestimated.

5. Angular and energy distribution

For every accepted particle the emission angle with respect to the incoming primary neutron was measured where $0°$ indicates the forward direction. The total range and thus the energy of these particles can not be determined without a second etching which would prolong the cones up to the point of interaction (see Fig. 1b). The sum R+L, however, provides a lower limit for the total range of the particles. Due to the chosen etching time specific regions for R+L are covered for the individual elements and some regions are lost. Therefore the integral range distributions from which energy distributions for the individual isotopes may be deduced need complex corrections which are not done up to now.

Thus we present preliminary two dimensional distributions for the emission angle and the sum R+L (Fig. 4, 5 and 6). Near $0°$ and $180°$ the scanning efficiency is reduced while around $90°$ the etching procedure decreases the detectable particle number. Preliminary lower limits for the forward to backward ratio are 3.2, 3.5 and 3.6 for He, Li and Be respectively. Concerning R+L it is to be seen that the number of large R+L values increases towards the forward direction $0°$. This is most obvious for Li and Be and indicates a strong energy dependence of the forward to backward ratio regardless of the required correction to particle losses. As already mentioned multi cone events caused by energetic fragments are detected only in the forward direction.

Fig. 4, 5, 6:

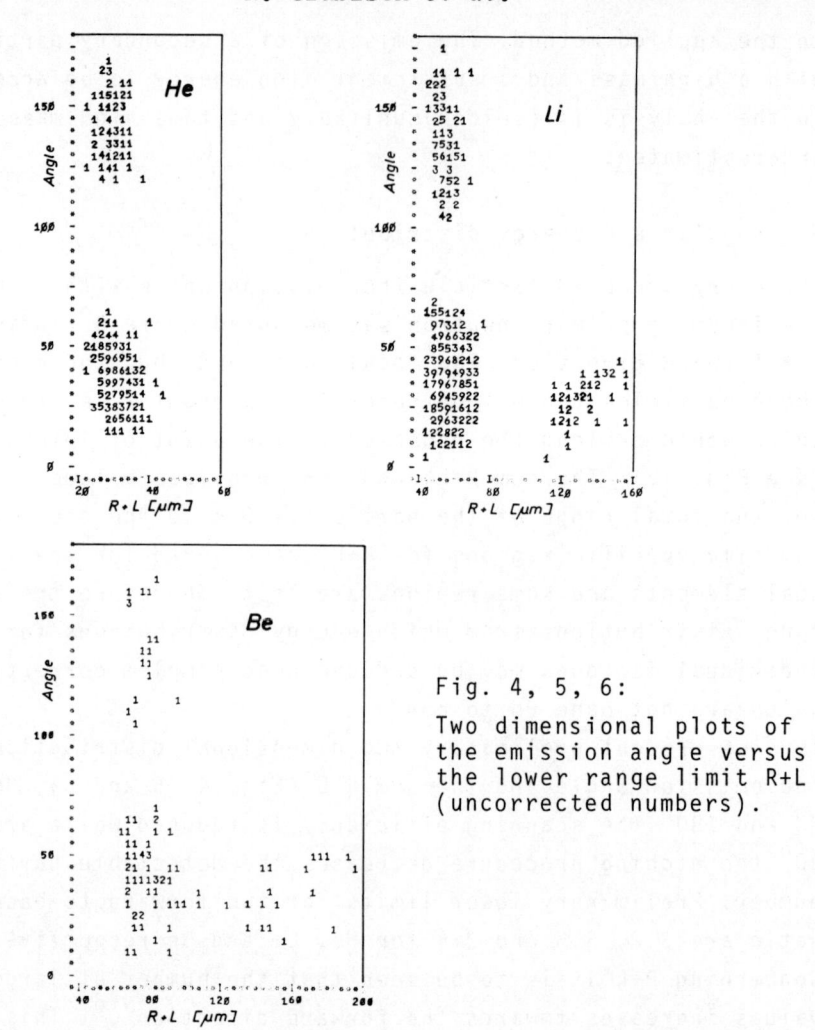

Two dimensional plots of the emission angle versus the lower range limit R+L (uncorrected numbers).

6. Conclusions

The mass of light nuclei which are produced in nuclear interactions of energetic neutrons with the Oxygen, Nitrogen and mainly Carbon nuclei of the cellulose nitrate detector is determined with a resolution of 0.4 amu. Further studies are in progress to correct the measured angular and range distributions for the individual isotopes. Shorter etching times will be applied to extend the measurements to lower energies of the fragments and re-etching will provide the identification of those tracks where the reagent had not reached the end of range in the first etching.

Acknowledgement

The authors like to thank Prof. Dr. E. Bagge for his support and Dr. Baarli, Dr. Höfert and the staff of the Cern accelerator for performing the exposure.

References

1. O. Skjeggestad, S.O. Sörensen,
   Phys. Review, 113, 1115 (1959)

2. S. Katcoff
   Phys. Review, 114, 905 (1959)

3. K. Fukui, W. Enge, R. Beaujean
   Z. f. Physik, A 277, 99 (1976)

4. R. Beaujean, W. Enge
   Fragmentation and Isotope Measurements on Accelerator Neon and Argon Particles of 280 MeV/Nuc.
   To be published.

5. R. Silberberg, C.H. Tsao
   Astrophys. J. Suppl., 220, 25, 315 (1973)

FISSION CROSS SECTIONS OF HEAVY NUCLEI INDUCED BY 300 GeV PROTONS WITH THE HELP OF PLASTIC DETECTOR

M. Debeauvais, J. Tripier and S. Jokic

SADVI, Centre de Recherches Nucleaires, 67037 Strasbourg Cedex

ABSTRACT

Thin targets of heavy elements are sandwiched between sheets of polycarbonate detectors. These makrofolstacks were irradiated with 300 GeV protons at FNAL and permitted the registration in 4π geometry of all the fission fragments in correlation. The incoming proton beam intensity were made by measuring ^{24}Na produced in Al foil with a calibrated proportional counter and so we have obtained the binary and ternary fission cross section of U, Th and Bi. A new kind of event, the break up of the nucleus in 4 fragments was found and studied.

The study (1) of binary and ternary fission of heavy nuclei induced by protons of 0.6 to 23 GeV energy, proved a dependance of the cross sections, binary as well as ternary, with the increase of energy.

When we succeeded in having irradiation at the FNAL with proton of 200 and 300 GeV, it seemed interesting to us to go on with that study.

In this work, we shall limit ourselves in giving the first results obtained on U, Th and Bi targets, without making a complete physical analyze.

EXPERIMENTAL PROCEDURE

We used the "sandwich technic" already put up for other experiments (2), which consist in sandwiching the target between two sheets of polycarbonate detectors of 200 microns thickness.

A mecanical system, allows after development, to reposition the sample, on its initial position, with a precision of a few microns

This way, we obtain a good correlation of the tracks in space.

The thickness of the targets, prepared by evaporation determined simultaneously by weighting and thickness a measure of interferential microscope , vary between 15 and 100 micrograms per cm^2.

The proton flux is measured by the activation of calibrated aluminium monitors and limitating the irradiated surface. The integrated flux measured by means of the ^{27}Al (p, 3pn) reaction is of 10^{11} protons per cm^2. The measurement of the beam intensities were made by measuring the ^{24}Na produced in the aluminium foils with a calibrated end - window proportional counter for which we are grateful to Dr Sam Baker in the department of Dr AWSCHALOM.

After having been developped in a NaOH solution 5 N, 60° C and 40 min, the sandwiches are put in a vacuumed polyethilene bag and observed with a magnification of 220 for the scanning and of 2500 for measurements (see Fig. 1).

RESULTS

See for the results concerning the binary cross sections on table 1.

TABLE 1

BINARY FISSION CROSS SECTIONS σ_b in mb

Target	Ep (GeV)				
	0.6	3	18	23	300
U	1200 ± 130	1330 ± 130	830 ± 120	780 ± 125	870 ± 130
Th	2300	870	850	910	980 ± 140
Bi	250	280	300		300 ± 40

We notice that the cross sections of the 3 studied targets do not change significantly between 18 and 300 GeV.

In a similar study by Hudis and Katkoff, but done on mica, we found a good agreement for U but not for Bi. Their values are far beneath ours: 180 mb at 300 GeV against 300 mb for us. This difference results probably from the various sensitivity of the mica and polycarbonates.

As pointed out by the authors, in the mica are visualized only tracks of $Z \geq 15$ and KE ≥ 10 MeV. The polycarbonates detects ion of $Z \geq 8$ with a length $> 5 \mu$ for the standard development ; and the difference will be greater on less heavy targets.

This effect is put into evidence by the measuring of the mean range of the binary fragments according to targets and the energy. We represent the mean range R and the standard deviation SD in microns. (Table 2)

TABLE 2

MEAN RANGE (R) AND STANDARD DEVIATION (SD) IN MICRON

Energy		U	Th	Bi	Pb
0.6 GeV	R	16.9			15.2
	SD	2.7			3
18 GeV	R	16.1			13.4
	SD	4			5.5
300 GeV	R	15.6	15.05	14.0	11.2
	SD	3.7	4.1	3.7	3.5

We are giving at the same time the corresponding values of the lead fission, to be able to compare with the values found before- and which are to be reported only to the U and Pb (5).

It shows that the range decreases when the energy of the incident proton increases but the same range decreases when the Z of the target decreases, on a similar energy.

These last points could agree with the hypothesis of coulombian repulsion of recoil nucleus, lighter than the targets nucleus and having emitted that more particles that the incident energy is greater.

We noticed, at high energy, ternary fissions which ratio T/B seem to depend on the energy of the incident particle as well as on the nature of the target.

So we measured this ratio at 200 and 300 GeV. The results are to be seen in table 3.

TABLE 3

RATIO OF TERNARY TO BINARY FISSION T/B

Ep GeV

Target	0,6	3	18	23	200	300
U	0,05 ± 0,03	0,8 ± 0,15	2,0 ± 0,3	1,7 ± 0,2	1,35 ± 0,2	1,53 ± 0,2
Th		0,3	2,9	1,6	1,7	2,7 ± 0,3
Bi		0,3		2,2		3,4 ± 0,3

Those values are by far superior than those found by Hudis and Katkoff and cannot be compared because of the difference of sensitivity of the detector.

For the Th and Bi, we only have a few results at 0,5, 3 and 23 GeV.

A complete study was essentially done on U and Pb targets. At 300 GeV, the ratio T/B increases when the Z of the targets decreases.

This increase seems to be in relationship with a new phenomena, that is appearence of breakup of 4 fragments (which we call quaternary events (Q).

The values given here have been found at 300 GeV. For Uranium we found 7 events Q on a total of 798 ternaries such as Q/T ∿ 0,9 ± 0,4 and 1 event with 5 break ups. For the thorium 23 Q on 1070 ternaries such as Q/T ∿ 2,1 ± 0,6 and 1 event with 5 break ups (Fig. 1).

The Q could not be found in mica because of the weaker sensitivity. This phenomena is more frequent when the Z of the target is smaller. There could be an analogy with the ternary events.

We shall continue this work by studying other irradiated targets and eventually in irradiation of proton of 500 GeV energy,

CONCLUSION

The binary cross sections do not seem to vary between 18 and 300 GeV. The ratio of binaries to ternaries (a characteristic of high energy interactions) presents a fundamental difference of behavior with the ratio T/B corresponding to a fission induced by heavy ions which increases with the parameter Z^2/A of the compound nucleus but decreases in our case.

On the other hand, this experiment allowed us to show a privileged mean of desintegration of the heavy nuclei in 4 fragments. This is specific with energies superior than 200 GeV and which calls high excitation energy transfers.

REFERENCES

(1) G. REMY.-
    These C.R.N. Strasbourg (janvier 1974).
(2) M. DEBEAUVAIS, R. STEIN, J. RALAROSY, P. CÜER.-
    Nucl. Phys. A 90, 186 (1967).
(3) J. HUDIS, S. KATCOFF.-
    Phys. Rev. C 13, 5, 1961 (1976).
(4) R. BRANDT, F. CARBONARA, E. CIESLAK, I. JARSTOFF, J. PIEKARZ, R. RINZIVILLO, J. ZAKRZEWSKY.-
    J. Phys., Paris, 31, 21 (1970).
(5) G. REMY, J. RALAROSY, R. STEIN, M. DEBEAUVAIS, J. TRIPIER.-
    Nucl. Phys. A 163, 583 (1971).

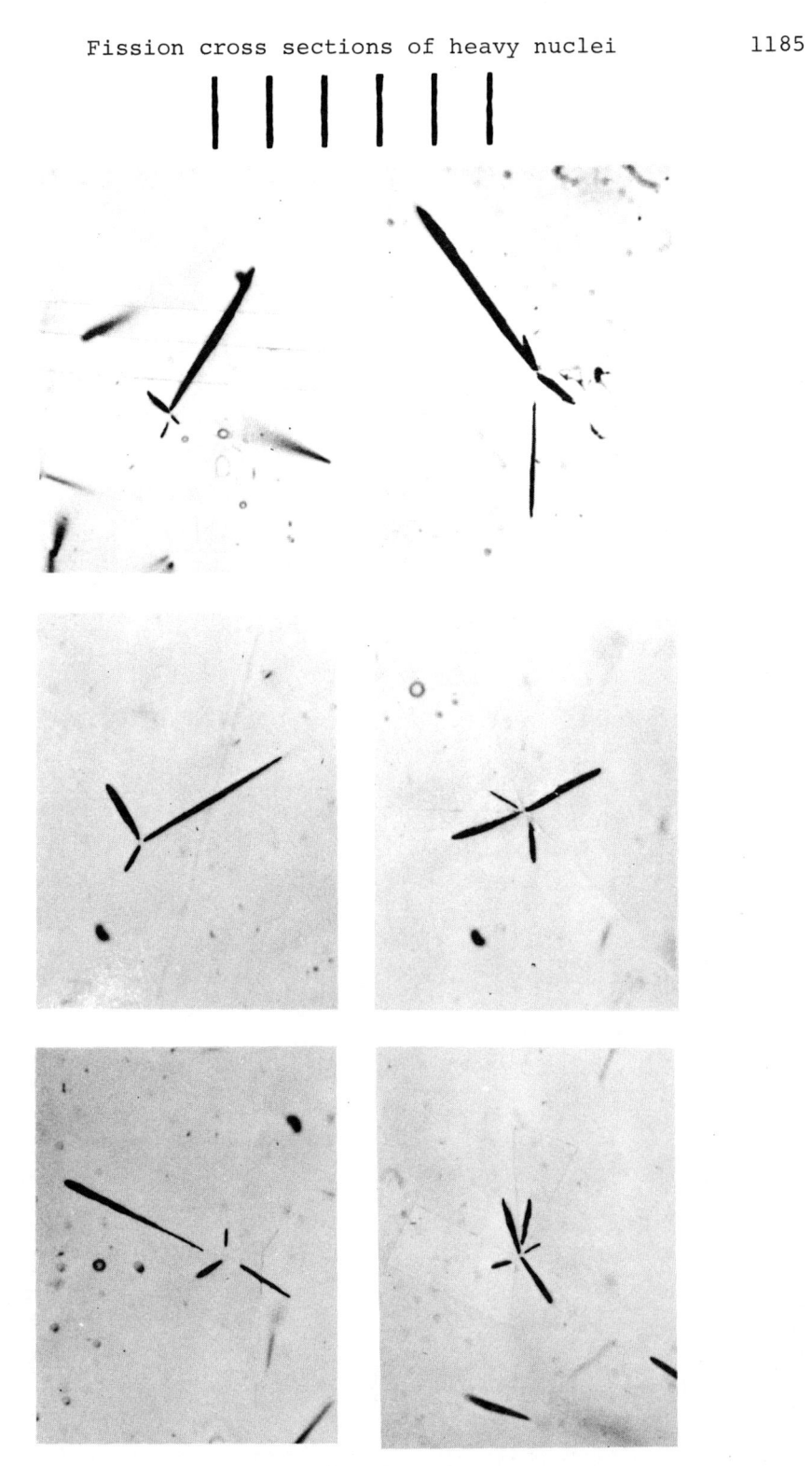

Fig. 1.

FISSION OF Bi, Pb AND Au INDUCED BY 0.65, 1.74 AND 4.12 GeV ALPHA PARTICLES

B. Grabež, Ž. Todorović and R. Antanasijević

Institute of Physics, 11001 Belgrade, SFR, Yugoslavia

Abstract

The binary and ternary fission of Bi, Pb and Au induced by high-energy α-particles has been analyzed using a polycarbonate detector. The fission characteristics have been studied and the corresponding cross-sections have been determined as a function of α-particle energy.

I. INTRODUCTION

In recent years many papers have been published which elucidate from various aspects the fission induced by α-particles of energies up to 150 MeV |1-5|. However, only scarce data are available on fission induced by helium ions of energies exceeding several hundred MeV. Therefore the object of the present study is α-induced fission in high-energy region. Using polycarbonate detectors we determined the cross-sections and fissionability of Bi, Pb and Au at incident α-particle energies of 0.65, 1.74 and 4.12 GeV. The values obtained for the cross-sections were compared with those obtained in the case of proton and deuteron projectiles in the same energy range.

II. EXPERIMENTAL

In the present experiment a polycarbonate detector (a macrofol as provided by the Bayer firm) having a detection limit of $Z \geq 8$ was used as track detector.

Targets were evaporated onto a macrofol foil of a thickness of 200 μm. Another foil, of the same size as that with target, was glued to the latter in such a way that the target was between them. The thickness of the targets was several hundred Å and was determined by weighing.

The sandwiches thus prepared were exposed at Saclay (France) to α-particles. The total flux at energies of 0.65, 1.74 and 4.12 GeV was 6×10^9, 4.7×10^{10} and 3.6×10^{10} respectively.

After exposure the targets were removed by dissolution in corresponding acids. Chemical development of tracks was carried out with a 20% NaOH solution for 45 min. An ultrasonic field of a strength of 1 W/cm^2 ensured reagent diffusion in the course of target dissolution and track development.

Scanning was made with a Zeiss optical microscope at a magnification of about 350 times.

III. RESULTS AND DISCUSSION

III.1. Binary fission

The results obtained in the present work for the binary fission cross-sections of all the three elements at the three energies quoted and those obtained by other authors for fission induced by protons and deuterons in the same energy region are presented in Table I. By binary fission is meant any event with two tracks in correlation having a range $R \geq 2$ μm. The cross-sections were corrected for track loss in target and in the dissolved layer of the detector. The evaluated errors comprise the uncertainty in flux determination (15%) and in target thickness (10%) as well as statistical errors.

As may be seen in Table I, the results obtained show that the cross-sections of all the three elements are constant within the limits of error in the given energy region.

Table I

	Project.	Kinetic energy (GeV)	Bi	Pb	Au
Binary fission cross section (mb)	$_2He^4$	0.65	437.20±138.05	253.78±77.33	200.80±67.92
		1.74	368.00±122.66	245.01±71.65	187.76±54.34
		4.12	383.28±108.41	269.79±78.24	203.43±58.26
	$_1d^2$	2.1	323 ± 60[a]	182 ± 40[c]	157 ± 20[a]
	p	0.6	216 ± 33[b]	134 ± 18[c]	59 ± 9[b]
		1	191 ± 30[b]		66 ± 10[b]
		2	165 ± 25[b]	139 ± 20[c]	76 ± 11[b]
		2.9	227 ± 33[d]	149 ± 23[d]	93 ± 15[d]
		3	161 ± 24[b]	135 ± 26[c]	76 ± 11[b]

[a]reference|9|; [b]reference|6|; [c]reference|7|; [d]reference|8|

Compared to the values obtained for the same elements in the same energy region for interaction with protons|6,7,8| and deuterons|9|, ours are larger by about a factor of 2 and by 15-35% respectively.

Figures 1a, 1b and 1c show the binary fission cross-section (σ_B) of Bi, Pb and Au as a function of the incident α-particle energy obtained in this experiment and those obtained by other authors at lower incident particle energies. From the figures it may be seen that the energy dependence of the Bi, Pb and Au binary fission cross-sections in the case where α-particles are projectiles is similar to that in the case where incident particles are protons and deuterons. In previous papers|5| it has been pointed out that there are differences in values between the cross-sections for binary fission induced by protons and by α-particles of energies up to 150 MeV, but that their behaviour with energy variation is analogous. According to the present results, differences in value and an analogy in behaviour of the binary fission cross-sections of Bi, Pb and Au are apparent also at high energies.

Table II

E_α (GeV)	σ_F/σ_R (%)		
	Bi	Pb	Au
0.65	15.5	10.6	8.5
1.74	18.2	10.4	8.1
4.12	15.7	11.2	8.5

Table II gives the values of the σ_F/σ_R ratio (fissionability). The total reaction cross-section for the He^4 + target reaction are calculated according to the soft sphere model|10|. The results obtained show that the correlation of fissionability and target atomic number in our case (where α-particles are projectiles) is similar to that in the case of target interaction with protons|8| and deuterons|9|. It should be noted that our values for fissionability are larger by several percentages than those obtained in experiments with incident protons|8| and deuterons|9|.

III.2. Ternary fission

Table III shows the values obtained for the ratio of ternary to binary events (T/B) and those obtained previously for other projectiles.

It may be seen that in the present experiment as well as in those with protons|11| the T/B ratio is found to increase with energy. At an incident particle energy of E_α = 0.65 GeV no ternary event was detected; the data on the interaction

Table III

	Project.	Kinetic energy (GeV)	Bi	Pb	Au
Ratio ternary to binary events (in percent)	$_2He^4$	0.65	–	–	–
		1.74	0.85*	–	0.93*
		4.12	1.4±0.51	2.54±0.79	3.67±0.87
	$_1d^{2\,a}$	2.1	1.83±0.24	1.97±0.25	1.96±0.3
	p^b	0.6		0.15*	0.8*
	p^c	2–3	0.16	–	0.22
	p^b	3		1.8±0.2	7.3±0.6

*poor statistics; [a]reference|9|; [b]reference|11|; [c]reference|12|

with protons|11| at an energy of E_p = 0.6 GeV show that the yield of ternary events is low. At an energy of E_α = 4.12 GeV our values for the T/B ratio show that it decreases with increasing atomic number of the target, which is in accordance with the behaviour observed in experiments with protons|11| and deuterons|9| as projectiles. The similarity in behaviour of the relative yield of ternary events as a function of target mass and projectile energy in the cases where α-particles, deuterons and protons are incident particles indicates the same mechanism of production of ternary events.

CONCLUSION

The fission of Bi, Pb and Au induced by 0.65, 1.74 and 4.12 GeV alpha-particles has been studied. The results obtained show that there are differences in value but a similarity in behaviour of the cross-sections for binary fission and fissionability in comparison with those for fission induced by protons and deuterons. The relative yield of ternary events increases with decreasing target mass and increasing projectile energy, which is in analogy to the behaviour observed in experiments with protons and deuterons.

REFERENCE

1. D.S.Burnett, R.C.Gatti, F.Plasil, P.B.Price, W.C.Swiatecki and S.G.Thompson, Phys.Rev., B134, 952 (1964)
2. J.Gindler, H. Münzel, J.Buschmann, G.Christaller, F.Michel and G.Rhode, Nucl. Phys., A145, 337 (1970)
3. J.R.Huizenga, R.Chaudry and R.Vendenbosch, Phys.Rev. 126, 210 (1962)
4. J.Ralarosy, These, Strasbourg (1972)
5. R.Bimbot et Y.Le Beyec, Journal de Phys., 32, 243 (1971)
6. J.Hudis and S.Katcoff, Phys.Rev. 180, 1122 (1969)
7. G.Remy, J.Ralarosy, R.Stein, M.Debeauvais and J.Tripier, Nucl. Phys., A163, 583 (1971)
8. R.Brandt, F.Carbonara, E.Cieslak, H.Piekarz, J.Piekarz and J.Zakrzewski, Rev.Phys.Appliquée, 7, 243 (1972)
9. F.Rahimi, D.Gheysary, G.Remy, J.Tripier, J.Ralarosy, R.Stein and M.Debeauvais, Phys.Rev. C8, 1500 (1973)
10. P.J.Karol, Phys.Rev., C11, 1203 (1975)
11. G.Remy, J.Ralarosy, R.Stein, M.Debeauvais and J.Tripier, Journal de Phys., 31, 27 (1970)
12. S.Kactoff and J.Hudis, Phys.Rev.Lett., 28, 1066 (1972).

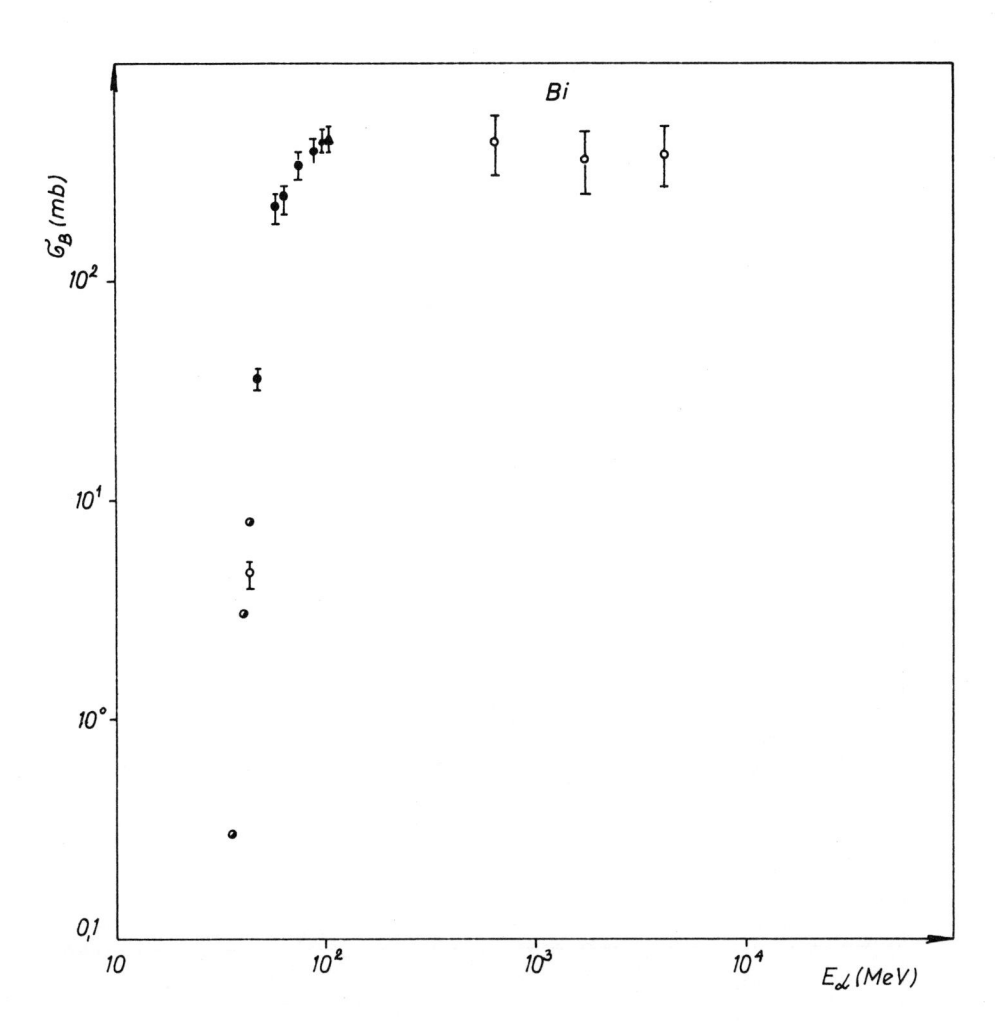

Fig. 1a. – Binary fission cross-section of Bi versus He^4 ion energy: mica detectors, ▲ Ref.|2| ; ◐ Ref.|3| ; macrofol ● Ref.|4| ; ○ present work.

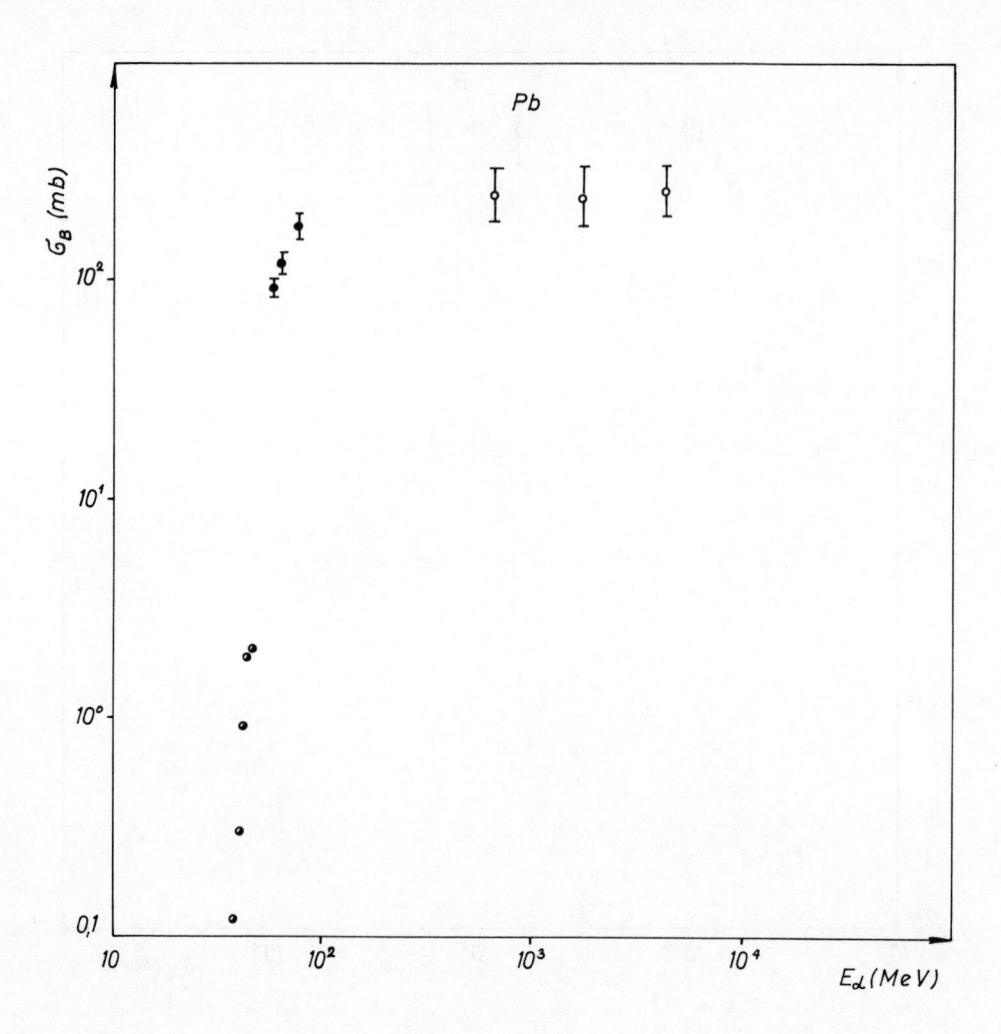

Fig. 1b. - Binary fission cross-section of Pb versus He^4 ion energy: notation the same as in Figure 1a.

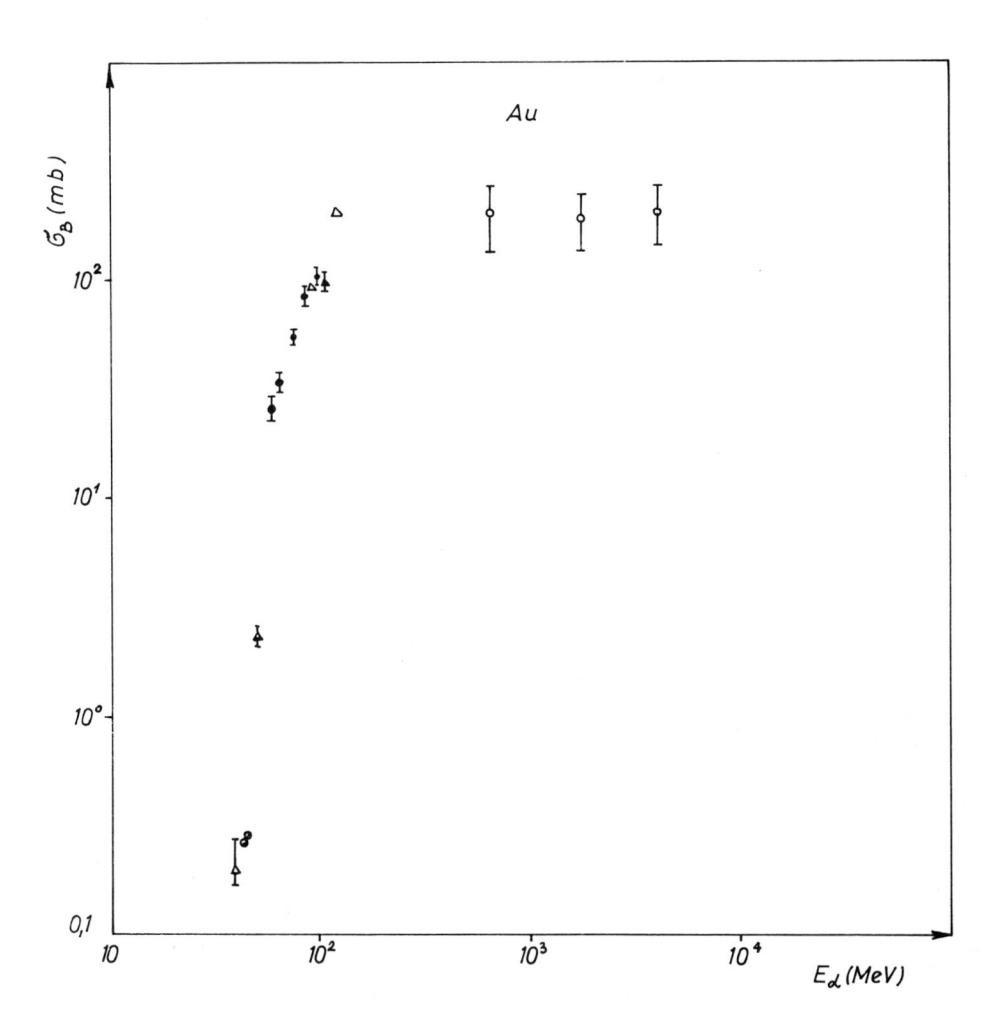

Fig. 1c. – Binary fission cross-section of Au versus He^4 ion energy: mica detectors, △ Ref.[1]; further notation the same as in Figure 1a.

FRAGMENTATION AND ISOTOPE MEASUREMENTS ON ACCELERATOR NEON AND ARGON PARTICLES OF 280 MeV/nuc

R. Beaujean and W. Enge

*Institut für Reine und Angewandte Kernphysik, University of Kiel,
23 Kiel, West Germany*

Abstract

Accelerator Argon and Neon particles of 280 MeV/nuc were stopped in stacks of 250 um Daicel cellulose nitrate visual track detector sheets. The nuclear charge and mass of 172 particles from the Neon exposure and of 23 particles from the Argon exposure was determined by means of the cone length versus residual range method using the REL criterion. A mass resolution of 0.41 amu for Ne^{20} and 0.25 amu for Ar^{40} was obtained. The influence of fragmentations within the registration range on the particle identification is discussed.

1. Introduction

Besides many other detector systems stacks of nuclear track plastic detector sheets can be used to measure the charge Z and the mass A of stopping ionizing particles. After developing the latent track by means of etching cone length L and residual range R can be measured under an optical microscope along the particle's track. One advantage of such a detector system is the fact that several L-R couples can be measured and thus a multiple dE/dx versus E analysis can be made. In this work the response and the resolution of this detector system are studied.

The stacks consist of 100 sheets of Daicel cellulose nitrate each of the size 250 µm x 5 cm x 5 cm, and they were exposed at the Princeton accelerator. The energy of the particles was about 280 MeV/nuc and appropriate absorbers were used to stop the Neon and Argon ions in stack A and Stack B respectively. For further details see Filz et al. (1).

In stack A an area of about 1 cm was cut from sheets 1 to 20 and etched for 240 min at 40 °C in a mechanically stirred 6n NaOH solution with 0.05 % Benax surfactant added. The etching of one quarter in area of stack B was performed in the same reagent but divided into three sets with different etching times according to the increasing track etching rate towards the stopping end. At residual ranges R < 4000 μm 4000 μm < R < 8000 μm and R > 8000 μm the etching time was 40 min, 120 min and 240 min respectively. About every third sheet remained unetched for future studies.

2. Calibration and mass determination

The cone length and range measurements were performed under a Leitz microscope with a 100x objective using inductive displacement transducers. The digital output was analysed with a computer and yielded the wellknown L-R plots.

The response of a plastic track detector can be expressed by the track etching rate v_t as a function of the Charge Z, mass A and velocity $\beta \cdot c$ of the incoming particle. The different track formation models try to reduce these three values to one particle independent expression. In this work we apply the model of the restricted energy loss REL according to Benton (2) with $\omega_0 = 1$ keV.

The track etching rate v_t, the total etching time T and the cone length L are linked by

$$L = \int_0^T v_t \cdot dt \quad \text{or} \quad T = \int_R^{R+L(R)} \frac{dR}{v_t(R)} \quad (1)$$

Eq. (1) are integral equations for the track etching rate $v_t(R)$ if the cone length versus residual range dependence L(R) is known. They may be solved either by an iteration procedure starting with $v_t(R+L/2) = L(R)/T$ or by a least square fit of a polynom $g(R) = 1/v_t(R)$ (3). Using the theoretical REL(R) function the calibrated response v_t(REL) is established. Hence, L(R) curves can be calculated for other ions using v_t(REL), eq. (1) and REL(R) functions. This is done for the most abundant and the adjacent isotopes of those elements which are apparent in the L-R plot. Excluding isotopes with a high neutron deficit or excess compared to the most abundant isotope each element can be clearly separated covering approximately three and three isotopes on the neutron poor and rich side of the reference isotope.

For every L-R couple a mass value is obtained by an interpolation between two adjacent isotope lines which are calculated following the above described method. Thus several mass values are obtained and the distribution of the individual cones to the mean value of the track may be calculated. The normal deviation equivalent to approximately 0.7 amu includes mainly the local changes in the response due to the manufactoring process and a subjective personal error of 1 - 2 μm which is equivalent to about 0.3 amu.

According to the difference ΔL of adjacent isotopes lines different weight factors ($\Delta L)^2$ are assigned to the individual cones. Then a 3σ cut-off procedure is applied which effects about 6 % of the cones in the utilized Daicel CN and thereafter the mean mass for the track is calculated.

The range measurements along a track are not independent but biased mainly with the error of the range R_1 of the largest cone (at $R < R_1$ the cones were etched together and no information on the dE/dx can be obtained). This error is included in all subsequent ranges R_2, R_3 etc. and is equivalent to about 0.15 amu. Therefore the error of a multi cone event is expected to be

$$\sigma^2 = \sigma_R^2 + \sigma_c^2/N \qquad (2)$$

where σ_R and σ_c are the range and individual cone length uncertainties and N is the number of included cones per track. As σ_c was found to be 0.6 to 0.8 amu in all experiments with Daicel CN the mass resolution depends mainly on the number of cones which increases with the nuclear charge of the stopping particle due to the increasing registration range.

3. Fragmentation within the registration range

A disadvantage of a long registration range is the increasing likelihood for fragmentations. Such events with proton and/or neutron losses may result in an incorrect particle identification when occuring within the registration range. Let us consider the influence of the two cases in more detail.

a) $(A,Z) \rightarrow A(A-1,Z)$: The loss of one neutron at the range R_o will shift all measurements at ranges $R > R_o$ to the left in the L-R plot compared to the undisturbed track. This range shortening is proportional to R_o/A and decreases the deduced mass value at the range R by R_o/R mass units. In the region $R > R_o$ the primary particle shows a weak tendency of slowly decreasing mass values with decreasing residual ranges but this will usually be concealed by the normal cone length fluctuations.

b) $(A,Z) \rightarrow (A-1,Z-1)$: The loss of one proton at the range R_o can easily be detected due to the suddenly decreased energy loss if the cones are still separated. If this is not the case the range extension shifts the measurements at $R > R_o$ to the right in the L-R plot and a particle with increasing mass at decreasing range seems to be detected. For one proton loss this tendency is rather large and the particle can be rejected. But in connection with the loss of several neutrons the range extension might be compensated and no unusual tendency or fluctuation may be seen.

4. Results for Neon

A total of 172 particles are analysed in stack A and 150 Neon, 7 Fluorine, 8 Oxygen and 1 Nitrogen particles are identified. 6 tracks are rejected due to large fluctuations and among those 150 Neon tracks only 118 are accepted for the final mass analysis. The rest of 32 particles has a regression of more than 1.4 amu in the mass to range relation.

Fig. 1a shows the mass distribution of the accepted Neon particles, Fig. 1b the mass histogram for the rejected ones (four particles fall outside the diagram). In Fig. 1a each particle covers a small histogram of unit area which represents a Gaussian distribution around the mean mass of the track with a standard deviation according to eq. (2).

The calibration procedure is performed in such a way that the peak of the distribution fits to Ne^{20}. A least square fit to the distribution yields a standard deviation of 0.41 amu and the indicated numbers for Ne^{19}, Ne^{20} and Ne^{21} (dashed line). The average number of cones is N=6, the mean standard deviation for the cones along a track is $\sigma_c = 0.67$ amu. It is obvious that the multi dE/dx method based on several cones improves the resolution although the obtained error is larger than the value 0.31 amu expected from eq. (2). As the measured particles were selected at random they are believed to be an unbiased representative set. Hence the measured number 0.06 ± 0.02 for the ratio Ne^{19}/Ne^{20} may be compared with the expected one. Taking a cross section of 50 mb for the reaction $Ne^{20} \rightarrow Ne^{19}$ in Hydrogen at 100-200 MeV/Nuc. and a target factor $A^{1/4}$(4,5) a rought estimate for the expected number of Ne^{19}/Ne^{20} in back of 8.6 g/cm^2 is 0.05 which is in good agreement with the measured value.

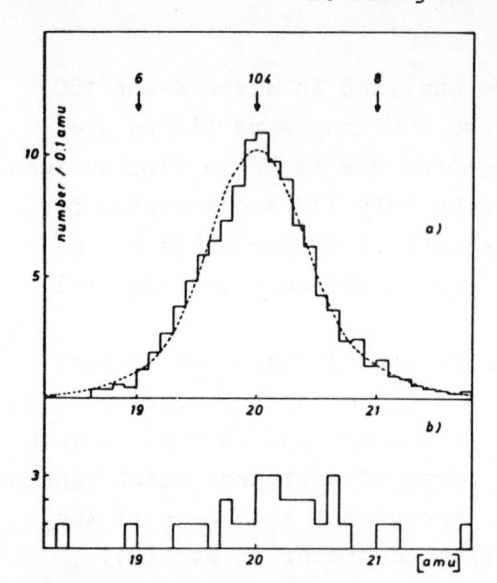

Fig. 1. a) Mass distribution of 118 Neon particles with a least square fit (dashed line).
b) Mass distribution of 32 rejected Neon particles.

5. Results for Argon

The resolution in that part of stack B etched for 40 min was found to be too poor for a mass determination due to the short etching and the saturation of the etching rate versus REL. Thus Fig. 2 shows the result for the 120 min (a) and 240 min etching (b) and the mass difference between these two measurements for the individual tracks. 19 particles are plotted among those 16 are identified as Ar^{40} whereas one or two entered the stack as Ar^{39} and the rest may undergo not identified fragmentations. Taking 60 mb for the neutron loss cross section the expected number of Ar^{39} in back of 3.4 g/cm^2 at the top of the stack is one particle.

The obtained resolution of 0.23 and 0.26 amu calculated for the above mentioned 16 Ar^{40} particles is in good agreement to the expected values. The average number of cones per track is N=9, the mean cone length fluctuation is equivalent to 0.82 and 0.68 amu for 120 and 240 min respectively. Although the distance ΔL of two adjacent isotope lines in the L-R plot for 240 min etching time is twice as much compared to 120 min, the resolution calculated from cone length fluctuations is not twice as good. Thus a relation of decreasing fluctuations with decreasing etching time may be deduced but the result for the short etching of 40 min claims a somewhat constant minimum value in the utilized Daicel CN.

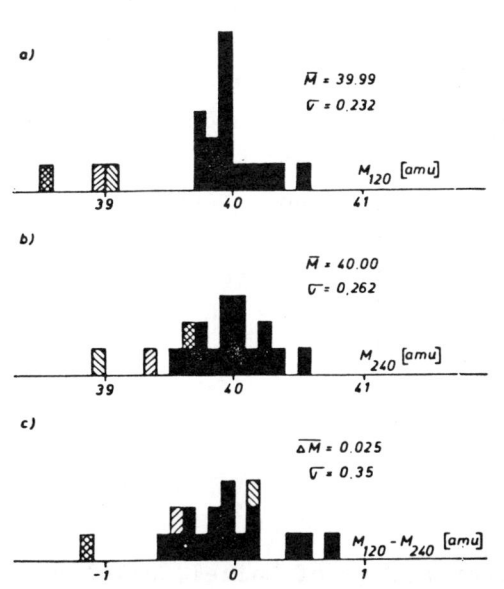

Fig. 2. Argon mass histogramm for 120 min (a) and 240 min (b) etching. The black area includes 16 Ar^{40} particles which were taken for the calculation of the mean mass and the standard deviation. The rest of three particles is individually shaded.

c) Mass difference between the 120 min and 240 min measurements for the individual tracks. The mean and the S.D. is calculated from 18 tracks.

Fig. 3 shows the fragmentation of four Argon particles within the registration range in stack B. Particle 1 survives as Cl^{35} and it is a good example for the compensation of neutron and proton loss which results in a rather small range change compared to the primary particle. Without the shortened etching time of 120 min this fragmentation would not be detected. Particle 2 reacts at a residual range of about 8700 μm and is not registrated down to 3700 μm. There two small cones are available in a 120 min etched sheet and the track is identified as Aluminum with mass 25 or 26. From the shift in the 240 min data a range extension of about 1500 μm can be estimated and this is the amount comming out for Al^{25} assuming a range relation proportional to A/Z^2. Particles 3 and 4 undergo not clearly identified events but they are separated from Ar^{40}.

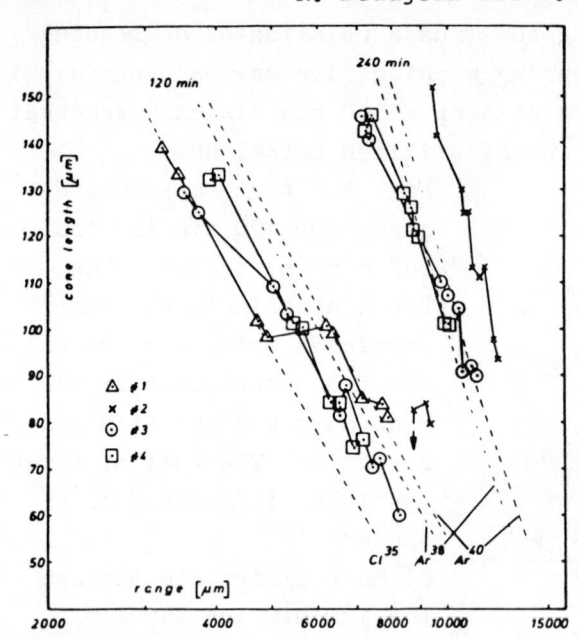

Fig. 3. L-R couples along tracks which undergo fragmentations. The plotted L-R curves are calculated using v_t(REL) and eq. (1). The corresponding measurements of the individual tracks are connected to guide the eye.

6. Conclusion

It can be stated that a detector system of Daicel cellulose nitrate sheets is capable to resolve isotopes with a resolution of 0.41 amu at nuclear charge $Z = 20$ and 0.25 amu at $Z = 18$. The utilized commercial Daicel material however shows a relative large fluctuation of the cone length which is compensated by the excellent steep v_t(REL) function. Local changes in the response mainly contribute to this fluctuation while the subjective personal error is less important but not negligible. A different response of the two surfaces was not detected.

The method of a multi dE/dx analysis improves the resolution with increasing number of cones. The residual range measurements along the track however are not independent and therefore the deduced mass of the particle strongly depends on the residual range R_1 of the largest cone. This range may be affected by fragmentations.

Acknowledgement

The authors like to thank Prof. Dr. E. Bagge for his support and R.C. Filz and the staff of the Princeton accelerator for the performance of the exposure. This work was supported by the Deutsche Forschungsgemeinschaft.

References:

1) R.C. Filz, P.J. McNulty and A.F. Davis, AF Cambridge Research Lab., Rep. AFCRL-72-0321 (1972)

2) E.V. Benton, U.S. Naval Radiological Defense Lab., Rep. USNRDL-TR-68-14 (1968)

3) G. Siegmon, K.-P. Bartholomä and W. Enge, Nucl. Instr. and Meth. 128 (1975) 461

4) R. Silberberg and C.H. Tsao, Astroph. J. Suppl., 220, 25 (1973) 315

5) P.J. Lindstrom, D.E. Greiner, H.H. Heckman, B. Cork and F.S. Bieser, LBL report 3650 (1975)

MULTI-FRAGMENT DECAY REACTIONS INDUCED BY HEAVY IONS AND STUDIED WITH MICA TRACK DETECTORS

P. Vater*, H.-J. Becker*, R. Brandt* and H. Freiesleben**

*Institut für Kernchemie, Philipps-Universität, F.G.R.
**Fachbereich 13 Physik, Philipps-Universität, D-3550 Marburg, F.G.R.

Abstract: The 2π-mica-technique was used to study binary, ternary, and quarternary decay in heavy ion (i.e. ^{56}Fe, ^{84}Kr, ^{136}Xe and ^{238}U) induced reactions. In the case of ^{84}Kr, the phenomena have been studied in some detail, whereas in the case of ^{136}Xe and ^{238}U the results are rather preliminary. In the reactions induced by ^{56}Fe and ^{84}Kr in U, ternary decay is the dominant decay mode. Quarternary decay is a rather rare decay mode.

Introduction:

Ternary decay reactions are considered in this paper as being defined by three rather massive break-up partners (≥ 30 u) in the exit reaction channel. (Sometimes, such reactions are also called "ternary fission", but such a term shall be reserved for those reactions, where all three partners are emitted from a compound nucleus). Binary or quarternary decay reactions are defined in an analogous manner: two or four rather massive reaction partners (≥ 30u) in the exit reaction channel. The lower limit of m=30u for

*Present address: GSI-Gesellschaft für Schwerionenforschung
D-6100 Darmstadt

the reaction partners is the threshold for registration of tracks in mica-however, this arbitrary limitation has been used widely and with some success.

The field of ternary fission and ternary decay has been reviewed[1], additional experimental results are given in Ref. 2-4 . The essential result is that ternary decay induced by ions as heavy as ^{40}Ar is a <u>small part</u> of the total reaction cross-section, and it is of the order of a few percent or less. This small contribution of ternary decay to the total reaction cross-section is somewhat surprising, as a rather simple estimation of the reaction energy released in a n-fold break-up of a composite system reveals the following: For systems heavier than about 200μ a multi-fragment (n > 2) decay releases more kinetic energy than the usually observed binary decay (Ref. are given in 1).

In this paper we describe recent experiments with ^{56}Fe, ^{84}Kr, ^{136}Xe and ^{238}U. The 2π-mica technique is always employed. The irradiations with 540 MeV ^{56}Fe and 800 MeV ^{84}Kr were carried out at the LINAC in Manchester. The experiments are nearly completed. The irradiations with 1080 MeV ^{136}Xe were performed at the SUPERHILAC at Berkeley, and irradiations with 1740 MeV ^{238}U at the UNILAC in Darmstadt. For the last two reactions only preliminary results are given.

II) Experimental technique

The 2π-mica technique is schematically shown in Fig.1. We irradiate with heavy ions (such as ^{84}Kr, ^{136}Xe, or ^{238}U) a mica-foil, covered with appr.1 mg/cm^2 UF$_4$. Only about 10^6 ions/cm^2 are used. After irradiation, the target material is dissolved and heavy ion tracks are developed by etching with HF. Afterwards, one observes with a microscope one-pronged events (mostly dots) due to the primary heavy ion, and also two-pronged, three-pronged and (very rarely) four-pronged events are observed (Figures are shown later). The technique has been described in some detail in Ref.5. It is possible to separate two-pronged events due to nuclear interactions from Rutherford scattering, as shown in Ref.5. In this paper, we only consider events due to nuclear interactions. Furthermore, this technique gives absolute cross-sections in a straight-forward manner.

In the irradiation with ^{56}Fe, we could use a much higher heavy-ion flux, as it is possible to anneal the tracks due to 540 MeV ^{56}Fe without annihilation of fission fragment tracks[3].

III) Experimental results

III.1) Binary decay reactions (BD) in the interaction of 800 MeV ^{84}Kr with ^{238}U

As the detailed study of nuclear binary decay reactions is quite elaborate with the 2π-mica-technique, we report here only results for the above-mentioned interaction. For all 389 two-pronged events observed the spatial coordinates of their two tracks are measured:

1) The angle α between the tracks in the plane normal to the ion beam.

2) The true track-lengthes (l_1, l_2) were calculated from the measurement of the projected length of the tracks and their depth.

3) The angles ξ_1, ξ_2 between the track and the beam direction were determined.

For 235 events a value of $\alpha \leq 165°$ was measured, and they are considered in Sect.III.2. 154 events with $165° < \alpha \leq 180°$ were considered as possible candidates for binary decay reactions of a compound system. For this, the experimental values l_1, l_2, ξ_1, and ξ_2 were used in trials to fit such an event to the kinematics of two-body decay of a composite system (Details are given in Ref.6,7). It was possible to achieve such a fit for 26 events. Fig.2 shows an example of such a two-pronged event due to a binary break-up. Fig.3 gives the distribution of track-lengths l and angles ξ.

The track-lengths are considerably larger than those originating from fission induced by thermal neutrons on uranium. Fig.5 gives a two-dimensional representation of TKE (total kinetic energy in the c.m. system) vs. the mass-ratio of the two fragments. The uncertainty in the mass-determination is appr. 15u, and in TKE it is appr. 20 MeV.

As one can see, all mass-ratios between a symmetric split of the composite system and $m_1/m_2 = 84/238 = 0.35$ are observed, indicating a rather broad mass-distribution in binary decay reactions. This agrees with radiochemical results[8]. As we study only the 2π-geometry, we can give only a lower limit for a binary decay cross-section $\sigma(BD)$:

(1) $\quad \sigma(BD) \geq (135 \pm 30)$ mb

III.2) Ternary decay reactions (TD) in the interaction of 800 MeV ^{84}Kr with ^{238}U

Three-pronged events are observed in large abundance. One of 97 events found is shown in Fig.5. From these calculate a cross-section for observed ternary decay $\sigma(T_{OBS})$:

(2) $\quad \sigma(T_{OBS}) = (381 \pm 50)$ mb

(3) $\quad \sigma(T_{OBS}) = (0.15 \pm 0.02)\sigma_R$

The total reaction cross-section σ_R has been calculated: $\sigma_R = 2.7$ b. Furthermore, the distribution of projected angles α in the plane normal to the beam for all two-pronged events is shown in Fig.6. One observes a rather broad distribution. For $\alpha < 180°$ one obviously needs for reasons of momentum conservation a third reaction partner. It has been estimated in Ref.6, that for $\alpha \leq 165°$ this third partner must be in most cases rather massive with $m > 30$ u; thus we consider events with $\alpha \leq 165°$ as being due to ternary decay $\left(B^{\alpha \leq 165°}\right)$. This yields as a lower limit for the total ternary decay cross-section $\sigma(TD)$

(4) $\quad \sigma(TD) \geq \sigma(T_{OBS}) + \sigma(B^{\alpha \leq 165°}) = (1.64 \pm 0.25)$ b

Thus, the limits for ternary decay can now be given as follows:

(5) $(0.60 \pm 0.15)\sigma_R \leqslant \sigma(TD) \leqslant (0.95 \pm 0.02)\sigma_R$

The upper limit is due to **binary decay** (BD), as mentioned before. Due to two effects we cannot define these limits narrower:

1) one-pronged events can originate from ternary decay with two fragments going backward in the laboratory. Such events cannot be analysed with our technique.

2) two-pronged events with $165° < \alpha \leqslant 180°$ can be due to ternary decay with one fragment going backward in the laboratory. Such events cannot be analysed with certainty so far, unless they have been identified as nuclear binary decays. (Sect.II,1)

Nevertheless, ternary decay has been established as a dominant reaction channel. This agrees with radiochemical results (Ref.8), here this decay mode has also been termed "quasi-ternary-fission" etc. Similar results are obtained by radiochemical methods and the mica-technique in an analysis of the reaction (540 MeV Fe+U), as shown in Ref.3,4,9. Table 1 summarizes the results obtained so far.

The reason for this large contribution of ternary decay in the reaction (800 MeV ^{84}Kr+U) will now be discussed: The angular distribution and track-length distribution for 97 T_{OBS}-events are given in Fig.7 and 8. There it is shown, that for each individual T_{OBS}-event one has one long track and two shorter tracks. Fig.7 shows that the longest tracks have a range between 18 μm and 62 μm, whereas the shorter tracks have a considerable shorter range between 2 μm and appr. 20 μm (this is approximately the range of fission fragments originating from thermal neutron induced fission in uranium). The angular distribution of the longest track is centered around the "quarter-point" angle, while the shorter tracks have a much wider angular distribution. These facts are plausibly explained as follows (Fig.9).

In a deep-inelastic collision (DIC) of the Kr-ion with the U-nucleus the Kr is scattered preferably near the "quarter-point" angle. The heavy partner becomes excited in this collision and fissions in the usual binary decay mode. This two-step mechanism in ternary decay is in agreement with other interpretations of such reactions[8,9].

Fig.10 shows a two-dimensional representation of the track-length of the longest tracks l vs. the angle ξ for such tracks. One observes that for large values of l the angular distribution is rather narrow, for lower values of l the angular distribution in ξ becomes wider. This again is in qualitative agreement with the "deep-inelastic-collission" (DIC) reactions: There is a smooth transition between DIC-reaction involving little energy transfer and a focussing to a grazing angle and large energy transfer (i.e. small value in l) and a wide angular distribution.

III.3) Quarternary break-up reaction in the interaction of ^{56}Fe, ^{84}Kr, ^{136}Xe, and ^{238}U with Uranium

Recently, a rather rare decay process has been observed in the interaction of heavy ions ($>50u$) with uranium. Fig.11a gives an example of quarternary decay (Q_{OBS}) in the interaction of 800 MeV ^{84}Kr with ^{238}U. As we see all four tracks in mica, we call this an "direct" quarternary decay event. Fig.11b gives an example of quarternary decay (Q_{IND}) in the interaction of 1080 MeV ^{136}Xe with ^{238}U. We see only three tracks in mica, however, we can infer from momentum conservation, that a fourth partner must have been emitted. Such events are defined as "indirect" events. Table 2 gives a summary of all examples of quarternary decay observed so far by us. This decay rate is rather rare, however, in the Kr-induced interactions one event represents 4 mb cross-section. For the sake of completeness, Fig.12 gives the distribution of track-lengths and angles for Kr- and Xe- induced quarternary decay events. No conclusion shall be drawn at this stage from the data, not even whether quarternary decay is a one-step or two-step or three-step process (analog to ternary decay as shown in Fig.9).

Fig.13 gives our first example of a four-pronged event seen in the interaction of 1750 MeV ^{238}U with ^{238}U. The irradiation has been carried out at the UNILAC (GSI, Darmstadt). We observe two pairs of tracks, each with $\alpha \sim 180°$. One pair of tracks is rather flat, the other rather steep. Here quarternary decay could be a rather trivial process: The two uranium nuclei interact to some extend, both nuclei get excited and they both fission into two fragments, yielding in total four fragments.

Obviously, this process has to be studied further, before any firm conclusions can be made.

Table 1 : Summary of observed cross-sections

Reaction	$\sigma(BD)$	$\dfrac{\sigma(T_{OBS})}{\sigma(T_{OBS})+\sigma(B_{OBS})}$	$\dfrac{\sigma(TD)}{\sigma(T_{OBS})+\sigma(B_{OBS})}$	radiochemical** results for $\sigma(TD)$
540MeV Fe + U	—	0.041 ± 0.005	$\geqslant (0.38 \pm 0.06)$	—
800MeV Kr + U	$\geqslant (135 \pm 50)$mb	0.16 ± 0.04	$\geqslant (0.68 \pm 0.12)$*	—
540MeV Fe + U[9]	—	—	—	$(0.46 \pm 0.14)\sigma_R$
605MeV Kr + U[8]	—	—	—	$(0.5 \pm 0.1)\sigma_R$

$\sigma(BD)$: cross-section for binary decay of the composite system
$\sigma(T_{OBS})$: cross-section for three-pronged events in the 2π-geometry
$\sigma(B_{OBS})$: cross-section for two-pronged events in the 2π-geometry
$\sigma(TD)$: cross-section for ternary decay, i.e. three particles in the exit channel

* limits for $\sigma(TD)$ in the reaction 800MeV Kr + U : $(0.6 \pm 0.15)\sigma_R \leqslant \sigma(TD) \leqslant (0.95 \pm 0.02)\sigma_R$

** as radiochemical results are obtained by the thick-target technique, they are not strictly comparable with the results from mica-experiments. $\sigma(TD)$ is considered to be equivalent with $\sigma(B/2)$ and $\sigma(D/2)$, as defined in Ref.8.

Table 2: Observed numbers of quarternary and ternary decay event

Reaction	T_{OBS}	Q_{OBS}	Q_{IND}	References
U + (305-380) MeV ^{40}Ar	500	0	0	Perelygin et al. Nucl.Phys. 127 (1969) 577
U + 540 MeV ^{56}Fe	112	0	1	Vater et al. Radioch. Radio.Lett.19 (1974) 87
U + 540 MeV ^{84}Kr	16	0	0	irradiation at Orsay
U + 800 MeV ^{84}Kr	97	1	1	" Manchester
U + 1080 MeV ^{136}Xe	60	1	3	" Berkeley
U + 1785 MeV ^{238}U	3	1	0	" Darmstadt

T_{OBS} : number of observed three-pronged events in 2π -geometry
Q_{OBS} : " " " four- " " " "
Q_{IND} : " " " three " " " " , where a heavy fourth partner must have been emitted during the decay (Fig.11b)

References

1) R.Brandt, Angew.Chemie, 83 (1971) 1000
2) P.Vater, R.Brandt, Radiochim.Acta, 21 (1974) 191
3) H.-J.Becker, P.Vater, R.Brandt, A.H.Boos, H.Diehl, Phys.Lett., 50B (1974) 445
4) P.Vater, H.-J.Becker, R.Brandt, Radiochem.Radioanal.Letters,19 (1974) 87
5) P.Vater, H.-J.Becker, R.Brandt, J.Radioanal.Chemistry, 32 (1976) 275
6) P.Vater, Doktorthesis, Marburg (in preparation)
7) P.Vater, H.-J.Becker, R.Brandt, H.Freiesleben (publication in preparation)
8) J.V.Kratz, A.E.Norris, G.T.Seaborg, Phys.Rev.Lett.,33 (1974) 302
9) U.Reus, A.M.Habbestad-Wätzig, R.A.Esterlund, P.Patzelt, and I.S.Grant (Marburg-Manchester, to be published).

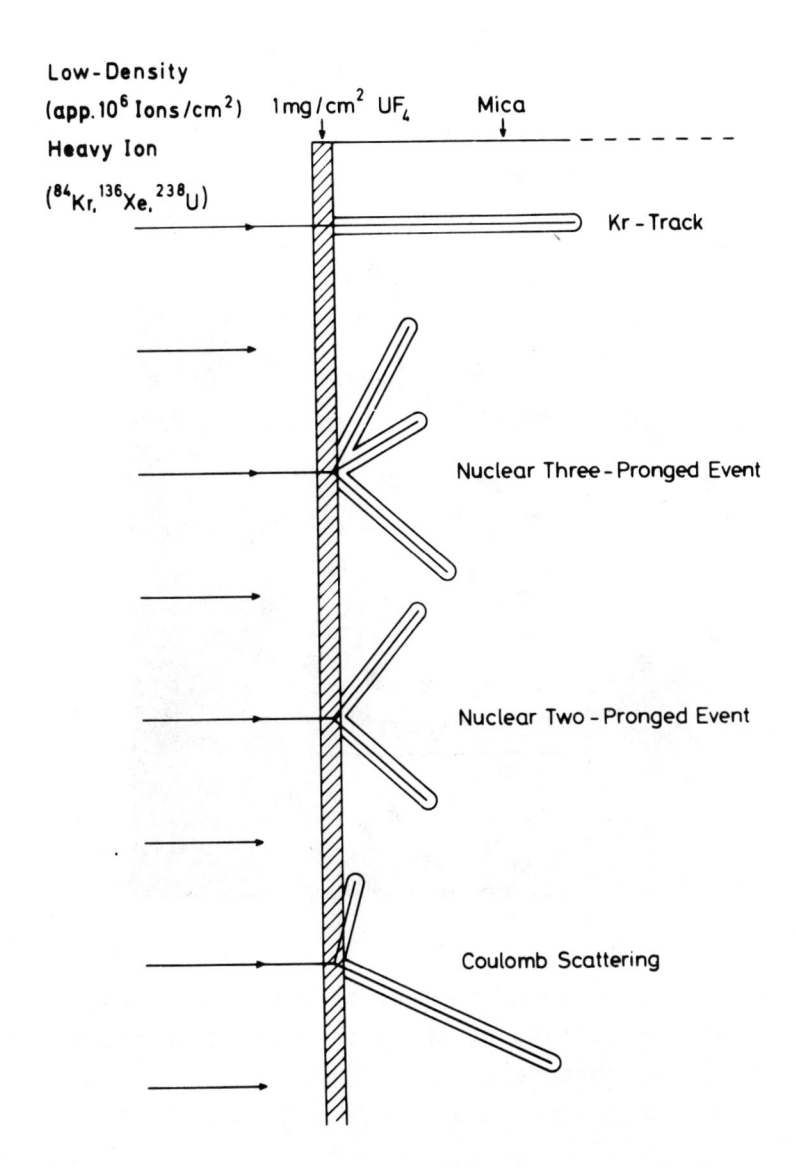

Fig.1: Schematic representation of the 2π-geometry technique using mica detectors.

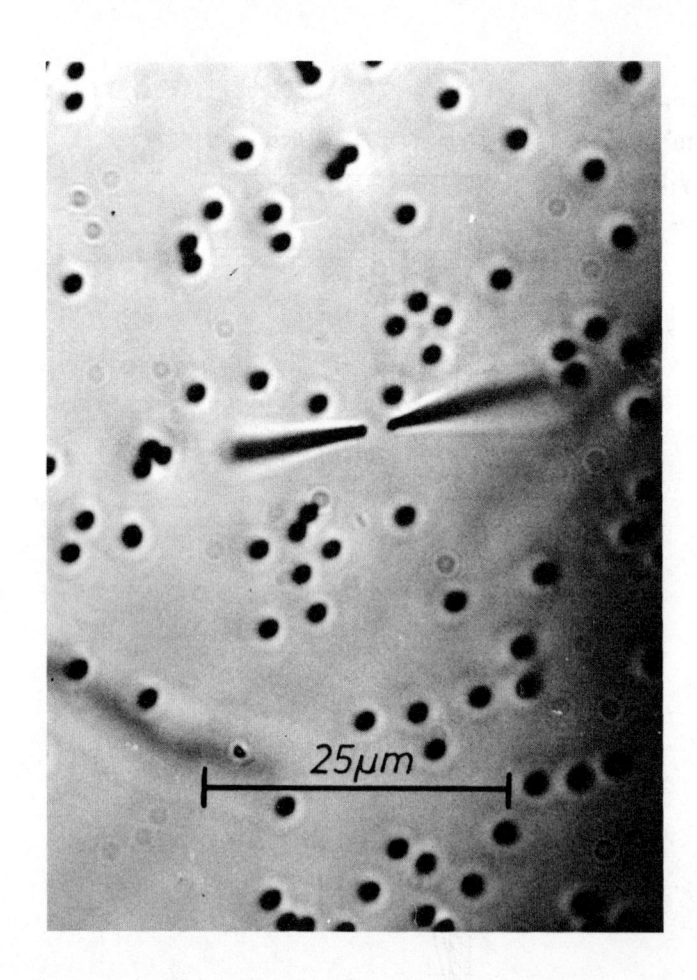

Fig.2: Typical example of nuclear binary decay (BD). (Only the surface of the mica foil is in the focus of the microscope).

26 BINARY DECAY EVENTS

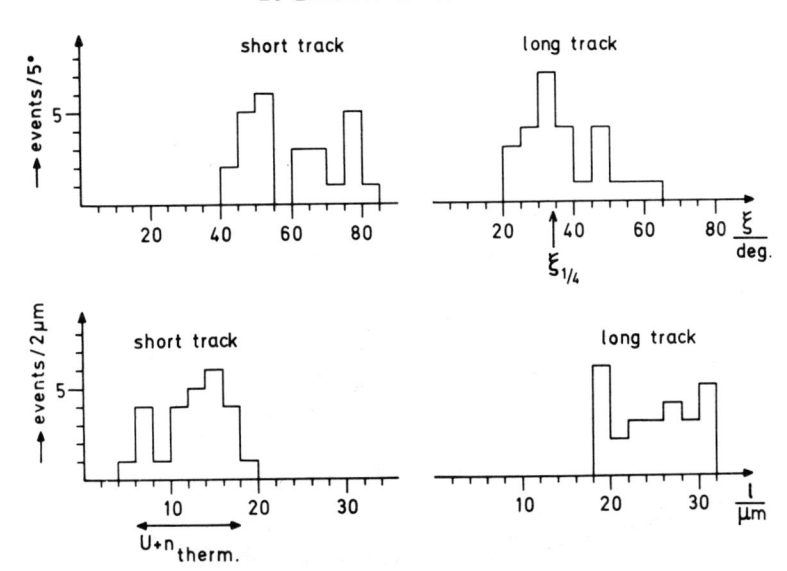

Fig.3: Distribution of track-lengths l and angles ξ between the track and beam direction for nuclear binary decay (BD)

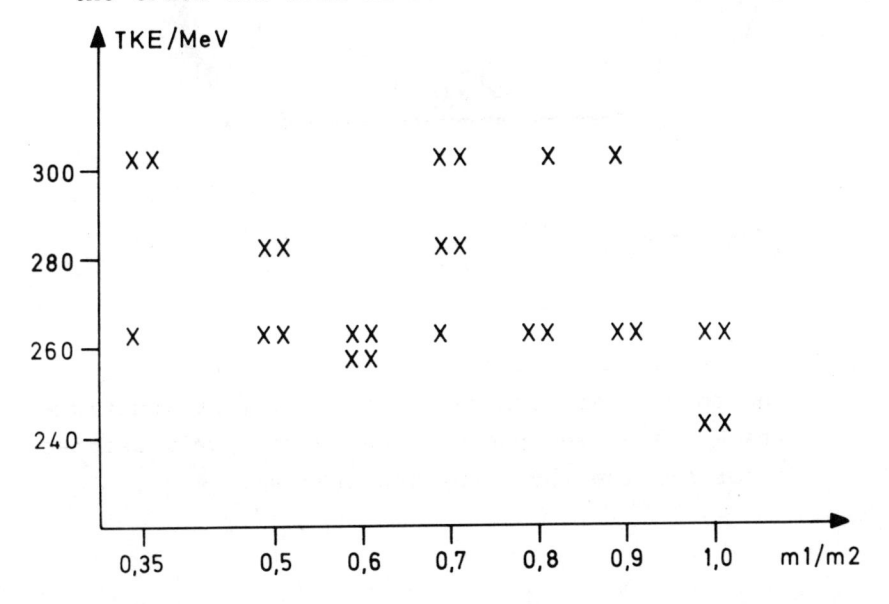

Fig.4: Two-dimensional representation of the total kinetic energy (TKE in the c.m. system) vs. the mass-ratio of two fragments in nuclear binary decay (BD) (Further details see text).

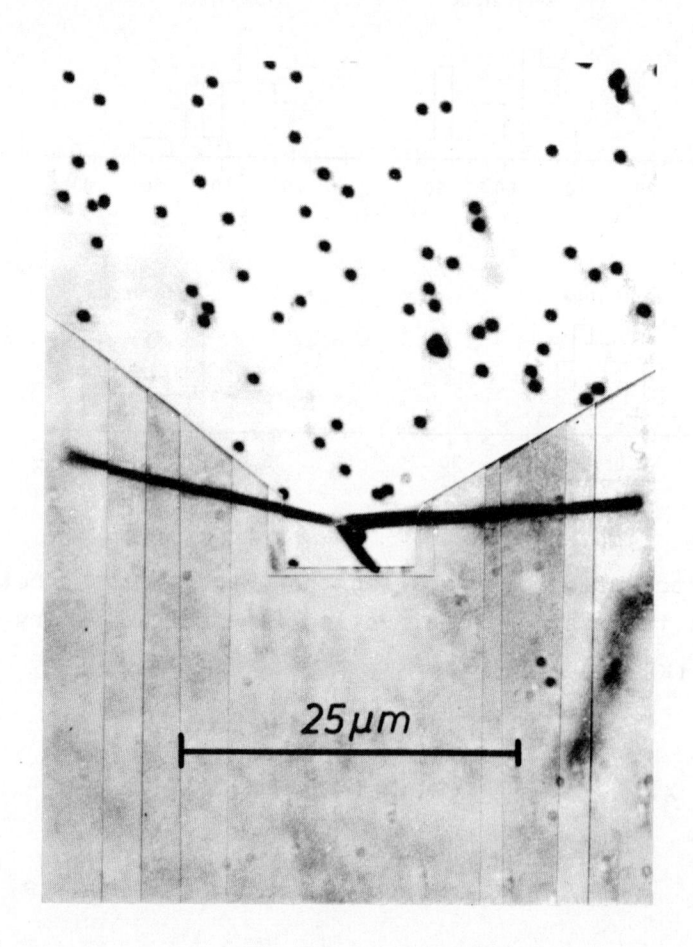

Fig.5: Typical example of a three-pronged event (T_{OBS}) due to nuclear ternary decay (TD). (The individual tracks have been photographed with a well defined focus all the way along the tracks).

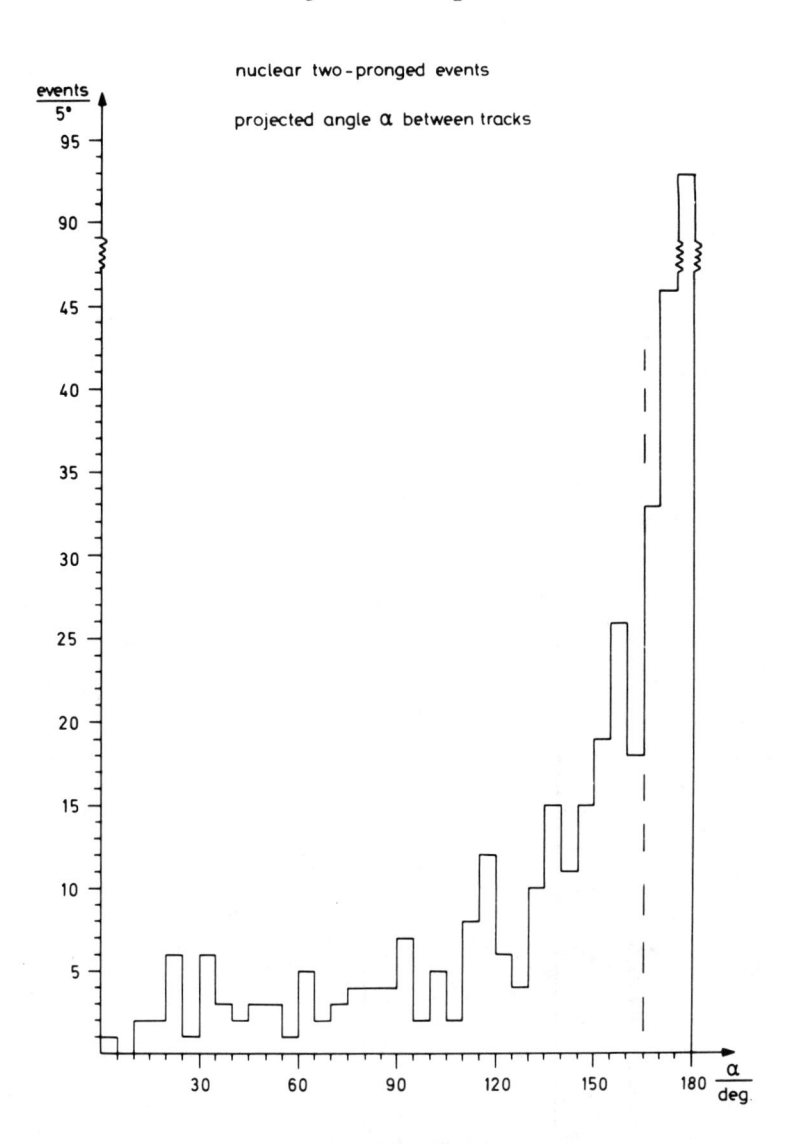

Fig.6: Distribution of the angle **α** between two tracks in the mica-plane (i.e. perpendicular to the beam direction) for two-pronged events (B_{OBS}).

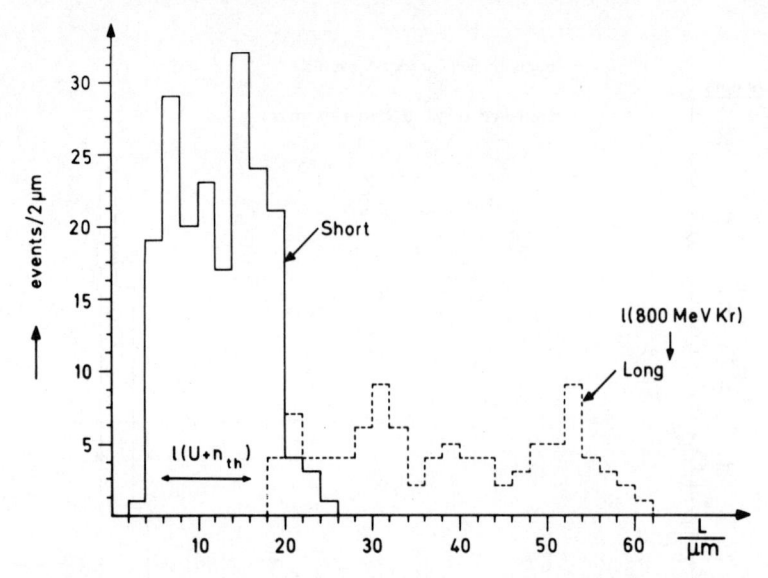

Fig.7: Distribution of track-lengths for 97 observed ternary decay events (T_{OBS}). The long track and the two short tracks for each event are shown in two different distributions (see text).

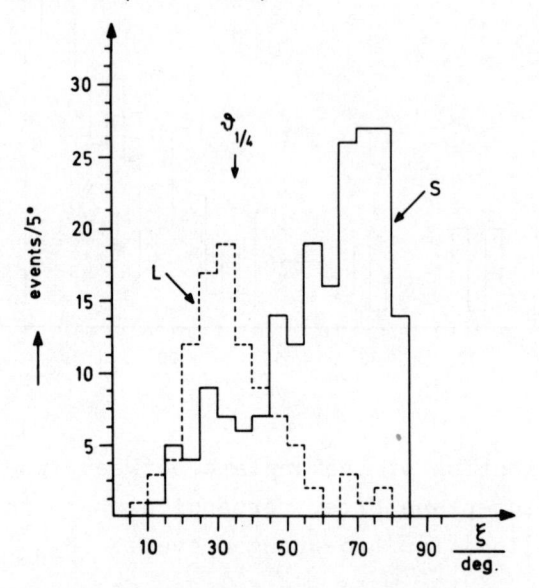

Fig.8: Distribution of the angles ξ between the track and beam direction for the same events as shown in Fig.7, also the long track- and (two) short track-distributions are indicated.

deep inelastic collision (DIC)

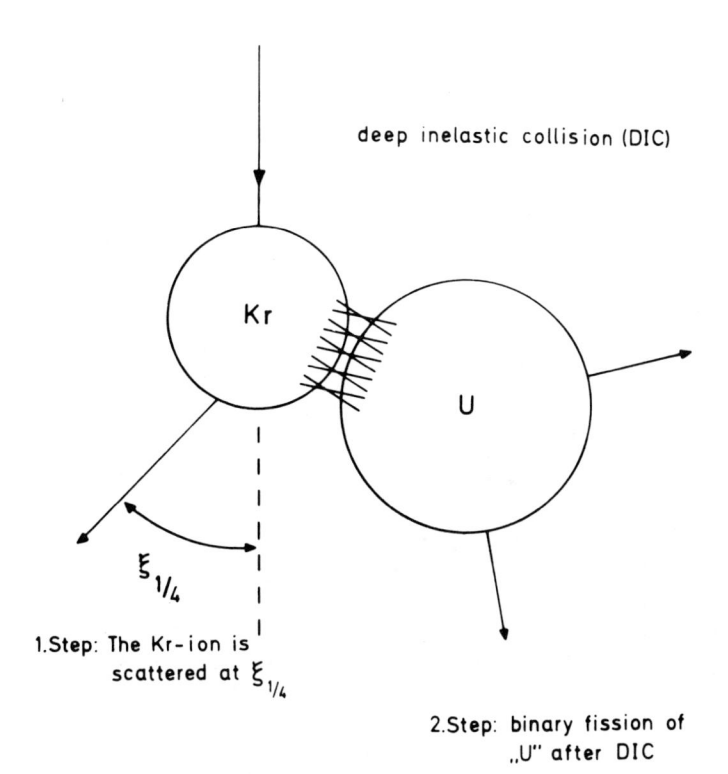

Kr

U

$\xi_{1/4}$

1.Step: The Kr-ion is scattered at $\xi_{1/4}$

2.Step: binary fission of „U" after DIC

Fig.9: Qualitative interpretation of the high amount of ternary decay in (800 MeV ^{84}Kr+U).
(Further details see text)

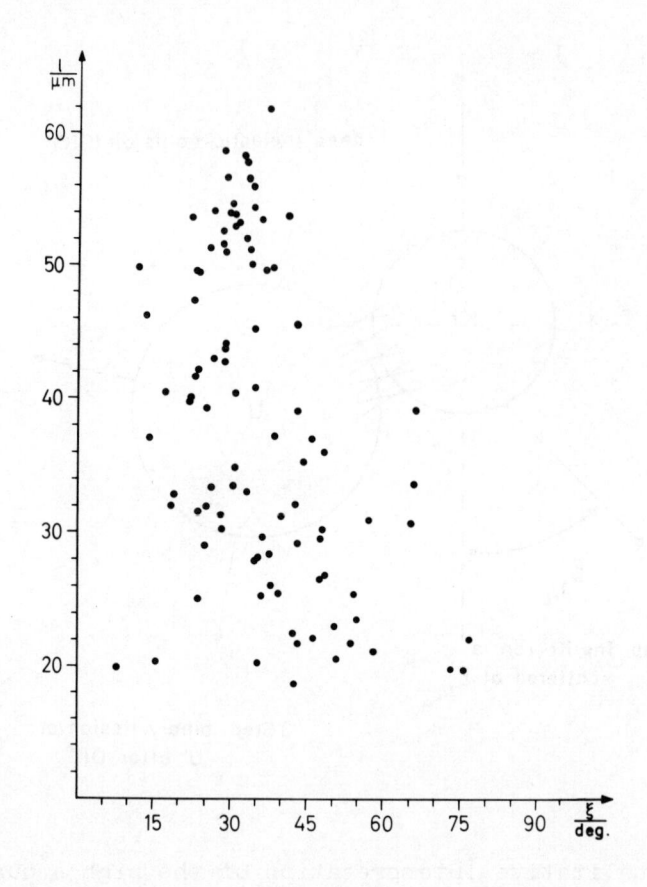

Fig.10: Two-dimensional representation of the track-lengths l and the angles ξ for 97 observed three-pronged events T_{OBS}. (Interpretation: see text).

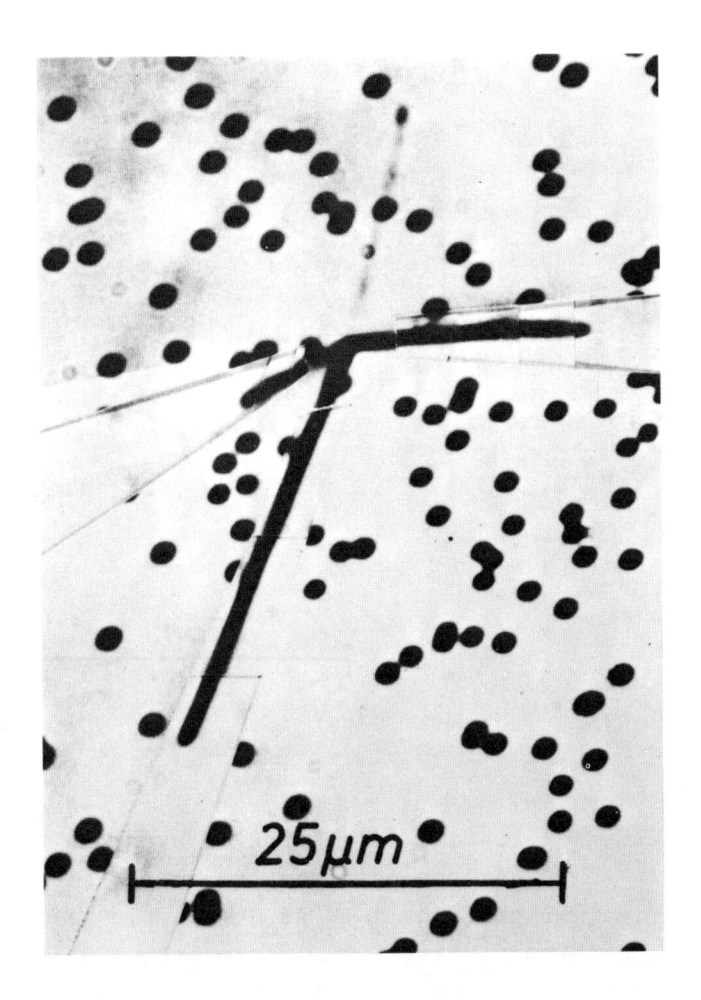

Fig.11: a) Observation of a four-pronged event in (800 MeV Kr+U). We consider this an example of "direct" quarternary decay (Q_{OBS}) (the photographic technique is described in Fig.5).

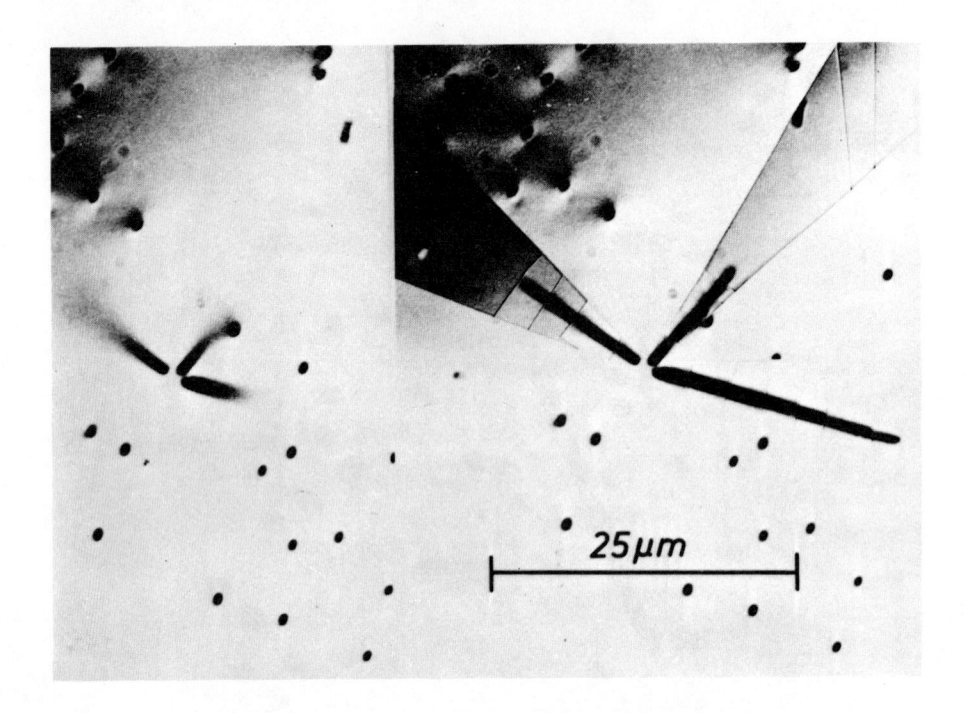

Fig.11: b) Observation of a three-pronged event in (1080 MeV ^{136}Xe+U). As one particle is not registered in mica, we consider this as an example of "indirect" quarternary decay (Q_{IND}) (the left-side photography is taken with the technique described in Fig.2, the right-side photography in Fig.5).

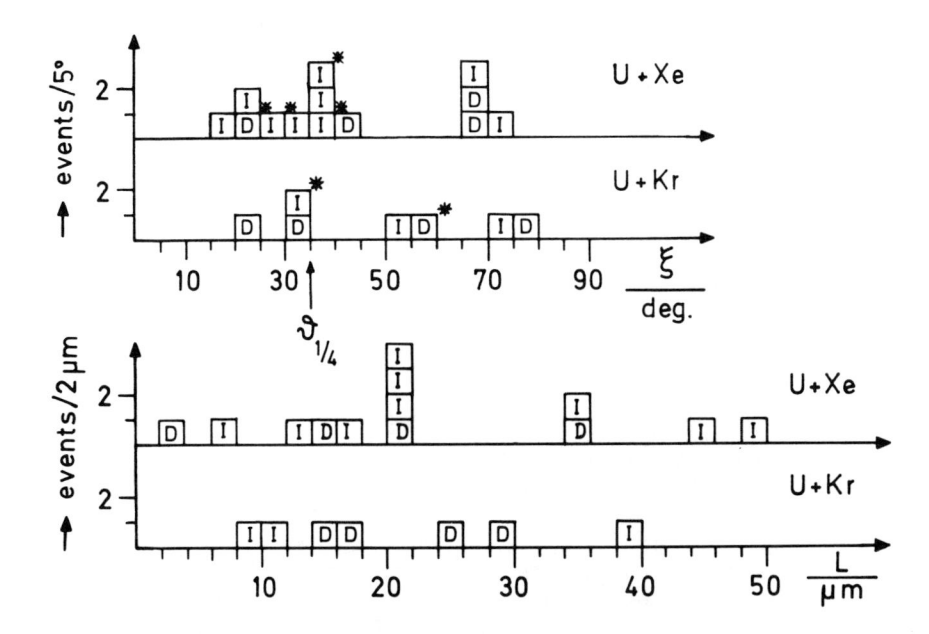

Fig.12: Distribution of track-lengths l and angles ξ for quarternary decays (Q_{OBS} and Q_{IND}) in ^{84}Kr- and ^{136}Xe-induced reactions. The stars in the angular distribution indicate the longest tracks (analog to Fig.7 and 8).

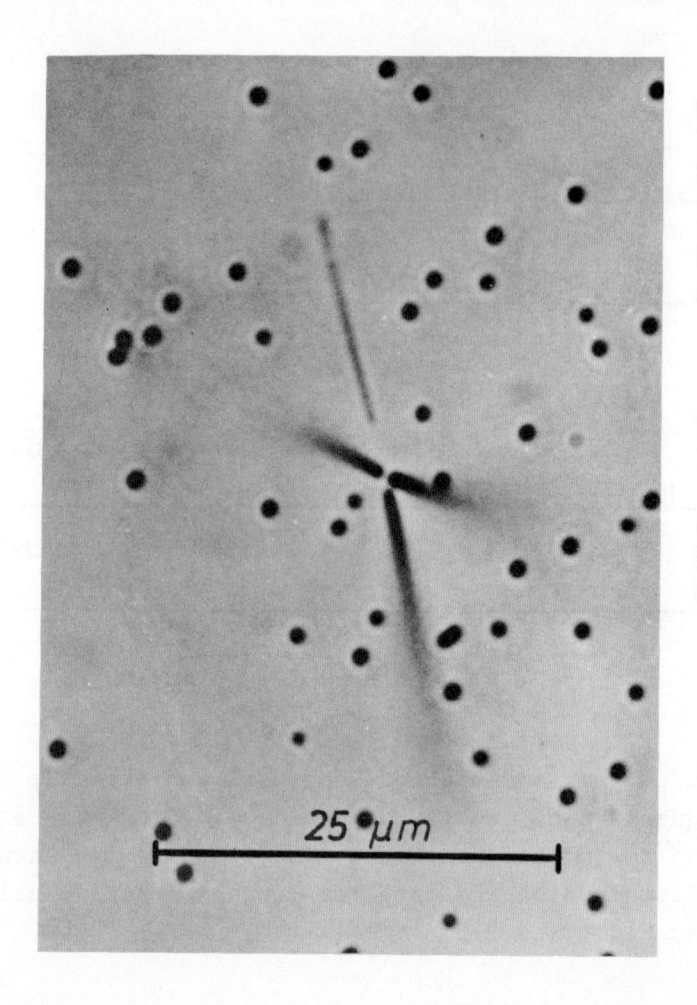

Fig.13: First example of a quarternary decay seen with mica-detectors in the interaction of (1785 MeV U+U).

THE MEASUREMENT OF HELIUM-ION-INDUCED FISSION CROSS-SECTION OF URANIUM BY GLASS TRACK DETECTORS

S. Mubarakmand*, Pervaiz Chaudhry*, K. Rashid**,
R. A. Akber** and H. A. Khan**†

*Nuclear Research Laboratory, Church Road, P.O. Box 1750,
Lahore, Pakistan
**Pakistan Institute of Nuclear Science and Technology (PINSTECH),
Nilore, Rawalpindi, Pakistan

ABSTRACT

The fission cross-sections of uranium induced by ^{3}He ions having energy, $E \leq 30$ Mev have been measured by using uranium loaded soda lime glass track detectors. The maximum cross-section for the reaction is 196 mb obtained at 30 Mev. It decreases rapidly with decreasing energy and reduces to about 2 mb at ^{3}He energy of 25.1 Mev.

---

† The person to whom all future correspondence may be addressed

1. INTRODUCTION

One of the main advantages of Solid State Nuclear Track Detectors is that with a careful selection, a 'Background Free' detection system for fission studies can be easily obtained. This is of particular importance for reactions having low fission yields.

Here, we describe application of uranium loaded soda lime glasses for the measurement of the ^{3}He ion induced fission cross-section of uranium.

2. EXPERIMENTAL DETAILS

In the present work, U-2 and U-3 uranium loaded references glasses (obtained from Corning Museum of Glass, U.S.A.) have been used. U-2 glass contains 43 parts per million of natural uranium while U-3 glass contains 43 ppm of depleted uranium ($^{235}U/^{238}U = 3.6 \times 10^{-3}$). The reference glasses were first given a fine polishing using 'Super Syndia' and 'Diadust' powders (manufactured by Van Hoppes & Sons, Basingstoke, England), then they were etched and scanned for background etch pits. Very low ($< 10/cm^2$) background track density was observed. This background disappeared after the reference glasses were annealed at $400°C$ for 10 min.

According to our previous experience with U-2 and U-3 glasses under these conditions all latent damage trails are completely erased. By performing separate experiments with these glasses it was observed that the uranium was distributed quite uniformly throughout the detectors and that their composition was such that identical etching properties could be expected from all parts of the detectors. At the end the detectors were given a fresh polishing and were fixed on aluminium backings. Plastic energy degraders were fixed on the surface of the glasses. The energy degradation for these plastics is accurately known. The detectors having thus been prepared were exposed to 10 MeV/nucleon ^{3}He ions obtained from the Nuffield Cyclotron at the Birmingham University, U.K. The integrated dose is kept to less than 10^{16} particles/cm^2. During exposure the temperature of the glasses was kept below $30°C$ by using an air ballast. Our annealing experiments have shown that no fading of latent damage trails occurs below this temperature.

After the exposure, the detectors were taken off the aluminium backings and the energy degraders were removed. After washing, the U-2 and U-3 glasses were etched in 4.8% hydrofluoric acid for 1 min at 22°C. Tracks were counted under different sections of the surface representing different regions of ^{3}He energy as defined by the energy degrading etchelon. The maximum track density was observed at 30 Mev ^{3}He energy where a total of 3492 area tracks were counted in the U-2 detector and 3215 area tracks were counted in the U-3 detector. The track density for the U-2 and U-3 detectors was $5.35 \times 10^5 /cm^2$ and $5.31 \times 10^5 /cm^2$ respectively.

TABLE I : FISSION CROSS-SECTION AS A FUNCTION OF INCIDENT ENERGY

$E_{^3He}$ (Mev)		25.1	26.4	27.5	28.1	28.85	29.5	30.0
σ_f (mb)	U-2	1.96	3.94	6.8	23.64	41.36	94.56	197.0
	U-3	0.98	2.94	4.90	19.6	35.28	103.88	196.0

3. RESULTS AND DISCUSSION

The fission cross-section as a function of incident energy for the reaction $U(^3He, f)$ is shown in Table 1. The fission cross-section was calculated from the expression

$$\sigma_f = N_f / J N_t \qquad (1)$$

where N_f the number of fissions/cm^3 is given by

$$N_f = S/R \cos^2\theta_c \, f(t) G, \qquad (2)$$

S = number of fission tracks/cm^2,
R = range of fission fragments in glass
   = 15.2×10^{-4} cm,
θ_c = critical angle = $31°45'$,
$f(t)$ = 1.25,
G = the geometric factor,
   = 0.5, for counts on the surface,
   = 1.0, for counts a few μm below the top surface,
N_t = number of target (uranium) nuclei/cm^3
   = 2.664×10^{17} nuclei/cm^3,
J = integrated incident flux
   = 7.3×10^{15} ions/cm^2.

The fission cross-section rises sharply from approximately 20 mb at 28.1 Mev to almost 196 mb at 30 Mev ^{3}He ion energy. The Coulomb barrier for singly ionized ^{3}He particles approaching a ^{238}U nucleus is 28 Mev. This barrier has been calculated by taking the sum of the radii of ^{3}He and uranium nuclei as the distance of closest approach. The rapid increase in the fission cross section above 28 Mev incident particle energy is apparently due to the fact that above the Coulomb barrier there is a great probability of direct interaction between ^{3}He and uranium. Above the Coulomb barrier the contribution to the fission yield could come from anyone or all of the following processes viz. U(^{3}He, f), U(^{3}He,nf), U(^{3}He, pf) or U(^{3}He, df). The fission cross-section for protons shows a similar trend[1], where the Coulomb barrier for U is about 14 Mev. The fission cross-section for the reaction ^{238}U(p, f) for E = 10 Mev is 35 mb[2,3]. The corresponding cross-section for the ^{238}U(^{3}He, f) reaction for E_3 = 30 Mev (10 Mev/nucleon) is 196 mb. The lesser value for the proton induced fission cross-section in uranium at 10 Mev can be largely due to the fact that at this energy the protons are below the Coulomb barrier and the proton barrier penetrability is calculated to be only about 28%. The value for σ_f in U for E_{3He} = 27 Mev (9 Mev/nucleon) is 5 mb. There is a 30% barrier penetrability for ^{3}He ions and U nucleus at this energy.

The corresponding penetrability for 9 Mev protons is 22% while the σ_f for 9 Mev protons in U is ~9 mb[1]. It is therefore observed that if the Coulomb barrier penetrability for the two types of incident particles is comparable, then their fission cross-sections are also comparable. For a more extensive comparison between proton and ^{3}He induced fissions in uranium, the present data need to be extended beyond 30 Mev ^{3}He energy.

4. REFERENCES

1. S. Baba, H. Umezawa and H. Baba. Nucl. Phys. A175 (1971) 177.

2. P.C. Stevenson, H.G. Hicks, W.E. Nervik and D.R. Nethaway. Phys. Rev. 111 (1958) 886.

3. C.B. Fulmer, Phys. Rev. 116 (1959(118.

4. G.R. Choppin and E.F. Meyer, Jr., J. Inorg. Nucl. Chem. 28 (1966) 1569.

APPLICATION OF THE TRACK DIAMETER MEASUREMENT METHOD IN NUCLEAR PHYSICS

V. A. Nikolaev

V. G. Khlopin Radium Institute, Leningrad, U.S.S.R.

Glass track detectors such as sodium-silicate, phosphate and quartz glasses proposed first in /1, 2/ have found wide application in nuclear physics as fission fragment recorders /3/. We and also other authors have undertaken investigations to find the possibilities for use of glasses as spectrometric devices. In this report a brief review of works on the track diameter measurement method and its application in nuclear physics is given, the regularities observed are explained, and some new data on the application of silicate and phosphate glasses for the measurement of symmetric radium fission contribution are presented.

We have investigated the dependences of track diameter in silicate glasses on etching time as well as track diameter distributions /4, 5/. It has been revealed that in weak acid solutions (~3%) the curves describing track development of various fragments up to a certain point overlap and then they diverge (see. Fig. 1a). A narrow one-humped diameter distribution broadens in a gradual manner and assumes a weakly pronounced double-humped character (first demonstrated in /6/

and then in /7/). (See. Fig. 1b). It has been shown /8, 14/ that when etching silicate glass in weak solutions the track takes in succession the form of a cone (until the acid reaches the limit of radiation damage zone), a conjugation of a cone part with a sphere, and a part of sphere, and that the breaks of the curve (Fig. 1a and 1C) are connected with transition to a spherical surface. The break appears the later, the greater the length of radiation damage zone is. Thus, the track diameter in the last stage of etching is connected with radiation damage zone length and fragment range. In silicate glasses this connection is weakly pronounced, therefore the distribution "humps" are only 10% apart /6, 7/ which practically does not permit spectroscopic measurements.

The studies of tracks from pre-moderated fragments /4, 5/ in silicate glasses have shown that in this case the curves of diameter growth diverge immediately for various fragments (see. Fig. 1c), and the track diameter distributions have a pronounced double-humped character. (See. Fig. 1d for ^{235}U). Later the double-humped character of track spectrum for moderated particles was also demonstrated in /9, 10/.

The comparison of track spectra with the data on dE/dx /11/ at various thickness of absorber has shown /5/ that in the first stage of etching the track diameter depends on dE/dx of fragments which are close for all fragments at the beginning, and differ greatly half-way the range. Since dE/dx of fragments half-way the range is more dependent on energy than on fragment mass /11/ it can be expected that the track spectra are similar to the energy spectra, which has been shown in /5, 8/ for well-studied isotopes ^{233}U, ^{235}U, ^{239}Pu and

^{252}Cf as well as in /7/ for separated fragments. Using these data a calibration curve has been plotted for determination of average kinetic energies of fragments; with the help of this curve for the first time the average kinetic energies of spontaneous and neutron-induced fission of ^{238}Pu /12, 13/ have been determined.

It should be emphasized that although moderated fragment track diameters permit to estimate fragment energies (see /4, 5, 7, 9, 10/), the connection between track diameter and fragment energy is indirect and reveals itself due to dE/dx dependence on energy half-way the range. The track diameter in glass depends directly on dE/dx in the first stage of etching. This is in agreement with the conclusions of /14/, where it is noted that the velocity of acid penetration into track zone (V_T) in the first stage of etching depends on dI/dx, i.e. on primary specific ionization of fragment, since the track diameter is defined by V_T, and dE/dx is close to dI/dx.

Last years we have used silicate glasses with the filters 1,5 mg/cm^2 Al for study of fission of ^{227}Ac and ^{226}Ra /16/. Fig. 2 shows the distributions of fragment track diameters in the reaction ^{226}Ra (γ, f) at energy E = 11, 18, 28 MeV. The use of these detectors has proved to be the only possible way due to a strong background of α-particles and γ-quanta. As the energy of γ-quanta increases, the "hump" of heavy gragments displaces gradually to the right owing to the increase of symmetric fission contribution. Up to energies of 30 MeV the "hump" of a light group can be separated by means of extrapolation from heavy group fragments and from symmetric fragments, and so the contribution of symmetric fission can be defined.

As the fission barrier approaches, the definition of symmetric fission contribution becomes difficult because of impurities of natural uranium in silicate glass and in aluminium filter and background track formation. Phosphate glasses have about 100 times less impurities of fissile material than silicate ones; therefore we have investigated the possibility of using phosphate glasses.

From /14/ it follows that the curve of connection of V_T with dE/dx for silicate glasses reaches rapidly plateau with increase in dE/dx of fragment. This is demonstrated also in the experiment with heavy ions /17/. Therefore, when using silicate glasses the filters are necessary. For phosphate glasses the connection of V_T with dE/dx is much more pronounced. As a result, for phosphate glasses even in the first stage of etching the diameter increase of various tracks is described by different curves (of type shown in Fig. 1c), and in the third stage, when the track diameter is affected also by fragment range, the divergence of curves becomes essentially greater than in the case of silicate glasses. Therefore, if phosphate glass is etched for sufficiently long time, the pronounced double-humped distribution can be obtained without pre-moderation of fragments, which is shown in /18/ for ^{252}Cf.

The measured distributions of track diameters in silicate glasses with a filter 1,5 mg/cm^2 Al and in phosphate glasses without filter for ^{226}Ra + n (E_n = 0 - 11 MeV) and also semiconductor spectrum are presented in Fig. 3. In Figures the positions of light peaks are brought into coincidence. The relation "light peak/depression" is seen to be similar for silicate and phosphate glasses, but the peak of heavy fragments in silicate glass is essentially deformed as related to semiconductor spectrum, whereas the spectrum shape in phosphate glass is quite similar to it. In this connection the use of phosphate glass is preferred to silicate one in those cases, where the shape of spectrum is of great importance and a detector with low background is required. However, the measurement of tracks by eyepiece micrometer is more time-consuming in the case of phosphate glasses where the size of tracks reaches 100 μm instead of $\sim$ 15-20 μm in silicate glasses. The limiting density of tracks in phosphate glasses is $\sim$ 10 times lower than in silicate ones. The track contrast also decreases greatly, which impedes the use of automatic devices of MAGMO-I-type /19/ for measurement and calculation of tracks. Keeping in mind all this, in most cases it is more convinient to use silicate glasses. At present we use just these glasses with filters for determination of asymmetrical component of fission product nuclei lighter than radium at the barrier.

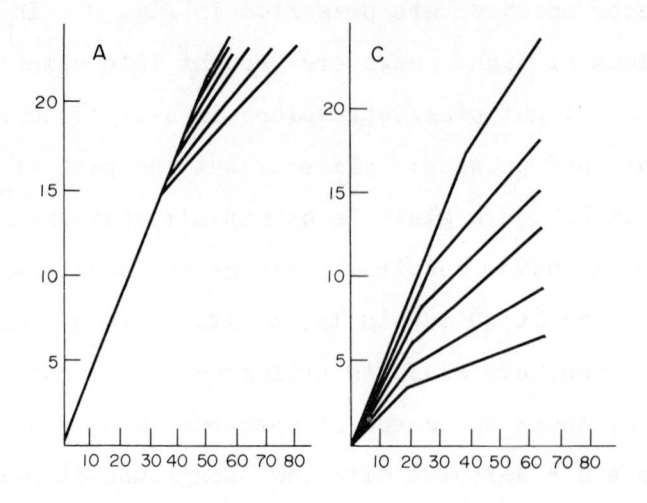

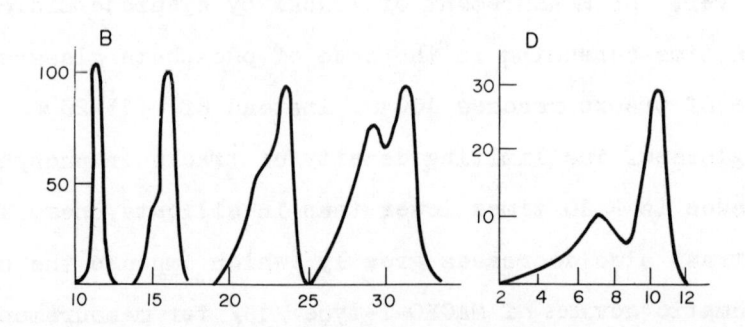

Fig. 1. Track diameters from fission fragments of ^{235}U in silicate glass depending on etching time for unmoderated (a) und moderated (c) fragments. Track diameter distribution for unmoderated (b) and moderated $\sim$ 1,5 mg/cm^2 Al fragments (d).

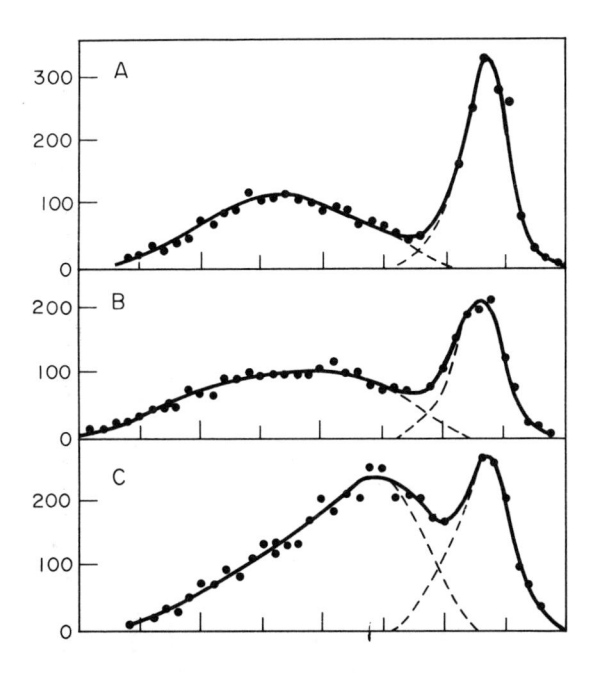

Fig. 2. Track diameter distribution in silicate glasses irradiated by fission fragments ^{226}Ra + γ:
a) E_γ = 11 MeV; b) E_γ = 18 MeV; c) E_γ = 28 MeV.

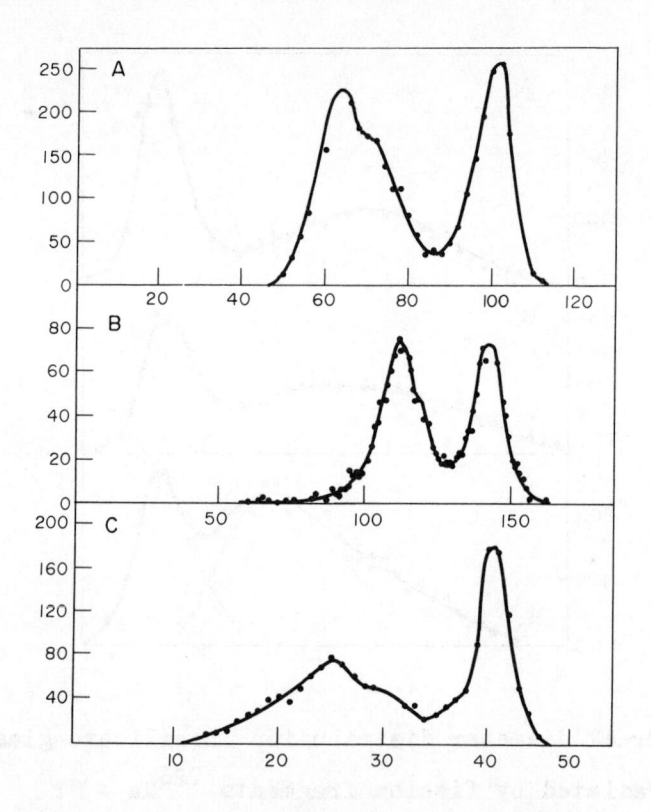

Fig. 3. Energy spectra of neutron-induced fission fragments of ^{226}Ra with energy 0 - 11 MeV measured by
a) solid-state detector; b) phosphate glass;
c) sodium-silicate glass with aluminium filter
$\sim$ 1,5 mg/cm^2.

REFERENCES

1. Fleischer R.L., Price P.B. J. Appl. Phys., 34, 2903, 1963.
2. Перелыгин В.П. и др. „Приборы и техника эксперимента", № 4, 70, 1964.
3. Proc. 8-th Intern. Conf. on Nucl. Phot. & SSTD, Ed. Nicolae, IFA, Bucharest, 1972.
4. Николаев В.А. Авт. свид. № 273008, приор. от 06.05.68. Б.И. № 19 07.1970.
5. Громов А.В., Николаев В.А. „Приборы и техника эксперимента", № 1, 245, 1970.
6. Somogyi G., Atomki Közl., 9, 77, 1967.
7. Höppner U. et al., Nucl. Inctr. & Meth., 74, 285, 1969.
8. Николаев В.А. Диссерт. на соиск. уч. степени кандидата физ.-мат. наук. Ленинград, Радиевый ин-т им. В.Г. Хлопина, 1972.
9. Горшков В.К. и др. „Атом. энергия". 28, 504, 1970.
10. Khan H.A., Durrani S.A., Nucl. Instr. & Meth., 109, 341, 1973.
11. Насыров Ф. и др. „Атом. энергия". 19, 244, 1965.
12. Николаев В.А. „Изв. АН СССР, сер. физ." 35, № 1, 180, 1971.
13. Николаев В.А., Протопопов А.Н. „Изв. АН СССР, сер. физ." 36, № 1, 215, 1972.
14. Fleischer R.L. et al., Phys. Rev., 188, 563, 1969.
15. Николаев В.А. и др. „Яд. физика". 19, 4, 751, 1974.
16. Жагров Е.А. и др. „Письма в ЖЭТФ". 20, 3, 220, 1974.
17. Капусцик А. и др. „Приборы и техника эксперимента", № 1, 43, 1968.
18. Aschenbach J. et al. Nucl. Instr. & Meth., 116, 389, 1974.
19. „Методы и техника машинного анализа биологических структур", под ред. Г.М. Франка и Г.Р. Иваницкого. Наука, 1972.

ON THE DETECTION OF LOW-ENERGY ^{4}He, ^{12}C, ^{14}N, ^{16}O IONS IN PC FOILS AND ITS USE IN NUCLEAR REACTION MEASUREMENTS

G. Somogyi, I. Hunyadi, E. Koltay and L. Zolnai

Institute of Nuclear Research of the Hungarian Academy of Sciences, Debrecen, Hungary

It is shown that by using a proper etching reagent the registration sensitivity of polycarbonate foils can be enhanced and they prove to be very suitable track recorders for alpha-particles emitted from nuclear reactions. At 6 MeV an energy resolution of 0.2 MeV can be achieved when using the track diameters as a measure of the particle energy. A theoretical way to calculate the track parameters important in nuclear reaction measurements involving alpha-particles recorded in polycarbonate foils is given. For this purpose the track etch rate vs residual range curve was determined by a parameter optimization procedure. The energy resolution of the track-diameter method as a function of the particle energy was predicted. In our earlier studies the track-diameter method was mostly used in angular distribution measurements of (d,α) nuclear reactions. In this work it is shown that with polycarbonate foils it can be well applied to excitation function measurements, as well. Such studies are presented for the α_0 and α_1 groups of the ^{27}Al$(p,\alpha)^{24}$Mg reaction in an energy interval between 1540 and 1920 keV. Finally, preliminary results on the track etching properties of low-energy O^+, N^+, C^+ and He^+ ions accelerated with a 5 MV Van de Graaff generator are given.

1. Introduction

In the application of plastic track detectors for the investigation of nuclear reactions the differentiation of reaction products according to type and energy is required. In the investigations performed with low-energy accelerators, which is the field of interest of the authors, the selection of reaction products according to type gives rise to no problems in the majority of the cases, since this question can be solved by properly choosing the detector type and the irradiation and etching conditions. In most investigations on nuclear reactions, the energy-dispersive measurement of identical nuclear particles is first of all necessary. The performance of such studies, based on the relation between the track diameter and particle energy [1] can be regarded today, in the knowledge of the theoretical and practical bases of the method [2,3,4], a routine task.

The energy-dispersive track diameter method is expedient for application primarily in experiments where it proves to be more advantageous as compared to that of the semiconductor detector technique. Such field is e.g. the investigation of very low cross section nuclear reactions /e.g. the nuclear processes of astrophysical interest/ or measurements carried out against intensive gamma or light particle backgrounds.

In the investigations performed in the above field, although in different degrees, CN, CA as well as PC detectors were applied. In these detectors, already, under special conditions the detection of proton tracks is also possible. This, however, has rather theoretical importance only. In low-energy charged-particle nuclear reaction studies plastic track detectors were used almost exclusively in cases where the detection of alpha particles was required. It is clearly shown in Table 1, in which a survey is given of the data reported in different publications related to this field.

Our first nuclear reaction investigations were performed in 1967 with CN sheets. The energy resolution of these sheets was rather poor, especially at low alpha energies. It was obvious that solution must be sought for in the application of plastic track detectors less sensitive to alpha particles. Then we began to use CA foils, together with the use of a basic etchant containing $KMnO_4$ [2], the energy resolution and track detecting properties of which proved to be more satisfactory. With these foils, on the other hand, the formation of uniform MnO_2 layer on the detector surface, which is one of the preconditions of good etching, required special care.

The etching difficulties were practically eliminated by the application of Makrofol-E detectors etched in a basic solution containing ethylic alcohol [5]. In our investigations, at present, these foils are used, since in view of energy resolution, track contrast, regularity of the pit contour, these show the best properties.

In the present work we give account of our recent studies on the etching properties of low-energy reaction products registered in PC foils. We shall present for the case of alpha particles registered in Makrofol-E foils a theoretical calculation related to the spectrometrical properties of the track-diameter method. Then we shall describe an experimental--technical development suitable for simultaneous angular distribution and excitation function measurements and present an investigation performed with high energy-resolution on the α_0 and α_1 groups of the $^{27}Al(d,\alpha)^{24}Mg$ reaction using plastic track detectors. In the end we present some preliminary experimental results obtained in Makrofol-E for the etching properties of the tracks of $^{4}He^+$, $^{12}C^+$, $^{14}N^+$ and $^{16}O^+$ ions accelerated with a 5 MV Van de Graaff generator. The direct aim of these measurements is: research into the detector technical details of nuclear reaction investigations based on residual nucleus registration.

2/ Theoretical basis

In our nuclear reaction studies discrimination between the various alpha groups was based on the correlation of the etch-pit diameters and the particle energy [1]. The basic principles of this method have been reported on in a previous communication [4]. Here we wish to present the results predicted by the etch-pit growth kinetics for the optimum etching times and the attainable energy resolving power of Makrofol-E sheets. This calculation gave valuable information for planning the irradiation, etching and track evaluation conditions of the sheets used in our experiments.

To solve the above task one had to determine the $d(h,E_0)$ function, for normally incident alphas i.e. the changes in the track diameter as a function of layer removal and energy. The calculation of this function of two variables was performed by breaking the problem down to the curve $d(h)$ at constant range and to the curve $d(R_o)$ at constant layer removal. Finally, for the calculations the following parametric relationships [4] were used:

$$\left.\begin{array}{c} d = 2\left[x_o - H(x_o)\right] \cdot \sqrt{\dfrac{V(x_o)+1}{V(x_o)-1}} \\ \\ h = \dfrac{x_o V(x_o) - H(x_o)}{V(x_o) - 1} \end{array}\right\} \quad \text{if } h \leq h_o \qquad /1/$$

$$d = 2\left(\left[R_o - H(R_o)\right] \cdot \left[2h - R_o - H(R_o)\right]\right)^{1/2}, \text{ if } h \geq h_o \qquad /2/$$

where $H(x_o) \equiv \displaystyle\int_0^{x_o} V^{-1}(x)\,dx$, /3/

$$h_o = \frac{R_o V(R_o) - H(R_o)}{V(R_o) - 1}, \qquad /4/$$

and $V(x_o) = V(x)\big|_{x=x_o}$, $H(R_o) = H(x_o)\big|_{x_o=R_o}$.

The key problem of the calculations with the above formulas is the knowledge of the $V(x)$ function, i.e. the etching rate ratio as a function of the distance measured along the track axis from the entrance point of the particle. It is expedient to choose the form of this function so that the integral /3/ be explicitly calculable with the introduction of a minimum number of free parameters. For alpha particles registered in Makrofol-E sheets applying 70°C PEW etchant /=15g KOH+40g C_2H_5OH+45g H_2O/ the function of form

$$V(x) = 1 + \exp\left[-A(R_o - x) + B\right] \qquad /5/$$

was found most suitable, and using this we get the integral /3/ in the form

$$H(x_o) = x_o + \frac{1}{A} \ln \frac{1 + \exp(-A\,R_o + B)}{1 + \exp(-A\,R_o + Ax_o + B)}. \qquad /6/$$

The A and B constants in the function $V(x)$ were determined from the best fit of the experimental and theoretical $d(h)$

curves. The results of fitting performed for 1,2,3,4 and 5.5 MeV alpha particles are presented on the upper part of Fig. 1, from which the values $A=0.22$ μm^{-1} and $B=0.45$ were obtained. /It is noted that in this fitting procedure the range-energy relation described in ref. [4] was applied./ On the lower part of Fig. 1 we present the theoretical $d(h)_{R_0}$ curves calculated with the above A and B values using the formulas /1/ and /2/, gradually increasing the value of the range by 2 μm.

The computer programme written for the above calculations was also suitable for the determination of the energy resolving power of the track-diameter method. For this aim the set of curves $d(R_0)_h$ had to be determined after re-writing eqs /1/ and /2/ into the parametric form of $d(x_0)$ and $R_0(x_0)$. The calculations were performed with an iteration procedure. The results obtained are given in Fig. 2. The best energy resolution can obviously be attained if the differential quotient of the track diameter vs range curves is the maximum. This condition is satisfied in the case of alpha energies higher than 2.5 MeV uniformly at track diameters round 10 μm. Consequently to assure the most favourable conditions for track evaluation it is expedient to etch the tracks so that they reach a diameter of 5-10 μm /see Fig.11/. The theoratically expectable energy resolution of the track-diameter method at these diameter values is shown in Fig.3. Here energy resolution was defined as the energy difference of two such alpha groups for which the difference of the track diameters is 2 μm. It should be noted, however, that in such case the alpha groups can be distinguished even with the unaided eye, without actually measuring the diameters.

If the energy resolution of the method is defined as the halfwidth of the experimentally obtainable track diameter distribution curves, then, about one quarter of the energy resolution shown in Fig. 3 is obtained, thus e.g. at an alpha energy of 6 MeV, the remarkable 60 keV. This corresponds to a track diameter distribution curve with halfwidth of about 0.5 µm. which is for PC, as the data in Fig. 4 show, a typical value. From Fig. 4 it can also be seen that the attainable energy resolution with CA detectors is about twice worse than with PC. This can be accounted for by the fact that the track contours of normally incident alpha particles are very regular in PC sheets /see Fig.11/, whereas they are more irregular spots in CA.

The above results all show that in low-energy nuclear reaction studies at present it is most expedient to use PC foils with favourable etching properties and excellent energy resolution. In the following part of our paper the technical circumstances and the results of a nuclear measurement performed with such foils are reported on.

3/ <u>Investigation of the ^{27}Al(p,α)^{24}Mg nuclear reaction</u>

From the energy and angular dependence investigations in charged particle reactions /excitation function, angular distribution/ performed with the high-stability and low energy--spread bombarding beams of Van de Graaff accelerators results highly valuable from the nuclear spectroscopic viewpoint can be derived. Such are the excitation energy, spin, parity, width etc. characterizing the ground and excited states of nuclei.

In such investigations if the target nuclei are bombarded with several MeV protons - with the exception of the quite light target nuclei - the scattered protons and the emitted alpha particles have approximately identical energies, whereas the intensity of the protons exceeds that of the alpha particles by several orders of magnitude. In such case the polycarbonate track detector possessing good energy resolution and particle discriminating power can be favourably applied in the spectroscopic study of low energy (p,α) reactions.

As an illustration, in Fig. 5 a charged particle spectrum is presented, which was measured with a semiconductor detector during the bombardement of an Al target with protons. It can be seen that the low-intensity α_1 group is covered by the proton peaks and in practice it cannot be separated from the spectrum. Although the α_0 group appears in the spectrum well-separated in energy, the scattered protons strongly growing in intensity towards the small scattering angles overload the semiconductor detector and the electronics, which prevents the observation even of the α_0 group at small angles. Therefore the α_0 and α_1 groups were registered in 200 μm thick Makrofol-E sheets. In the E_p=1500-1900 keV bombarding energy interval at $\Theta = 90°$, the energy of the alpha-groups varied between E_{α_0} = 2.6-3.0 MeV and E_{α_1} = 1.4-1.8 MeV. Since the aim of the measurement was to study the resonance states of the ^{28}Si compound nucleus, the excitation function had to be measured in fine steps with high energy resolution. To attain this we had to solve the quick change of the detector sheets in vacuum. To this end a special irradiation unit was constructed.

a/ Experimental technique

The bombarding proton beam was obtained from the 5 MV Van de Graaff generator of the Institute of Nuclear Research in Debrecen. The beam of $\sim$ 1 keV energy spread outgoing from the 90° analyser magnet was deflected by another magnet into one of the measuring channels, in which the experimental arrangement shown in Fig. 6 was set up. The detailed spatial view of the irradiation unit holding the plastic detectors is given in the lower part of the Figure.

The experimental arrangement consists of two adjoining cylindrical scattering chambers in the common axis of which the proton beam is running. The first chamber serves for the measurement of the yield of the (p,γ) reaction with a scintillation detector. In the second chamber the yield and angular distribution of the (p,α) reaction is measured with plastic track detectors. The (p,γ) and (p,α) measurements were performed alternately. In each chamber a Faraday cup with evertible bottom was used and target pairs of identical thickness made of spectral purity Al by vacuum evaporation were placed. In the ~ 5 μg/cm^2 Al layer evaporated onto a 40 μg/cm^2 carbon backing the energy loss of the bombarding protons was ~ 1.5 keV. In the investigation of the (p,α) reaction the bombarding beam passed through the first chamber and hit the target placed at an angle of 45° to the direction of the beam in the centre of the irradiation unit holding the plastic track detectors.

The irradiation unit consists of three coaxial cylinders. The detection and solid angles are determined by the openings in the inner cylinder. The Makrofol-E sheets are fixed onto the inner wall of the intermediate double-walled cylinder in 8 rows one above the other. The outer cylinder serves for the protection of the detectors against background.

The double-walled cylinder can be vertically shifted and turned round by an angle of 90°. This allows us to perform irradiation at 16 energies around a resonance and at every energies simultaneously at 16 angles between 10°-80° and 100°-170°, without opening the vacuum system.

To stop the nuclei recoiled from the target and carbon backing, in front of the plastic sheets at forward angles 0.5 mg/cm^2, at backward angles 0.2 mg/cm^2 Al foils were used. First, in the energy range under study the (p,γ) excitation function was measured in the first chamber, then knowing this for the (p,α) measurements in the second chamber the optimum energy steps were decided. The irradiation time was varied according to the strength of the resonance studied. The charge collected in one irradiation varied between 0.4-4.2 mCoul.

For track revealing the PEW solution at 70°C was used [5]. In this case the irradiated surface of the detector is marked out by the high density proton tracks as opalescent spots, in which the very contrasty alpha-tracks become visible. In the reaction under study the 1.37 MeV energy difference between the $α_o$ and $α_1$ groups was sufficient for separating the tracks at a magnification 500x without measuring diameters.

b/ Excitation function and angular distribution measurements

In order to find the resonances the excitation functions were determined at $80°$ and $150°$. The results obtained at $80°$ are shown in Fig. 7. In the lower part of the Figure the γ-yield curve is also presented. In the energy range studied about 250 irradiations were performed with 0.5-1 keV steps on the resonances and 5-10 keV steps between the resonances. The experimental energy resolution was better than 1.8 keV. At the resonances observed in the excitation functions at the 11 different energies denoted by numbers in Fig. 7, angular distributions were measured with PC detectors in the angular ranges of $20°-80°$ and $100°-170°$. From these some are presented in Fig. 8 as an illustration.

At last without going into fuller details of nuclear relations we should like to emphasize two important results of ours obtained by applying the track detector technique. The first one is that it was the first time that the $^{27}Al(p, \alpha_1)^{24}Mg^*$ reaction was investigated in detail with the direct detection of α_1 particles.

The other result is that while with the semiconductor detector technique, because of the proton background, angular distribution could be measured only at the backward angles, with the plastic track detectors data were obtained for the whole angular range. This permits a more exact theoretical evaluation than the measurement of half angular distribution, where it is customary to assume symmetric distribution for $90°$. A characteristic example is the case of the angular distributions denoted by 4 in Fig. 8, which were found to be asymmetric.

5/ Detection of accelerated low-energy light ions

In the end we shall give account of some of our preliminary experimental results obtained for the detection of the tracks of low-energy accelerated O^+, N^+, C^+, and He^+ ions in Makrofol-E foils. One of the aims of these investigations was to get direct information on the etching properties of the tracks of low-energy residual nuclei produced in nuclear reactions as well as of the tracks of recoils produced in the detector by neutrons.

The singly-charged ^{12}C, ^{14}N and ^{16}O ions were accelerated by a 5MV Van de Graaff generator. The desired ion beam was chosen with magnetic deflection, and after scattering on thin gold foil it reached the detector surface at right angles /in the irradiation unit shown in Fig.6/. The energy spread due to the scattering foil was less than 0.1 MeV.

One of the typical results of the track etching experiments is summarized in Fig. 9, from which it can be established that on the basis of the track diameters the light ions of different energies can be well discriminated. At the same time in this low-energy region the track diameter is hardly sensitive to the particle type.

On the right side of Fig. 9 a comparison is made between the responses of the Makrofol-E sheets to ^{16}O and ^{4}He ions with two different etchants. It can be seen that the PEW and 30 % KOH solutions give identical curves for the growth of the diameters for ^{16}O ions, but not for alpha particles. It is obvious from this that if we wish to apply PC foils as neutron detectors, the use of the PEW solution is more advantageous since it gives higher registration sensitivity and efficiency. In nuclear reaction investigations, however, where the residual nucleus is to be detected in alpha-particle background it is recommended to use 30 % KOH.

In another experiment we studied the response of the Makrofol-E foils to very low-energy ($<$0.2 MeV/amu) accelerated ^{4}He^{+} ions, too. Here the investigations performed with a monoenergetic beam are very interesting in order to decide whether etchable tracks can be produced in this energy loss region. In this region measurements, so far performed does not allow us to make the decision whether there exists a so-called lower registration threshold, since in these measurements alpha-particles of greatly degraded energy from radioactive sources were used, where the energy straggling is high.

In our experiments the response of PC sheets to alpha particles was investigated in the energy range 1 MeV - 0.2 MeV. With normally entering particles and using PEW etchant the d(h) track diameter versus layer removal curves were determined in small etching time steps. Then, according to the procedure described in ref. [4], from the straight portion of the $d^2 = f(h)$ curves we determined the values of the etchable track lengths. The range-energy curve obtained is presented in Fig. 10 /left/. It can be seen that the extrapolation of this curve towards the lower energies allows the assumption of a lower registration threshold of 0.1 MeV at most. Our range data as shown by the right-side curve in Fig. 10, are systematically lower than the theoretical values obtainable from the range-energy table of Henke and Benton. However, this table does not claim to give realistic data for the range at such low energies, thus, the divergence between the experimental and theoretical ranges must be attributed to this fact and not to the existence of a non-etchable track end.

References

[1] G. Somogyi: Nucl. Instr. Meth. 42 /1966/ 312 and ATOMKI Közl. 8 /1966/ 218
[2] G. Somogyi, B. Schlenk, M. Várnagy, L. Meskó, A. Valek: Nucl. Instr. Meth. 63 /1968/ 189
[3] G. Somogyi, B. Schlenk: Rad. Effects 5 /1970/ 61
[4] G. Somogyi, A.S. Szalay: Nucl. Instr. Meth. 109 /1973/ 211
[5] G. Somogyi, J. Gulyás: Radioisotopy 13 /1972/ 549
[6] L. Meskó, B. Schlenk, G. Somogyi, A. Valek: Nucl. Phys. A130 /1969/ 449
[7] I. Hunyadi, B. Schlenk, G. Somogyi, D.S. Srivastava: Acta Phys. Hung. 30 /1971/ 73
[8] I. Hunyadi, G.M. Osetinsky, B. Schlenk: ATOMKI Közl. 16 /1974/ 289
[9] M.H.S.Bakr, I. Hunyadi, B. Schlenk, G. Somogyi, A. Valek: ATOMKI Közl. 16 /1974/ 123
[10] I. Hunyadi, E. Koltay, L. Zolnai: ATOMKI Közl. 17 /1975/ 371
[11] J. Szabó, J. Csikai, M. Várnagy: Nucl.Phys. A195 /1972/ 527
[12] S. Szegedi: Acta Phys. Hung. 34 /1973/ 215
[13] M. Várnagy, J. Csikai, J. Szabó, S. Szegedi, J. Bánhalmi: Nucl. Instr. Meth. 119 /1974/ 451

[14] V. Zoran, Gh. Popescu, Stud. Cercet. Fiz. 23 /1971/1127
[15] H. Nakayama, M. Ishii, K. Hisatake, F. Fujimoto, K. Komaki: Nucl. Phys. A208 /1973/ 545
[16] R.P. Sharma, J.U. Andersen, K.O. Nielsen: Nucl. Phys. A204 /1973/ 371
[17] W. Dolak, D. Lehmann, K. Lindner, G. Otto, D. Reich, H.J. Treffkorn: Jahresbericht, Rossendorf, ZfK-283 /1974/ 20
[18] H.J. Treffkorn, G. Otto, V. Geist, Jahresbericht, Rossendorf, ZfK-283 /1974/ 103.

Table 1. Use of plastic track detectors for studying low-energy nuclear reactions emitting alpha-particles

NUCLEAR REACTION STUDIED	ENERGY	QUANTITY MEASURED	DETECTOR USED	REFERENCES		LABORATORY
$^{19}F(d,\alpha)^{17}O$	E_d=459,407,355 303 keV	$\sigma(\theta)$ for $\alpha_0, \alpha_1, \alpha_2, \alpha_3$	CA	/6/	1969	INSTITUTE OF NUCLEAR RESEARCH, DEBRECEN, HUNGARY
$^{27}Al(d,\alpha)^{25}Mg$	E_d=650,585, 540 keV	$\sigma(\theta)$ for α_0, α_1	CA	/7/	1971	
$^{15}N(^{3}He,\alpha)^{14}N$	E_{3He}=2-4 MeV	$\sigma(E), \sigma(\theta)$ for $\alpha_0, \alpha_1, \alpha_2$	CA	/8/	1974	
$^{14}N(d,\alpha)^{12}C$	E_d=640,510, 350 keV	$\sigma(\theta)$ for $\alpha_0, \alpha_1, \alpha_2, \alpha_3$	PC	/9/	1974	
$^{27}Al(p,\alpha)^{24}Mg$	E_p=1540-1930 keV	$\sigma(E)$ for α_0, $\sigma(\theta)$ at resonances	PC	/10/	1975	
$^{10}B(p,\alpha)^{7}Be$	E_p=60-180 keV	$\sigma(E)$	CA	/11/	1972	INSTITUTE OF EXPERIMENTAL PHYSICS, DEBRECEN HUNGARY
$^{9}Be(d,\alpha)^{7}Li$	E_d=500,400, 300,200 keV	$\sigma(\theta)$ for $(\alpha_0 + \alpha_1)$	CA	/12/	1973	
$^{6,7}Li(p,\alpha)^{3,4}He$	E_p=100-180 keV	$\sigma(E)$	CA	/13/	1974	
$^{27}Al(p,\alpha)^{24}Mg$	E_p=7.4 MeV	$\sigma(\theta)$	CA	/14/	1971	INST. OF ATOMIC PHYS. BUCHAREST, RUMANIA
$^{27}Al(p,\alpha)^{24}Mg$	E_p=1183,937, 731,633 keV	lifetime from blocking	CN	/15/	1973	TOKYO INST. OF TECHNOLOGY, JAPAN
$^{31}P(p,\alpha)^{28}Si$	E_p=642 keV	lifetime from blocking	CN	/16/	1973	INST.OF PHYS. UNIV. OF AARHUS DENMARK
$^{27}Al(p,\alpha)^{24}Mg$	E_p=633 keV		CN	/16/	1973	
$^{27}Al(p,\alpha)^{24}Mg$	E_p= 936 keV	$\sigma(\theta)$	CA	/17/	1974	K.MARX UNIV. LEIPZIG GDR
	E_p=1183 keV	lifetime from blocking	CA	/18/	1974	

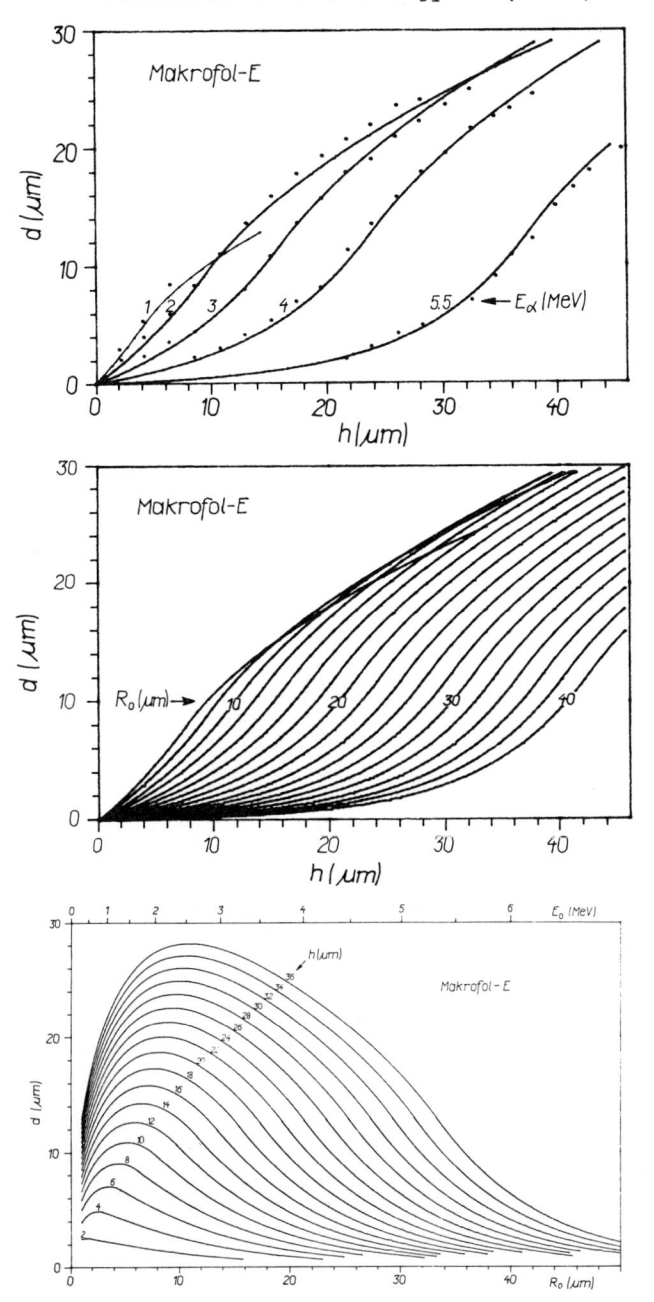

Fig.1 Upper figure: experimental /dots/ and least-squares fitted theoretical /solid lines/ track diameter versus layer removal curves for different alpha-energies in Makrofol-E foil etched in PEW solution /see in the text/ at 70°C. Lower figure: Theoretical track diameter versus layer removal curves when using the relation $V(x_o) = 1+\exp/-A(R_o-x_o)+B/$ with $A=0.22$ /μm^{-1} and $B=0.45$ in the formulas /1/ and /2/

Fig.2 Theoretical track diameter versus particle range curves at different layer removals for alphas entering Makrofol-E foil at rigth angles, calculated with $V(x_o)$ mentioned in the text of Fig.1.

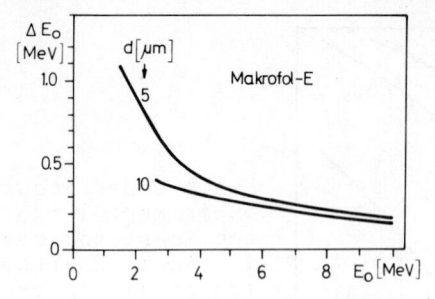

Fig.3 Theoretical energy resolving power for separating the alpha-groups $\alpha(E_0)$ and $\alpha(E_1)$ with the unaided eye under optical microscope where ΔE_0 is defined as E_0-E_1 if for the mean diameters the $d(E_0)-d(E_1)=2$ μm condition is fulfilled.

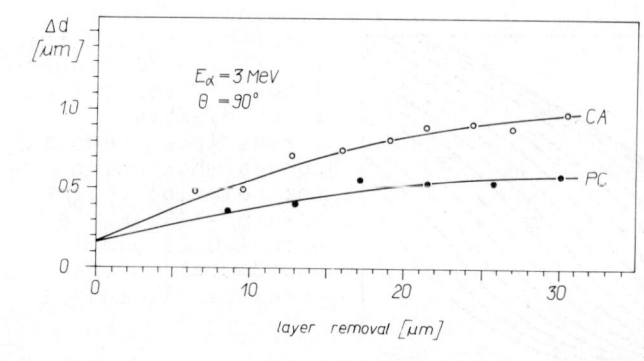

Fig.4. Halfwidth of the track diameter distribution curves for normally incident 3 MeV alphas as a function of layer removal in Cellit-T etched in B solution [2] and in Makrofol-E etched in PEW solution /see in the text/ at 70°C.

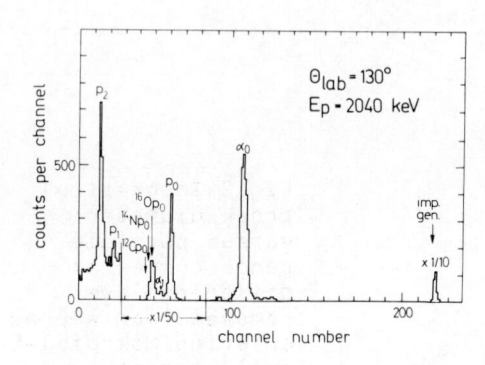

Fig.5 Charged particle spectrum measured by semiconductor detector in the case of bombarding an Al target with protons.

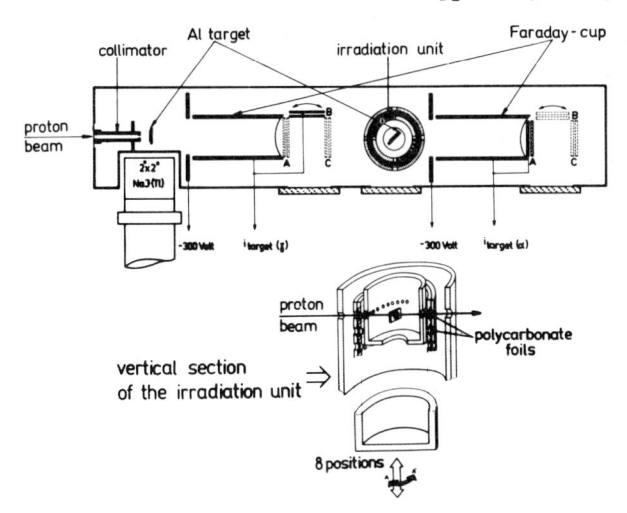

Detection of low-energy ^{4}He, ^{12}C, ^{14}N, ^{16}O ions   1259

Fig.6. Schematic view of the experimental arrangement used for excitation function and angular distribution measurements in (p,α) nuclear reactions. The lower drawing shows an enlarged sectional picture of the unit in which the irradiation and quick replacement of the plastic foils in vacuum can be performed.

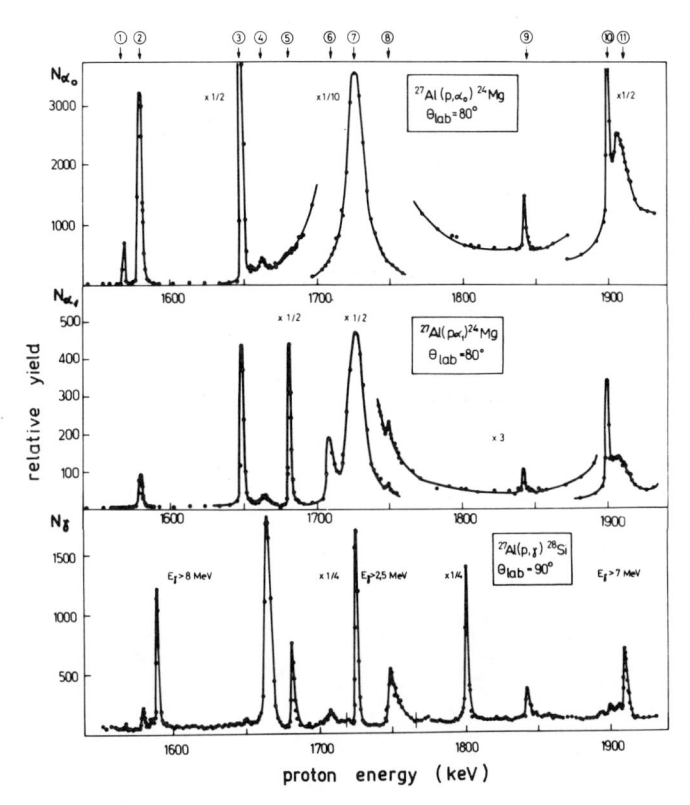

Fig.7 Excitation functions measured with Makrofol-E foils at 80° for the α_0 and α_1 groups in the ^{27}Al(p,α)^{24}Mg reaction. In the lower part of the figure the excitation function of the ^{27}Al(p,γ) reaction measured with a scintillation crystal is shown.

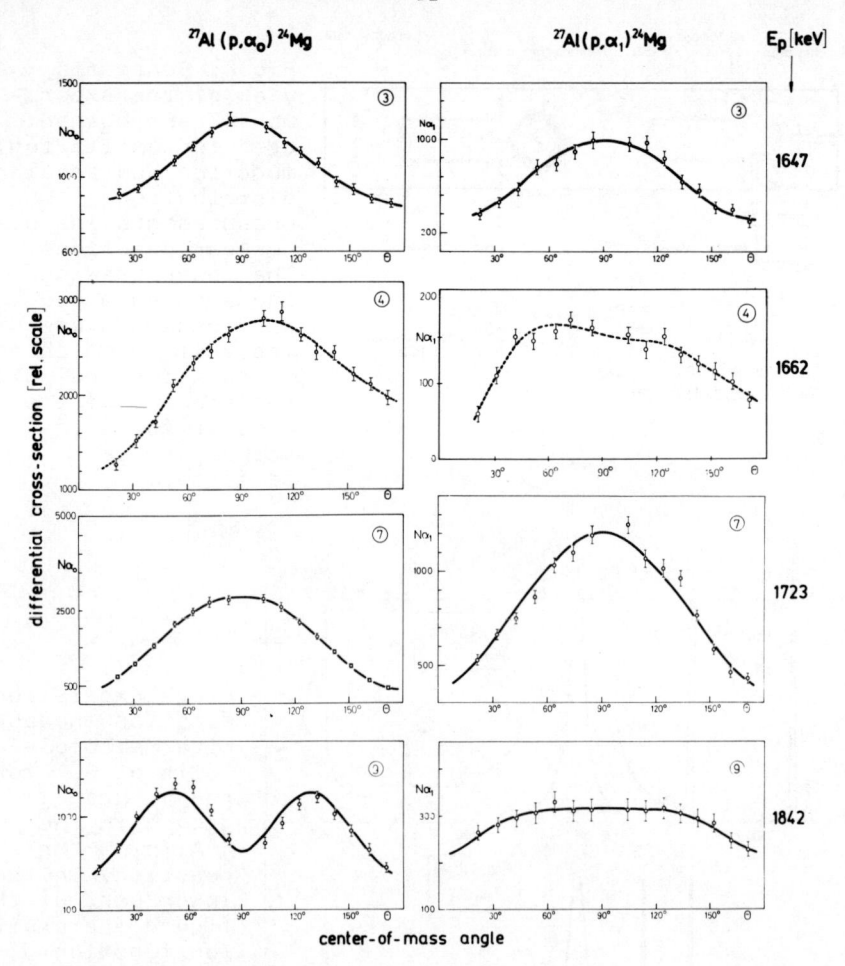

Fig. 8 Curves showing α_0 and α_1 angular distributions measured at different resonances, plotted in the centre of mass system.

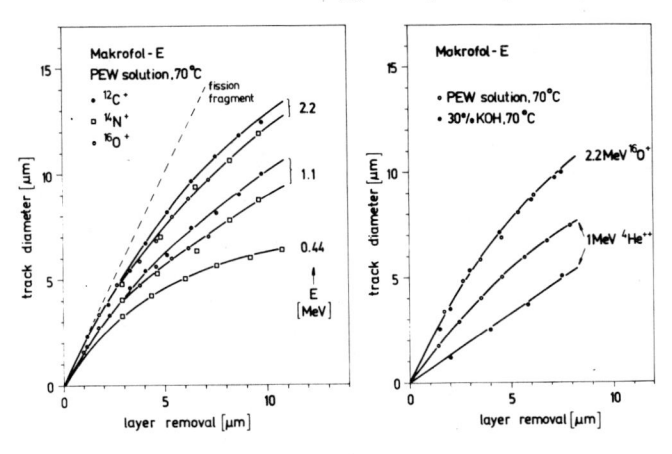

Fig. 9 Track diameter versus layer removal curves of low-energy C^+, N^+ and O^+ ions registered in Makrofol-E foil etched in PEW solution /left/ and those of O^+ and 1 MeV He^{++} for two different etching conditions /right./

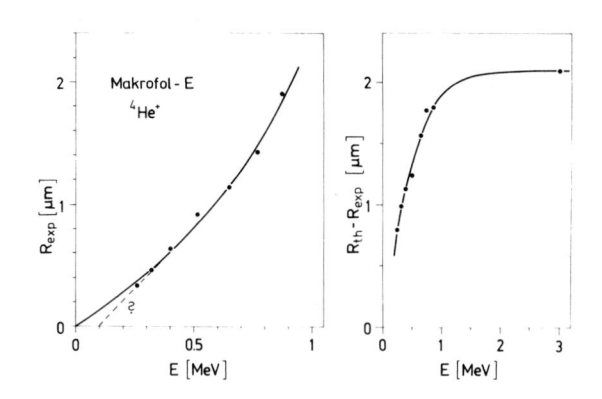

Fig. 10 Experimental range-energy relation of low-energy accelerated He^+ ions in Makrofol-E /left/ and the difference between the theoretical and experimental ranges as a function of particle energy /rigth/.

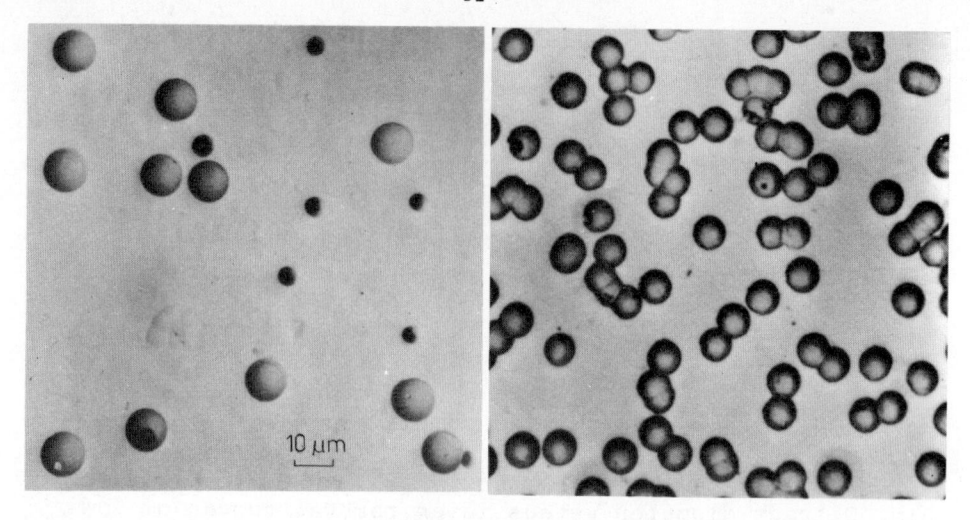

Fig.11 Microphotos of the tracks of 1 MeV and 3 MeV alpha particles /left/, and 2,2 MeV accelerated $^{16}O^+$ ions /right/ entering Makrofol-E at right angles. The alpha and $^{16}O^{+9}$ ion tracks were etched at 70°C in PEW and 30 % KOH solutions, respectively.

Combined Session 11

Space Biophysics

Chairmen: J. P. Massue
O. C. Allkofer

DOSIMETRIC SIGNIFICANCE OF COSMIC RADIATION IN THE ALTITUDE OF SUPERSONIC TRANSPORTS AND IN FREE SPACE

O. C. Allkofer

Institut für Reine und Angewandte Kernphysik, University of Kiel, 2300 Kiel, West Germany

The integral cosmic-ray-flux and hence the dose rate increases with altitude. At the cruising altitude of the subsonic jets, about 10 km, the dose rate is already about a factor 70 higher than at sea level. At the higher altitudes of SST the situation is different because the composition of the galactic component differs from that at the subsonic level, the solar flares are more efficient, and a small number of heavy nuclei are still present. In free space an additional radiation hazard appears when the radiation belts have to be crossed.

1. Introduction

When cosmic-ray particles interact with tissue, they produce a radiation dose similar to a radioactive source. At sea level, and at mountain altitudes, this dose is negligible compared with doses from environmental radioactivity. The integral cosmic-ray flux and hence the dose rate increases with altitude, the latter much more rapidly, due to the changing composition of cosmic radiation. At the cruising altitude of the subsonic jets, about 10 km, the dose rate is already about a factor of 70 higher than at sea level. However, the cosmic radiation from solar flares is almost completely absorbed higher in the atmosphere, and heavy nuclei play a minor role at this level, because they are more or less completely fragmented in higher atmospheric layers.

At the higher altitudes of the Supersonic Transports (SST), the situation is different because the composition of the galactic component differs from that at the subsonic level, the solar flares are more efficient, and a small number of heavy nuclei are still present. The dose rate is about a

factor 5 more than at the subsonic level, and about 400
times higher than at sea level (20 km).

In free space, there is no longer any shielding against solar
flares, and an additional radiation hazard appears when the
radiation belts have to be crossed. For long, manned space
flights, the effects of all the cosmic ray components have
to be considered.

A useful sub-division of the radiation hazards due to cosmic
radiation is as follows:

-- radiation dose produced by galactic radiation
-- local radiation injuries due to heavy nuclei of the
   galactic component
-- total radiation hazard due to solar flares
-- occasional very high dose rates due to strong solar flares.

In the following a short survey about galactic and solar
cosmic radiation will be given, then the main features about
geomagnetic effects and interaction in the atmosphere will be
treated. In the further paragraph the main points about
dosimetry relevant to this item will be mentioned and finally
the radiation hazard at the level of supersonic transport
and in free space will be treated.

2. Galactic Cosmic Radiation

The primary cosmic radiation is incident isotropically at the
top of the atmosphere and is composed of 86 % protons, 13 %
alpha-particles, and 1.4 % heavy nuclei. Table 1 shows the
charge spectrum with the relative abundance, relative mass,

Table 1 Composition of heavy nuclei

charge	integral flux $(m^2 \, s \, sr)^{-1}$	relat. frequency %	relat. mass %	relat. total energy %
1	610 ± 30	86.0	54.6	70.7
2	90 ± 2	12.7	32.2	21.0
3 – 5	2 ± 0.2			
6 – 9	5.6 ± 0.2	1.3	13.2	8.3
10 – 19	1.4 ± 0.2			
20 – 29	0.4 ± 0.1			

and relative total energy of the primary particles. Figure 1 shows a diagram of the abundances of nuclei relative to C6 at the top of the atmosphere /1/. The primary flux is dependent on the geomagnetic latitude and is modulated by solar events. A mean number for the flux of primary particles as measured with a counter telescope during a rocket flight is
$I = 0.03$ cm^{-2}s^{-1}sr^{-1} at a geomagnetic equator and
$I = 0.3$ cm^{-2}s^{-1}sr^{-1} at a geomagnetic latitude of 60°. This is because the cutoff-energy varies with geomagnetic latitude;

Fig. 1.

the cutoff-energy has its maximum value at the geomagnetic equator and decreases to zero for particles incident in the polar region. The protons of the primary component have an energy spectrum which covers the energy range from some MeV to 10^{20} eV. It can be described by a power law

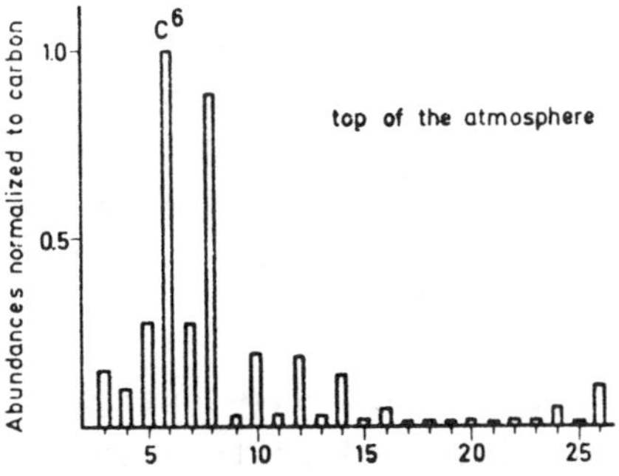

$$I = A \cdot E^{-r}$$

E   energy (eV)
A   constant
I   integral particle flux (cm^{-2}s^{-1}sr^{-1})

The exponent r has in the low energy region the value r = 1.6, at intermediate energies the spectrum steepens to r = 2.2.

The energy spectra of the heavy nuclei are of similar shape. Figure 2 shows the spectra for different charge groups /2/.

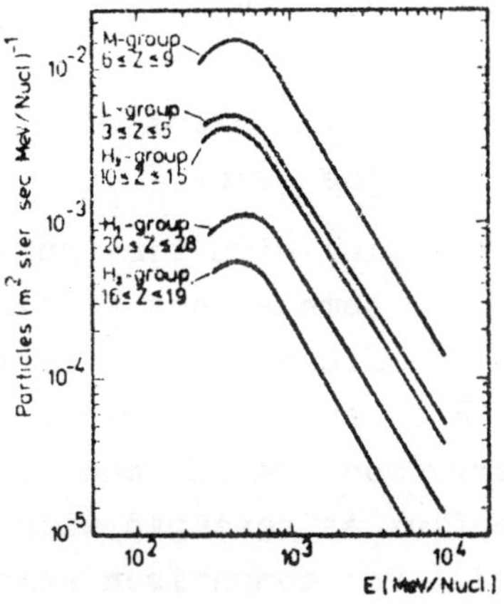

Differential energy spectrum of different charge groups of heavy primaries [2]

Fig. 2.

3. Solar Cosmic Radiation

Occasionally large eruptions of solar gas from the chromosphere take place on the sun the periods of which are typically one day. These solar flares consist of about 95 % protons. The information about the frequency of these events is of statistical kind. The number of eruptions decreases with increasing size. For events with remarkable fluxes a low rate of only 2 up to 3 per year is observed. The correlation of number and fluxes of flares with the number and size of sunspots have been observed since the last century. These events occur with a period of 11 years.

Most of the particles of a solar flare event have energies below 100 MeV, but some rare events involved protons up to some GeV. The individual events vary widely in intensity and energy distribution. Their time dependence can be approximated by an exponential law

$$N(>E,t) = I_o(t) \exp\left[- E/G(t)\right]$$

$N(>E,t)$ Number of protons with energy greater than E at the time t (particles $cm^{-2} s^{-1} sr^{-1}$)

$I_o(t)$ Characteristic parameter describing the flux of an individual event

$G(t)$ Characteristic parameter for the slope of the energy distribution of the event

Both characteristic parameters are changing with time during a flare. Figure 3 shows the spectra for some events at fixed times /3/. At the top of the atmosphere fluxes of 10^5 particles $cm^{-2} s^{-1} sr^{-1}$ with energies $E > 40$ MeV can be observed. Figure 4 shows the differential energy spectra of some flare events at the time of their maximum flux as measured with an ESRO satellite during 1968-69 /4/. For comparison some flare spectra along with the spectrum of the galactic component are shown in Figure 5.

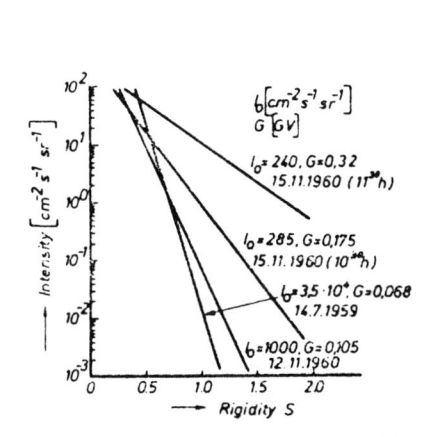

Some typical Flares Spectra [3]
Fig. 3.

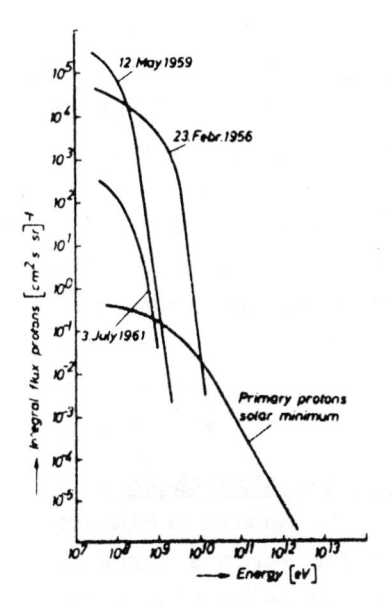

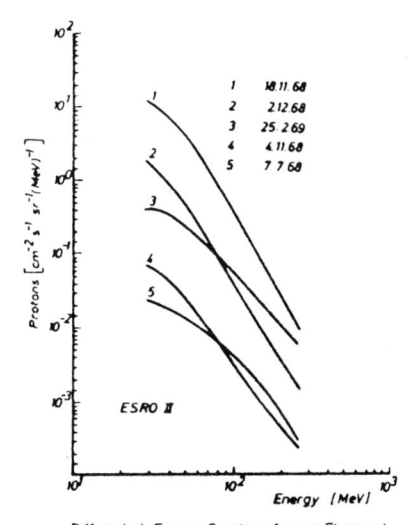

Fig. 4. Differential Energy Spectra of some Flares at the Maximum Intensity during 1968-69 as Measured with an Esro Satellite [4]

Comparision of Flare Spectra with the Galactic Component

Fig. 5.

4. Geomagnetic Effects and Modulation

Before entering the atmosphere the primary particles have to traverse the magnetic field surrounding the earth. The trajectories of the charged particles are bent by the field, leading to cut-off energies which depend on geomagnetic latitude, direction of incidence, and the charge of the particle. This leads to a latitude dependence of the primary as well as the secondary components. Fig. 6 shows the latitude effect of the primary flux for different years /5/. The latitude variation is also present in the secondary particles including neutron component /6/ (Fig.7). The solar flares enlarge the particle flux of the galactic component. But in addition to this effect there is in correlation with the solar 11-year-cycle a modulation effect imposed on

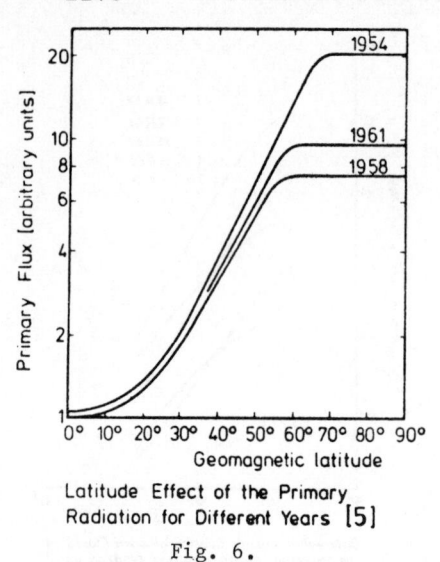

Latitude Effect of the Primary
Radiation for Different Years [5]
Fig. 6.

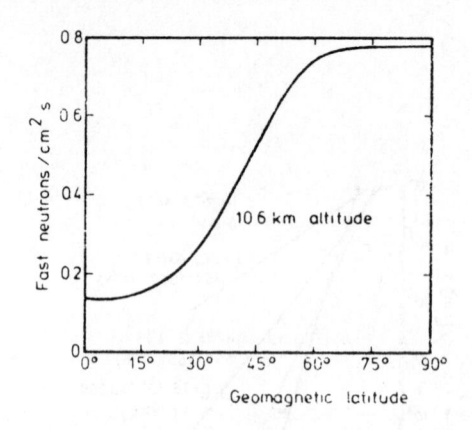

Latitude Profile of Fast Neutron Flux in
the Atmosphere [6]
Fig. 7.

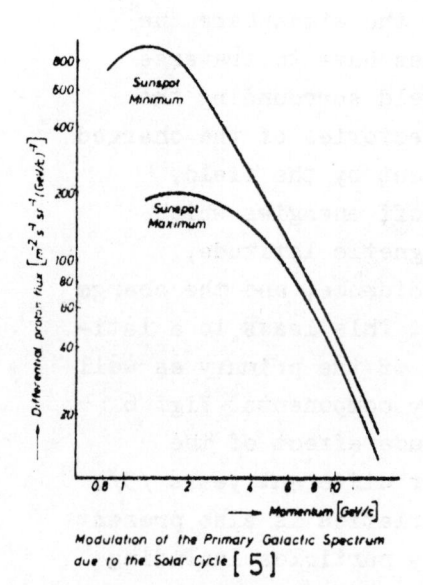

Modulation of the Primary Galactic Spectrum
due to the Solar Cycle [5]
Fig. 8.

the galactic radiation. The interplanetary space is filled with the solar wind, a plasma stream with frozen-in magnetic fields. High energy particles penetrate the plasma clouds with little difficulty, but below about 20 GeV the galactic spectrum is quite strongly modulated. So the differential proton flux at 1 GeV can vary by about a factor 5 over the solar cycle. In Fig. 8 the low momentum region of the primary spectra are shown at sun spot minimum and sun spot maximum, from which the maximum modulation can be taken.

5. Interaction in the Atmosphere

The primary galactic cosmic radiation is influenced by the earth's magnetic field, its intensity is modulated by solar activity, and it is transformed

by the interaction with the atmosphere. The energy loss due
to ionization and excitation for a vertical passage through
the atmosphere is about 2 GeV for single charged relativistic particles. But the primary protons produce in strong
interactions with the nuclei of the air mesons and hyperons
with an interaction length of about 80 g cm^{-2} and a mean
inelasticity of about 50 %. Furthermore nucleons and
fragmentation products are released form the nuclei in this
type of interaction. The instable hadrons decay directly
or via other particles into leptons mostly. The variety of
these processes leads to different particle compositions
and different energy spectra at different altitudes.
Fig. 9 shows roughly the altitude dependence of the main
components such as electrons, muons, and protons /7/.
The atmospheric neutron flux profile at a geomagnetic latitude
of 42° can be seen in Fig. 10 /8/.
The heavy nuclei are fragmented in interactions with air
nuclei producing thereby lower charged nuclei. In addition
they have a far higher energy loss due to ionization.

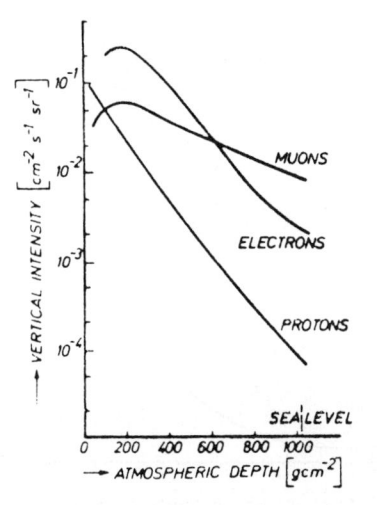

Fig. 9. Variation of Vertical Fluxes of the Main Components with Altitude[7]

Fig. 10. Altitude Profile of Fast Neutron Flux in Atmosphere[8]

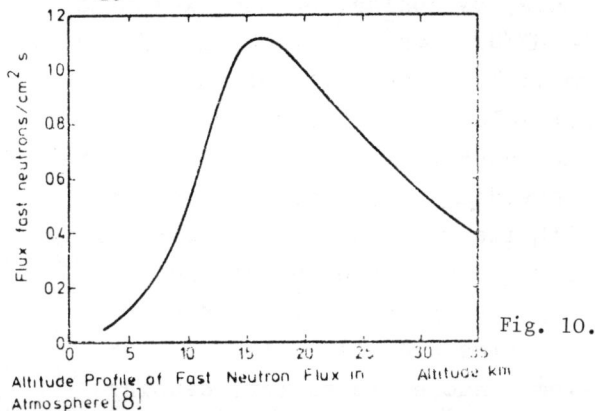

The mean free path for fragmentation is
dependent also on the charge of the nuclei.
Using the overlap model this dependence
on the charge has been calculated for air
based on the measured quantities in
nuclear emulsions /9/. Fig. 11 shows the
dependence of the mean free path for
fragmentation in air on the charge of the

nuclei. Thus a nucleus with a charge around Z = 20 has a four times shorter interaction length than the protons in the atmosphere. With a stack of plastic detectors the number of thin-down hits as a function of the atmospheric depth has been measured for nuclei with Z ≧ 10 /10/; this is shown in Fig. 12.

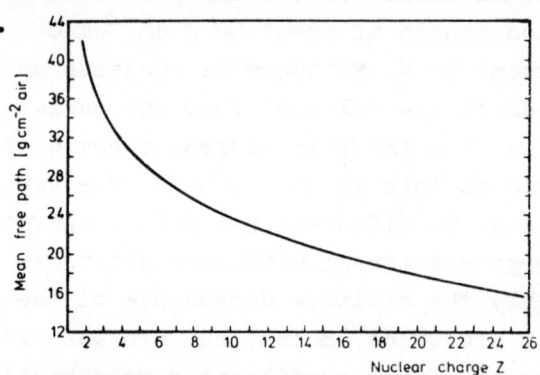

Fig. 11. Mean Free Paths of Nuclei in Dependence on the Mean Charge in Air [11]

6. Dosimetric Background

When a charged particle traverses material such a tissue, it loses energy by ionization and excitation, $(- dE/dx) \propto z^2/v^2$ (z charge of the particle, v velocity of the particle). The radiation damage increases with the charge Z, of the bombarding particle, and decreases with its velocity v. Thus, stopping heavy ions at the end of their range are particularly dangerous. Particles can also interact with atomic nuclei in matter, causing breakup or emission of mesons and hyperons. Low-energy secondary particles also have a high ionization density. Since neutrons undergo nuclear interactions and produce secondary particles, they too have a similar effect.

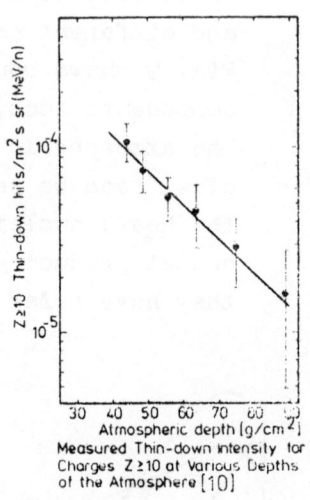

Measured Thin-down Intensity for Charges Z ≧ 10 at Various Depths of the Atmosphere [10]

Fig. 12.

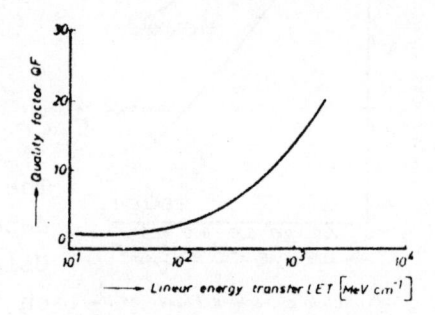

Variation of the quality factor QF, with linear energy transfer LET

Fig. 13.

The toxicity of a specific type of particle increases with Linear Energy Transfer, LET, i.e. more ions are produced within a given distance. A dose unit which takes into consideration the degree of toxidy is the rem (radiation equivalent man). If two different kinds of charged particles produce the same rad dose, they have different radiotoxidities if they differ in LET, since toxicity depends on the distribution of ionization along the particle trajectory. A quality factor QF, is introduced, which increases with LET (Fig. 13). The dose unit which takes into consideration the differences in LET is the rem

$$rem = QF \cdot rad.$$

This is the most convenient unit for describing the biological radiation hazard.

Heavy nuclei have high QF values, because of their high charges, whereas stopping particles have high QF values, because of their low velocities. Strongly interacting particles, such as protons and neutrons produce secondaries with high charge and low velocity, and consequently also have high QF values. Heavy nuclei have the highest QF values with $QF \leq 20$. It is useful to know for comparison, what radiation dose people receive from the environment. A natural component is present due to radioactivity and an artificial one due to medicin and technical applications. Table 2 gives a summary of average yearly doses. The mean radiation dose per year is seen to be 250 to 300 mrem. The amount of

Table 2   Natural and artificial environmental radiation doses

type of irradiation	mean dose rate (mrem a^{-1})
external irradiation	50
cosmic radiation	30
internal irradiation	50
medicine (diagnostics and therapy)	130
technical applications	2
radioactive fallout from nuclear explosions	20
professional exposure (reactors, x-ray therapy, radionuclides)	2
total	250 - 300

radiation damage is strongly dependent upon which organ of
the body is exposed to radiation. There are different
tolerance levels for different organs. A differentation is
made between somatic effects, genetic effects and effects to
a foetus or embryo in the early stages of development.

7. Radiation Hazard at the Level of Supersonic Transport(SST)

The cruising altitude of SST flights is now 16 km and could
be in future between 17 and 23 km. In order to evaluate the
radiation hazard, the particle fluxes of the galactic component, in particular the flux of heavy nuclei, and the
radiation produced by solar flares have to be considered
separately. Fig. 14 gives the fluxes of the main cosmic-ray
components present at subsonic and
SST levels at high geomagnetic lati-
tudes.

In principle hadronic and electro-
magnetic cascade multiplication can
take place in the wall of the aircraft.
However, because the characteristic
lengths for these processes in aluminum (interaction length λ_{Al} = 32,3 cm

Flux composition at the level of SST and at subsonic level [3]   Fig. 14.

and radiation length X^o_{Al} = 9.1 cm) exceed the wall thickness, they can be neglected for dose rate evaluation. The
dose rates at SST level have been measured and calculated.
Variations in the results have been obtained due to the
latitude effect, altitude changes, change with the 11-year
cycle, and possibly due to different measuring methods.
Values between 0.8 and 4 mrem h^{-1} have been measured at
various SST-levels. At a cruising altitude of 20 km, a dose
rate of 2 mrem h^{-1} is a reasonable average figure. A duty
time of 20 hours per month at SST altitude, leads to a dose
rate of 480 mrem a year.

Dosimetric significance of heavy nuclei. The intensities of
heavy nuclei are very low, because the primary flux is only
about 1 % of its initial value, and because fragmentation

occurs in the upper atmosphere. However, it is generally believed that single hits of low-energy or stopping nuclei in tissue, can produce a high local damage because of the large amount of energy deposited. Such stopping tracks of heavy nuclei are called enders or thin down hits. The number of stopping nuclei for different charge groups have also been both, measured and calculated. Fig. 15 shows the omnidirectional flux of stopping nuclei per cm^3 of tissue, for different atmospheric depths, at the solar minimum. From these values, the number of stopping nuclei within a sphere of tissue of diameter 12 cm, simulating a human body, has been calculated for different charge groups at an altitude of 20 km. For nuclei with charge $Z \geq 6$ a value of 5.6×10^{-3} stopping nuclei per cm^3 of tissue per hour was obtained. The radiation hazard caused by these low fluxes can barely be evaluated, although a possible collision with an embryo during organogenesis could have a serious result. Taking a value of 1 cm^3 for the embryo, and a duty time of 40 h per month, the probability of such a hit for a pregnant crew member is 20 % per month, at an altitude of 20 km. At the altitude of 16 km the probability is reduced by a factor of 10. For passengers the probability is lower, because less time is spent at SST level, although the total risk increases with the number of passengers.

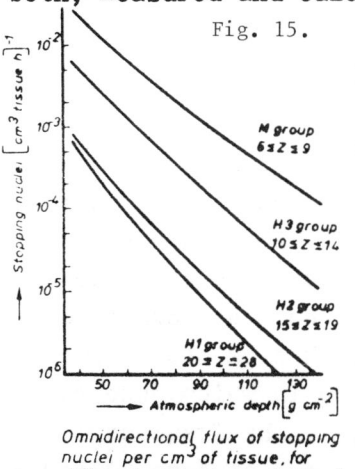

Fig. 15. Omnidirectional flux of stopping nuclei per cm^3 of tissue, for different atmospheric depths at solar minimum [10]

Solar flares. There is at present no direct experimental information on the time profile of the radiation surge at SST altitude during large solar flares. Only those flares which are able to produce extraordinarily high intensities at SST altitude are important. Such flares occur during the time of the active sun, with a frequency of about 12 per year.

Fig. 16. Dose rate versus atmospheric depth at solar minimum [9]

Fig. 17. Depth dose distribution produced by average flare events at SST level

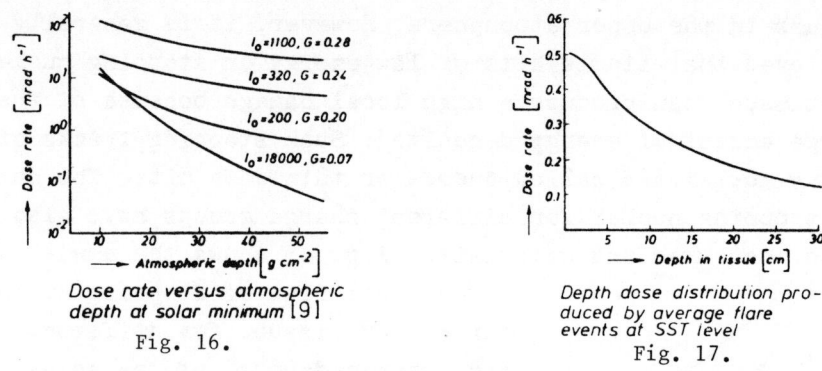

Fig. 16 shows the variation of proton dose rate near the polar region, at different altitudes, for some typical flares. Flares produce no effects in the equatorial region because of the high geomagnetic cut-off. Secondary neutrons produced in the atmosphere by flare protons, contribute 20 to 50 % of the dose rate, depending on the flare parameters. Occasionally events of very high flux occur, such as the event of the 23 rd February 1956 in which dose rates of some rems were produced at SST level. The frequency of high-intensity events producing dose rates of some 100 mrem h^{-1}, is about 2 per year. In addition to the high-intensity flares, there are other flares which produce an average radiation dose of about 1 mrem h^{-1}, the same order of magnitude as the dose rate from the galactic component. To give an idea of the distribution of the dose rate in tissue, the depth-dose distribution is shown in Fig. 17, for a typical flare events. It is seen that the calculated, interior dose rates are only slightly smaller than those on the surface.

8. Radiation Doses in Free Space

In free space as at SST level, there exist the galactic component containing protons and heavy nuclei, in the solar flares, here producing higher dose rates because there is no atmospheric shielding, and the radiation belts. In order to evaluate the dose rates in free space, several kinds of

particles (electrons, protons, heavy nuclei), with various
energies must be considered. Dose rates in free space are
important for astronauts, particularly on long-term missions.
Some dose rates for different sources are given in Table 3.
The reason for the wide range of values is that there are
modifications due to the influence of the magnetosphere, the
solar cycle and the geomagnetic cut-off.

Table 3 Radiation doses in free space [12]

radiation	dose rate	remarks
galactic radiation	0.01 - 0.05 rad d^{-1} 0.07 - 0.30 rem d^{-1}	a 10-day Apollo mission ~0.5 rad
radiation belts	protons: 1 - 10 rad h^{-1} (behind a shielding of 1 cm Al) electrons: 10^2 - 10^3 rad h^{-1} (at surface)	traversing the radiation belts ~0.5 rad almost completely absorbed by the walls of the spacecraft
solar flares	12 - 350 rad per event at surface 1 - 15 rad at 2 cm depth of tissue	the very strong event of 23.2.56, ~1000 rad in tissue
heavy ions ($Z \geq 10$; LET $\geq$ 550 MeV cm^{-1})	(0.9 - 1.7) x 10^3 particles m^{-2} d^{-1}	1.5 hits cm^{-2} for a 10-day Apollo moon mission

Galactic radiation. Possible hazards due to the galactic
radiation are a shortening of the lifespan and an increased
possibility of contracting cancer. The only results available
are for mice. Applying these values to man, the values for
radiation damage given in table 4 are obtained. The conse-
quent shortening of life for astronauts is on average about
0.25 % of the mission time. Using observational data obtained
at clinics, where patients have been treated by radiotherapy,
is is predicted that the probability for contracting leukaemia

Table 4 Shortening of lifespan due to irradiation by the galactic component in free space [13]

radiation	LET (MeV cm^{-1})	lifetime shortening (d rad^{-1})	
		accute irradiation	chronic irradiation
primary protons, alpha particles and high-energy secondary particles	35	12	3
heavy ions (3 $\leq Z \leq$ 28), low-energy secondary particles and neutrons	35	24	24

is doubled in space missions.

Solar flares. At the advent of a solar flare, astronauts would be irradiated for a period of about one day. The walls of the space vehicle would provide some shielding. Fig. 18 shows the integral percentage of flare events behind 2 cm and 40 cm of aluminum. Because most flare particles are of relatively low energy, they do not penetrate far into the human body. Nevertheless, very strong flares produce appreciable doses at a depth of several cm of tissue; at 5 cm the dose is reduced about 20 % of that on the surface. The blood-producing organs situated near the skin, might receive as much as 30 rem during a strong flare event, and so in addition to the chronical injuries produced by galactic radiation, further radiation injuries can occur. The very strong flare of 23 rd February 1956, produced a radiation dose in free space of about 1 000 rem.

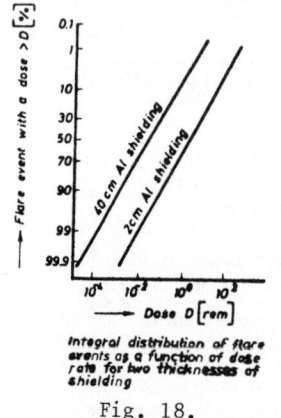

Fig. 18. Integral distribution of flare events as a function of dose rate for two thicknesses of shielding

Radiation belts. The potential hazards of irradiation in the radiation belts are also difficult to evaluate, because high fluxes of low-energy particles are present. At various locations within the radiation belts, protons produce dose rates of between 1 and 10 rad h^{-1}, behind 1 cm Al, an appreciably higher dose rate than that of the galactic radiation. During a space mission, the radiation belts are usually crossed by the spacecraft in less than one hour, and so the resultant dose rate is less than the accumulated dose of galactic radiation, during a 10-day mission to the moon. Because the energy of the electrons in the radiation belts is very low, most of them are absorbed in the walls of the space vehicles.

Heavy ions. The radiation hazard due to heavy nuclei has to be evaluated in terms of single events as at SST level. A comparison of the number of hits in space and at an SST

level of 20 km, finds that the crew members of an aircraft
during 20 years of service will receive the same number of
hits as astronauts on a 10-day trip to the moon. The
influence of heavy ions was studied during the Apollo
mission to the moon where astronauts saw light flashes with
closes eyes. Laboratory experiments found that such flashes
were due to hits on the retina of the eye. When a low-energy
heavy ion hits a cell nucleus it is very probable that the
cell nucleus will be destroyed because of the large amount
of energy deposited. Furthermore, because the region of
ionization around the particle path increases with the
charge of an ion, it is probable that every cell struck by
a heavy ion will be destroyed. Table 5 shows the number of
sensitive organ cells destroyed during the moon mission of
Apollo 12 and for a hypothetical long-term mission of two
years. The lower limit is calculated under the assumption
that a hit on any part of a cell destroys it. Under normal
environmental conditions, nerve-cells also die, making it
difficult to evaluate real radiation damage. Certainly, for
future long-time missions in space, such effect must be
taken into consideration. Experiments have even shown that
roots of hairs become grey, if they are hit by a heavy ion.

Table 5 Number of nerve-cells destroyed due to hits by heavy ions [13]

cells	number of cells destroyed per 10^6 cells	
	Apollo mission (12 days)	2 year space mission
cerebellum	0.50 - 0.65	40 - 50
netskin ganglies cells	0.64 - 5.7	50 - 500
cerebrum	2 - 14	16 - 120
shell of cerebrum (BETZ cells)	18 - 83	1050 - 6600
spinal cord	26 - 200	2000 - 16000

In order to obtain direct information about the biophysical
significance of the radiation hazard of heavy ions, the biostack experiment was performed. During the manned Apollo 16
and 17 missions to the moon, and the Apollo-Soyus-manoeuvre
a biostack was carried in the capsule /14/. Such an object

consists of several layers of biophysical specimens, alternatively stacked between track-sensitive detectors. Some of the specimens were shrimp eggs (Artemia Salina). Fig. 19 shows eggs under magnification, together with the track of heavy ion, which has penetrated one of the eggs and been measured in a plastic detector. Preliminary results found a high reduction in larval emergence and hatching, plus the beginnings of developmental anomalies in the struck eggs.

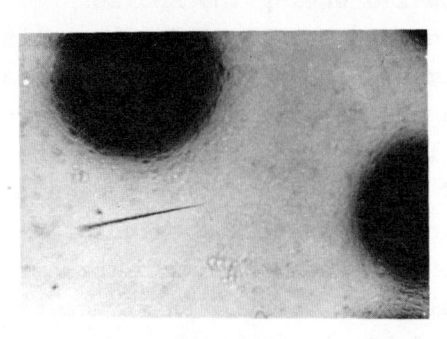

Fig. 19.

References

/ 1/ Shapiro, M.M., R. Silberberg: Ann. Rev. Nucl. Sci. 20 (1970) 323

/ 2/ Webber, W.R., I.F. Ormes, Journ. Geophys. Res. 72 (1967) 5971

/ 3/ Allkofer, O.C., M. Simon, Atompr. 16 (1970) 1

/ 4/ Engelmann I, Proceedings of the International Congress on Protection against Accelerators and Space Radiation, CERN 11-16, Vol.1 (1971) 463

/ 5/ Schopper, E., Handb. Phys. Springer-Verlag, Vol. X LVI/2 (1967) 372

/ 6/ Korff, S.A., R.B. Mendell, M. Merker, W. Sandie, Canad. J. Phys. 46 (1968) S 1023

/ 7/ Peters, B., Handbook of Physics, McCraw-Hill, New York (1958) 9-201

/ 8/ Haymes, R.C., J. Geophys. Res. 69 (1964) 841

/ 9/ Cleghorn, T.F., P.S. Freier, C.I. Waddington, Canad. J. Phys. 46 (1968) S 572

/10/ Allkofer, O.C., W. Enge, W. Heinrich, H. Röhrs, Proceedings of International Congress on Protection against Accelerators and Space Radiation, CERN 71-16, Vol. 1 (1971) 512

Allkofer, O.C., W. Heinrich, Nucl. Phys. B 71 (1974) 429

/11/ Allkofer, O.C., W. Heinrich, Health Physics 27(1974) 543

/12/ Wilson, I.J., Proceedings of the International Congress on Protection against Accelerators and Space Radiation, CERN 11-16, Vol. 1 (1971) 575

/13/ Schäfer, H.J., Biophysics 5 (1969) 315

Tobias, C.A. et al., AGARD Conference Preprint 95 (1971) C 6

/14/ Bücker, H., C. Horneck, E. Reinholz, W. Scheuermann, W. Rüther, E.H. Graul, H. Planel, J.P. Soleilhavoup, P. Cüer, R. Kaiser, J.P. Massúe, R. Pfohl, R. Schmitt, W. Enge, K.P. Bartholomä, R. Beaujean, K. Fukui, O.C. Allkofer, W. Heinrich, H. Francois, G. Portal, H. Kühn, H. Wollenhaupt, and C.H. Bowman

Part A. Biostack Experiment, Apollo 16, Preliminary Science Report, NASA SP-315 (1972) 27-1

Bücker, H., G. Horneck, E. Reinholz, W. Rüther, E.H. Graul, H. Planel, J.P. Soleilhavoup, P. Cüer, R. Kaiser, J.P. Massué, R. Phohl, W. Enge, K.P. Bartholomä, R. Beaujean, K. Fukui, O.C. Allkofer, W. Heinrich, E.V. Benton, E. Schopper, G. Henig, J.U. Schott, H. Francois, G. Portal, H. Kühn, D. Harder, H. Wollenhaupt, and G. Bowman

Biostack Experiment, Apollo 17, Preliminary Science Report, NASA SP-330 (1973) 26-1

Bücker, H., R. Facius, D. Hildebrand, G. Horneck, G. Reitz, U. Scheidemann, M. Schäfer, C. Thomas. B. Toth, A.R. Kranz, E.H. Graul, W. Rüther, M. Delpoux, H. Planel, J.P. Soleilhavoup, C.A. Tobias, T. Yang, E. Schopper, J.U. Schott, E. Obst, O.C. Allkofer, K.P. Bartholomä, R. Beaujean, W. Enge, W. Heinrich, H. Francois, G. Portal, R. Kaiser, J.P. Massue, R. Pfohl, C. Jacquot, E.V. Benton, and D.D. Peterson

Biostack III-Experiment MA-107, Apollo-Soyuz Test Project, Preliminary Science Report NASA TM X-58173 (1976) 14-1

RADIOBIOLOGICAL INVESTIGATIONS OF COSMIC HZE-PARTICLES WITH VISUAL TRACK DETECTORS IN THE BIOSTACK EXPERIMENT

R. Facius, G. Hölz, B. Toth and H. Bücker

*Arbeitsgruppe Biophysikalische Raumforschung, Universität Frankfurt,
D-6000 Frankfurt, Kennedy Allee 97, West Germany*

Abstract: The aim of the Biostack namely to contribute to a dosimetry of the HZE-particles as the highly structured component of the cosmic radiation leads to a variety of physical and biological requirements in order to achieve that goal. The temporal, spatial and atomic characteristics of that radiation field rendered visual track detectors with the capability of individually recording and correlating single hit events as the most suitable physical detectors. One of the objects selected had been the spores of Bacillus subtilis, which were flown in the missions of Apollo 16,17 and the Apollo-Soyuz Test Project (ASTP). A substantial biological improvement was the development of routine methods which allowed the investigation of various biological endpoints of one individual spore together with a more precise determination of its impact parameter with respect to the path of the heavy ion. Preliminary results of the ASTP experiment are presented which strongly support the findings of the previous Biostack experiments.

Introduction: Like the previous Biostack experiments I and II on board of Apollo 16 and 17 the current ASTP Biostack experiment contained spores of Bacillus subtilis as one of several other specially selected biological test organisms /1/. The analysis of especially the Biostack II results on Bacillus subtilis justified the anew inclusion of this object, in view of the since then advanced methods of localization and biological evaluation /2,3/ Among the questions raised by these results /4,5/ were the apparent indepence of the biological response in a target area of ≈ 6 $\mu m \emptyset$, and the question why effects of the observed magnitude could be correlated with the properties of the passing/penetrating particles at all - considering the known dose effect curve for Bacillus subtilis spores under low LET irradiation. The results so far obtained from the still progressing evaluation of the ASTP Bacillus subtilis subunit indicate that the pains taken with the advanced methods may have payed out.

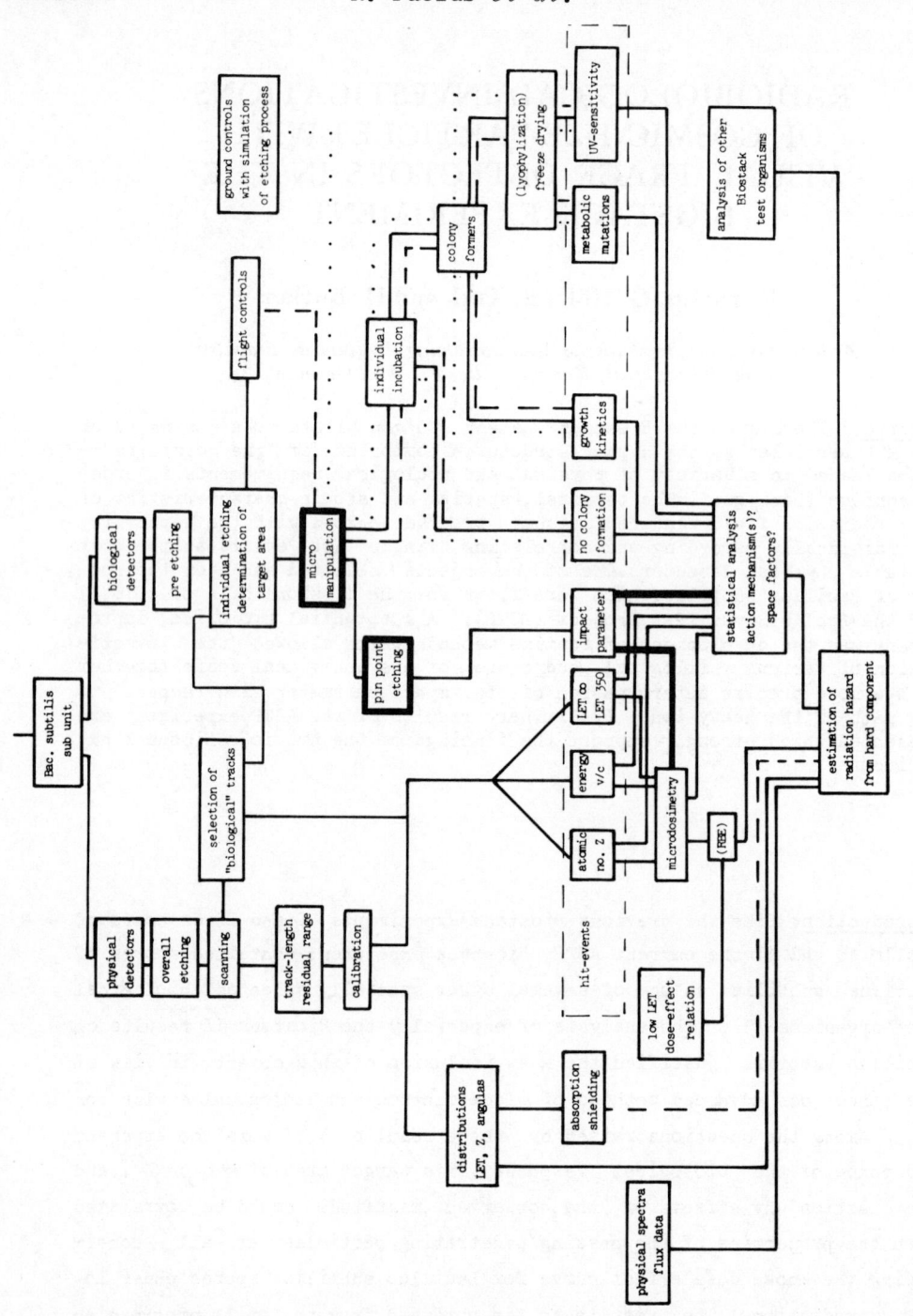

fig.1 : Concept for the ASTP Bacillus subtilis evaluation

Methical survey: In order to demonstrate the importance of the above mentioned experimental improvements, their position within a flow chart is given which represents the concept according to which the primary problem of the Biostack experiments is tackled, - namely of estimating the radiation hazards from the hard component of the cosmic radiation (fig. 1). The possibility of individually removing and incubating single spores (micromanipulation) allowed to use standard microbiological procedures as e.g. test of colony formation and of induced mutations, rendering their results comparable with other radiobiological findings. Thereby a variety of independent data on significant biological endpoints can be obtained. This increases the chances of the statistical analysis for the detection of correlations between the physical particle properties and the biological response if they do exist in reality. Taking into account the accuracy of localization attainable in the Biostack II experiment the distance to the particles trajectory was measured in terms of hit probability instead of using the impact parameter /5/. This impact parameter occupies a crucial position in the course of analysis, in so far as it links the particles characteristics and the biological endpoints in both the microdosimetric and the statistical analysis. The method of pinpoint etching /2/ rendered possible its sufficiently accurate determination and as a welcome side effect it significantly increased the fraction of evaluable particle tracks /3/. This experimental approach of individually correlating the biological effects with the properties of the passing HZE-particle is a mandatory prerequisite when the biological effectiveness of the cosmic radiation is to be investigated /8/. In the ASTP experiment the total absorbed dose was measured in thermoluminescence dosimeters as about 1 mJ/Kg /1/, a value where no biological effect at all is to be expected.

Results: So far only the target spores attached to the Daicelcellulose nitrate (CN_D) detectors are available for a preliminary analysis which comprises less than half of the total expected sample size. The biological endpoint to be discussed is mainly the colony forming ability (CFA) of the target spores. First of all the results for the CFA of the flight and ground controls - these already containing spores attached to the Kodak detectors (CN_K) - are compared under various aspects. Since the advanced procedures - especially the micromanipulation - are rather time consuming, table I shows the joint asymptotic χ^2 test of the hypothesis that during the time of analysis no decline of CFA developed among the spores. The re-

Table I: Test of temporal stability of controls

	Colony formation			Time of sampling
	(-)	(+)	total	
flight controls Daicel	6	44	50	Aug. 75 - Dec. 75
ground controls Daicel	10	79	89	Dec. 75 - Jan. 76
ground controls Kodak	9	80	89	Mar. 76 - Apr. 76
flight controls Kodak	11	64	75	Jun. 76 - Aug. 76
	36	267	303	

Null hypothesis:
All samples are taken from the same population

$\chi_3^2 = 2.96 \quad p(\chi^2 > 2.96, 3) = 0.40$

Table II: Colony formation of ground and flight controls

		Colony formation		$p[n_-(I) \geq n_-(II)]$	$p[n_-(I) \leq n_-(II)]$
		(-)	(+)		
ground controls Daicel	I	10	79		
flight controls Daicel	II	6	44	.55	.67
ground controls Kodak	I	9	80		
flight controls Kodak	II	11	64	.26	.87
ground controls Daicel	I	10	79		
ground controls Kodak	II	9	80	.69	.50
flight controls Daicel	I	6	44		
flight controls Kodak	II	11	64	.44	.75
ground controls	I	19	159		
flight controls	II	17	108	.27	.83

sulting test value for χ^2 is far beyond the (one-sided) critical value of 7.82 which would allow to refute this assumption on a 5 % level. Table II contains a detailed comparison between ground and flight controls and between spores from CN_D and CN_K detectors. None of the given exact multinomial error probabilities (FisherYates test) indicate any nonrandom discrepancy among the control samples. So the controls are pooled into one sample resulting in 89 % of CFA for the controls which have undergone the same treatment as the target spores. The value of 89 % compares pretty well with the value obtained for spores immediatly incubated. So the treatment as such does not influence the CFA. In fig. 2 the CFA of target spores and controls is given. Although the target spores comprise cells of up to 18 μm off the particles trajectory (see also fig.3 right ordinate) this time a highly significant decrease of their CFA is indicated by the error probabilities. The integral distribution of the number of evaluated target spores against their impact parameter B is given in fig. 3 (right ordinate). It shows that up to 4 - 5 μm the sampling can be considered to be unbiased. For impact parameters beyond 5 μm obviously some kind of necessary subjective selection of the spores to be manipulated became effective. The points refering to the left ordinate are derived as follows: Given an im-

Radiobiological investigations of cosmic HZE-particles

[Figure 2: Bar chart showing colony formation 100% for ground-controls, flight-controls, and target spores, with accompanying table:]

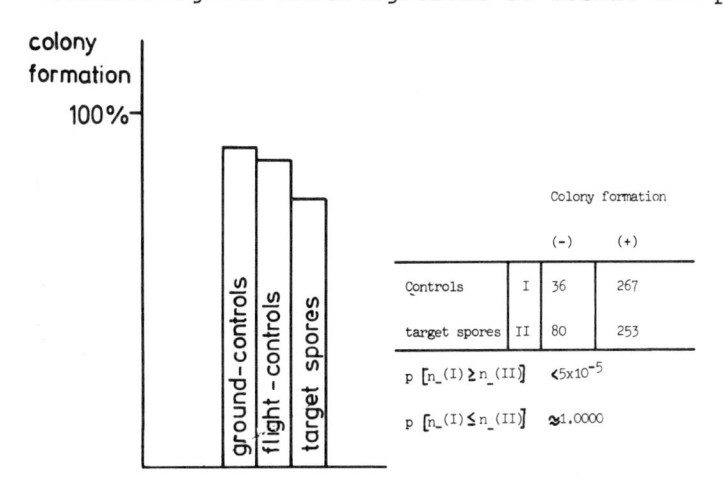

		Colony formation	
		(−)	(+)
Controls	I	36	267
target spores	II	80	253

$p\,[n_-(I) \geq n_-(II)]$ $< 5 \times 10^{-5}$

$p\,[n_-(I) \leq n_-(II)]$ ≈ 1.0000

fig.2

pact parameter value B the number N_B of target spores is counted which were lying closer to a particle track than B. Substraction of the expected number of noncolony formers according to the observed control value and division by N_B yields the integral net fraction of inactivated target spores. The points of fig. 4 are similarly derived by a method of moving averages with additional division by the size of the impact parameter interval. Fig. 4 may be considered as an attempt to determine a somewhat smoothed (negative) slope of the general trend of the data points in fig. 3 without destroying completely any possibly existing superimposed structure. Fig. 5 finally gives the partition of the so far evaluated target spores according to the atomic number of the HZEparticles. Concerning the additional biological endpoints the preliminary results are as follows: Up to now no antibiotic or auxotrophic mutations have been detected among about 200 investigated target spores. An increased sensitivity against UV-irradiation (more than a factor of 2 in the D_{37}) was detected for 2 target spores. For a comparison 2000 controls were irradiated, where no such change in the UV sensitivity was found. The probability for this finding to be a random event is 8.4×10^{-3}.

Discussion: Leaving aside the possibility of indirect effects fig.4 may be considered as an estimate of the probability density $k(B)$ that a viable spore will be killed if passed by an HZE-particle with impact parameter B. It indicates, that between 1 to 4 μm $k(B)$ decreases less steeply than B^{-2} as approximately it has to be expected from the conjecture that inactivation is caused by the radiation dose delivered by δ-rays. Even $k(B) \sim$

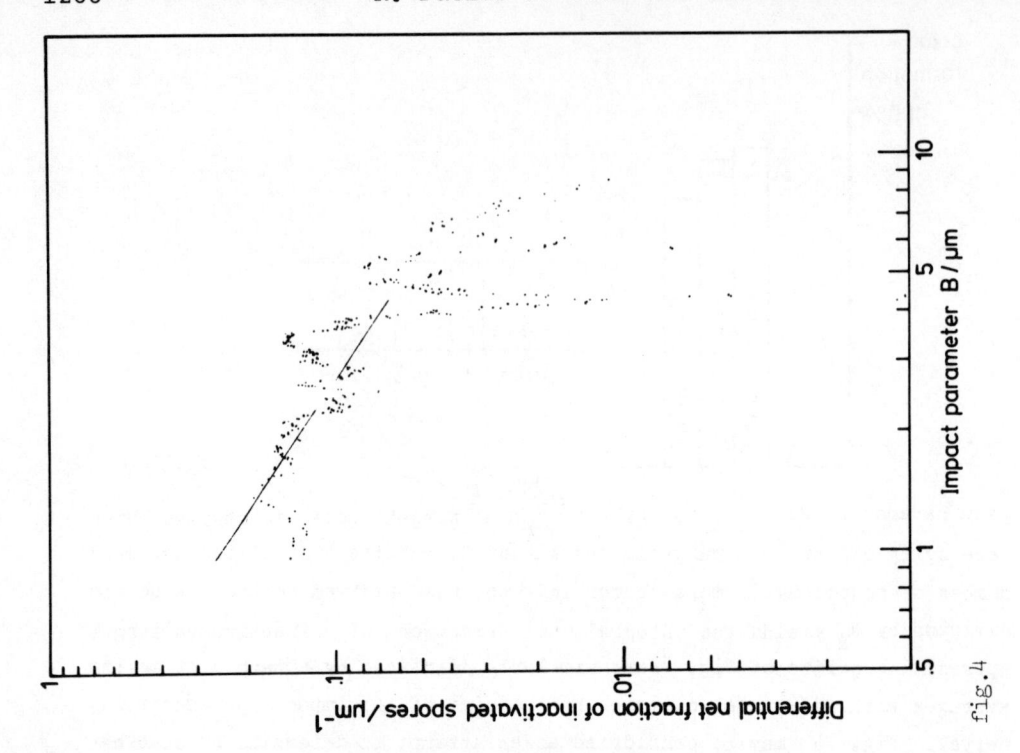

Integral fraction of spores with b ≤ B

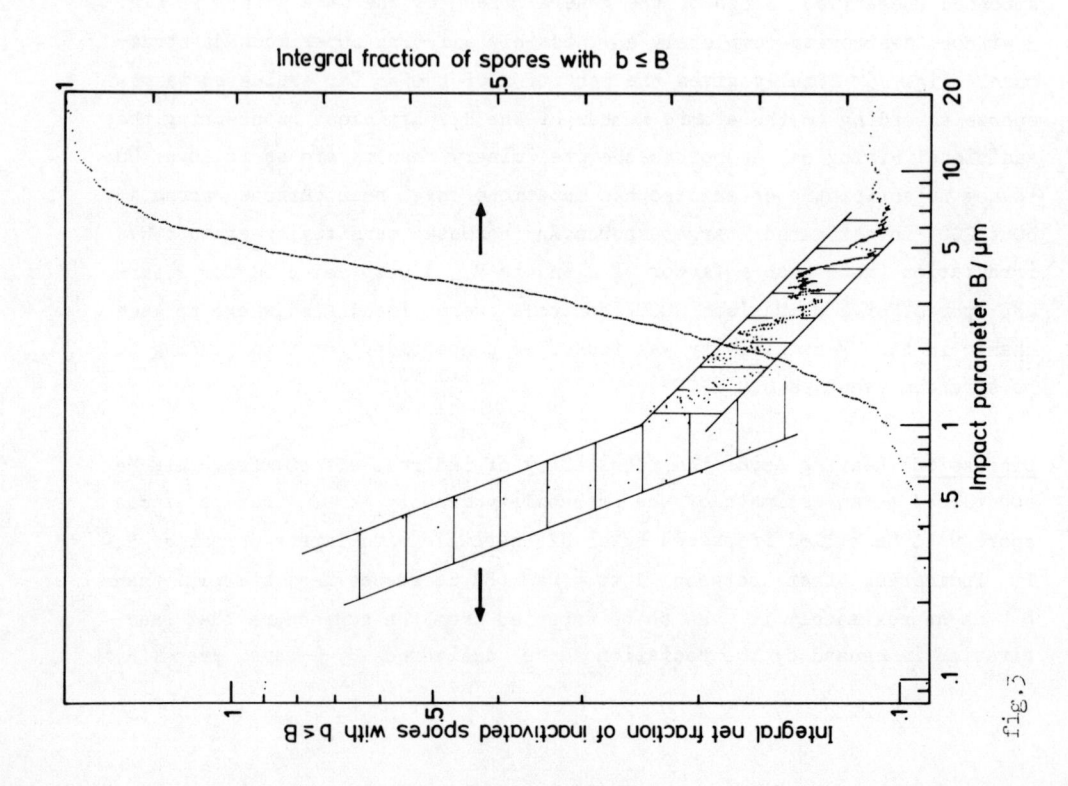

fig. 3

Differential net fraction of inactivated spores/μm⁻¹

fig. 4

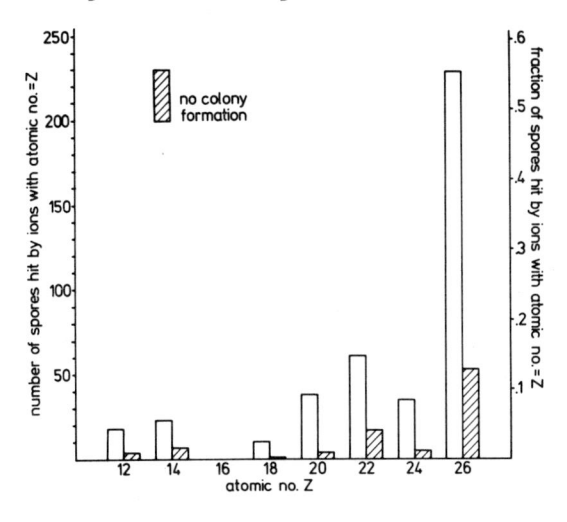

fig.5

B^{-1} as represented by the (manually adjusted) straight line seems to overestimate the decrease of effectiveness as shown by the data between 3 and 4 μm. This superimposed structure is also present in the data points of fig.3, where the trend of the data between 1 and 5 μm is delimited by (again manually adjusted) two parallel straight lines. So far no attempt was made to determine how far this structure may be related to the distribution of atomic numbers (fig.5) nor to check, whether it is significant at all. The distribution in fig.5 is somewhat discourageing as far the possible establishment of these correlations is concerned. The trend of datapoints below 1 μm in fig.3 (again indicated by two parallel lines) more closely corresponds to the behaviour expected from an inactivation by the δ-ray dose considering both slope and range.

Conclusions: The preliminary analysis of only one endpoint for approximately half of the expected total number of target spores already produced significant results. They confirm in a more direct way the conclusions derived as statistical inferences from the multifactorial analysis of the Biostack II experiment. Especially the conjecture of two superimposed different inactivation mechanisms seems to be corroborated. One short-ranged component may be traced back to the δ-ray dose. The second one extends at least to somewhere between 4 - 5 um off the particles trajectory. The detailed microdosimetric and multivariate analysis of the complete set of both the biological and physical data hopefully will add to a discrimination between these mechanisms.

Speculations: Concerning the longer-ranged component we still consider the (adhoc) hypothesis of (acoustic) shock waves as mentioned in /6,7/ as consistent with the observed approximate dependence of the inactivation effectiveness proportional to B^{-1} /5/.

References:

/1/ Bücker H. et al. in: Life Science and Space Research XV,
    Akademie Verlag Berlin (to be published)

/2/ Hildebrand D., Bücker H., Facius R., Schäfer M., Toth B., Rüther W., Pfohl R. and Kaiser R. in: Proc. 5th Symp. on Microdosimetry
    Commission of the European Communities, Luxembourg, EUR 5452, p.929

/3/ Schäfer M., Bücker H., Facius R. and Hildebrand D., presented at the
    9th Intern. Conf. on Solid State Nuclear Track Detectors
    München, Sept. 1976

/4/ Bücker H. et al., Apollo-Soyuz Test Project Preliminary Science Report
    NASA Rep.No. TM X-58173, p. 14-1 (1976)

/5/ Facius R. in: Life Science and Space Research XV,
    Akademie Verlag Berlin (to be published)

/6/ Budinger T.F. et al., Apollo-Soyuz Test Project Preliminary Science
    Report, NASA Rep.No. TM X-58173, p.13-1 (1976)

/7/ Chattarjee A. and Tobias C.A. in: Abstracts 5th Intern. Congress
    Radiat. Research, Academic Press, New York 1974 (p.297)

/8/ Bücker H., Facius R. and Schäfer M. in: Life Science and Space
    Research XIV, Akademie Verlag Berlin (in press)

HIGH PRECISION LOCALIZATION METHODS FOR HZE-PARTICLES

M. Schäfer, H. Bücker, R. Facius and D. Hildebrand

*Arbeitsgruppe biophys. Raumforschung, Universität Frankfurt/M.
D 6000 Frankfurt/M Kennedy Allee 97, West Germany*

Abstract:

For experimental investigations in the field of microdosimetry of heavy ions a high precision localization method for biological objects relativ to path of the penetrating ion is mandatory. New methods are described for corn seeds of Arabidopsis thaliana and spores of Bacillus subtilis. In the latter case the actual penetration point of the particle in the spore layer is determined by an individual microetching technique with an accuracy of about 0.2 um. This is done after removing the spores from the hit area by "break-through"etching. With the help of reference spores the impact parameter is determined for correlating the biological effects with the physical event. Besides the aspect of water during the track development is described.

Introduction:

The Biostack program was designed to study the biological effects of individual HZE-particles of galactic cosmic radiation, to study the influence of additional space flight factors, to obtain knowledge on the mechanism by which HZE-particles damaged biological materials. For these purposes the Biostack experiment includes a wide spectrum of biological objects like bacterial spores, protozoa cysts, plant seeds, shrimp eggs and insect eggs. The biological objects were sandwich-packed together with different track detectors as nuclear emulsions, plastics, silver chloride-crystals and lithium fluoride dosimeters (fig. 1).

During the space flight of Apollo 16 and 17 the dose contribution of the HZE-particles was only 0.5 percent of the total space radiation dose of about 900 mrad and 600 mrad respectively measured with integrating dosimeters. But the biological damage by single HZE-particles has been shown to be significant. This means that the average absorbed dose is obviously an inadequate term to describe the biological effects of such high structured radiation (1).

The configuration and the methods of the Biostack are considered as one of the most suitable ways to investigate the radiation effects of individual particles. It allows to evaluate these objects, which are affected with a high probability by a hit. The observed biological effects are then correlated with the physical quantities determined by continuous model calculations.

A critical requirement under these microdosimetric aspects is therefore a high precision localization for the biological objects relative to the path of the penetrating particles. Going out from the methods used in the Biostack I and II experiments the aim was to increase the accuracy of the localization method.

For small objects like the spores of Bacillus subtilis with about 1 μm in diameter an experimental procedure for the determination of the actual penetration point of the particle trajectory in the biological layer was developed, which allows the calculation of the impact parameter of each spore. An additional essential improvement is given by the fact that no more constraints like minimum LET and dip angle as they have been necessary in Biostack I and II reduced the sample size of evaluable spores. That is any etchable track can now be evaluated in regard to possible biological effects. This is an advantage in comparison to the experimental constraints of Biostack I and II. The preparation methods are the same used in Biostack I and II (2,3).

Arabidopsis thaliana:
An entirely different aspect of the improvement of localization methods in the ASTPBiostack is demonstrated by the seeds of Arabidopsis thaliana. For test objects of that size no more the accuracy of localization poses the predominant problems. However the course of the HZEparticle within the already differentiated seed becomes the biological important feature (4). To improve the localization of Arabidopsis thaliana corn seeds in this respect the seeds have been treated in a good approximation as rotational ellipsoids. Their size and orientation in three dimensions is measured together with the etch cone of the particle obtained by overall one side etching. If a particle penetrates a given corn the coordinates of the points of intersection with its surface are given in the system defined by the seeds main axis (fig. 2). This allows to determine immediately whether internal structures as for instance the cotyledons, root or stem-meristems have been crossed by a particle. This is possible since these structures can be recognized within the partly translucent seeds. This on-line evaluation greatly facilitates the later judgement of the observed biological effects.

Bacillus subtilis:
In the case of Bacillus subtilis spores pieces of 6 mm in diameter are cut

out of the pre-etched CN-foil carrying the spore layer. It contains the etch cones of single particle tracks at the spore free side. These tracks will be further etched individually in a special frame (fig. 3) under microscopical observation. The etching conditions are 6n NaOH at 40°C for CN(Daicel) and 3n NaOH at 40° C for CN(Kodak) respectively. The development of the etch cone is stopped with destilled water a few microns before the tip of the etch cone reaches the spore layer side. An extrapolation measurement with photographical registration (fig. 4) marks a preliminary penetration point, which determines a hit area of about 8 μm in diameter. Therefrom single spores are removed by a micromanipulation method (4) for individual biological evaluation. The etch cone is etched a second time by so called "break-through" etching until the tip of the cone impinges on the PVA-layer resulting in a circular spot, which is also photographically registrated (fig. 4b). The centre of this spot, which is comparable with the size of the spore, has to be determined as the penetration point. The accuracy of its determination is in the limits of about 0.2 μm.

With the help of the reference spores around the hit area the x,y-position of each manipulated spore in relation to the penetration point is measured. The origin of the coordinate system is placed in the penetration point and the track direction is identical with the x-axis (fig. 4d) This geometry allows the calculation of the impact parameter by a simple equation:

$$b = f\sqrt{x^2 \sin^2 \delta + y^2}$$

x,y are the coordinate values in the spore layer plane, δ is the dip angle and f is a conversion factor of the magnification of the photographes mentioned above. The accuracy of the impact parameter depends on the dip-angle of the track and the coordinates of the spores. The individual error components have been estimated as Δx= 0.35 μm, Δy= 0.36 μm and $\Delta \delta$ = 0.2 degrees. With these estimates the propagated errors for the impact parameter have been calculated for the spores so far evaluated. Their frequency distribution are shown in fig. 5.

Track development:

Throughout the development of this track etching procedure some devices similar to those described in the literature (5, 6) did not produce satisfying results. It can be pointed out that in comparison to an etching bath the track etching rate decreases. The etching with higher temperatures at 50°C or 60°C makes this effect much more evident, so that the track etching rate seems to be reaching the bulk etching rate resulting in "round" etch cones (fig. 6b).

Because of the spores the etching time at 40°C has to be kept as short as possible. Therefore these etching chambers could not be used. Instead it was attempted to simulate the same conditions as they are available in a bath. This has been attained by sealing the spore layer side with a cover glass against the surrounding air. This closed volume is filled with a high relative air humidity due to the water diffusion through the detector foil. The cover plate prevents the loss of humidity within the detector foil, which implies a stable profile of water content through the etching process. Under these conditions the etching process occurs with track etching rates comparable with those obtained in an etching bath (fig. 6c, d).

The development of carbon ion tracks in CN(Daicel) with 6n NaOH at 40°C is shown in fig 7. The only difference between both curves is the etching with and without a cover glass. All other etching parameters are constant. This experimental result demonstrates that the water is an important etching parameter. This has been found also in earlier investigations (7).

References:

(1) H. Bücker, R. Facius, M. Schäfer, Life Science and Space Research XIV, (in press)

(2) G. Horneck, R. Facius, W. Enge, R. Beaujean and K.-P. Bartholomä, Life Science and Space Research XII, 75 (1974)

(3) H. Bücker, R. Facius, D. Hildebrand and G. Horneck, Life Science and Space Research XIII, 161 (1975)

(4) H. Bücker, et al., Apollo-Soyuz Test Project Preliminary Science Report, NASA Rep. No. TM X-58173 p. 14 - 1 (1976)

(5) M. Monnin, These presentee a la faculte des Sciences de l'Universite de Clermont, Serie No. d'Ordre: 108, (1969)

(6) F. H. Ruddy, H. B. Knowles, G.E..Tripard, "Etch induction time in cellulose nitrate track detectors", Department of Physics, Washington State University, Pullman, Washington 99163

(7) D. Hildebrand, G. Reitz, H. Bücker presented in: Proc. 9th Intern. Conf. on Solid State Nuclear Track Detectors, München 1976

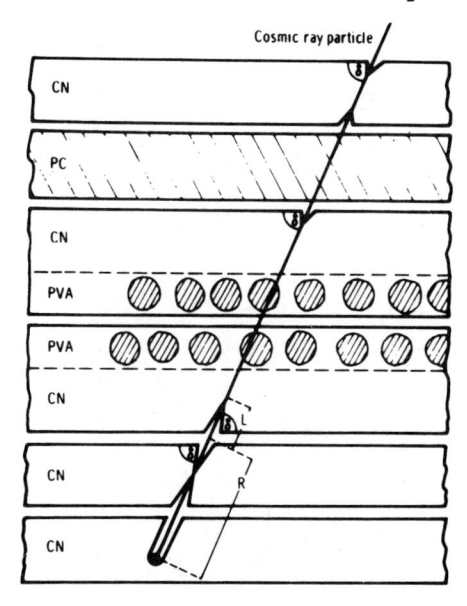

Fig. 1 Schematic configuration of the BIOSTACK
        Biological objects in connection with plastic detectors

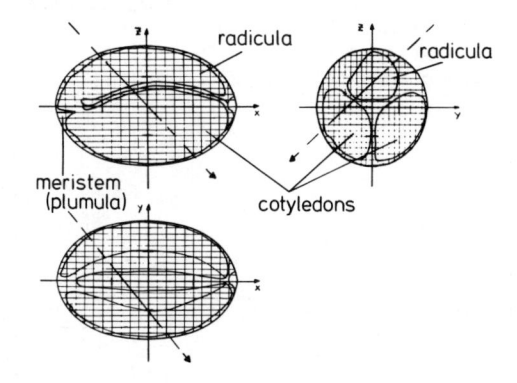

Fig. 2 Schematic of orthogonal projections of Arabidopsis thaliana
        seeds used for localization of HZE-particle penetration

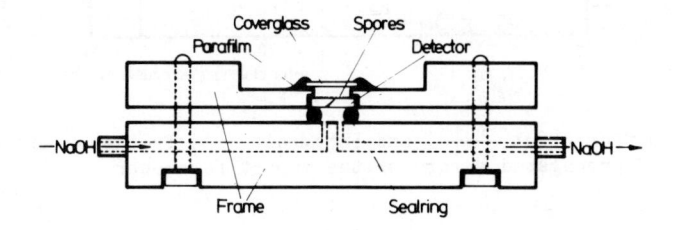

Fig. 3 Schematic of microetching device

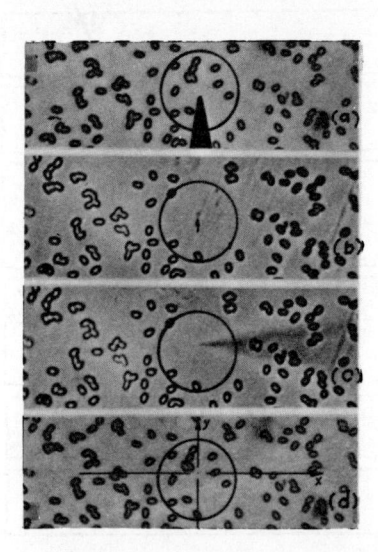

Fig. 4   Demonstration of the microetching procedure
         a) Determination of a preliminary penetration point
         b) "break-through" etching
         c) cone at high aperture
         d) x,y-coordinate determination

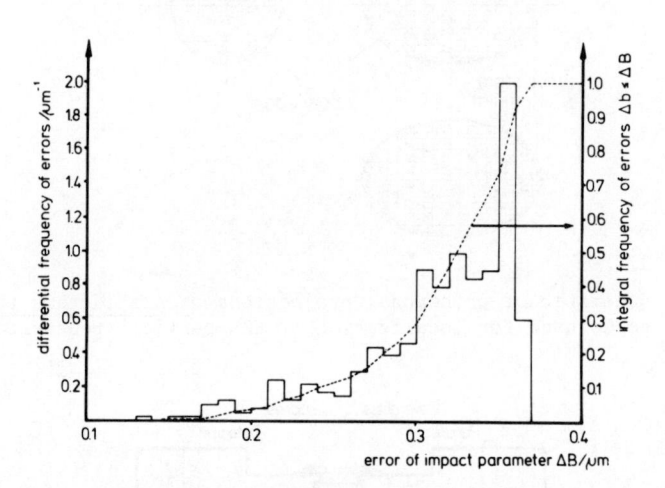

Fig. 5   Differential(———) and integral(- - - -) frequency distribution
         of propagated errors of the impact parameter

Localization methods for HZE-particles 1297

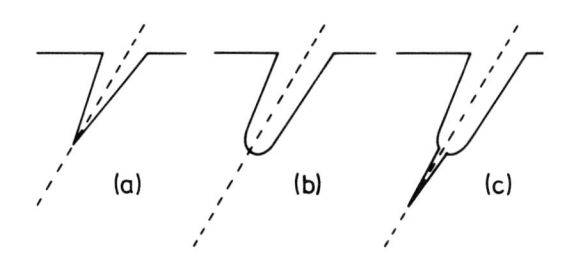

Fig. 6 Track development
   a) in the bath
   b) by one side etching with low humidity
   c) continued etching with high humidity
   d) same conditions as under c)

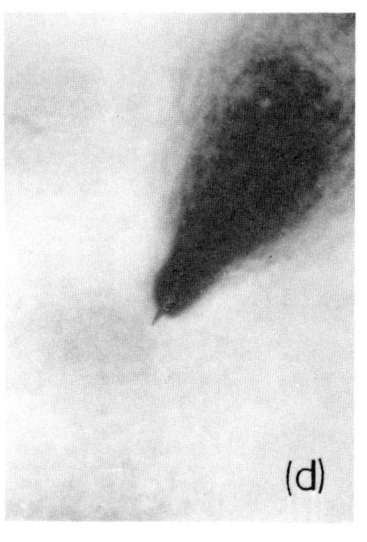

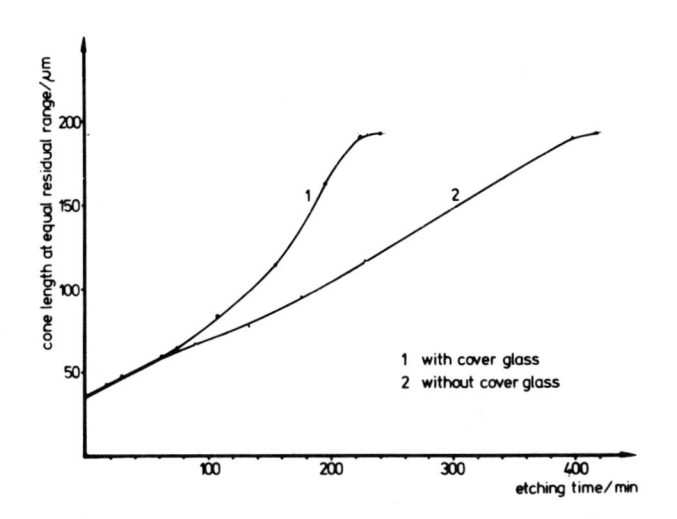

Fig. 7 Track development for Carbon ions with a high (1) and a low (2) humidity

DETERMINATION OF THE TRAJECTORIES OF HZE PARTICLES IN SEEDS OF *ARABIDOPSIS THALIANA* BY USE OF PLASTIC DETECTORS

U. Scheidemann, H. Bücker, R. Facius and C. Thomas

Arbeitsgruppe für Biophysikalische Raumforschung, Universität Frankfurt, Kennedyallee 97, 6000 Frankfurt/Main, West Germany

The study of the effects of individual HZE particles on biological subjects requires a correlation between the particle's path, determined by measurement of the etch cone in the plastic detector and the biological objects suspicious to be hit by the particle. A differentiation between hit and non hit subjects is not sufficient for highly organized biological objects such as seeds of Arabidopsis thaliana with differentiated tissues and repair possibilities.

In seeds of Arabidopsis thaliana three well differentiated organs can be distinguished, namely cotyledons, root and stem-meristem, which are clearly localized at well determined places in the seed and may show different radiation sensibilities.

A method shall be presented that allows to determine the trajectory of a particle within a plant seed. The conditions for the application of this method are that the seeds are in fixed contact with the plastic detector, that the plastic detector can be etched at one side, while the other side containing the biological samples are protected from the etch solution, and the seeds can be treated as well defined geometrical bodies.

A seed of Arabidopsis thaliana may be treated in a good approximation as a rotational ellipsoid. The seeds can be brought in fixed contact with the plastic detector using polyvinyl alcohol and methods for one side etching have been developed.

The size and orientation of the seed is determined together with the data of the etch cone. If a particle penetrates a given seed, the coordinates of the points of intersection with its surface are given in the system defined by the seeds main axis. Since the internal morphology and the position of the seed are known from the microscopical observation during the measurement, the hit region within the seed can be determined.

RADIOBIOLOGICAL STUDIES ON BIOLOGICAL SYSTEMS OF ANIMALS EXPOSED TO THE HEAVY NUCLEI OF COSMIC GALACTIC RADIATION

E. H. Graul and W. Rüther

*Klinik und Poliklinik für Nuklearmedizin, Universität Marburg,
355 Marburg (Lahn) Lahnstraße 4a, F.R.G.*

Introduction

In considering the biological problems which experience so far has shown to appear in longer space flights, and which must be considered as risk factors, the two main points are weightlessness and the threat posed by ionizing radiation. Naturally, there are other factors as well, such as ozone intoxication, decreases in air pressure, and temperature fluctuations.

The second, hitherto unsolved problem during longer space flights is the radiation load imposed by ionizing cosmic radiation, especially hits by so-called heavy primaries. This problem has a qualitative and a quantitative aspect.

The radiobiological relevancy of such radiation components cannot be satisfactorily estimated at present, since their quantitative and chronological appearance is still largely unknown from a spatiological point of view, and their relative radiological effect has not yet been satisfactorily explained. In addition, the radiation flow is relatively slight, except in cases of excessive proton showers from solar flares, so that the rem loads which appear lie within the range of natural (terrestrial) radiation, and are thus difficult to evaluate. In order to clarify these problems, certain basic radiobiological experiments are being carried out at present.

As mentioned above, galactiv radiation contains approximately 1% heavy atomic nuclei, the so-called heavy primaries, i.e. high-energy nuclei, especially of elements of the iron-nickel group. When these strike matter, including human tissue, they produce orbital tracks of high ionization density with a correspondingly strong biological effect, especially at the end of the track. Even though the intensity and quantity of this radiation is slight (approximately 3 nuclei/cm^2 per day), it can still lead to strong localized hits which under certain conditions, if there were to be coincidental summation in a particular area, could cause serious complications in biologically important control centers (e.g. the respiratory center in the brain). These hits were the cause of the flashes of light seen during recent Apollo flights by the astronauts, even when their eyes were closed. Laboratory investigations revealed that this phenomenon results when individual heavy ions strike the retina, or when Crenkov radiation is induced in the vitreous body. In addition, it has been found that such hits on hair roots can cause the hair to turn gray. Furthermore, it is suspected that hits by heavy ions can cause the turbidity of the eye lens.

Material and Method

After numerous preliminary experiments, we found three different organisms in which eggs met the requirements we had set up, such as the ability to survive several months of fixation in polyvinyl alcohol, laboratory breedability, and other criteria.

These animals were:
1. Artemia salina, the brine shrimp
2. Tribolium confusum, the flour beetle
3. Carausius morosus, stick insect

The eggs of these three animals shows fig. 1. After the flight the biological objects which had been hit by heavy primaries were found with the aid of the nuclear-track detector layers and the network of squares. Thus it was possible to carry out our study of the biological radiation effects of HZE particles within a relatively short time after the mission was over.

The biological processes which we studied most closely were hatching, development, and growth.

Results

Since the experiments with Artemia produced the most impressive results, I shall begin with this organism.

Artemia belongs to the order Branchiopoda, suborder Euphyllopoda, and lives in sea water (Fig.2). The adult animal is around 12 mm long. The eggs of the brine shrimp, Artemia, in the blastula stage consist of about 4200 cells and can be stored for long periods in a dried state. These dried cysts are about 200 µm in diameter and covered with thick, hard shells. In this condition, metabolic activity and respiration are arrested. The inert nature of the dry eggs allows the study of radiation-induced events. The eggs also withstand exposure to high or low temperatures, various gases, and high vacuum. Most of these conditions can be varied between wide limits; for example, viability of eggs was not impaired after storage in high vacuum (10^{-6} Torr) for 6 months.

Under favorable conditions in 1.5% salt water and a temperature of 25°C, the larva will normally hatch within 24 to 48 hours; it is called a nauplius. During hatching, the eggs shell breaks open and the nauplius emerges from the egg still covered by the membrane. Then, with the aid of its swimming organs, the nauplius tears this membrane off and begins to swim in the salt water. Assuming that there is sufficient algae for nutrition, the developmental cycle lasts around three to four weeks.

Under the conditions described above, the hatching rate of normal eggs was around 90%. For all the experiments we used specially selected Artemia eggs from Great Salt Lake in Utah.

In order to investigate radiation damage caused by HZE particles, we removed the hit eggs from the plastic plate with a micromanipulator glass capillary tube or an eye scalpel, and cultivated them in salt water (1.5%) at 25°C. [2.3]

The frequencies of the appearance of these developmental stages in around 900 eggs hits by heavy primaries, in around 1200 eggs which had received only background radiation, and in around 1600 eggs from the ground controls are shown in Fig.3.

The eggs hit by heavy primaries show the lowest hatching rate (14%) if one compares their development up to the stage of the freely swimming nauplius with that of the control eggs (90%). In addition, the viability of these nauplii is slight, since only a few of them attained the stage of sexual maturity around four weeks later. However, the eggs that had received background radiation also produced only around half as many freely swimming nauplii (45%) compared with the ground controls.

The number of nondeveloping eggs was highest (43%) in the eggs hit by heavy primaries, but there was also a relatively high number of nondeveloping eggs (20%) among those which had received background radiation. This result clearly shows that heavy nuclei cause severe damage to the blastula during the quiescent period. Depending upon the spatially varying radiation damage to certain blastula cells during the quiescent phase, only relatively few larvae develop into freely swimming nauplii during the following period of organogenesis.

The extreme damage done by heavy nuclei is further shown by an abnormality of embryonic development, in which the egg shells rupture, but with no visible organ development (34%) inside the transparent membranes. To a slighter extent (16%), this was also observed in the egg culture that received only background radiation in space. In the opinion of the NASA biophysicists this was the result of the influence of high-energy protons and the stars they form. A good criterion here is the single median eye. Although the basic physical conformation was visible in 9% of the animals from eggs hit by heavy primaries, their extremities did not have sufficient energy to tear open the egg membrane, thus indicating a decrease in vitality.

Anomalies of the body and extremities were found more frequently (6%) in the nauplii which hatched from eggs that had been hit by heavy primaries. Changes of this kind were relatively rare (0.5%) in the ground and flight controls. Here, the main changes observed were a shortening of the extremities, and anomalies in the thorax or abdomen.

Histological examination of these animals with anomalies showed that the externally visible radiation damage could also be recognized on the histo-cellular and organ level; this was also true of the radiation chimeras.

Similar results were also found in the flour beetle (<u>Tribolium confusum</u>) after irradiation with heavy primaries.

Figure 4 shows the life cycle of Tribolium confusum, the egg, the larva stages (1-11d), the pupa and the adult beetle. They live in flour-yeast medium [1,4,5].

Some permanent damage due to HZE irradiation was observed in hatching, early embryo development, metamorphosis of the larvae and in the structure of adult beetle. Figure 5 demonstrates the results of the three groups: earth control (normal development) flight control (cosmic background radiation) and HZE particle (cosmic radiation). The first group, the earth control group (400 eggs) shows a normal development. The hatchingrate was 87% and from these hatched larvae develop 84% to pupae and from these 82% to beetles. On the flight control group (cosmic background radiation) 69% of the 200 flown eggs hatched. From these larvae 62% reached the pupa stage and 59% developed into beetles. In the last group of 122 eggs, which were hit by cosmic HZE particles hatched 66%. Only 34% of these reached the pupa stage and 32% grew to beetles, a high mortality was found in this group during the larva stages. The decrease of pupation also indicates, that HZE irradiation injuries remain during Tribolium development through the larvae stages. Presently further tests are carried out in order to determine anomalies of the abdomen, wings and the antennae. The results of the investigation of those eggs not damaged by HZE particles confirm, that there are also other factors, like vibration, acceleration, weightlessness, which can have a negative influence on the development in connection with irradiation by heavy ions or the background irradiation.

The third animal we proved was the stick insect Carausius morosus. Figure 6 shows the life cycle, the egg, the larva stage and the adult insect, bred in the laboratory on raspberry leaves.

147 eggs were found to be hit by HZE particles. The isolation was executed and the development from these eggs lasted 6-9 month.

Figure 7 shows the results compared with the earth control group and the flight control group. In the experiment 83% of the normal eggs hatched and from these larva 71% grewed into adult insects. The hatching rate in the two flight group with background irradiation and HZE irradiation was 63% or 61%, that means a decrease of about 20% compared with the earth control group. In the flight groups nearly the same amount developed into adult animals. In these first experiments we have observed body anomalies like curved abdomen and deformed antennae.

In summary, it can be said of our investigation of Artemia that there was a clear difference between the development into nauplii of eggs which had been irradiated with heavy nuclei, eggs which had received background radiation, and eggs which had remained on earth. Our studies indicate that the radiation damages the encysted blastula, and that this damage continues to a certain extent in the gastrula and on up to the hatching of the nauplius. However, many nauplii which had been irradiated with cosmic HZE particles not survive the developmental phase (1-4 days) of initial development, with its high mitosis rate. Studies carried out with heavy primaries (He^{2+}, C^{6+}, A^{17+}) have shown that the dose-effect curves take an exponential course. If we assume that with sigmoid dose-effect curves a part of the irradiated cells retain their function, and thus enable the organism to survive, then the effect of heavy primaries on the Artemia blastula must be to destroy an essential funcional unit, thus negatively influencing the course of embryogenesis. However, it must be pointed out here that other stress factors, such as vibration, acceleration, etc., can likewise have a negative influence on development in connection with irradiation with heavy primaries or background radiation. This is indicated by the results of studies made on eggs which had not been damaged by cosmic HZE particles. We have seen analogous results in other insect eggs (Tribolium and Carausius). Only the time of development in the eggs of the three species of animals differs: artemia salina 1-2 days, Tribolium confusum 4-8 days and Carausius morosus 75-110 dys. Therefore it is obvious, that the radiobiological effect in the determined Artemia egg blastomes cells caused by a cosmic radiation particle is higher than in the other two animals, Tribolium confusum and Carausius morosus.

Conclusion

The biophysical space experiments Biostack I, II and III were carried out in order to study the biological effects of HZE particles in cosmic radiation and their significance for longterm sojourns in space.

Bombardment with heavy primaries from cosmic radiation caused varying reactions in the various biological objects. The range of biological damage was wide: Complete destruction, deformities, delays in development, and possible mutations were observed, as well as processes which remained completely unaffected, such as germination in plant seeds and bacterial spores that were also included in the test program. In most cases there is a long chain of widely varying biological processes between the directly caused radiation damage, the destroyed cell, and finally, the externally visible changes. In addition, penetration by a heavy primary does not have the same effect on all cells. Here it depends especially upon the function and the activity of the cell; this is naturally of particular importance for manned space flight.

References

(1) Buckhold, B., J.V., Slater, I.L. Silver, T. Yang and C.A. Tobias: Some effects of spaceflight on the flour beetle, Tribolium confusum, EXP P-1+39, in: The Experiments of Biosatellite II, NASA SP-204 (JF. Saunders, ed.), (1971), P. 79

(2) Easter, S. and F. Hutchinson: Effects of radiations of differend, LET on Artemia eggs. Radiat. Res. 15 (1961) 333

(3) Hutchinson, F. and S. Easter: A difference between the biological effects of gamma rays and of heavy ions. Science 132 (1960) 1311

(4) Slater, J.V., B. Buckhold, and C.A. Tobias: Effect on a Flour Beetle of irradiation During Space Flight. BioScience 18: 622-632 (1968).

(5) Yang, C.H., B.D. Heinze, I.L. Silver, and C.A. Tobias: The Combined Effect of Radiation and Compensated Gravity on the Wing Development in Tribolium confusum. Lawrence Berkeley Laboratory Report No. 596, (1972).

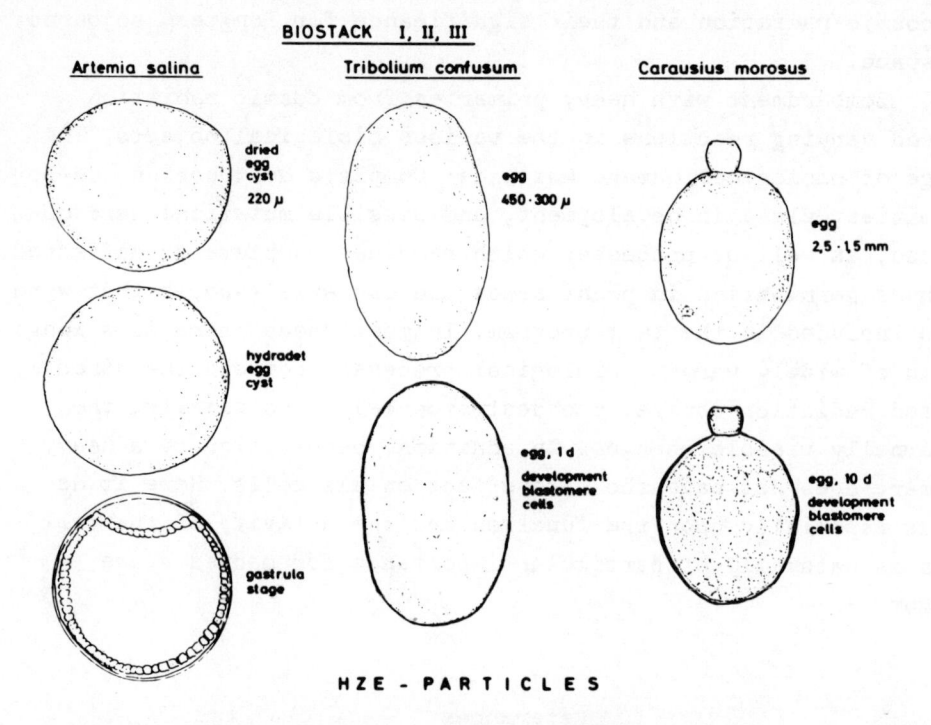

Figure 1. Eggs of three organism: Artemia salina, Tribolium confusum and Carausius morosus

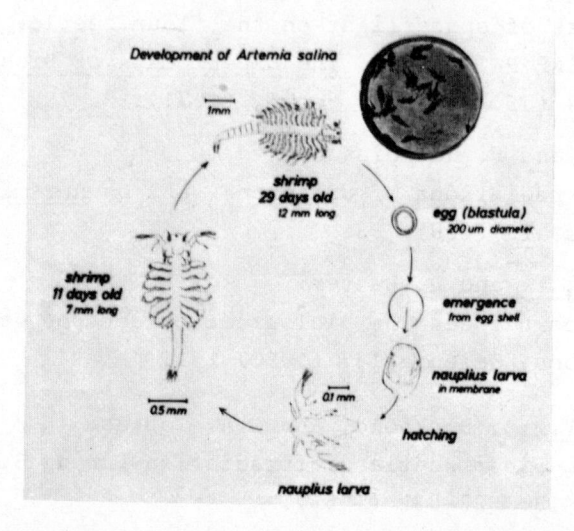

Figure 2. Life cycle of Artemia salina

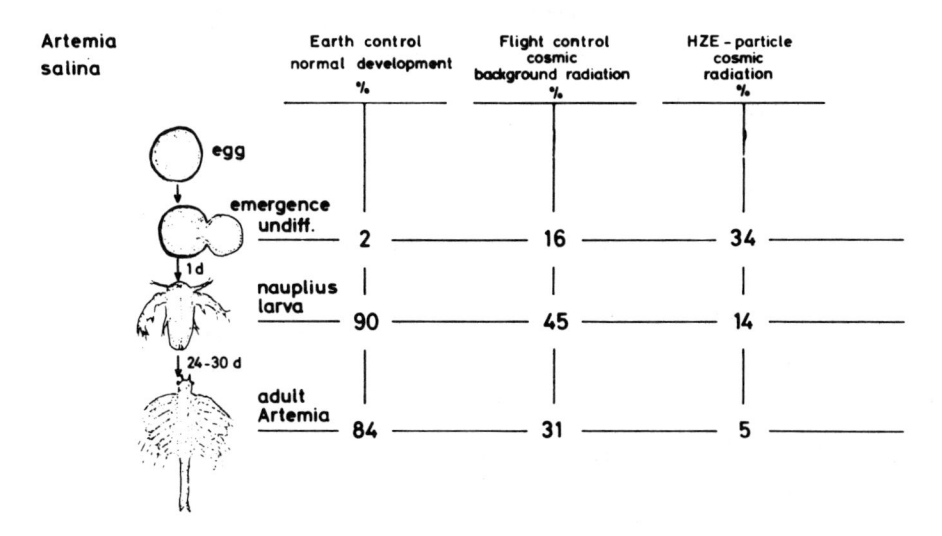

Artemia salina	Earth control normal development %	Flight control cosmic background radiation %	HZE - particle cosmic radiation %
egg → emergence undiff. (1 d)	2	16	34
nauplius larva (24-30 d)	90	45	14
adult Artemia	84	31	5

Figure 3. Results of Artemia salina, Biostack I, II and III

Life cycle of Tribolium confusum

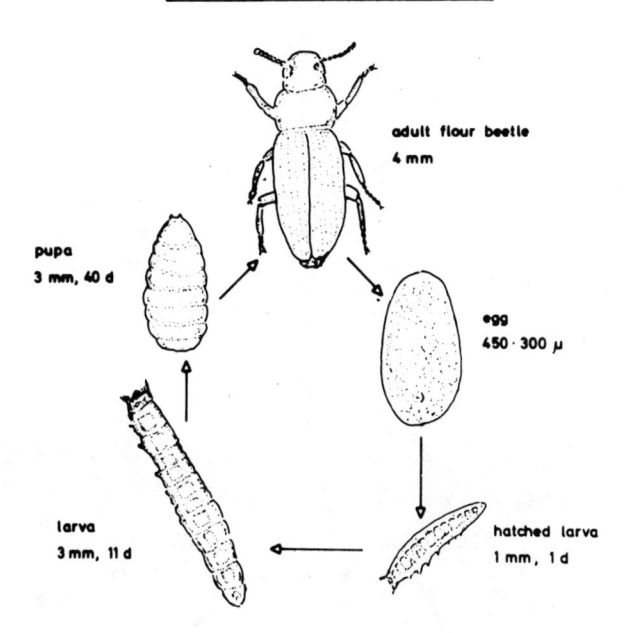

- adult flour beetle, 4 mm
- egg, 450·300 μ
- hatched larva, 1 mm, 1 d
- larva, 3 mm, 11 d
- pupa, 3 mm, 40 d

Figure 4. Life cycle of Tribolium confusum

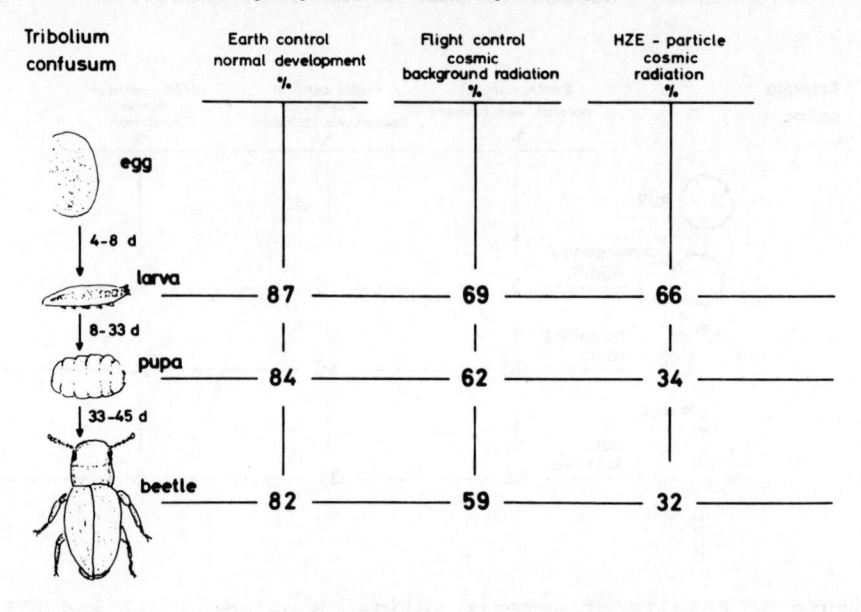

Tribolium confusum	Earth control normal development %	Flight control cosmic background radiation %	HZE - particle cosmic radiation %
egg			
↓ 4-8 d			
larva	87	69	66
↓ 8-33 d			
pupa	84	62	34
↓ 33-45 d			
beetle	82	59	32

Figure 5. Results of Tribolium confusum, Biostack II and III

Life cycle of Carausius morosus

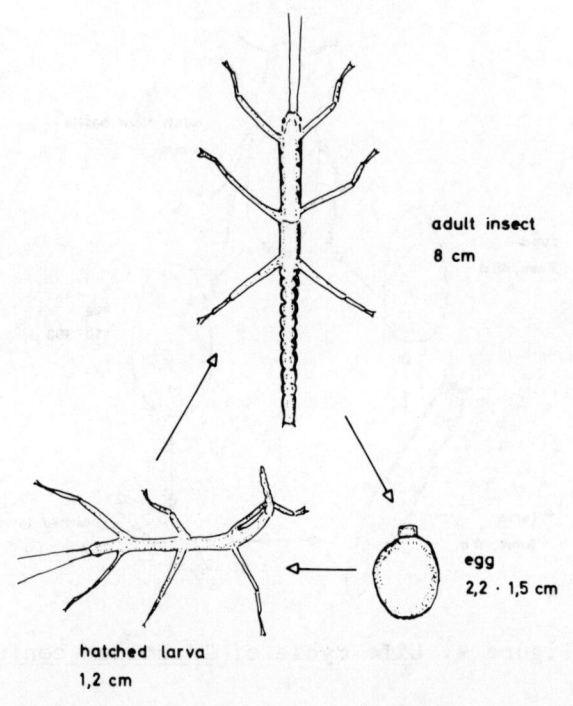

adult insect
8 cm

egg
2,2 · 1,5 cm

hatched larva
1,2 cm

Figure 6. Life cycle of Carausius morosus

Carausius morosus	Earth control normal development %	Flight control cosmic background radiation %	HZE - particle cosmic radiation %
egg ↓ 75-110 d			
hatched larva	83	63	61
↓ 4-5 m			
adult carausius	71	39	34

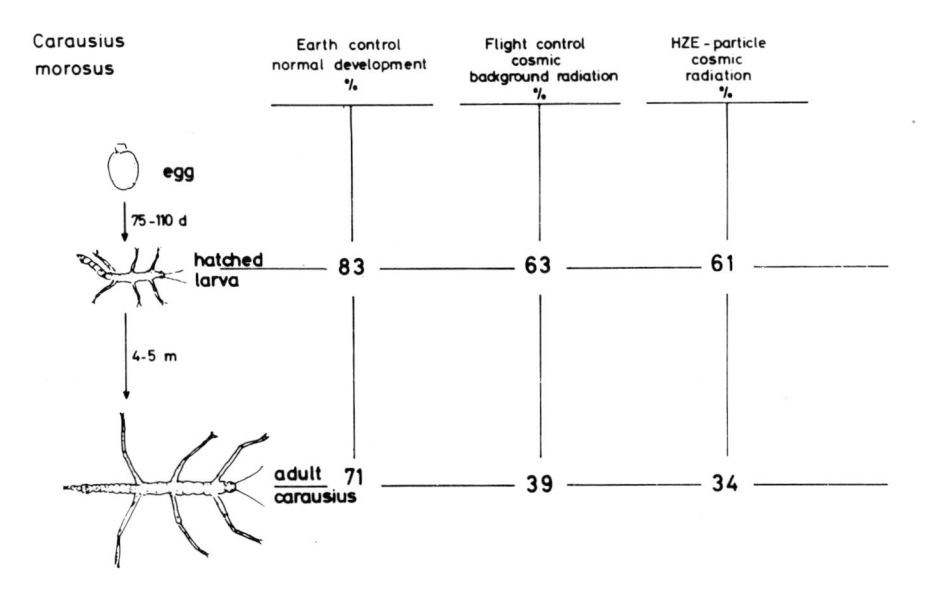

Figure 7. Results of Carausius morosus, Biostack II and III

CALCULATION OF LET-SPECTRA OF HEAVY COSMIC RAY NUCLEI AT VARIOUS ABSORBER DEPTHS

W. Heinrich

*Gesamthochschule Siegen, Fachbereich Physik, Hölderlinstr. 3,
D 5900 Siegen 21, F.G.R.*

The calculation is based on a factorization of fragmentation cross sections into a projectile depending and a target depending part, which was found at Bavalac for high energy C and O ions. With this the known cross sections for the fragmentation of heavy ions in collisions with protons can be extrapolated to heavier targets. Considering the fragmentation and the energy loss of heavy nuclei the energy spectra of individual elements are calculated for different depths of absorber. This energy spectra can easily be converted to LET (linear energy transfer)-spectra. The changes of the isotopic composition with absorber depth and its influence on the shape or the LET-spectra are discussed. The calculated results are compared with experimental data.

1) Introduction

In the past decade the radiation risk caused by heavy primary nuclei of the cosmic radiation during space flights and flights at supersonic transport (SST) altitude (16-22 km) gained growing interest. Investigations of radiobiological effects by single hits of heavy nuclei were performed in the Biostack experiments, flown aboard Apollo 16, Apollo 17 and Apollo Soyuz (ASTP) (1,2,3). These experiments showed a serious radiation damage of insect eggs during space flights (4,5). Parallel to these investigations the total flux of the high-LET particle radiation was monitored with plastic track detectors(6,7,8,9). Also at SST-level the intensity of heavy nuclei was measured for different periods of solar modulation (10,11).

Beside these experimental activities calculations of cosmic ray nuclei fluxes behind shielding are necessary to predict the flux of heavy nuclei for planned space missions under different conditions of solar modulation and earth magnetic field cut-off. In the past such calculations were performed for charge groups of nuclei using the fragmentation probabilities measured in nuclear emulsion (10,12,13). Recently an empirical formula for the calculation of fragmentation cross sections for collisions with heavy target nuclei was developed by C.H.Tsao and R.Silberberg (14). In this paper the energy spectra of individual elements behind absorbing material are presented, which were calculated using this relation. From these energy spectra the LET spectra were derived.

2) Calculation of energy spectra

The calculation starts with the differential energy spectra in free space, that is the flux of nuclei per square meter, second, energy interval in MeV/nucleon and steradiant as a function of the energy. Figur 1 shows in the uppermost curves ($0 g cm^{-2}$) the differential energy spectrum of C-, O-, F- and Fe-nuclei. These spectra for individual elements from carbon to iron at the flight time of Apollo 16 and Apollo 17 were derived from the primary carbon spectrum measured by G.M.Mason (15) in 1969 - 1970 with the IMP-5 satellite. An identical shape of the spectra for all nuclear charges was assumed and the spectra for the other elements were deduced according to their relative abundances.

Behind absorbing material the flux of heavy nuclei with energies extending over a broad spectrum, like for cosmic ray nuclei, is reduced because the low energy particles are stopped by ionization losses. The isotopic composition of heavy nuclei that pass through the absorber will change due to nuclear collisions in which the heavy nuclei break up into lighter fragments. Furthermore these nuclei with sufficient penetration power have lower energies behind the absorber.

In the present calculation the energy loss of the nuclei is determined from the range-energy relation for heavy nuclei after Barkas and Berger (16) in the modified form given by E.V.Benton (17). For the calculation of the fragmentation of the nuclei the partial cross sections σ_{PT}^{F} for the production of a fragment with mass number F from a projectile nucleus P in an interaction with a target nucleus T are needed. These were derived from an empirical relation suggested by C.H.Tsao and R.Silberberg (14) for heavy target nuclei. This relation is based on a factorization of the cross sections into a projectile and a target depending part, as measured by P.J.Lindstrom et al.(18) for C and O nuclei. This factorization makes it possible to calculate the unknown cross sections from the data for proton targets, determined in accelerator experiments with interchanged roles for projectile and target (19,20).

Before applying the relation of C.H.Tsao and R.Silberberg in the calculation reported here, it was tested making use of cosmic ray data. The cross sections calculated with this relation were used to compute the fragmentation probabilities of heavy nuclei and the absorption of heavy nuclei in the earth atmosphere. There is no contradiction between the results of this calculation and the existing experimental data (13).

To calculate the energy spectra of heavy nuclei behind absorbing material the numbers N_i of nuclei of charges $6 \leq Z_i \leq 26$ for fixed energy intervals ΔE of the primary spectra are determined. These energy intervals can be seen as steps in the curves of Fig.1. Then the variations $\Delta N_i/\Delta x$ caused by fragmentation of the nuclei in an absorber of Δx gcm^{-2} thickness are calculated for each energy interval. For this purpose a system of diffusion equations is used (13). The energy loss in Δx gcm^{-2} is considered for each charge Z_i by correcting the boundaries of the energy intervals ΔE. The calculation is then repeated for the next increment Δx. The intervals used in the calculation are $\Delta E = 1$ MeV/nucleon and $\Delta x = 1$ gcm^{-2} of absorber.

3) Shape of the energy spectra behind shielding
As results of the calculation we get differential energy spectra for shielding thicknesses of 1 gcm^{-2} to 60 gcm^{-2}. Fig.1 shows as an example the calculated spectra for C-, O-, F- and Fe-nuclei.

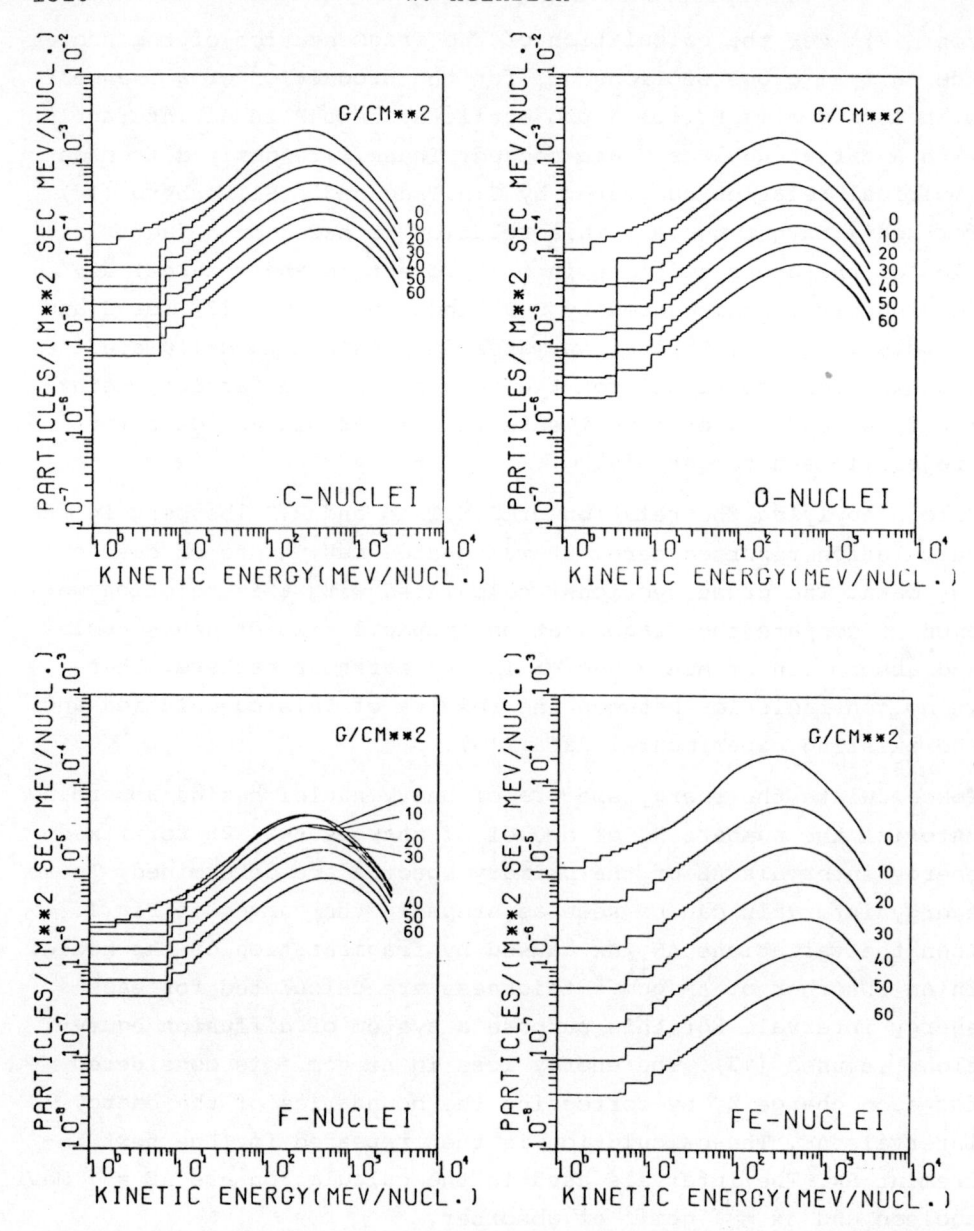

Figure 1: Energy spectra of cosmic ray heavy nuclei behind various shielding

The maximum intensity decreases by about 330-fold for iron, 25-fold for oxygen, 15-fold for carbon and 5-fold for fluorine after penetration of 60 gcm^{-2}. The decreased intensity depends on the number of nuclei lost in fragmentation processes and on the number of new nuclei produced as fragments of heavier particles. The number of lost nuclei is roughly proportional to the total cross section, which is nearly the geometric cross section. Therefore the flux of iron nuclei is reduced much faster than that of carbon, oxygen or fluorine nuclei. These last two elements of nearly equal geometric dimensions show differences in absorption because a significant number of carbon nuclei at greater absorber depths are fragments of primary oxygen.

For fluorine the flux of high energy particles is first enhanced in the absorbing material because the fragments of heavier elements contribute significantly to its flux. Behind about 30 gcm^{-2} of absorbing material a steady state of the ratio of fluorine to heavier nuclei develops. From this point the intensity of fluorine decreases in constant fractions like for the other nuclei.

The position of the maximum of the energy spectra is shifted towards higher energies after penetration of absorbing matter. In free space this maximum for iron (Fig.1) is found at about 300 MeV/nucleon whereas at absorver depths of 60 gcm^{-2} its position is at about 800 MeV/nucleon. This shift in the maximum is caused by the energy loss of the particles and is less significant for lighter elements. E.g. for carbon nuclei the flux maximum after penetration of 60 gcm^{-2} matter is found at 500 MeV/nucleon.

4) Calculation of LET-spectra

The LET-Spectra behind various shielding thicknesses can be calculated from the differential energy spectra. This is demonstrated in Fig.2 for iron nuclei. The right side of this figure shows in the back plane the differential energy spectrum. The ground plane shows a curve giving LET as a function of energy for iron nuclei. Above this curve in a three dimensional plot

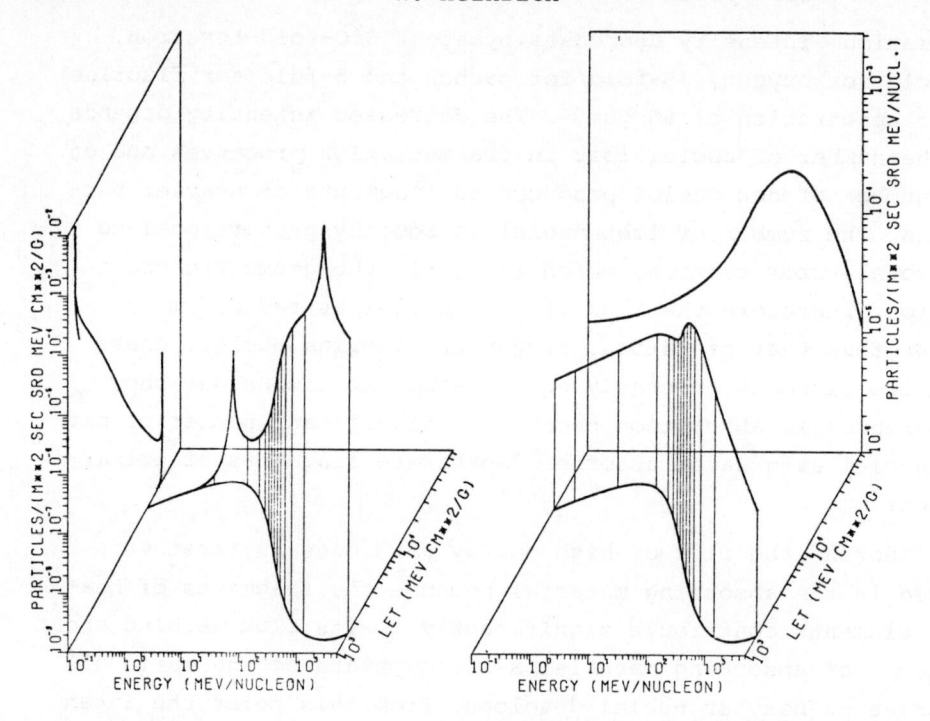

Figure 2: Derivation of the LET-spectrum from the energy spectrum (for details see text)

the number of iron nuclei per m² sec srd and energy interval can be seen. This particle flux per interval of energy must be converted into the particle flux per LET-interval, to get the differential LET-spectrum for iron. This was done by calculating the number of particles in an energy interval $\Delta E = E_2 - E_1$. Then the interval of $\Delta LET = |LET_2 - LET_1|$ corresponding to ΔE was determined from the LET(E) relation and the number of particles was divided by ΔLET. At LET-regions where LET(E) is a slowly varying function with energy, i.e. at the maximum and at the minimum of the LET(E) curves, a very narrow interval ΔLET corresponds to a broad interval ΔE. Therefore the differential LET-spectrum shows pronounced peaks at these positions.

The left side of Fig.2 shows a three dimensional plot of the differential LET-spectrum of iron nuclei. In the left side plane a projection of this spectrum is shown in a two dimensional plot with a high-LET and a low-LET peak. Between these peaks a roughly linear decrease of the intensity can be recognized in a double logarithmic presentation of the differential LET-spectrum. This region corresponds to energies between the LET-maximum and LET-minimum.

For energies above the LET-minimum the curve in the left side plane turns back to higher values of LET, but with an intensity which rapidly decreases by orders of magnitude. A similar behavior can be recognized for the high LET-peak. For each value of LET between maximum and minimum LET there exist three values for differential intensity. The differential LET-spectrum can be determined by summing up these three intensity values. A contribution to the LET-spectrum by particles with energies below the LET-maximum and above the LET-minimum is only significant in the vicinity of these extrema.

5) Form of LET-spectra

Figure 3 shows the differential LET-spectra of iron and carbon nuclei, which were derived from the energy spectra behind $2 gcm^{-2}$ shielding. The steps in the spectra originate from the ΔE steps of the numerical calculated energy spectra. The LET-spectra of all nuclei are of the typical form discussed above. For light nuclei like carbon in Figure 3 the high-LET peak is suppressed by the broadness of the LET intervals, whereas for heavier nuclei like iron this peak can be recognized. The upper curve in Figure 3 shows the differential LET-spectrum for all nuclei

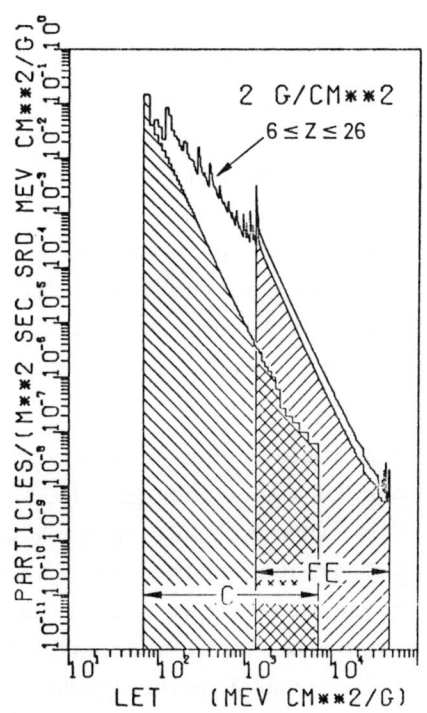

Fig.3: Differential LET-spectrum at 2 gcm^{-2} for C-nuclei, Fe-nuclei and nuclei with $6 \leq Z \leq 26$

from carbon to iron, which results as the sum of the spectra of the individual elements. The low LET peak can be distinguished for each element. The height of this peak represents the relative abundance of the elements.

In figure 4 the differential LET-spectra in free space (0 gcm^{-2}) and behind a shielding of 50 gcm^{-2} are shown. Comparing the low LET-peaks of both spectra it can be seen that behind 50 gcm^{-2} absorber the relative abundance of nuclei which are rare in the primary radiation like nuclei with odd charges and nuclei with charges $15 \leq Z \leq 25$ has increased. For nuclei with charges $15 \leq Z$ also the high-LET peaks can be recognized at 50 gcm^{-2} shielding. Furthermore the relative height of the low-LET iron peak is diminished. In the part of the spectrum for LET$>10^3$ MeV cm^{2}g^{-1}, which is governed by the iron nuclei, the flux is reduced about 50-fold in an absorber depth of 50 gcm^{-2}. For lower LET values the flux is only reduced about 10-fold to 20-fold in the same absorber.

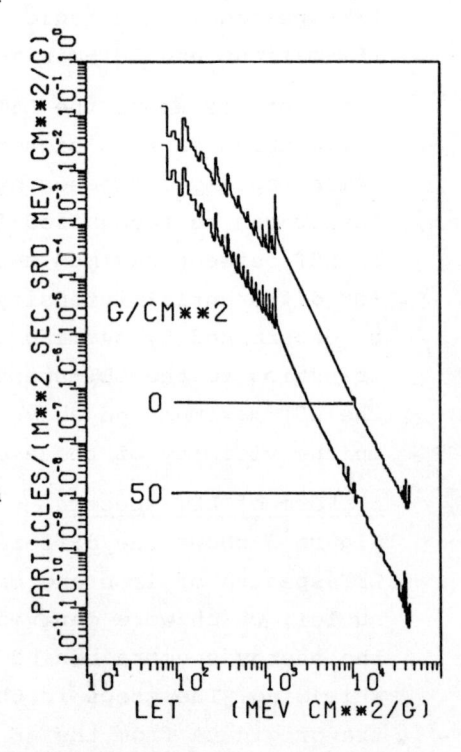

Fig.4: Differential LET-spectra in free space and behind 50 gcm^{-2} shielding

The differences of the differential LET-spectra cause differences in the slope of the integral LET-spectra shown in Figure 5. The integral LET-spectra give the flux of all particles with an LET greater than a threshold LET$_o$. For LET$_o > 10^3$ MeV cm^{2}g^{-1} these spectra show in a double logarithmic plot an approximately constant slope, independent of the shielding. For greater absorber depths a gentle bump can be noticed at about 10^4 MeV cm^{2}g^{-1}, which originates from the high-LET peaks in the differential LET-spectra. At about 10^3 MeV cm^{2}g^{-1} the flux of the integral LET-spectra increases rapidly, caused by the low-LET iron peaks of the differential spectra. This increase is more significant for low shielding where the relative abundance of this peak is greatest.

For LET_o < 10^3 MeV cm^{2}g^{-1} the differential spectrum can also roughly be approximated by a straight line, but with a slope which depends on the shielding. These differences of the slope can be understood very well from the changes of relative abundances of the elements.

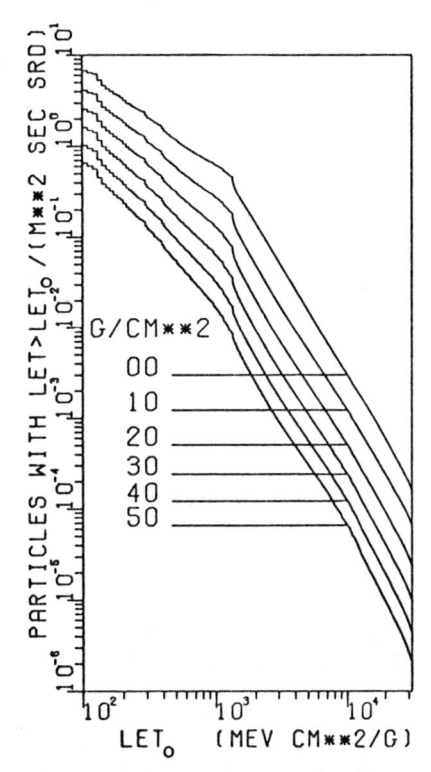

Figure 5: Integral LET-spectra behind various shielding

6) Comparison with experimental results and conclusion

The following features can be drawn from these calculations:
1) For LET>10^3 MeV cm^{2}g^{-1} the exponent of the slope of the integral spectrum is roughly constant in a double logarithmic plot and independent of the shielding. Slight deviations from a constant exponent can be noticed for greater absorber dephts around LET $\sim 10^4$ MeV cm^2 g^{-1}.
2) For LET < 10^3 MeV cm^{2}g^{-1} the slope of the calculated spectrum is quite different from that for higher LET's and depends on the shielding.

For comparison with the calculations REL (restricted energy loss)-spectra measured with plastic track detectors are available. REL includes the energy loss by δ-electrons with energies below a certain cut-off. In order to compare these measurements with the present calculations the REL-spectra were derived from the energy spectra.

To determine the particle flux at a position inside the space craft the flux must be add up for all possible angles of incidence, considering the different amount of absorbing matter for each direction. The distribution of matter is generally rather complicated. For the Biostack experiments(1,2,3) which were shielded only by the wall of the Apollo space ship, this problem is less serious. Our calculated REL-spectra inside the Biostack (21) at a depth of 4 gcm^{-2} and 20 gcm^{-2} respectively agree within a factor of two or better with the experimental results of R.Beaujean et al.(9). This agreement between experimental and calculated data for a spectrum that varies over three orders of magnitude shows that the cross section formula of C.H.Tsao and R.Silberberg (14) is quite valuable to calculate the energy spectra of individual elements behind shielding.

Acknowledgement
I want to thank M.Simon for valuable discussions.

Literature
1) H.Bücker et al.,Life Science and Space Research XI, 295, Akademie Verlag Berlin (1973)
2) H.Bücker et al., NASA Apollo 17 Preliminary Science Report, Order No.:NASA SP-330 (1974)
3) H.Bücker et al.,19th Cospar Meeting Philadelphia(1976)
4) W.Rüther, E.H.Graul, W.Heinrich O.C.Allkofer, R.Kaiser, P.Cüer, Life Science and Space Research XII, page 69, Akademie Verlag Berlin (1974)
5) W.Rüther, E.H.Graul, W.Heinrich, O.C.Allkofer, R.Kaiser, R.Pfohl, H.Bücker, Life Science and Space Research XIII, Akademie Verlag Berlin (1975)
6) E.V.Benton, R.P.Henke, J.V.Bailey, Science 187 (1975) 263
7) E.V.Benton, R.P.Henke, J.V.Bailey, Health Phys.27 (1974)79
8) D.D.Peterson, E.V.Benton Health Phys.29 (1975) 125
9) R.Beaujean, W.Enge, W.Herrman, K.P.Bartholomä, Paper presented at the Cospar Meeting, Karma (1975)
10) O.C.Allkofer, W.Heinrich, Health Physics 27(1974)593

11) K.Fukui, Y.K.Lim, P.S.Young, Nuovo Cimento B61 (1969) 210
12) O.C.Allkofer, W.Heinrich, Journal of Spacecraft and Rockets 12(1975) 119
13) O.C.Allkofer, W.Heinrich, Nuclear Physics B71(1974) 429
14) C.H.Tsao, R.Silberberg, Proc.14th Int.Cosmic Ray Conf., München, page 516 (1975)
15) G.M.Mason, Astrophys.Journ.171(1972) 139
16) US National Academy of Sciences - National Research Council Publication - 1133(1964)
17) E.V.Benton, Naval Radiol.Defense Laboratory, San Francisco, Report USNRDL - TR - 68 - 14 (1968)
18) P.J.Lindstrom, D.E.Greiner, H.H.Heckman, B.Cork, F.S.Bieser, Preprint LBL - 3650 (1975)
19) R.Silberberg, C.H.Tsao, Naval Research Laboratory, Washington D.C., Report 7593(1973)
20) R.Silberberg, C.H.Tsao, Astrophys.Journ.Suppl.25(1973) 315
21) W.Heinrich, 19th Cospar Meeting, Philadelphia (1976), in press in Life Science and Space Research.

DOSIMETRY OF COSMIC PARTICLES IN NUCLEAR EMULSIONS FOR THE APOLLO 16, 17 AND APOLLO-SOYOUZ-TEST-PROJECT EXPERIMENTS (1972-1975) SESSION 11

R. Pfohl, R. Kaiser, J. P. Massue and H. Francois

SADVI, Centre de Recherches Nucleaires, 67037 Strasbourg Cedex, France

ABSTRACT

In the Biostack I, II, II experiments flown on the Apollo 16 and 17 and ASTP missions, nuclear emulsion plates have been used for the first time for heavy ions identification in connection with biological layers. After the calibration step using the heavy ions available at the Bevalac at Berkeley, the nuclear emulsion can give dosimetric results for the heavy ions hitting biological materials. The results have been sent for exploitation to the biological laboratories of Toulouse, Frankfurt and Marburg. We summarize all the dosimetric results in the field of heavy ion tracks and nuclear stars induced by cosmic-ray particles. The last results from Biobloc (Dec. 1975) are presented.

INTRODUCTION

Nuclear emulsion presents many advantages for dosimetric studies in space missions. A small volume of this detector makes it possible to register the trajectories of charged particles, keeping the tracks for further investigations.

In space research, volume and weight constrainsts still remain, nuclear emulsion can register a large spectrum of particles from relativistic protons with tracks of some grains per 100 μ up to the heaviest nuclei whose tracks are tens of microns thick.

I - STUDY OF THE HEAVY IONS

1°) Aspect

The heavy ion tracks recorded in nuclear emulsion have a specific aspect, especially at the end of the track. The thinning down of these tracks is known. Picture (1) shows such a cosmic track from the Apollo 16 experiment.

2°) Calibration

The problem of the calibration of these tracks is now easier since the Bevalac in Berkeley has accelerated ions up to the charge 18 with an energy of 2.1 GeV/nucleon. Fig. (1) shows the thickness of the track versus the residual range.

It is essential to know the energy-loss along the track ($\frac{dE}{dx}$) as well for the computing the energy lossed inside biological material as for dosimetry calculations. The basic measure is a photometry using the MPV Leitz Photometer. It consists in measuring the light intensities through the emulsion on the track and beside it. This method is precise, because our apparatus was calibrated for a large energy-interval, from 0 to 2.1 GeV/nucleon. Our calibration is valid for energies up to 2.1 GeV/N and charges between 1 and 18. The relation between photometry and energy-loss is unique, that is to say independant from the heavy ion considered.

A similar result was obtained by the Lund Laboratory using more heavy ions but lower energies. We have developed acurate measurement processes which enable us to reliable results on energy-loss for biologists. Fig. (2) is the calibration curve for a K2 emulsion.

3°) Dosimetry

With this calibration curve it is possible to make dosimetric studies. On one hand the number of tracks gives us the total incoming flux, on the other - as explained - we measure the lineic energy loss (LET). We know the energy lossed in a given volume and calculate the corresponding dose. Table (1) summarizes the results for the ions.

But it is necessary to print out that in fact a global dose for heavy ions does not reproduce exactly the feature of the interactions. The action of the energy deposited by the heavy ions along their trajectory is restricted to a short cylinder along the path of the ion. If we assume that the total energy is deposited in a cylinder of 10 µm diameter, the local dose (for A16) is not 3.3 mrad but 20 rads.

4°) Application of dosimetry to biological studies

If we know the energy loss $\frac{dE}{dx}$ of a heavy ion near a biological object and the length of the track through it, we can calculate the energy lossed inside it. Fig. (3) is the distribution of $\frac{dE}{dx}$ of 227 heavy ions (Apollo 17) which hit artemia salina eggs. Fig. (4) is the distribution of the energy deposited by the same ions. These results are the base for further sudies done by the biologist team from the Biostack Experiment.

II - NUCLEAR STARS IN THE EMULSION

The interactions between the incident radiations and the nuclei of the emulsion detector give nuclear stars whose amoint is related to the intensity of the radiation.

In the Biostack III A experiment we made an experimental study about the dose coming from these stars.

The stars with a number of tracks ≥ 8 are related to the interactions with Ag and Br.

We have statistically a mean number of 12.2 tracks 1.6 black tracks, - 5.4 grey tracks - 5.2 white tracks.

In a first approximation we assume an energy loss 20-2-0,5 KeV/μ for respectively black, grey, white tracks and a mean range of 500 μ per track.

This gives a mean energy deposition of 22.7 MeV per AgBr nuclei star. For the stars with a number of tracks < 8 (the most of these related to interactions with C, N, O) we have a mean number of 4.6 tracks.

1.4 black tracks, - 1.8 grey, - 1.4 white tracks.

This gives a mean energy deposition of 16.2 MeV per star. In an object (cube of 10 μm length) the local dose would be 600 mrad.

Table (2) summarizes the results for the different experiments in the last years.

III - DOSIMETRY OF LOW CHARGE (1.2) PARTICLES

If we lood at a Kz emulsion in the microscope we see a lot of low charge tracks protons, α. With a few assumptions we try to estimate the energy lossed in the emulsion. By these particles, the mean number of tracks measured in a cm^2 is 8400 for protons with a $\frac{dE}{dx}$ about 2.5 KeV/μ and 400 tracks with a $\frac{dE}{dx}$ of about 10 KeV/μ. With these assumptions the energy deposition is 2×10^{10} eV = $3,1 \times 10^{-2}$ erg. (Fig. 5).

The corresponding dose calculated for the volume of 6×10^{-5} dm^3 is 1.4 mrad (about 1 % from the total dose measured by LiF dosimeter).

IV - CONCLUSION

This method of passive dosimetry is useful and must be continue in the future.

TABLE I

HEAVY IONS	TOTAL nb/cm^2	ENDERS/cm^2	TOTAL DOSE LiF mrad	DOSE HEAVY IONS mrad
A 16	130	2,1	650	3,3
A 17	140	2,7	730	3,5
ASTP	41		130	
BIOBLOC	125		450	2,0
BALLON 1974 GRBS 35000 m	12			0,26

TABLE II

STARS	TOTAL/cm^3	Nb/cm^3/day	TOTAL DOSE mrad	FLUX (p,α)/cm^2	TOTAL DOSE mrad
A 17	12.700	950	1,0		
ASTP	15.000	1.650	1,3	8.400 p 400 α	1,4
BIOBLOC	25.000	1.250	2,0	20.500 p 1.500 α	11,0
BALLON GRBS 35.000 m	310	670	0,06	2.200 p 630 α	0,6

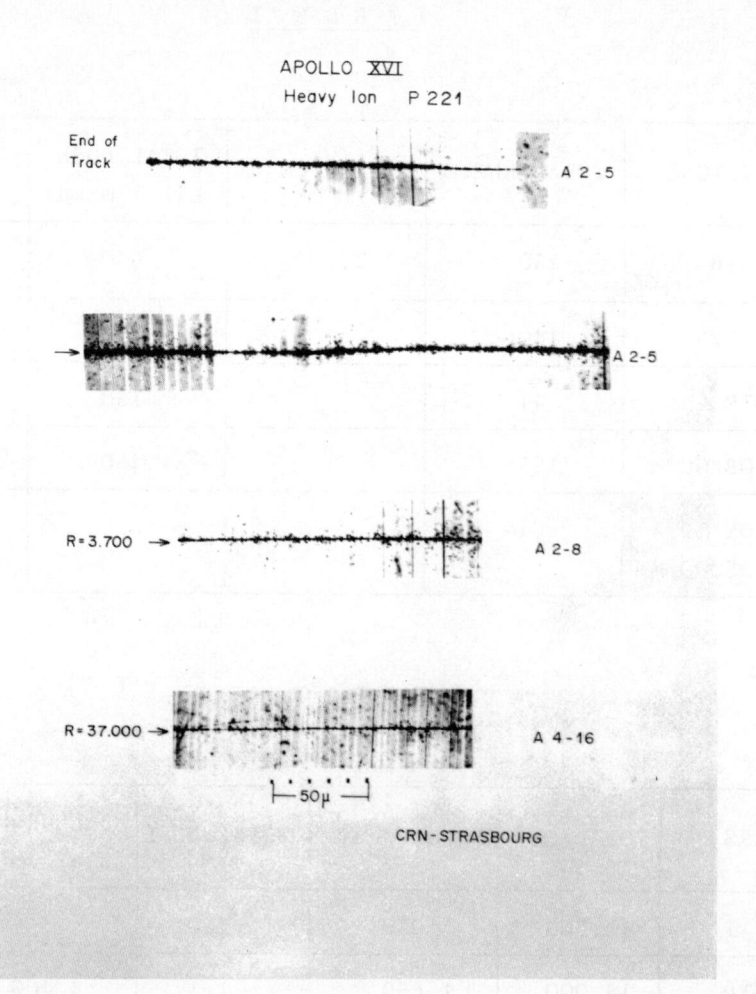

Picture 1.

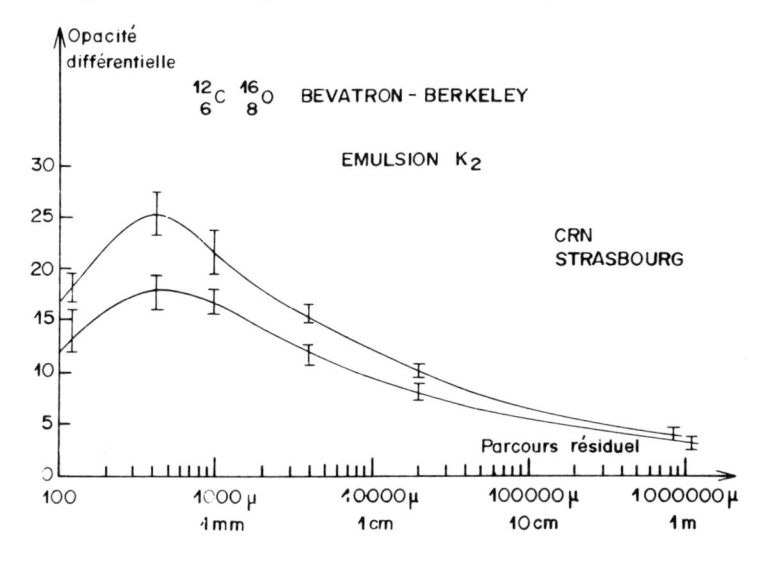

Fig. 1.

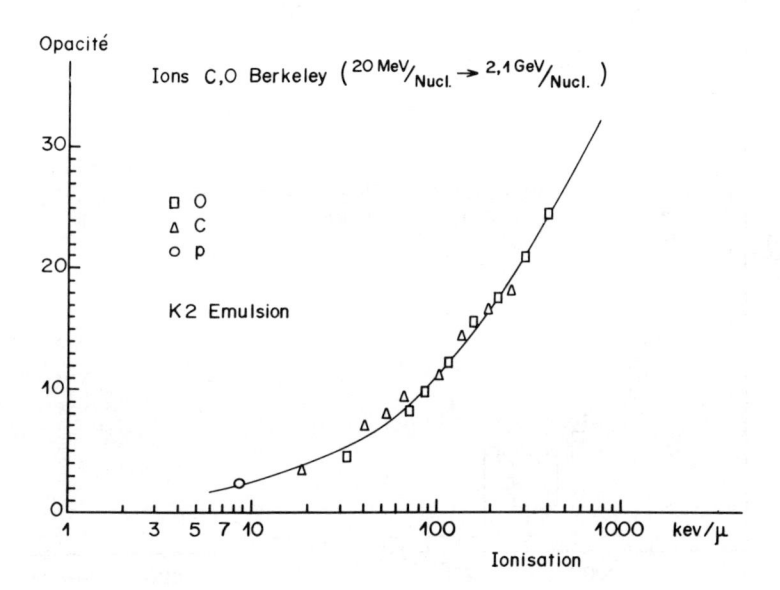

Fig. 2.

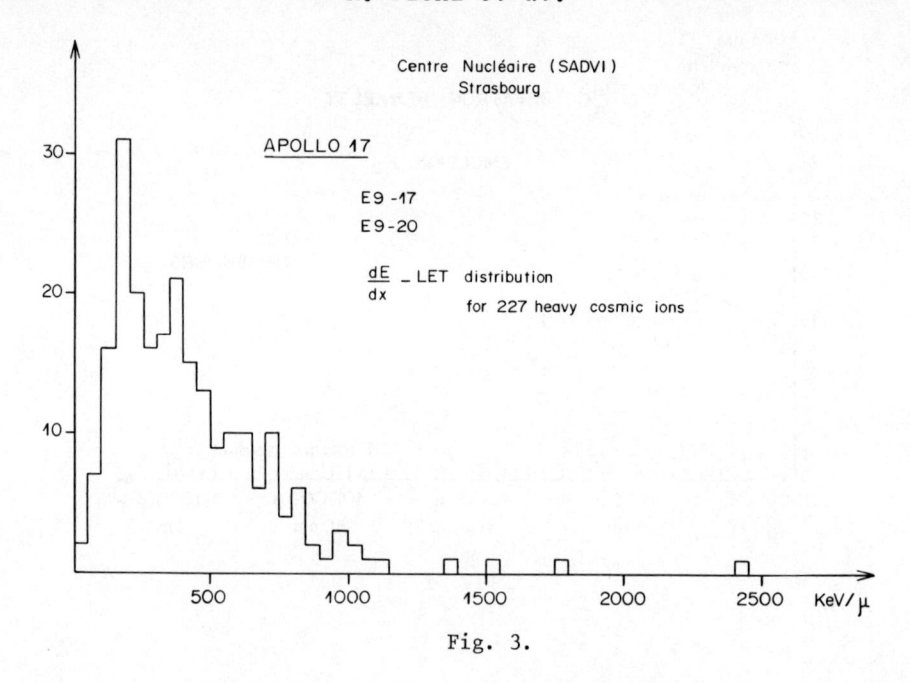

Fig. 3.

Centre Nucléaire (SADVI) Strasbourg

APOLLO 17

E9-17
E9-20

$\frac{dE}{dx}$ — LET distribution for 227 heavy cosmic ions

Fig. 4.

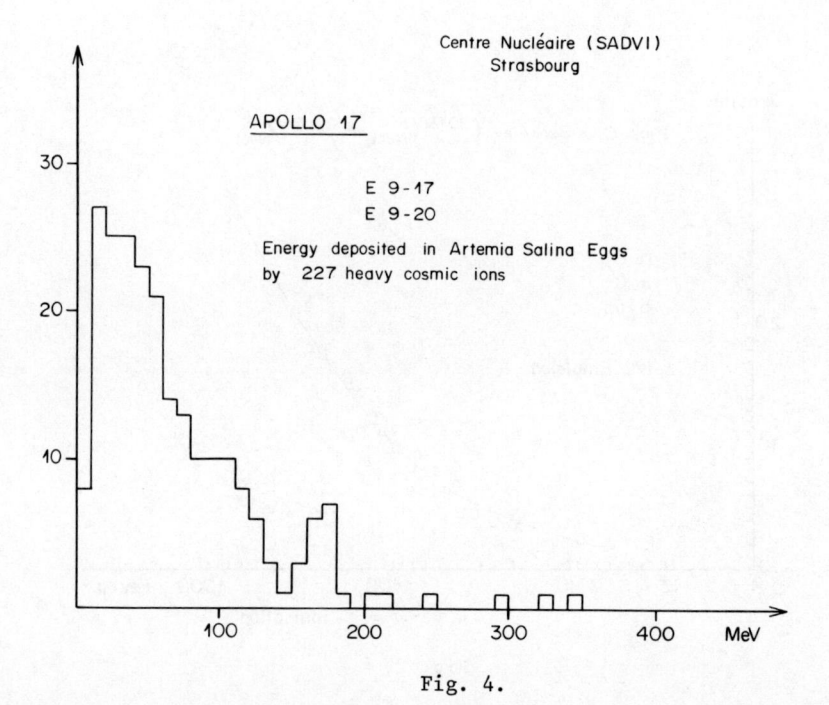

Centre Nucléaire (SADVI) Strasbourg

APOLLO 17

E 9-17
E 9-20

Energy deposited in Artemia Salina Eggs by 227 heavy cosmic ions

APOLLO XVI

Back ground: protons, muons, electrons, gamma

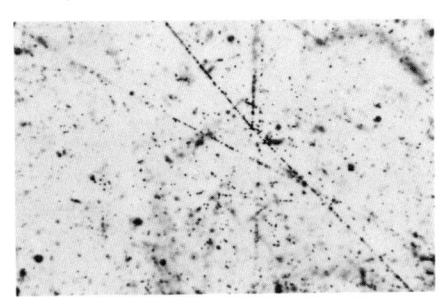

K 2

K 5

50 μ

CRN - STRASBOURG

Fig. 5.

PARTICLE INDUCED VISUAL PHENOMENA IN SPACE

P.J. McNulty*, V. P. Pease*, V. P. Bond**, R. C. Filz***
and P. L. Rothwell***

*Clarkson College of Technology, Potsdam, New York 13676, U.S.A.
**Brookhaven National Laboratory, Upton, New York 11973, U.S.A.
***Air Force Geophysics Laboratory, U.S.A.

There have been a large number of laboratory experiments on particle induced visual sensations which have resulted in a variety of visual phenomena that are similar in appearance to the so-called "light flashes" described by astronauts on Apollo missions 11 through 17 and Skylab 4. Unfortunately, no direct comparison of the laboratory and space observations have been made by observers who have experienced both. More than one physical mechanism has been shown to be involved in the laboratory phenomena and presumably in the space observations also. A number of models for particle induced visual phenomena are described and a quantitative estimate of their contribution to the space observations is attempted.

Presented at the 9th International Conference on Solid State Track Detectors, Munich, September 30-October 6, 1976.

INTRODUCTION

Astronauts on Apollo and Skylab missions have reported observing a variety of visual phenomena when their eyes are closed and adapted to darkness (1,2). These observations were studied under controlled conditions during a number of sessions on board Apollo and Skylab spacecraft. The data were obtained in the form of descriptions of the phenomena and the frequency of occurrence. Visual phenomena which are similar in appearance have been demonstrated at a number of accelerator facilities (3-11) by exposing the eyes of human subjects to beams of various types of radiation.

More than one physical mechanism is known to be involved. In some laboratory experiments (3-7), Cerenkov radiation was shown to be the basis for the flashes observed while other experiments (8-11), were such that Cerenkov radiation could be ruled out.

The types of flashes described by Apollo and Skylab astronauts are listed in Table 1. Only point-like flashes are included as stars in Table 1 and not the large "nova-like flashes." The crescent shaped flashes appear in the far periphery of the upper temporal field of view both in space and the Cerenkov experiments. The crucial question is whether the laboratory flashes are in fact the same in appearance as those observed in space. Or are they only similar in description. Unfortunately, there has not been a direct comparison of the flashes observed in space with either the Cerenkov or non-Cerenkov flashes by an observor who has experienced both. Nor have the Cerenkov and non-Cerenkov flashes been compared by the same observors, although such a comparison is planned for the near future. Until such direct comparisons are carried out, one is limited to a comparison of subjective descriptions by different observors. To date, the large area flashes have only been observed in the laboratory experiments with Cerenkov radiation and the star flashes have only been observed in the non-Cerenkov studies.

NON-CERENKOV FLASHES

In at least one of the observing sessions carried out in space, Cerenkov radiation can be eliminated as a dominant mechanism. When Skylab 4 passed through the portion of the earth's radiation belts known as the South Atlantic Anomaly

(SAA) Lt. Col. William Pogue reported flash rates as high as 20 per minute (2). These rates are too high to be accounted for by the number of particles in the SAA that would produce detectable flashes of Cerenkov radiation (12). On the other hand, the situation is different in deep space. Calculations (13,14) show that there are more than enough Cerenking particles passing through an astronaut's eye to explain the observed rates that can run as high as 1-2 per minute. Furthermore, Table 1 shows that the deep space observations include a variety that is absent in the SAA and the non-Cerenkov experiments.

The contribution Cerenkov radiation makes to the flashes in deep space has been discussed previously and the interested reader is referred elsewhere (5, 14-16) for details. The SAA flashes take on a particular significance in what follows just because Cerenkov radiation can be ruled out and alternative mechanisms can be tested.

Quantitative statements regarding the non-Cerenkov contribution to the flashes in space have not been possible because the mechanism is not known. The available data are not sufficient to determine whether the concept of threshold which is so useful in modeling the detection of optical (17,18) and Cerenkov (7, 9-11) flashes is even valid for the non-Cerenkov case. There is no agreement as to which of the particle's characteristics determines detection efficiency. Budinger, et al conclude that linear energy transfer LET, is the relevant parameter and that particles with LET values above 10 keV/μm will be detected at the retina with an efficiency of 40% at high exposure rates and 5% at low rates(10,11).

The difficulty with LET is that it provides no distinction between tracks that maintain high LET values over long trajectories and those that maintain their LET above 10 keV/μm for only a few microns. Pinsky et al (2) suggest that two requirements must be satisfied simultaneously for a non-relativistic particle to be detected. First, the particle must enter the sensitive layer of the retina with some minimum LET and, secondly, maintain that value over some critical path length. If both conditions are satisfied the particle is detected and not otherwise. After analyzing the Apollo observations and those on Skylab outside the SAA, they proposed a threshold requirements of 37 keV/μm for the LET and 40 μm for the path length.

PROPOSED MODEL FOR NON-CERENKOV FLASHES

Because the dark-adapted retina is known to integrate signals over distances as great as 300 μm, it seemed more appropriate to express threshold in terms of a minimum deposition of energy within a retinal area or unit of summation 300 μm in diameter (12). If more than a threshold amount is deposited, the unit is triggered and if at least one unit is triggered on the retina the particle is "seen." A rough estimate of threshold can be obtained from the available data. The alpha and nitrogen exposures (10,11) imply 5% detection efficiency for particles with 10 keV/μm LET incident tangentially to the posterior retina. Such particles would deposit energies up to 3 MeV within a 300 μm unit. At normal incidence to the sensitive layer (assuming it to be 30 μm thick-corresponding to the layer of rod-cell outer segments) the energy deposited would be only 0.3 MeV but it is not clear whether these particles would be detected at normal incidence. Similarly, the twin threshold requirements of 37 keV/μm and 40 μm path result in particles that deposit at least 1.5 MeV in a summation unit.

DISCUSSION

The flash rates predicted by our model for the SAA is plotted in Fig. 1 as a function of the energy requirement for threshold. The astronauts observed rates of one every few seconds. The dashed curve represents the contribution from the passage of the protons of the radiation belt through the sensitive layer of the retina and the solid curve represents the contribution from nuclear stars. There is no direct proton contribution to the flashes for thresholds above 0.3 MeV as shown in Fig. 1 primarily because they reach LET values of 10 keV/μm or more only in the last few microns of range. No trapped protons have a LET of 37 keV/μm over a 40 μm path length.

The solid curve represents the contribution from nuclear stars. The details of the calculation are given elsewhere (12) and we restrict ourselves here to a few brief comments. The nuclear star contribution is model dependent. Few nuclear stars emit particles that maintain a LET of 37 keV/μm for over 40 μm of path. But more than enough stars occur close enough for two or more prongs to enter the sensitive layer within 300 μm and deposit the same amount of energy (1.5 MeV).

As more energy is needed the stars must occur closer to the sensitive retinal layer for enough prongs to enter within 300 μm. This is illustrated in Fig. 2 where the fraction of flash producing nuclear stars that occur within the sensitive layer is plotted versus threshold.

The SAA flux at the location of Skylab is not well enough known to justify a careful comparison to Lt. Col. Pogue's observations. Rates of 20 per minute correspond in Fig. 2 to a threshold of 1-2 MeV. This agrees quite well with the 1.5 estimate earlier from Apollo and Skylab outside the SAA (2) and the 3.0 MeV estimated from the stopping alpha and nitrogen experiments (10,11).

TABLE I

VISUAL PHENOMENA OBSERVED BY DARK-ADAPTED SUBJECTS

	Space Flight		Non-Cerenkov			Cerenkov		
	Deep Space	SAA	Neutron	Stopping Alpha	Stopping Nitrogen	Nitrogen	Muon Bursts	Pion Bursts
Star	X	X	X	X	–	–	–	–
Streak	X	X	X	X	X	X	–	–
Cloud	X	–	–	–	–	–	X	X
Crescent	X	–	–	–	–	–	X	X
Band	X	–	–	–	–	–	–	X
Annulus	X	–	–	–	–	–	X	–

REFERENCES

1. L.S. Pinsky, W.F. Osborne, J.V. Bailey, R.E. Benson and L.F. Thompson, Science 183, 957 (1974).

2. L.S. Pinsky, W.Z. Osborne, R.A. Hoffman and J.V. Bailey, Science 188, 928 (1975).

3. P.J. McNulty, Nature 234, 110 (1971), Air Force Cambridge Research Laboratories Report No. 71-0377 (1971).

4. P.J. McNulty, V.P. Pease, L.S. Pinsky, V.P. Bond, W. Schimmerling and K.G. Vosburgh, Science 178, 166 (1972).

5. P.J. McNulty, V.P. Pease and V.P. Bond, Science 189, 453 (1975).

6. P.J. McNulty, V.P. Pease and V.P. Bond, J. Opt. Soc. Am. 66, 49 (1976).

7. P.J. McNulty, V.P. Pease and V.P. Bond, Radiation Research 66, 519 (1976).

8. J.H. Fremlin, New Scientist 47, 42 (1970).

9. C.A. Tobias, T.F. Budinger, and J.T. Lyman, Nature 230, 596 (1971).

10. C.A. Tobias, T.F. Budinger and J.T. Lyman, Proc. Natl. Symp. on Natural and Manmade Radiation in Space, 1971, E.A. Warman, ed. (NASA-TMX-2440, 1972), p. 1002.

11. T.F. Budinger, H. Bischel and C.A. Tobias, Science 172, 868 (1971).

12. P. Rothwell, R. Filz and P.J. McNulty, Science 193, 1002 (1976).

12. W.N. Charman, J.A. Dennis, G.G. Fazio and J.V. Jelley, Nature 230, 522 (1971).

14. R. Madey and P.J. McNulty, Proc. Natl. Symp. Natural and Manmade Radiation in Space, 1971, E.A. Warman, ed. (NASA-TMX-2440, 1972), p. 757.

15. P.J. McNulty, in Proc. 8th Internatl. Conf. on Nuclear Photographic and Solid State Track Detectors, Vol. 1, M. Nicolae (ed.) (Institute of Atomic Physics, Bucharest, 1972) p. 598.

16. P.J. McNulty, V.P. Pease, and V.P. Bond, L. feSciences and Space Research 13, (in press).

17. E.J. Denton and M.H. Pirenne, J. Physiol. 123, 417 (1954).

18. M.H. Pirenne and F.H. Marriott in Psychology a Study of Science, Vol. 1, S. Koch, ed. (McGraw-Hill, New York, 1959).

PREDICTED FLASHRATE AS A FUNCTION OF THE RETINA THRESHOLD SENSITIVITY
$\ell = 300\,\mu$

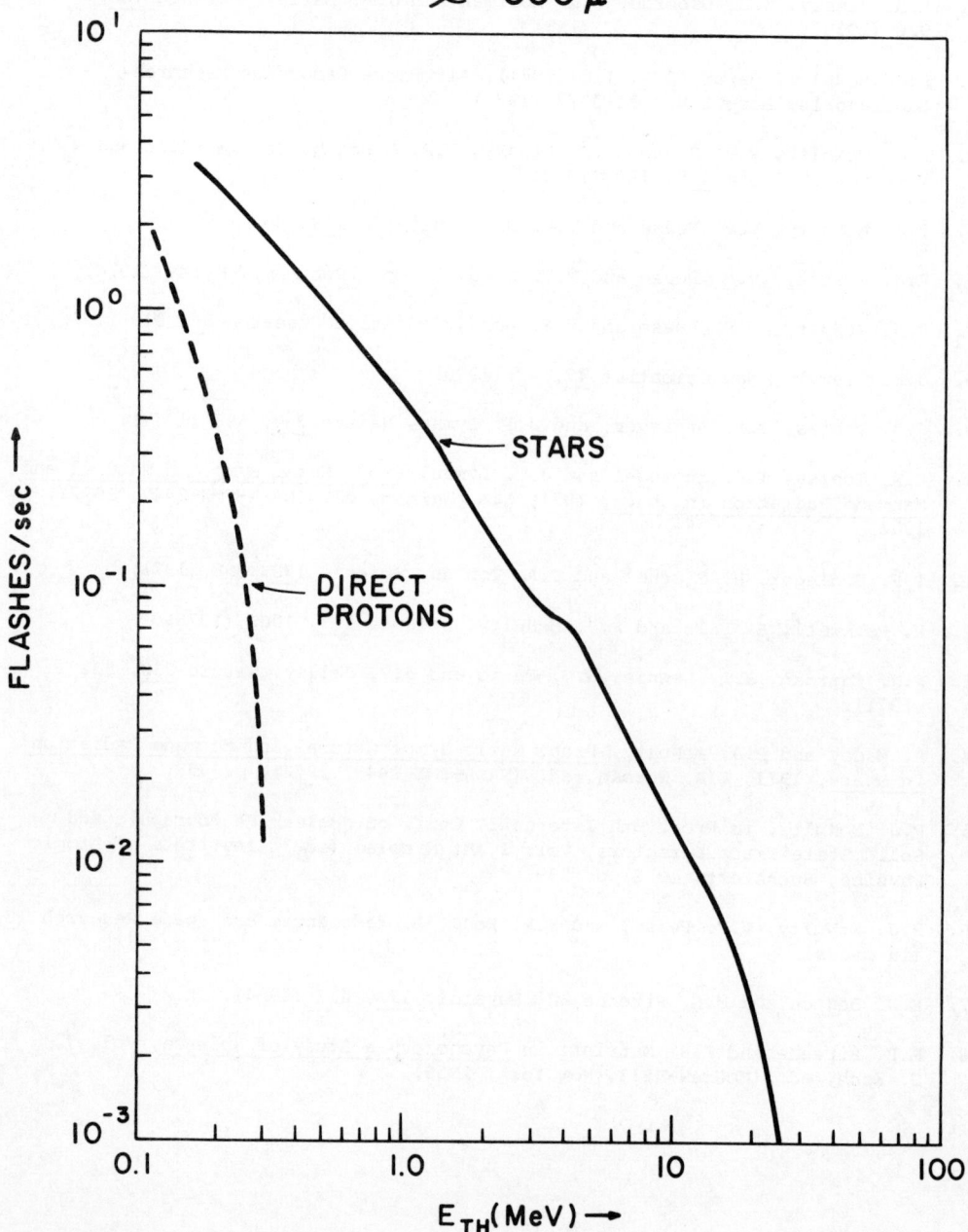

Figure 1. Flash rates calculated for the SAA as a function of threshold.

RATIO (R) OF RETINAL STARS TO TOTAL STARS

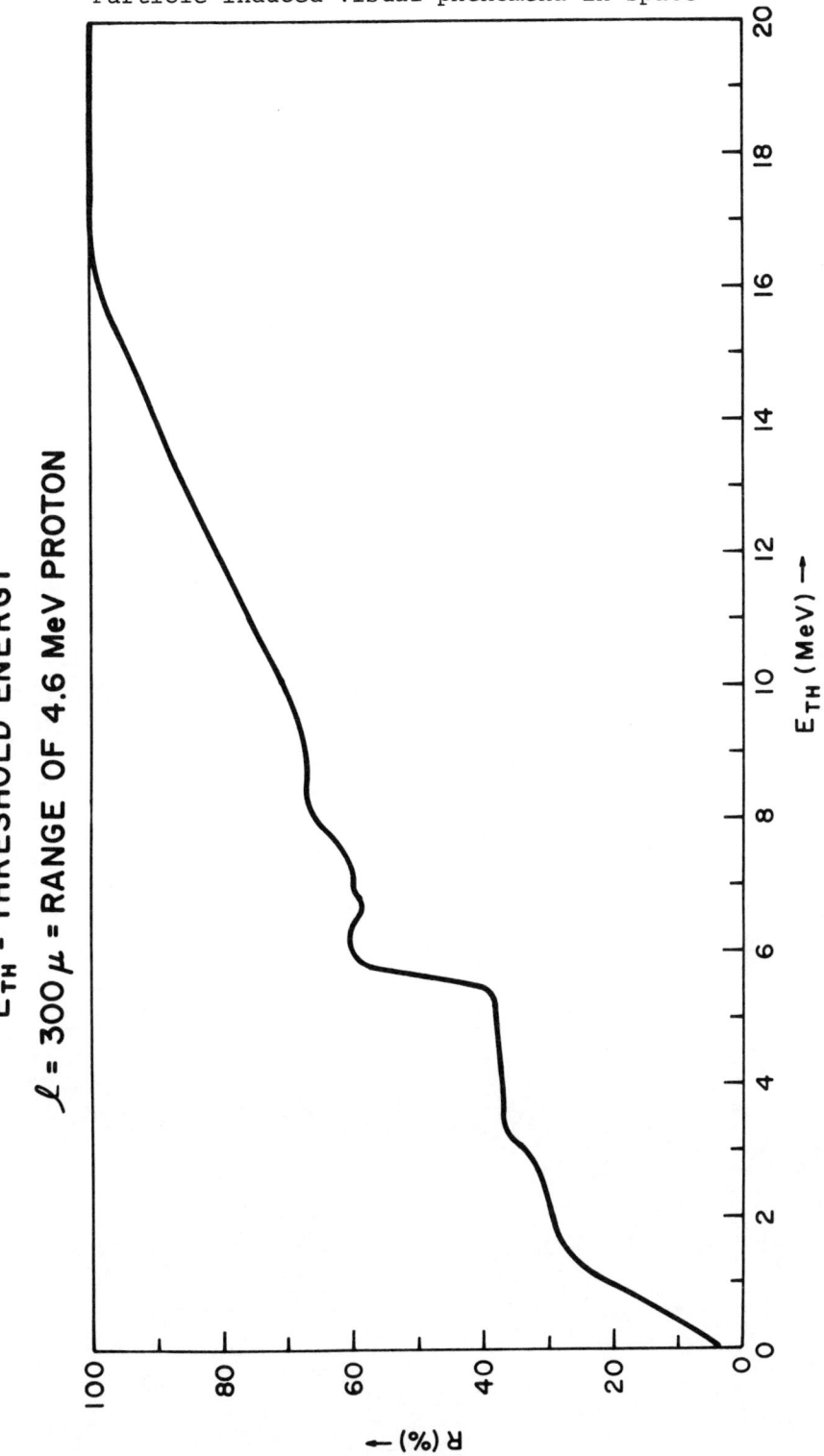

E_{TH} = THRESHOLD ENERGY

ℓ = 300 μ = RANGE OF 4.6 MeV PROTON

Figure 2. Percentage of eye-flash producing stars that occur within the 30 μm thick layer of photoreceptors.

AUTHOR INDEX

Abmayr, W. 643
Aframian, A. 447, 651
Afzal, M. 815
Akber, R.A. 403, 439, 803, 815, 931, 1231
Akopova, A.B. 547
Alexander, C. 1007
Al-Haddad, I.K. 883
Ali, A. 101
Allkofer, O.C. 1265
Annoni, H. 1153
Antanasijevic, R. 831, 835, 1187
Apanasenko, A.V. 553
Apostol, A. 967
Aschenbach, J. 417
Avdonina, E.D. 221
Azimi-Garakani, D. 625

Badea, E.G. 715, 719
Bagge, E. 753
Balcazar-Garcia, M. 351
Balzer, R. 387
Baranov, D.G. 1081
Baranov, V.I. 567
Bartholomä, K.-P. 137, 1059
Basha, M.A. 943
Bashir, S. 677
Beaujean, R. 1069, 1075, 1171, 1197
Becker, H.J. 1207
Belous, V.M. 221
Benton, E.V. 175, 349, 739
Bernas, A. 301
Besant, C.B. 363
Bichsel, H. 891
Bogomolov, C.S. 511, 523
Bogomolov, K.S. 553
Bond, V.P. 1335

Bozin, S. 251
Brandt, R. 1207
Brun, M. 1153
Bücker, H. 325, 1283, 1291, 1299
Bull, R.K. 1031
Chambaudet, A. 301, 307
Chaudhry, P. 439, 815, 1231
Chikunova, E.I. 553
Clapham, V.M. 1007
Coppens, R. 677
Cruty, M.R. 739

Damm, G. 729
Danis, A. 715, 719, 725
Dartige, E. 395
Debeauvais, M. 1091, 1179
Decossas, J.L. 317
Delaunay, B. 317
DiLiberto, S. 635
Ditlov, V.A. 511
Dmitriev, V.D. 217
Dragu, A. 967
Dran, J.C. 707
Drndarevic, S.B. 1145
Dühmke, E. 753
Dupuy, J. 229
Dura, A. 1087
Duraud, J.P. 395, 707
Durrani, S.A. 101, 351, 375, 651, 1031
Dutrannois, J. 905, 953

Eenmaa, J. 891
El-Fiki, M. 943
El-Konsol, S. 943
Enge, W. 103, 119, 137, 697, 753, 1039, 1059, 1069, 1075, 1171, 1197
Erlenkeuser, H. 697

Facius, R. 1283, 1291, 1299
Fadel, M.A. 943
Fain, J. 243
Fernandez, F. 1087
Fetisov, V.V. 523
Fiedler, G. 417
Filz, R.C. 501, 1023, 1335
Fleischer, R.L. 3, 663
Fowler, P.H. 983, 1007, 1017
Francois, H. 977, 1325
Freiesleben, H. 1207
Fukui, K. 1155, 1163

Gadiullin, R.S. 553
Gagarin, Yu. F. 1081
Gais, P. 643
Ginobbi, P. 635
Gisclon, J.L. 229
Gourcy, J. 243
Grabez, B. 1187
Grabisch, K. 103, 119, 1171
Granzer, F. 199, 425
Graul, E.H. 1301
Green, P.F. 375

Haase, G. 199, 237
Hasegan, D. 967
Hassib, G.M. 905
Heilmann, C. 977
Heinrich, W. 1313
Henke, R.P. 739
Henshaw, D.L. 1007, 1017
Hertzmann, S. 1075
Hildebrand, D. 325, 1291
Hölz, G. 1283
Hunger, W. 753
Hunyadi, I. 599, 1245
Hussain, G. 931

Ivanov, V.O. 553
Ivanova, N.S. 1081

Jacquot, C. 977
Jal, J.F. 229
Jasiak, J. 587
Jensen, J.M. 697
Jokic, S. 1179
Juric, M. 1129, 1137, 1145

Kafalenos, V.P. 669
Kaiser, R. 1325
Kartuzhanski, A.L. 221
Kashkarov, L.L. 393, 401
Katz, R. 27, 145, 175
Kehva, T.E. 221
Kenawy, M.A. 943
Khadduri, T.Y. 761, 883
Khan, H.A. 403, 439, 803, 815, 931, 1231

Khilyuto, I.G. 1081
Khruleva, L.S. 523, 553
Klein, N. 571
Kligerman, M.M. 787
Knowles, H.B. 333, 787, 891
Kocherov, N.P. 217
Köhnen, H.J. 137
Koltay, E. 1245
Koshkin, V.L. 401
Kotel'nikov, K.A. 553
Kozinets, G.I. 523
Krätschmer, W. 1025
Kristiansson, K. 1075
Kulinkov, V.N. 1081
Kuznetzova, G.G. 553

Langevin, Y. 395, 707
Larsson, L. 145, 175
Liehu, A.E. 689
Lotz, U. 875
Luckstead, S.C. 333

Magradze, N.V. 547
Massue, J.P. 1325
McNulty, P.J. 501, 1335
Medina, J. 1087
Medveczky, L. 599
Melkumyan, L.V. 547
Miocinovic, D. 831
Moliton, J.P. 317
Monnin, M. 243
Mubarakmand, S. 439, 815, 1231
Myl'zeva, V.A. 553
Myshkin, V.E. 1081

Nagi, F.I. 815
Nakanishi, T. 705
Nicolae, M. 771, 967
Nikolaev, V.A. 1235
Novikova, N.R. 217, 533
Nyako, B. 599

O'Ceallaigh, C. 1007, 1017
Oncescu, M. 715, 719
Ortega, M. 1087
O'Sullivan, D. 1007, 1017
Otterlund, I. 1107
Otto, W. 417
Ouseph, P.J. 215

Paretzke, H.G. 87, 643, 821, 917
Pease, V.P. 1335
Perfilov, N.A. 217, 533
Pestov, V.S. 393
Peterson, D.D. 349
Petit, J.C. 707
Pfohl, R. 1325
Piesch, E. 587
Pinkerton, F.E. 145, 175
Pitt, E. 875

Author Index

Plachenov, B.T. 221
Prokhorenko, Y.P. 547
Purica, I. 715, 719

Qaqish, A.Y. 363

Raabe, O.G. 663
Rao, Y.V. 1155, 1163
Rashid, K. 1231
Rautenberg, T. 417
Razorenova, I.F. 523
Rechenmann, R.V. 463
Reitz, G. 325
Richard, P. 677
Ristic, R. 251
Roberts, J.H. 669
Rodenacker, K. 643
Rohrs, H. 1089
Rössler, K. 99
Romanovskaya, K.M. 541, 1081
Romary, P. 307
Roncin, J. 301
Roose, U. 753
Rothwell, P.L. 1335
Ruddy, F.H. 333, 787, 891
Ruditskaya, I.A. 523
Rüther, W. 1301

Sagebiel, H. 1069
Sakanoue, M. 705
Sansoni, B. 705
Savateeva, J.P. 541
Schäfer, M. 1291
Scharmann, A. 875
Scheidemann, U. 1299
Scherzer, R. 103, 119, 753, 1075, 1171
Schmidt, H. 237
Schopper, E. 199, 615
Schott, J.U. 615
Schraube, H. 917
Schwarzkopf, G. 643
Senger, B. 463
Sequeiros, J. 1087
Siegert, G. 417
Siegmon, G. 137, 1059
Sigrist, A. 387
Simonovic, J. 835
Smathers, J.E. 891
Smorodin, J.A. 553
Söderström, K. 1075
Sojka, B. 1089
Sokolovskaya, I.N. 553
Solomon, P. 571
Somogyi, G. 103, 119, 255, 285, 599, 1245
Spurny, F. 839, 863
Stamatovic, A. 831
Staudte, R. 615
Steinhauser, U. 417

Teyssier, J.L. 317
Thiel, K. 729
Thomas, C. 1299
Thompson, A. 1007, 1017
Thorne, R.T. 1087
Tobias, C.A. 739
Todorovic, Z. 831, 1187
Tommasino, L. 571
Toth, B. 1283
Tran, M. 349
Tripard, G.E. 333, 787, 891
Tripier, J. 1091, 1179
Troshin, A.N. 401
Turek, K. 829, 863
Tuyn, J.W.N. 905, 953

Vareille, J.C. 317
Varga, B.B. 363
Varyukhin, V.V. 1081
Vater, P. 1207
Vidal-Quadras, A. 1087
Vitt, B. 875
Vukovic, J. 835

Waheed, A. 439, 815, 1129, 1137
Werba, T. 425
West, G.M. 787
Williams, J.G. 625
Willkomm, H. 697
Wittendorp, E. 463

Yakubovsky, E.A. 1081
Young, P.S. 1155, 1163
Yule, T.J. 669

Zakharov, V.I. 533
Zhdanov, G.B. 553
Zizić, B. 251
Zlatarov, V. 1137
Zörgiebel, F. 237
Zolnai, L. 1245

LIST OF PROCEEDINGS OF FORMER CONFERENCES

Photographie Corpusculaire
Proc. of the 1st International Colloquium on Corpuscular Photography, Strassbourg (France), 1957.
(Centre National de la Recherche Scientifique, Paris 1958)

Photographie Corpusculaire II
Proc. of the 2nd Int. Colloqu. on Corpuscular Photography, Montreal (Canada) 1958
Ed.:P.Demers;(Les Presses Universitaires de Montréal,1959).

Photographie Corpusculaire III
Yadernaya Fotografia; Ed.: Academy of Science of the USSR (Moscow 1962)
Proc. of the 3rd Int. Colloqu. on Corpuscular Photography, Moscow (USSR) 1960
Ed.:P.Demers;(Les Presses Universitaires de Montréal, 1964)

Korpuskularphotographie IV
Proc. of the 4th Int. Colloqu. on Corpuscular Photography, Munich (Germany), 1962.
Ed.: H. Frieser, G. Heimann (Institut für Wiss. Photographie der Techn. Hochschule München, 1963)

Proceedings of the Vth International Conference on Nuclear Photography, Vol.I and II
CERN, Geneva (Switzerland) 1964;
Ed.: E. Dahl-Jensen (CERN 65 - 4; 1964)

VIth International Conference on Nuclear Photography, Florence (Italy) 1966.
Ed.: M. della Corte (C.E.P.I., Rome, 1966)

7^e Colloque International de Photographie Corpusculaire et des Detecteurs Visuels Solides; Barcelone (Espagne) 1970.
Ed: P. Cüer, R. Schmitt (Lab. de Physique Corpusculaire, Strassbourg 1971).

List of proceedings of former conferences

Proceedings of the 8th International Conference on Nuclear Photography and Solid State Track Detectors, Bukarest, Rumania, 1972.
Ed.: M. Nicolae; (Institute of Atomic Physics, Bukarest, 1972)